Studies in Classification, Data Analysis, and Knowledge Organization

T0140458

Springer
Berlin
Heidelberg
New York
Hong Kong
London
Milan
Paris
Tokyo

Titles in the Series

Manfred Schwaiger · Otto Opitz (Eds.)

Exploratory Data Analysis in Empirical Research

Proceedings of the 25th Annual Conference
of the Gesellschaft für Klassifikation e.V.,
University of Munich, March 14–16, 2001

With 93 Figures and 60 Tables

Springer

Professor Dr. Manfred Schwaiger
University of Munich
Munich School of Management
Institute of Corporate Development
and Organization
Kaulbachstraße 45/I
80539 Munich
Germany
schwaiger@bwl.uni-muenchen.de

Professor Dr. Otto Opitz
University of Augsburg
Department of Mathematical
Methods in Economics
Universitätsstraße 16
86159 Augsburg
Germany
otto.opitz@wiwi.uni-augsburg.de

ISBN 3-540-44183-2 Springer-Verlag Berlin Heidelberg New York

Cataloging-in-Publication Data applied for
Die Deutsche Bibliothek – CIP-Einheitsaufnahme
Exploratory data analysis in empirical research: University of Munich, March 14–16, 2001; with 60
tables / Manfred Schwaiger; Otto Opitz (ed.). – Berlin; Heidelberg; New York; Hong Kong; London;
Milan; Paris; Tokyo: Springer, 2003
    (Proceedings of the ... annual conference of the Gesellschaft für Klassifikation e.V.; 25)
    (Studies in classification, data analysis, and knowledge organization)
    ISBN 3-540-44183-2

Springer-Verlag Berlin Heidelberg New York
a member of BertelsmannSpringer Science+Business Media GmbH

http://www.springer.de

© Springer-Verlag Berlin · Heidelberg 2003
Printed in Germany

Softcover-Design: Erich Kirchner, Heidelberg

SPIN 10891974   43/2202-5 4 3 2 1 0 – Printed on acid-free paper

Preface

The 25th annual conference of the Gesellschaft für Klassifikation e.V. (German Classification Society) took place at the University of Munich between March 14-16, 2001. Within the scope of **Exploratory Data Analysis in Empirical Research** scientists and practitioners discussed recent developments in this field and established cross-disciplinary cooperations in their own fields of research. The scientific program of the conference included 18 plenary or semiplenary lectures and about 130 presentations in more than 20 special sections. Furthermore, publishing companies informed the participants about current software products and books at several exhibition booths.

The peer–reviewed papers are presented in 4 chapters as follows:

1. Classification, Data Analysis and Statistics

2. Web Mining, Data Mining and Computer Science

3. Medicine, Biological Sciences, Health Care

4. Marketing, Finance and Management Science

In contrast to the alphabetical order in each of the four chapters, the following overview is arranged according to textual considerations.

Several problems of **Classification, Data Analysis** and **Statistics** are considered in 23 papers of chapter 1.

The paper of F. DOMENACH and B. LECLERC deals with the notion of Galois connection which is proved to be a fundamental tool in various situations of classification theory. A new clustering method based on the notion of biclosed relation is presented by the authors. E. GODE-HARDT and J. JAWORSKI describe two models which use random bipartite graphs to provide clusters and classifications. The authors derive mathematical properties, characterizations and asymptotic results for their models. P. BRITO and F.A.T. DE CARVALHO introduce an approach useful for clustering purposes, taking into account constraints on probabilistic data. In this context they propose new generality measures in order to compare couples of constrained probabilistic objects, together with an appropriate extension of generalization operators to get new constrained probabilistic objects. The main focus of the contribution of A. GRAUEL, I. RENNERS and E. SAAVEDRA is the presentation of a general methodology for structure optimization of fuzzy classifiers in classification techniques. The paper of M. CSERNEL and F.A.T. DE CARVALHO is embedded in the field of symbolic data analysis. The authors propose a special decomposition method of a symbolic data table which splits the table into subtables. It is shown that the decomposition reduces computation time from exponential to polynomial with bounded

factors for memory space. M. KREUSELER, T. NOCKE and H. SCHU-
MANN investigate the combination of numerical and visual exploration
techniques focused on cluster analysis of multidimensional data. They
describe new visualization approaches and selected clustering techniques
and discuss the major features of them. The paper of D. WISHART addresses practical issues in k-means cluster analysis or segmentation with
mixed types of variables and missing values. A more general k-means
clustering procedure is developed that is suitable for use with very large
datasets.

The paper of C. WEIHS and U.M. SONDHAUSS considers the interpretability of partitions generated by classification rules with severe concern for the adequacy for human understanding, the mental fit. The
authors discuss various criteria introduced in relation to mental fit in
literature and derive a general criterion for the interpretability of partitions generated by classification rules. Using hierarchical methods, the
main focus of S. UEDA'S and Y. ITOH'S contribution is to show that
languages can be mainly divided into prepositional and postpositional
or adpositional languages with numerals and nouns as second most important clustering variables.

J. KRAUTH considers clusters resulting from connecting the nearest
neighbours of points generated on the real line by a Poisson process. His
particular interest is the distribution of the number of nearest neighbour
clusters which can be used to test the hypothesis of no clustering. Furthermore, an approximation of the exact distribution of the test statistics
is given. The paper of M.T. GALLEGOS is dealing with clustering objects in the presence of outliers. The author develops an estimator which
simultaneously detects outliers and partitions. Another algorithm is proposed which approximates the estimator. G.J. MCLACHLAN, S.K. NG
and D. PEEL consider several problems using mixture models for clustering multivariate data. They report some recent results on speeding
up the fitting process by an EM–algorithm. Furthermore, the problem
of clustering high–dimensional data by use of the factor analyzers model
is considered. A. DOUGARJAPOV and G. LAUSEN study the problem of
matching time series. They present a technique for evaluating a classifier
which allows a reasonable choice of attributes to describe time points.
As an evaluation function of the classifier they present the domain error.

Based on the generalized correlation coefficient introduced by Kendall
K. JAJUGA, M. WALESIAK and A. BAK present a general distance measure for ordinal, interval and ratio scaled data. Y. TANAKA, F. ZHANG
and W. YANG discuss a sensitivity analysis for principal component and
canonical correlation analyses based on Cook's local influence. They
compare their results with those obtained by procedures based on the
influence function approach. Using an additive conjoint measurement

A. OKADA analyzes social network data. In contrast to many other approaches his procedure applies not only to binary but also to discrete or continuously valued data as well as to asymmetric relations. In this context, an application to interpersonal attraction among students is presented. P.J.F. GROENEN and J. POBLOME propose a constrained correspondence analysis for seriation of archaeological artefactual assemblages. The method is applied to sagalassos ceramic tablewares data.

Using computer intensive Bayesian methods such as Gibbs sampling or Markov Chain Monte Carlo R. WINKELMANN discusses in a survey paper advances in analyzing complex count data models. The paper of L. FAHRMEIR, C. GÖSSL and A. HENNERFEIND considers alternatives of spatial magnetic resonance imaging (MRI) priors in functional MRI's which are expected to have better edge preserving properties. The authors study the performance of functional MR priors with random weights using simulated and real data. They show improved edge preserving properties for the fitted functional MRI, however the edge preserving properties are lost for the fitted standardized functional MRI. An extension of the so-called Sliced Inverse Regression (SIR) to dynamic models is discussed by C. BECKER and R. FRIED. They discuss the general application of dynamic SIR, the detection of relevant directions and relations among the variables as well as the comparison with other methods in this context like graphical modeling. TH. OTTER, R. TÜCHLER and S. FRÜHWIRTH–SCHNATTER present a Bayesian analysis of the latent class model using a new approach towards MCMC estimation in the context of mixture models. Conjoint data from the Austrian mineral water market serves to illustrate the method. S. LANG, P. KRAGLER, G. HAYBACH and L. FAHRMEIR provide a Bayesian semiparametric approach which permits to simultaneously incorporate effects of space, time and further covariates within a joint model for analyzing the claims process in non–life insurances. The authors apply the approach to analyze costs of hospital treatment and accommodation using a large data set from a German health insurance company.

Chapter 2 includes 9 contributions dealing with **Web Mining**, **Data Mining** and **Computer Science** in a wide sense.

In a survey paper J.R. PUNIN, M.S. KRISHNAMOORTHY and M.J. ZAKI propose the use of a graph description language, XCMML and LOGML, especially developed as a tool for representing web navigation for discovering navigational patterns. F. SÄUBERLICH and K.-P. HUBER suggest a combination of different data mining techniques to analyze common log files of web users to find out structural problems in web design, typical navigation paths and to improve the web design. TH. LIEHR discusses several methods for the imputation of missing values in the course of data preparation in large real–world data mining projects.

Before clustering T.A. RUNKLER and J.C. BEZDEK transform considered text–data to relational data using Levenstein distance. These relational data can be clustered by relational ACS (RACE) providing keywords as the RACE cluster centers for further use. In a survey paper D. SAAD uses methods adopted from statistical physics to analyze the dynamics of online learning in multilayer neuronal networks. The analysis is based on monitoring a set of macroscopic variables from which the generalisation error can be calculated. Two approaches for document retrieval which group pre–processed documents are presented in the paper of A. NÜRNBERGER, A. KLOSE, R. KRUSE, G. HARTMANN and M. RICHARDS. The authors apply self–organizing maps which help the user to navigate through similar documents.

One of the major problems in developing educational software is the appropriate way to visualize complex data structures for learners; the paper of K. FRIESEN and H. SCHMITZ deals with the semantic modelling issues in cost accounting and describes the software environment of the solution.

Finally the consumption patterns for information goods are studied by W. BÖHM, A. GEYER–SCHULZ, M. HAHSLER and M. JAHN in the context of an information broker at a virtual university. The authors investigate whether Ehrenberg's repeat–buying theory can also be applied in electronic markets for information goods. In the field of automatic documentation and the combination of different documentation systems U. RIST presents a metadata management system with mapping tables to link the different documentation systems.

Applications of data analysis and data mining in **Medicine**, **Biological sciences** and **Health Economics** are discussed in 9 contributions of chapter 3.

Several applications of unsupervised neural networks or self–organizing maps in medicine are considered in the paper of G. GUIMARÃES and W. URFER. Especially, the authors focus on sleep apnea discovery, protein sequence analysis and tumor classification. For the analysis of cancer mortality data U. SCHACH makes an intrapolation from coarse time sequential data to the belonging small area data. The small area estimation uses special and temporal dependence structures based on a Bayesian hierarchical modelling approach. C. VOGEL and O. GEFELLER discuss the effects of misclassification on the estimation of measures of association in epidemiologic studies. The effects on the attributable risk and on the relative risk are compared. The paper of T. HOTHORN, I. PAL, O. GEFELLER, B. LAUSEN, G. MICHOLSON and D. PAULUS presents first results of the development of an automatic classification scheme of optic nerve head topography images for Glaucoma screening. Different linear and tree based discriminant techniques as well as stabilized methods are evaluated for the classification tasks. C. ERNST, G. ERNST and A. SZCZESNY estimate a learning curve related to knee replacement surgery from the data of surgeries performed in a German hospital. For this problem a regression model is applied. For constructing survival trees M. RADESPIEL-TRÖGER, T. RABENSTEIN, L. HÖPFNER and H.T. SCHNEIDER compare various split criteria for a special case of censored data. They use a likelihood model and Magee's R^2 as measures for the explained variation and estimate bootstrap confidence intervals of the explained variation.

The paper of A. ZIEGLER, O. HARTMANN, I.R. KÖNIG and H. SCHÄFER discusses recently developed techniques which might be alternatives for the identification of disease susceptibility genes and their function. They illustrate the absolute need for new statistical methods by means of two areas of current research, genome wide association analysis and gene expression groups.

To detect quality deficiencies in health care, J. STAUSBERG and TH. ALBRECHT propose the use of a common data mining tool. To receive useful results they found out that the underlying data have to be especially prepared for the OLAP–model. M. STAAT considers the question to provide physicians with appropriate information which improves the cost–efficiency relation of their work. He uses a bench marking method based on data envelopment analysis.

Many problems of **Marketing**, **Finance** and **Management Science** make use of models and methods of numerical and statistical data analysis. The 12 papers of chapter 4 are dealing with these issues.

M. MEYER considers the problem of sequence mining from a computational and a marketing point of view. He explains differences between these and points out the necessity to develop new algorithms, focusing stronger on marketing aspects.

C. BORNEMEYER and R. DECKER identify key success factors of city marketing projects by using structural equation modeling and discriminant analysis. The paper of H.H. BAUER, M. STAAT and M. HAMMER-SCHMIDT offers an analytical framework for an integrated treatment of market positioning and benchmarking based on data envelopment analysis. Furthermore, an exploratory data mining approach is used. For predicting market shares of competing products in assumed market scenarios D. BAIER and W. POLASEK compare empirical–traditional estimation procedures of conjoint analysis to a Bayesian approach. M. LÖFFLER presents a modification of the wellknown „tandem approach" for segmentation which takes into account certain shortcomings. Subsequently he applies his approach to the German automobile market to reveal additional insights.

Within the framework of the „default–mode" credit portfolio risk model R. KIESEL and U. STADTMÜLLER analyze the importance of factor correlation and granularity for the total risk of credit risky portfolios. S. HÖSE and S. HUSCHENS deal with the problem of estimating stable default probabilities from individual credit scores without loosing information by aggregating individual to rating grades. For this purpose they propose a maximum likelihood estimator first for the binomial, second for the poisson model. The paper of N. WAGNER focuses on the estimation of extreme parts of some special distributions with fat tails which exist in financial return distributions. In this context the problem of optimal subsamples is discussed. M. POJARLIEV and W. POLASEK use a VAR–GARCH model to predict the monthly returns of a portfolio on a stock market to find an appropriate portfolio as well as the kind of forecasts which improves the portfolio performance.

An optimization method is presented by R. BENNERT and M. MISSLER–BEHR to find promising portfolio positions for a reorganizing company. The method is based on a genetic algorithm using a cashflow index and a potential of success index for the overall fitness criterion. D. KWIATKOWSKA–CIOTUCHA, U. ZALUSKA and J. DZIECHCIARZ study the attractiveness of manufactoring branches in Poland from the investor's point of view. They analyze several descriptive variables obtained from statistical reports.

The main results of the investigation on the cognitive representation of trait–descriptive terms, considered by S. KROLAK–SCHWERDT and B. GANTER clearly show that the choice of a particular measurement technique determines which organizational aspects of cognitive relations between traits become visible and whether subjects attention is directed either to the semantic components of the trait terms or to stereotypes and self concept features.

The editors of these proceedings are very indebted to all colleagues who chaired a session during the conference and/or reviewed some papers for this volume. We gratefully acknowledge the help and support given by the members of the scientific program committee as well as the active cooperation of all participants and authors. Finally we would like to emphasize the excellent work of all assistants and secretaries involved in the organization of the conference and preparation of this proceedings volume.

We hope that the presented volume will find interested readers and may encourage further research.

Augsburg and Munich, June 2002

O. Opitz, M. Schwaiger

Acknowledgements

Among our principal supporters we would like to highlight the role of **Bain & Company**, who had a major stake in ensuring the success of our Munich conference. **Bain & Company** is known to be one of the leading strategic management consultancies worldwide. Since its foundation nearly thirty years ago, it has successfully focused on helping its diverse client base of over 2000 organizations formulating and implementing business strategies. The capacity to output relevant, functioning and lasting business solutions demands thorough and continuous development of knowledge, which **Bain & Company** successfully complied with by generously supporting our conference. The company's patronage underlines its strong and credible interest in both the development of innovative approaches in business research as well as the exchange of ideas between theory and practice. In respect of the above we would once more like to express our true appreciation and gratitude to **Bain & Company**.

Furthermore, we gratefully take the opportunity to acknowledge the support by:

- Bayerische Hypo-und Vereinsbank AG
- SAS Deutschland
- NFO Infratest
- Booz · Allen & Hamilton
- DaimlerChrysler AG
- Siemens AG, München
- HUK–COBURG Versicherungen Bausparen
- H.F. & Ph.F. Reemtsma GmbH
- EADS Deutschland GmbH
- Gesellschaft von Freunden und Förderern der Universität München
- Bavarian State Ministry of Sciences, Research and the Arts

Contents

Part 4: Marketing, Finance and Management Science . 411

Classification, Data Analysis and Statistics

Sliced Inverse Regression for High-dimensional Time Series

C. Becker, R. Fried

Fachbereich Statistik,
Universität Dortmund, D-44221 Dortmund, Germany

Abstract: Methods of dimension reduction are very helpful and almost a necessity when analyzing high-dimensional time series since otherwise modelling affords many parameters because of interactions at various time-lags. We use a dynamic version of Sliced Inverse Regression (SIR; Li (1991)) as an exploratory tool for analyzing multivariate time series. Analyzing each variable individually, we search for those directions, i.e. linear combinations of past and present observations of the other variables which explain most of its variability. This also provides information on possible nonlinearities. An application to time series representing the hemodynamic system is given.

1 Introduction

Modern technical possibilities allow for simultaneous recording of many variables at high sampling frequencies. Possibly there are interactions between the observed variables at various time lags and we have to treat the data as multivariate time series. Often, the data contain strong dynamic structures which are unknown in advance. In intensive care medicine for instance, physicians have to deal with a flood of high-dimensional data describing the patient's state. Vital signs of critically ill patients are recorded in the course of time by clinical information systems (CIS). Altogether, up to 2000 variables are reported. Thus, methods for compressing the information provided by the CIS in a suitable way would be very helpful. A model-based approach to dimension reduction which has been used successfully in this context (Gather et al. (2001)) is the application of dynamic factor models. However, specification of a factor model needs a lot of assumptions. Methods such as graphical modelling for multivariate time series (Dahlhaus (2000)) and a dynamic version of Sliced Inverse Regression (SIR; Li (1991)) can be used at an early stage to explore the dynamic structure between the observed variables.

After briefly reviewing multivariate time series analysis in Section 2, we describe in Section 3 how SIR can be applied to dynamic process data. In Section 4, we give an application to multivariate time series of vital signs measured in intensive care and compare the results with those obtained by using graphical models.

2 Multivariate Time Series Analysis

Consider a multivariate time series $\{x(t) = (x_1(t), \ldots, x_m(t))' : t = 1, \ldots, T\}$, generated by a stochastic process $\{X(t) : t \in \mathbb{Z}\}$. In classical time series analysis, the (time-lagged) relations between the variables are usually assumed to be linear because of computational convenience and since the theory of linear time series models is well established. Vector Autoregressive Moving Average VARMA(p, q) models probably form the most common model class. They formally resemble multiple regressions. In a VARMA(p, q) model for a stochastic process $\{X(t) : t \in \mathbb{Z}\}$ the influence of past observations and past shocks on the current observation is modelled using weight matrices $\Phi_1, \ldots, \Phi_p, \Theta_1, \ldots, \Theta_q \in \mathbb{R}^{m \times m}$:

$$\begin{aligned} X(t) &= \Phi_1 X(t-1) + \ldots + \Phi_p X(t-p) \\ &\quad + \Theta_1 \varepsilon(t-1) + \ldots + \Theta_q \varepsilon(t-q) + \varepsilon(t), \quad t \in \mathbb{Z}. \end{aligned}$$

The unobservable shocks $\varepsilon(t), t \in \mathbb{Z}$, are assumed to form a (Gaussian) white noise process. The VARMA representation of a process is not necessarily unique, several distinct VARMA equations can describe the same stochastic dependencies (cf. Reinsel (1997)). In practice, often VAR(p) models $(q = 0)$ are used since we get tractable formulae for estimation and prediction, and uniqueness of the representation is guaranteed.

A basic interest in time series analysis is the prediction of future outcomes. In intensive care, the primary task is the online detection of interesting patterns of change in the data. The large number of unknown parameters needed for modelling multivariate time series can cause problems since the estimates have to be based on a large number of observations. Therefore, methods for dimension reduction could be very helpful. However, for fitting an appropriate model in some lower-dimensional space we have to understand the possibly time-lagged relations between the variables. For instance, in Gather et al. (2001) a factor model with four latent factors is constructed for a ten-dimensional time series of vital signs. The observed multivariate time series is assumed to be a linear transform of a lower-dimensional latent factor process which progresses according to a VARMA model (see Peña and Box (1987) for details). More exactly, the analysis of Gather et al. assumes the factor process to follow a VAR(2) model with independent components. Another approach to factor analysis is suggested by Molenaar (1985). He assumes the observed time series to be a moving average of a lower dimensional factor process which is modelled using a state-space model. This is a simple example where a-priori information would be useful to select a suitable model form.

Assuming all interactions to be linear means a severe restriction which certainly is not always appropriate. Linear time series models are not sufficient to describe all real-world phenomena. For this reason a couple of nonlinear models has been suggested (Tong (1990)). A multivariate nonlinear autoregressive model of order p can be written as

$$X(t) = h(X(t-1)', X(t-2)', \ldots, X(t-p)', \varepsilon(t)'), \quad t \in \mathbb{Z}.$$

Here, the transfer function $h : \mathbb{R}^{(p+1)m} \to \mathbb{R}^m$ links the past and the present of the process. It specifies the nonlinear influence of past observations on the present observation. The specification of h is difficult. For $p = 3$ and a 10-dimensional time series we get a 30-dimensional regressor space even if simultaneous interactions are not considered. Hence, finding a suitable dimension-reducing subspace would be very helpful.

In the following we select one variable, denoted by Y, and concentrate on the influence of the other variables, denoted by X, on this variable. Then we get a (possibly nonlinear) transfer function model

$$Y(t) = g(X(t), \ldots, X(t-p), Y(t-1), \ldots, Y(t-p), \varepsilon(t)) \tag{1}$$

where the innovations $\varepsilon(t)$ form a white noise process. It is usually assumed that feedback is not present in the system, meaning that Y does not influence the explanatory variables X, and the processes $\{\varepsilon(t), t \in \mathbb{Z}\}$ and $\{X(t), t \in \mathbb{Z}\}$ are assumed to be independent. Even in case of a single explanatory variable, i.e. univariate $X(t)$, finding a suitable transfer function and estimation of its parameters is difficult (Karlsen et al. (2000)). Hence, appropriate dimension reduction is crucial to get an impression of g. Sometimes a small number of linear combinations ("directions") of time lagged observations of the influential variables may be sufficient to explain most of the variability in $\{Y(t) : t \in \mathbb{Z}\}$.

3 Sliced Inverse Regression for Dynamic Data

Li (1991) proposes sliced inverse regression (SIR) as a tool for dimension reduction in non-dynamic regression problems. The basic idea is that instead of taking $Y = g(X, \varepsilon)$, $X = (X_1, \ldots, X_d)'$, one assumes that it is sufficient to consider a lower-dimensional space – the so-called central dimension reduction (dr) subspace (Cook (1996)) $\mathcal{B}$ of dimension $r < d$ – such that there exists $f : \mathbb{R}^{r+1} \mapsto \mathbb{R}$ with

$$Y = f(\beta_1'X, \ldots, \beta_r'X, \varepsilon), \text{ where} \tag{2}$$

$$\mathcal{B} = \text{span}[\beta_1, \ldots, \beta_r],$$

and $\mathcal{B}$ is a space of smallest possible dimension r such that (2) is satisfied. Hence, a reduction of the regressor space from d to r dimensions is supposed to be possible. SIR estimates $\mathcal{B}$ using information contained in an estimate of the inverse regression curve $E(X|Y)$. For given data $(y_i, x_i')'$, $i = 1, \ldots, n$, $x_i \in \mathbb{R}^d$, $y_i \in \mathbb{R}$, SIR is performed in five steps:

1. Standardization: $z_i = \widehat{\Sigma}^{-1/2}(x_i - \bar{x})$, $i = 1, \ldots, n$, where $\widehat{\Sigma} = \sum_{i=1}^{n}(x_i - \bar{x})(x_i - \bar{x})'/n$, $\bar{x} = \sum_{i=1}^{n} x_i/n$.

2. Slicing: split $z_1, \ldots, z_n$ into H slices S_h, $h = 1, \ldots, H$, according to the order of the corresponding values of $y_1, \ldots, y_n$; let n_h denote the number of observations in S_h. Hence, observations belonging to the n_1 smallest y_i are sorted into the first slice, etc.

3. Slice means: $\widehat{m}_h = \sum_{S_h} z_i/n_h$, $h = 1, \ldots, H$.

4. PCA of slice means: $\widehat{SIR} = \sum_{h=1}^{H} n_h \widehat{m}_h \widehat{m}_h'/n$ with eigenvalues $\widehat{\lambda}_1 \geq \ldots \geq \widehat{\lambda}_d$ and respective normalized eigenvectors $\widehat{\eta}_1, \ldots, \widehat{\eta}_d$.

5. Estimate dr directions: $\widehat{\beta}_k = \widehat{\Sigma}^{-1/2} \widehat{\eta}_k$, $k = 1, \ldots, r$.

The number H of slices has to be chosen in advance. The reduced dimension r of $\mathcal{B}$ can be estimated simultaneously by performing consecutive tests of $H_0^j : r = j$ vs. $H_1^j : r > j$, starting with $j = 0$, until H_0^j is not rejected for the first time, cf. Li (1991).

For applying this procedure to multivariate time series, it has to be developed further to take the dynamic structure of the data into account. Let $(Y(t), X(t)')'$ denote the observation of $(Y, X')'$ at time t. In the following, a transfer function model (1) for the measurements of Y is assumed. Becker et al. (2001) suggest to search for a dr subspace of the regressor space using a modified version of equation (2):

$$Y(t) = f(\beta_1' \widetilde{X}(t), \ldots, \beta_r' \widetilde{X}(t), \varepsilon(t)) \ \forall t,$$

where $\widetilde{X}(t) = (X(t)', Y(t-1), X(t-1)', \ldots, Y(t-p), X(t-p)')'$ if we take p time lags into account. The idea is to bind various lagged measurements of the variables together to construct the regressor space, forming higher dimensional observations, and then to apply the original SIR method to $(Y(t), \widetilde{X}(t)')'$. This allows to consider time-lagged effects up to the order p. An appropriate order p can be searched for by applying dynamic SIR for $p = 1, 2, \ldots$ until the measurements at time lag p do no longer contribute to the dr directions. Sometimes this approach may fail because of e.g. seasonal effects. Hence, we should use preliminary information if possible. Experience shows that for online monitoring data observed in intensive care $p = 2$ is sufficient (Gather et al. (2000)).

4 Application

In a given data situation, where we do not know beforehand about the association structure between the variables but expect that there is some structure, we can apply dynamic SIR as an explanatory tool. In our application we are concerned with 10-variate time series describing the hemodynamic system as measured in intensive care, consisting of arterial and pulmonary artery blood pressures (diastolic, systolic, mean), denoted by APD, APS, APM, PAPD, PAPS, PAPM, central venous pressure (CVP), heart rate (HR), blood temperature (TEMP), and pulsoximetry (SP02). In Becker et al. (2001) a simulation study is performed for several linear and nonlinear VAR and dynamic regression models with 10 variables. It turns out that dr subspaces can be discovered by applying dynamic SIR. To get reliable results, for linear models time series with about 250 observations seem to be sufficient. For nonlinear models 3000 or even more observations are needed since we found about 30 slices with at least 100 observations appropriate then. For a retrospective analysis of multivariate time series observed in intensive care this does not pose a severe restriction since usually several thousands of observations are available.

In Gather et al. (2001), graphical models for multivariate time series are used to find relations between the variables. However, such undirected graphical models based on the partial spectral coherences (psc) accumulate the effects for all time lags and do not mirror causalities for instance. Directed graphical models giving information on possible causalities have been developed recently (Dahlhaus and Eichler (2001)), but so far no software is commonly available.

Using dynamic SIR for multivariate time series gives an impression of the time-lagged relations between the variables. We apply graphical models and the dynamic SIR procedure to a 10-dimensional time series of vital signs observed for a critically ill patient. A total of 360 observation times is included in the analysis. Figure 1 shows the graphical model for all vital signs derived from analyzing the partial spectral coherences. We classify a partial correlation as strong if the average squared psc is at least 0.4, as medium if it is between 0.2 and 0.4, and as weak if it is smaller, but significantly distinct from zero at any frequency.

In order to provide just one illustrative example, we concentrate on the relations of APD to the other vital signs further on. This variable is typically neglected by the physician, who might select APM instead for monitoring the arterial pressures. In Gather et al. (2001) APD, APM and APS are replaced by a common factor. Therefore we are interested in knowing which directions provide important information on the course of APD. Applying dynamic SIR to the standardized variables, we find a

Figure 1: Graphical model for the hemodynamic system of one patient.

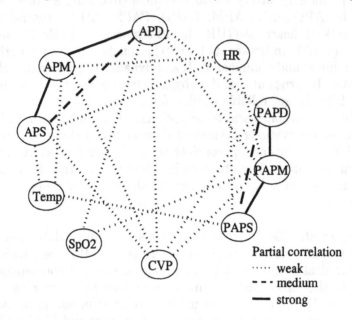

three-dimensional reduced subspace which is spanned by

$$d_t^1 \approx 1.31 APM_t - 0.57 APS_t + 0.47 APD_{t-1} - 0.43 APM_{t-1}$$

$$d_t^2 \approx 0.53 APM_t + 0.59 SPO2_t + 0.45 APM_{t-1} - 0.52 APS_{t-1}$$
$$-0.62 PAPD_{t-1} - 0.72 TEMP_{t-1} - 0.52 SPO2_{t-1}$$
$$-0.57 PAPD_{t-2} + 0.60 PAPM_{t-2}$$

$$d_t^3 \approx -0.65 PAPD_t + 0.73 PAPM_t + 0.79 APD_{t-1} - 1.04 APM_{t-1}$$
$$-0.54 PAPM_{t-1} + 0.68 PAPS_{t-1}$$

In Figure 2 we provide a graphical illustration of these findings. We connect APD to any of the other vital signs by a line if the simultaneous observation of the latter has a large weight on one of the directions, and draw an arrow from the vital sign towards APD if this is true for a past observation. The strength of the relation (weak, medium, strong) is interpreted according to the weight and the number of the corresponding direction. This is according to the fact that the first dr direction corresponds to the largest amount of variability in the relation between response and explanatory variables. The exact classification done here is subjective, of course, we use dynamic SIR as an exploratory tool only. As we can see, the results are rather similar to the results obtained by graphical models. Both methods identify APM and APS as the most important variables when we want to explain APD. Using dynamic SIR we additionally get some information about the dynamics since strong

Figure 2: Relations of APD to other vital signs for one patient derived from graphical model (left) and from dynamic SIR (right).

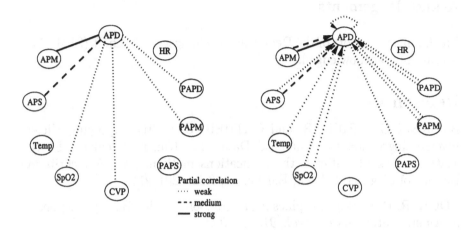

simultaneous effects and also somewhat weaker influences of past values seem to exist. Moreover, we could use dynamic SIR to get an impression of the functional form of the relations using plots of APD against each dr direction.

There are some weak relations of APD to other variables identified by graphical modelling but not by dynamic SIR and vice versa. This is due to some fundamental differences between both approaches: Undirected graphical models treat both variables symmetrically analyzing their partial correlations, while dynamic SIR uses an asymmetric regression approach based on the assumption that one of the variables can be explained using a structural equation model. Wermuth and Lauritzen (1990) discuss some differences between these frameworks in the non-dynamic case.

Finding a three-dimensional reduced subspace points at nonlinearities since for a linear link function only one direction is needed. Indeed, for linear link functions SIR detects one direction only even in the presence of feedback, cf. Becker et al. (2001). In our application, such nonlinearities can be due to therapeutic interventions influencing all variables in different ways. A linear factor model as used in Gather et al. (2001) can therefore only provide a local description of the relations in the steady state. This can be totally satisfactory if detection of relevant changes like intervention effects is the main goal.

Altogether, we consider dynamic SIR-type methods to be promising for dimension reduction of long multivariate time series from dynamic systems. Yet, the theoretical properties of dynamic SIR as well as the interactions found within the physiological time series measured in intensive care should be further validated.

Acknowledgements

The financial support of the Deutsche Forschungsgemeinschaft (SFB 475) is gratefully acknowledged.

References

BECKER, C., FRIED, R. and GATHER, U. (2001): Applying Sliced Inverse Regression to Dynamical Data. In: Kunert, Trenkler (Eds.): Mathematical Statistics with Applications in Biometry. Festschrift in Honour of Siegfried Schach, Eul-Verlag, Köln, 201–214.

COOK, R. D. (1996): Graphics for Regressions With a Binary Response, J. Amer. Statist. Assoc., Vol. 91, 983–992.

DAHLHAUS, R. (2000): Graphical Interaction Models for Multivariate Time Series. Metrika, 51, 157-172.

DAHLHAUS, R. and EICHLER, M. (2001): Causality and Graphical Models in Time Series Analysis. Preprint, Department of Mathematics, University of Heidelberg, Germany.

GATHER, U., FRIED, R. and IMHOFF, M. (2000): Online Classification of States in Intensive Care, in Gaul, Opitz, Schader (Eds.): Data Analysis. Scientific Modeling and Practical Application, Springer, Berlin et al., 413–428.

GATHER, U., FRIED, R., LANIUS, V. and IMHOFF, M. (2001): Online Monitoring of High Dimensional Physiological Time Series - A Case-Study. Estadistica, in press.

KARLSEN, H. A., MYKLEBUST, T. and TJOSTHEIM, D. (2000): Nonparametric Estimation in a Nonlinear Cointegration Type Model. Discussion Paper, SFB 373, Berlin, Germany.

LI, K.-C. (1991): Sliced Inverse Regression for Dimension Reduction (with discussion). J. Amer. Statist. Assoc., Vol. 86, 316–342.

MOLENAAR, P. C. M. (1985): A Dynamic Factor Model for the Analysis of Multivariate Time Series. Psychometrica, Vol. 50, 181–202.

PEÑA, D. and BOX, G. E. P. (1987): Identifying a Simplifying Structure in Time Series. J. Americ. Stat. Assoc., Vol. 82, 836–843.

REINSEL, G. C. (1997): Elements of Multivariate Time Series Analysis. Second Edition. Springer, New York et al.

TONG, H. (1990): Non-linear Time Series. A Dynamical System Approach. Clarendon Press, Oxford.

WERMUTH, N. and LAURITZEN, S. L. (1990): On Substantive Research Hypotheses, Conditional Independence Graphs and Graphical Chain Models. *J. R. Statist. Soc. B, Vol. 52, 21–50.*

Symbolic Clustering of Constrained Probabilistic Data

P. Brito[1], F.A.T. de Carvalho[2]

[1]Faculdade de Economia da Universidade do Porto,
Rua Dr. Roberto Frias, 4200 Porto, Portugal
mpbrito@fep.up.pt

[2]Centro de Informática - CIn / UFPE,
Av. Prof. Luiz Freire, s/n - Cidade Universitária,
50740-540 Recife-PE, Brasil
fatc@cin.ufpe.br

Abstract:

In previous work (Brito and De Carvalho (1999)) we have considered the presence of dependence rules between variables in the framework of a symbolic clustering method. In another paper Brito (1998) has addressed the problem of clustering probabilistic data. The aim of this paper is to bring together the two issues, that is, to take into account dependence rules on probabilistic data. This is accomplished by introducing new generality measures with an appropriate generalization operator. This approach allows for the extension of a symbolic clustering method to constrained probabilistic data.

1 Introduction

Symbolic data are more complex than standard data as they contain internal variation and they are structured. They come from many sources, for instance in summarizing huge relational data bases or as expert knowledge. The need to extend standard data analysis methods to symbolic data is increasing in order to get more accurate information and summarize extensive data sets (Bock and Diday (2000)).

This paper addresses the problem of clustering constrained probabilistic data. Probabilistic data occur when individual data has been aggregated and frequency distributions on discrete variables are taken into account. This is the case, for instance, when survey data must be aggregated for confidentiality issues. If moreover, as it is often the case, there are dependence rules between some variables, then we are in the presence of constrained probabilistic data. In previous work, Brito and De Carvalho (1999) have considered the presence of dependence rules in the framework of a clustering method which clusters Boolean symbolic data; Brito (1998) has defined appropriate measures and operators for clustering probabilistic symbolic data. In the present work, the more general problem of clustering constrained probabilistic data is addressed. Generalization operators are suitably extended to constrained probabilistic data, and new measures of generality for such data are defined. This allows extending the previously developed hierarchical and/or pyramidal symbolic clustering method to constrained probabilistic data. An example illustrates the presented algorithm.

2 Constrained Probabilistic Data

In this section, we start by recalling which kind of constraints we are concerned with, what is meant by probabilistic symbolic data, and finally we define constrained probabilistic symbolic data.

2.1 Hierarchical Dependence

A variable Z is said to be dependent from another variable Y if the range of values for Z depends on the value recorded for Y (Bock and Diday (2000)). We will consider in this text only a kind of dependence: hierarchical dependence.

We will say that a variable Z is hierarchically dependent of a variable Y if Z makes no sense for some categories of Y, and hence becomes "non-applicable".

Let $\mathcal{Y}$ be the range of Y and $\mathcal{Z}$ the range of Z, and $\mathcal{Y}' \subseteq \mathcal{Y}$ the set of values for which Z makes no sense. Then, the hierarchical dependence may be expressed by the rule (De Carvalho (1998)):

$$Y \text{ takes values in } \mathcal{Y}' \iff Z \text{ is not applicable}$$

When such a dependence exists, the special code NA (non-applicable) is used to express the fact that no value can be recorded for Z, this corresponds to consider an enlarged domain which contains NA as a category, $\mathcal{Z}' = \mathcal{Z} \cup \{NA\}$.

Example 1

Let Y ="sex" $\mathcal{Y} = \{\text{male, female}\}$ and Z = "nb. of births", $\mathcal{Z} = \{0, 1, 2, 3, ...\}$. Clearly, the applicability of Z depends on the value of Y:

if $Y(x) = $ "male" for some x then Z is not-applicable for x, $Z(x) = $ NA.

2.2 Probabilistic Data

A modal variable Y, with domain $\mathcal{Y}$, is a categorical multi-valued variable, such that, for each object $x \in E$ we are given not only the category set $Y(x) \subseteq \mathcal{Y}$, but also, for each category $y \in Y(x)$ a frequency, probability or weight which indicates how frequent or typical the category is for the object x.

So, a modal variable Y on a set of objects E with domain $\mathcal{Y}$ is a mapping $Y(x) = (U(x), \pi_x)$, $x \in E$, where π_x is a measure distribution on $\mathcal{Y}$ and $U(x) \subseteq \mathcal{Y}$ is the support of π_x in $\mathcal{Y}$ (Bock and Diday (2000)).

Example 2

Let Y = "Sex". A group of employees of a company may be described by a modal variable as $[Sex = \{(0.7)female, (0.3)male\}]$, expressing the relative frequencies of the categories.

2.3 Constrained Probabilistic Data

When we are in presence of hierarchically dependent variables, for which probability / frequency distributions are known, we say that we have constrained probabilistic data. In such cases, the distributions present some special features, which naturally follow from the exigence of coherence of the descriptions.

Example 3

Let again Y= "sex" and Z = "nb. of births". An example of a probabilistic constrained object could be:

$$[Y = \{(0.7)female, (0.3)male\} \wedge [Z = \{(0.3)0, (0.3)1, (0.1)2, (0.3)NA\}].$$

Notice that the frequency of *NA* of the dependent variable Z must equal the frequency of the value of the category 'male' of variable Y, which triggered off the rule. In fact, if in a sample there are 30% of males then, of course, for 30% of the individuals the variable "number of births" is not applicable.

In general, if $Y(x) \in \mathcal{Y}'$ implies that Z is not applicable, then, the frequency of the *NA* category must equal the sum of the frequencies of all categories in $\mathcal{Y}'$. Another alternative would be to use conditional probabilities in the description of the dependent variable Z. In this work, taking into account the generalization operators used and the corresponding interpretation (see §3) we prefer to use absolute probabilities.

3 Generalization

Generalization of two symbolic objects s_1 and s_2 consists in determining another object that is more general than both s_1 and s_2, that is, whose extension contains both the extensions of s_1 and s_2.

Let $s_1 = [y_1 \in V_1] \wedge \ldots \wedge [y_p \in V_p]$ and $s_2 = [y_1 \in W_1] \wedge \ldots \wedge [y_p \in W_p]$.

Generalization is performed variable wise:

$s_1 \cup s_2 = [y_1 \in V_1 \cup W_1] \wedge \ldots \wedge [y_p \in V_p \cup W_p]$.

In the case of existence of hierarchical rules, NA is treated as any other category.

For probabilistic data, two possibilities have been presented (Brito (1998)):

- take for each category the **Supremum** of its frequencies

$e := e_1 \cup e_2 = [Y \in \{(p_1^1)m_1, \ldots (p_k^1)m_k\} \cup [Y \in \{(p_1^2)m_1, \ldots (p_k^2)m_k\} :=$
$[Y \in \{(p_1)m_1, \ldots (p_k)m_k\}$ with $p_i = \mathrm{Max}\{p_i^1, p_i^2\}, i = 1, \ldots, k$.

We define the extension of e by $\mathrm{Ext}(e) := \{s \mid p_i^s \le p_i, i = 1, \ldots, k\}$.

Thus, $\mathrm{Ext}(e)$ consists of the set of all symbolic objects s which contain, for the variable Y, an event $[Y \in \{(p_1^s)m_1, \ldots (p_k^s)m_k\}$ such that each category m_i of Y has at most frequency p_i.

By defining the generalization and the extension in this way, we do generalize: the extension of e contains the extensions of e_1 and e_2.

If moreover there are hierarchical rules of the form $Y(x) \in \mathcal{Y}'$ implies Z is not applicable, the maximum is computed for all categories except NA and then the weight of NA in the generalized object is given by $\mathrm{Min}\{1, \sum_{m_i \in \mathcal{Y}'} p_i\}$.

Example 4

Let Y = "Education", with $\mathcal{Y}$ = {basic, secondary, superior} and Z = "University diploma", with $\mathcal{Z}$ = {BSc, MSc, PhD}. Clearly, if $Y(x)$ = basic or $Y(x)$ = secondary for some x then Z is not applicable for x, $Z(x)$ = NA.

Let us consider a town whose education profile may be described by

$s_1 = [Y \in \{(0.3) \text{ basic}, (0.5) \text{ sec.}, (0.2) \text{ sup.}\} \wedge$

$[Z \in \{(0.1) \text{ BSc}, (0.1) \text{ MSc}, (0.0) \text{ PhD}, (0.8) \text{ NA}\}$

and another town with education profile given by

$s_2 = [Y \in \{(0.1) \text{ basic}, (0.8) \text{ sec.}, (0.1) \text{ sup.}\} \wedge$

$[Z \in \{(0.0) \text{ BSc}, (0.05) \text{ MSc}, (0.05) \text{ PhD}, (0.9) \text{ NA}\}$

The generalized profile is

$s_1 \cup s_2 = [Y \in \{(0.3) \text{ basic}, (0.8) \text{ sec.}, (0.2) \text{ sup.}\} \wedge$

$[Z \in \{(0.1) \text{ BSc}, (0.1) \text{ MSc}, (0.05) \text{ PhD}, (1.0) \text{ NA}\}$

whose extension is composed by towns with *at most* 30% people with basic education, *at most* 80% people with secondary education, *at most* 20% people with superior education, etc., and so, the variable Z = University diploma may be non-applicable in up to 100% of the cases.

- take for each category the **Infimum** of its frequencies

$e := e_1 \cup e_2 = [Y \in \{(p_1^1)m_1, \ldots (p_k^1)m_k] \cup [Y \in \{(p_1^2)m_1, \ldots (p_k^2)m_k] = [Y \in \{(p_1)m_1, \ldots (p_k)m_k]$ with $p_i = \text{Min}\{p_i^1, p_i^2\}, i = 1, \ldots, k$.

We define the extension of e by $\text{Ext}(e) := \{s \mid p_i^s \geq p_i, i = 1, \ldots, k\}$

Thus, $\text{Ext}(e)$ consists of the set of all symbolic objects s which contain, for the variable Y, an event $[Y \in \{(p_1^s)m_1, \ldots (p_k^s)m_k]$ such that each category m_i of Y has at least frequency p_i.

By defining the generalization and the extension in this way, we do generalize: the extension of e contains the extensions of e_1 and e_2.

In the presence of hierarchical rules the minimum is computed for all categories except NA and then the weight of NA in the generalized object is given by $\sum_{m_i \in \mathcal{Y}'} p_i$

4 Generality Measures

In the framework of clustering analysis and related methods it is necessary to compare couples of symbolic objects as concerns generality. For this purpose, we compute the generality degree of a symbolic object, as the proportion of the description space that it covers. In the case of boolean symbolic objects, a relative description potential has been introduced (Brito and De Carvalho (1999)).

For probabilistic data, two measures are presented to evaluate the generality of an object, according to the generalization procedure. If generalization is performed by the maximum, and p_{ij} denotes the weight of category i of variable j ($j = 1, \ldots, p, i = 1, \ldots, k_j$) then the generality degree is defined as:

$$G_1(s) := \prod_{j=1}^{p} \frac{1}{\sqrt{k_j}} \sum_{i=1}^{k_j} \sqrt{p_{ij}} \qquad (1)$$

while, if generalization is performed by the minimum, the generality degree is:

$$G_2(s) := \prod_{j=1}^{p} \frac{1}{\sqrt{k_j(k_j - 1)}} \sum_{i=1}^{k_j} \sqrt{(1 - p_{ij})} \qquad (2)$$

If, for the modal variable j, $p_{1j} + \ldots + p_{k_j j} = 1$, notice that

$$\frac{1}{\sqrt{k_j}} \sum_{i=1}^{k_j} \sqrt{p_{ij}} = \sum_{i=1}^{k_j} \sqrt{p_{ij} \frac{1}{k_j}} \qquad (3)$$

which is the **affinity coefficient** (Matusita (1951)) between$(p_{1j} \ldots p_{k_j j})$ and the uniform distribution.$G_1(e_j) = 1$ is maximum when we are in presence of the uniform distribution : $p_{ij} = \frac{1}{k_j}, i = 1, \ldots, k_j$.
Again, if $p_{1j} + \ldots + p_{k_j j} = 1$ for the modal variable j, $G_2(e_j) = 1$ is maximum when we are also in presence of the uniform distribution. This means that we consider an object the more general the more similar it is to the uniform distribution. $G_2(e_j)$ is the affinity coefficient between $(1 - p_{1j}, \ldots, 1 - p_{k_j j})$ and $(\frac{1}{k_j} \frac{1}{k_j - 1}, \ldots, \frac{1}{k_j} \frac{1}{k_j - 1})$.

If, for each modal variable j, $p_{1j} + \ldots + p_{k_j j} = 1$ then we have:

$$\prod_{j=1}^{p} \frac{1}{\sqrt{k_j}} \leq G_1(s) \leq 1 \text{ and } \prod_{j=1}^{p} \frac{\sqrt{k_j - 1}}{\sqrt{k_j}} \leq G_2(s) \leq 1.$$

Moreover, in any situation, the following properties hold:

- $G_1(s \cup s') \geq \text{Max}\{G_1(s), G_1(s')\}$ if the union is performed by the maximum;

- $G_2(s \cup s') \geq \text{Max}\{G_2(s), G_2(s')\}$ if the union is performed by the minimum;

- $\prod_{j=1}^{p} \frac{1}{\sqrt{k_j}} \leq G_1(s) \leq \prod_{j=1}^{p} \sqrt{k_j}; \prod_{j=1}^{p} \frac{\sqrt{k_j - 1}}{\sqrt{k_j}} \leq G_2(s) \leq \prod_{j=1}^{p} \frac{\sqrt{k_j}}{\sqrt{k_j - 1}}.$

In the presence of hierarchical rules r_1, ..., r_t, the generality degree has to be adjusted by subtracting the generality of the non-coherent descriptions. Let $G(s \mid r_1 \wedge \ldots \wedge r_t)$ be the generality of a symbolic object taking into account rules $r_1, \ldots, r_t$, then

$$G(s \mid r_1 \wedge \ldots \wedge r_t) = G(s) - G(s \mid \neg(r_1 \wedge \ldots \wedge r_t)) \qquad (4)$$

Example 5

Let us consider again the town whose education profile is described by

$s_1 = [Y \in \{(0.3) \text{ basic}, (0.5) \text{ sec.}, (0.2) \text{ sup.}\} \wedge$

$[Z \in \{(0.1) \text{ BSc}, (0.1) \text{ MSc}, (0.0) \text{ PhD}, (0.8) \text{ NA}\}]$

In this example, the hierarchical dependence between the variables "Education" and "University diploma" is expressed by the rule

$r : Y(x) \in \{basic, secondary\} \Longleftrightarrow Z(x) = NA$

The non-coherent descriptions are those for which there is a contradiction with the rule r, i.e., they are those for which if $Y(x) =$ basic or secondary and $Z(x) =$ BSc, MSc or PhD, or if $Y(x) =$ superior and $Z(x) =$ NA. Then,

$$G_1(s_1) = \left(\frac{\sqrt{.3} + \sqrt{.5} + \sqrt{.2}}{\sqrt{3}}\right)\left(\frac{\sqrt{.1} + \sqrt{.1} + \sqrt{0} + \sqrt{.8}}{\sqrt{4}}\right) = 0.75$$

$$G_1(s_1 \mid \neg r) = \left(\frac{\sqrt{.3} + \sqrt{.5}}{\sqrt{3}}\right)\left(\frac{\sqrt{.1} + \sqrt{.1} + \sqrt{0}}{\sqrt{4}}\right) + \left(\frac{\sqrt{.2}}{\sqrt{3}}\right)\left(\frac{\sqrt{.8}}{\sqrt{4}}\right) = 0.34$$

is the generality corresponding to the non-coherent descriptions ({basic, sec}, {BSc, MSc, PhD}) and ({sup},NA).

Therefore, $G_1(s_1 \mid r) = G_1(s_1) - G_1(s_1 \mid \neg r) = 0.41$.

5   The Clustering Algorithm

Having defined the generalization procedure and the generality degree for constrained probabilistic data, we can now outline the main steps of the clustering algorithm. As concerns the cluster structure, hierarchical and pyramidal models are considered. Pyramids (Diday (1986), Bertrand and Diday (1985)) extend hierarchical clustering by allowing non-disjoint clusters which are not nested, but it imposes a linear order on the set being clustered, such that all formed clusters are intervals of this order. So, while a hierarchy is a set of nested partitions, a pyramid is a set of nested overlappings.

Let $E = \{x_1, \ldots, x_n\}$, be the set of objects we wish to cluster and let a_i be the symbolic description of $x_i, i = 1, \ldots, n$. We consider a bottom-up agglomerative algorithm, which is initialized with n concepts, $(x_i, a_i), i = 1, \ldots, n$. At each step, the algorithm forms a new concept (p,s), by merging two existing concepts, (p_1, s_1) and (p_2, s_2) such that

1. p_1 and p_2 can be merged together, that is:
   - if the structure is a hierarchy : none of them has been aggregated before ;
   - if the structure is a pyramid : none of them has been aggregated twice, and p is an interval of a total order on E.

2. $s = s_1 \cup s_2$ fulfills the condition $Ext_E(s) = p$

3. The generality degree of s, $G(s)$, is minimum among objects fulfilling condition 2.

When there is no pair (p_1, s_1) and (p_2, s_2) fulfilling the above conditions, the algorithm proceeds by trying to merge more than two concepts $(p_1, s_1) \ldots (p_t, s_t)$ at a time, for some $t > 2$, and such that suitable generalizations of the above conditions are met. The algorithm stops when the concept (E, s), for suitable s, is formed. Depending on how generalization is performed - by the maximum or by the minimum - the generality degree is computed according to G_1 or G_2.

Example 6

Consider a set $E = \{x_1, x_2, x_3, x_4\}$ where each x_i is described by a symbolic object a_i as follows:

a_1 = [Education = (0.4) basic, (0.3) secondary, (0.3) superior] $\wedge$

[Univ. diplom = (0.2) BSc, (0.05) MSc, (0.05) PhD, (0.7) NA]

a_2 = [Education = (0.5) basic, (0.3) secondary, (0.2) superior] $\wedge$

[Univ. diplom = (0.1) BSc, (0.05) MSc, (0.05) PhD, (0.8) NA]

a_3 = [Education = (0.1) basic, (0.2) secondary, (0.7) superior] $\wedge$

[Univ. diplom = (0.4) BSc, (0.2) MSc, (0.1) PhD, (0.3) NA]

a_4 = [Education = (0.1) basic, (0.1) secondary, (0.8) superior] $\wedge$

[Univ. diplom = (0.3) BSc, (0.3) MSc, (0.2) PhD, (0.2) NA]

The objective is to build a hierarchy. We choose the maximum option to generalize, and, accordingly, compute the generality degree by using G_1. We obtain the following generality degree matrix, where the entry (i, j) is the value of $G_1(a_i \cup a_j \mid r)$, $1 \le i \le j \le 4$):

$$\begin{pmatrix} 0.4265 & 0.4654 & 0.6222 & 0.6834 \\ & 0.4226 & 0.6611 & 0.7223 \\ & & 0.4579 & 0.5409 \\ & & & 0.4800 \end{pmatrix}$$

where r is the rule which expresses the hierarchical dependence between the variables "Education" and "University diploma" as in the example 5. Let,

$s_1 = a_1 \cup a_2$ = [Education = (0.5) basic, (0.3) secondary, (0.3) superior]

$\wedge$ [Univ. diplom = (0.2) BSc, (0.05) MSc, (0.05) PhD, (0.8) NA]

As the extension of s_1 is $p_1 = \{x_1, x_2\}$ and since $G_1(s_1 \mid r)$ is minimum, we form the concept (p_1, s_1). In the next step,

$$\begin{pmatrix} & a_3 & a_4 & s_1 \\ a_3 & 0.4579 & 0.5409 & 0.6611 \\ a_4 & & 0.4800 & 0.7223 \\ s_1 & & & 0.4654 \end{pmatrix}$$

(x_3, a_3) and (x_4, a_4) are merged to form (p_2, s_2) with $p_2 = \{x_3, x_4\}$,

$s_2 = a_3 \cup a_4 = $ [Education = (0.1) basic,(0.2) secondary, (0.8) superior]

$\wedge$ [Univ. diplom = (0.4) BSc, (0.3) MSc, (0.2) PhD, (0.3) NA]

and $G_1(s_2 \mid r) = 0.5409$. Finally, (p_1, s_1) is merged with (p_2, s_2) to form (p_3, s_3) with $p_3 = \{x_1, x_2, x_3, x_4\}$,

$s_3 = s_1 \cup s_2 = $ [Education = (0.5) basic, (0.3) secondary, (0.8) superior]

$\wedge$ [Univ. diplom = (0.4) BSc, (0.3) MSc, (0.2) PhD, (0.8) NA]

and $G_1(s_3 \mid r) = 0.7442$. The values of $G_1(s_i \mid r), i = 1, 2, 3$ are used for indexing the hierarchy.

6 Final Comments and Conclusion

We have introduced an approach to take into account constraints on probabilistic data which is useful for clustering purposes. This has been accomplished by introducing new generality measures, to compare couples of constrained probabilistic objects, together with an appropriate extension of generalization operators, to get new constrained probabilistic objects. The validation, by applying the method to real data, is being carried out.

Acknowledgements. This paper is supported by grants from the joint cooperation FEP (Portugal) and UFPE (Brasil) and by grants from CNPq (Brasil).

References

BERTRAND, P. and DIDAY, E. (1985): A Visual Representation of the Compatibility between an Order and a Dissimilarity Index: The

Pyramids. *Computational Statistics Quarterly, Vol. 2, Issue 1, 31-42.*

BOCK, H. H. and DIDAY, E. (Eds.) (2000): Analysis of Symbolic Data, Springer, Berlin.

BRITO, P. (1998): Symbolic Clustering of Probabilistic Data, in Rizzi, Vichi, Bock (Eds.): Advances in Data Science and Classification, Proceedings of the 6th Conference of the International Federation of Classification Societies, University of Rome (La Sapienza), Springer, Berlin.

BRITO, P. and DE CARVALHO, F. A. T. (1999): Symbolic Clustering in the Presence of Hierarchical Rules, in Studies and Research, Proceedings of the Conference on Knowledge Extraction and Symbolic Data Analysis (KESDA'98), Office for Official Publications of the European Communities, Luxembourg.

DE CARVALHO, F. A. T. (1998): Extension based Proximities between Constrained Boolean Symbolic Objects, in Hayashi, Ohsumi, Yajima, Tanaka, Bock, Baba (Eds.): Data Science, Classification and Related Methods, Proceedings of the 5th Conference of the International Federation of Classification Societies, Springer, Tokyo.

DIDAY, E. (1986): Orders and Overlapping Clusters by Pyramids, in J. De Leeuw et al. (Eds.): Multidimensional Data Analysis Proc., Dswo Press, Leiden

MATUSITA, K. (1951): Decision rules based on distance for problems of fit, two samples and estimation. *Ann. Math. Stat. 3, 1–30.*

On Memory Requirement with Normal Symbolic Form

M. Csernel[1], F.A.T. de Carvalho[2]

[1]INRIA - Rocquencourt
Domaine de Voluceau - Rocquencourt - B. P. 105
78153 le Chesnay Cedex - France, Email: Marc.Csernel@inria.fr

[2]Centro de Informatica - CIn / UFPE
Av. Prof. Luiz Freire, s/n - Cidade Universitaria
CEP: 50740-540 - Recife - PE - Brasil, Email: fatc@cin.ufpe.br

Abstract: Symbolic objects can deal with domain knowledge expressed by dependency rules. However this leads to exponential computation time. In a previous paper we presented an approach called Normal Symbolic Form (NSF) which is applicable when the dependency rules form a tree or a set of trees. This approach leads to a polynomial computation time, but may sometimes bring about an exponential explosion of space memory requirement. The goal of this paper is to study this possible memory requirement and to see how it is bounded according to the nature of the variables and the nature of the rules. We will see that the memory growth is bounded, and that in most cases, instead of a growth, we will obtain a reduction.

1   Introduction

Symbolic Data Analysis (*SDA*) aims to extend standard data analysis to more complex data, called *symbolic data*, as they contain internal variation and they are structured. The *SDA* methods have as input a Symbolic Data Table. The columns of this Symbolic Table are the symbolic variables. A symbolic variable is *set-valued*, i.e., for an object, it takes a subset of categories of its domain. The rows of this data table are the symbolic descriptions of the objects, i.e., the symbolic data vectors whose columns are symbolic variables. A cell of such a data table does not necessarily contain, a single value, but a set of values.

Symbolic descriptions can be constrained by some domain knowledge expressed in the form of rules. Taking this rules into account in order to analyse the symbolic data, requires usually an exponential computation time. In a previous work (Csernel and De Carvalho (1999)) we presented an approach called Normal Symbolic Form (NSF). The aim of this approach is to implement a decomposition of a symbolic description in such a way that only coherent descriptions (i.e., which do not contradict the rules) are represented. Once this decomposition is achieved, the computation can be done in a polynomial time (Csernel 1998), but in some cases, it can lead to a combinatorial explosion of space memory requirement.

The aim of this paper is to study the possible memory space requirement and to see how it is bounded according to the nature of the variables and the nature of the rules, and finally to show that in most cases, we will obtain a reduction rather than a growth.

In the following, we first introduce the symbolic data table and then we provide the definition of the kind of rules we are able to deal with. Then we briefly recall what the NSF is, and, after discussing the time complexity generated by the NSF, we consider the memory growth both locally and globally.

In this paper, we mostly focus on categorical multi-valued variables, but the particular case of continuous data is also mentioned. Before concluding we will look at a real example and see that space reduction can be from 55% to 80%.

2 Symbolic Data and Normal Symbolic Form

2.1 Symbolic Data

In classical data analysis, the input is a data table where the rows are the descriptions of the individuals, and the columns are the variables. One cell of such data table contains a single quantitative or categorical value.

However, sometimes in the real world the information recorded is too complex to be described by usual data. That is why different kinds of symbolic variables and symbolic data have been introduced (Bock and Diday (2000)). A quantitative variable takes, for an object, an intervals of its domain, whereas a categorical multi-valued variables takes, for an object, a subset of its domain.

2.2 Constraints on Symbolic Descriptions

Symbolic descriptions can be constrained by dependencies between couples of variables expressed by rules. We take into account two kinds of dependencies: hierarchical and logical. We will call premise variable and conclusion variable the variable associated respectively with the premise and the conclusion of each rule.

In the following $\mathcal{P}^*(D)$ will denote the power set of D without the empty set. Let y_1 and y_2 be two categorical multi-valued variables whose domains are respectively $\mathcal{D}_1$ and $\mathcal{D}_2$. A hierarchical dependence between the variables y_1 and y_2 is expressed by the following kind of rule:

$$y_1 \in \mathcal{P}^*(D_1) \Longrightarrow y_2 = NA$$

where $D_1 \subset \mathcal{D}_1$ and the term NA means *not applicable* hence the variable does not exist. With this kind of dependence, we sometimes speak of mother-daughter variables. In this paper will deal mostly with hierarchical rules.

A logical dependence between the variables y_1 and y_2 is expressed by the following kind of rule

$$y_1 \in \mathcal{P}^*(D_1) \Longrightarrow y_2 \in \mathcal{P}^*(D_2).$$

Both of these rules reduce the number of individual descriptions of a symbolic description, but the first kind of rule reduces the number of dimensions on a symbolic description, whereas the second does not. We have seen in De Carvalho (1998) that computation using rules leads to exponential computation time depending on the number of rules. To avoid this explosion we introduced the Normal Symbolic Form.

$$Wings \in \{absent\} \Longrightarrow Wings\_colour = NA \qquad (r_1).$$
$$Wings\_colour \in \{red\} \Longrightarrow Thorax\_colour \in (\{blue\}) \quad (r_2)$$

	Wings	Wings_colour	Thorax_colour	Thorax_size
d_1	{absent,present}	{red,blue}	{blue,yellow}	{big,small}
d_2	{absent,present}	{red,green}	{blue,red}	{small}

Table I: original table.

In the symbolic data table presented above there are two symbolic descriptions d_1, d_2, and three categorical multi-valued variables. The values are constrained by rules r_1 and r_2.

2.3 The Normal Symbolic Form

The idea of the NSF is slightly related to Codd's normal form for relational databases (Codd (1972)). The aim is to represent the data in such a way that only coherent descriptions (i.e., which do not contradict the rules) are represented, and the rules will no longer be needed any more. As a consequence, any computation made using these data will not have to take the rules into account. In order to achieved such a goal the original table is decomposed into several tables (according to the number of different premise variables) it is done in databases.

The variables not concerned by the rules remain in the original table. Each of these new tables contains variables to which the rules apply: a premise variable and all linked conclusion variables. It is easy to check wether for each line of these tables . the premise and conclusion variables doe not contradict the rules If a contradiction appears, it is easy to split the value into two parts: one where the premise is true, and one where the premise is false. We will decompose the original data table according the dependance graph between the variables, induced by the rules. We only carry out this decomposition if the graph is a tree. Then we will obtain a tree of data tables. We call the root of the data tree the *Main table*, we call the other tables *secondary tables*. We call two consecutive table repsectively *mother table* and *daughter table*

We say that a set of Boolean SO is conform to the NSF (Csernel and De Carvalho (1999)) if the following conditions hold:

First NSF condition: Either no dependence occurs between the variables belonging to the same table, or a dependence occurs between the first variable (premise variable) and all the other variables.

Second NSF condition: all the values taken by the premise variable for one table line lead to the same conclusion.

The following example shows the result of the NSF transformation of Table I. It can be seen that we now have three tables instead of a single one, but only the valid parts of the objects are represented: now, the tables include the rules.

	wings_r	Thorax_size
d1	{ 1, 3}	{big,small}
d2	{2, 4}	{small}

main table

wings_t	wings	colour_r
1	absent	4
2	absent	5
3	present	{ 1, 2 }
4	present	{ 1, 3 }

secondary table 1

colour_t	wings_colour	Thorax_colour
1	{ red }	{blue }
2	{ blue }	{ blue, yellow }
3	{ green }	{blue, red }
4	NA	{ blue, yellow }
5	NA	{ blue, red }

secondary table 2

The data form a unique (degenerate) data tree where each node is a table. To refer from one table to another, we need to introduce some new variables, called reference variables, which introduce a small space overhead. In the example these variables are denoted by '_r' at the end and the corresponding table by a '_t' at the end. The corresponding values refer to a line number within the table. The first column of *secondary tables* contain the name of the table in the first line and the line number in the other lines. The initial symbolic description can be found from the *Main table*. All the tables, except the *main table*, are composed in the following way:

1) the first variable in a table is a premise variable, all the other variables are conclusion variables;

2) in each line the premise variable leads to a unique conclusion for each of the conclusion variables;

The second NSF condition has two consequences:

1) we have to decompose each individual in a table into two parts: one part where the premise is true, and one part where the premise is false. In order to have an easier notation we will denote this consequence $CQ1$.

2) when we want to represent different conclusions in one table, we represent each description by as many lines as we have conclusions.

The main advantage of the NSF is that after this transformation the rules are included in the data and there are no longer any rules to be taken into account, and so no more exponential growth in computation time. Instead, computation time required to analyse symbolic data is polynomial as if there were no rules to be taken into account.

The $CQ1$ can induce a memory growth. In the next section, we will study this growth more closely according to the kind of rules we use.

3 Hierarchical Dependencies

In the following section, we will consider the possible memory growth, using only categorical set-valued variables. We will use two indices S_n and S_d. S_n indicates the maximum possible size of the table due to the $CQ1$. S_d indicates the maximum possible size of the table according to the domains of the different variables used.

In order to describe the growth more accurately, we will generally distinguish two cases :

1) the local case: where the computation is done comparing a mother table to one of its daughters.

2) the global case: where the comparison is done comparing the original table with a leaf table.

3.1 When All the Conclusions are Determined by the Same Set of Premise Values

Locally

In this case we have one premise variable y and m conclusion variables $x_1, \ldots, x_m$. Each conclusion variable x_j is defined on a domain $\mathcal{X}_j$.

Let the domain $\mathcal{Y}$ of y be partitioned on two sets: A and $\overline{A}$. A is the set of values which makes the conclusion variables inapplicable. Let N_m be the number of lines of the mother table.

According to $CQ1$, the size N_d of the daughter table is at most : $S_n = 2 \times N_m$.

According to the domain of the variables, the size N_d is at most :

$$S_v = (2^{|A|} - 1) + (2^{|\overline{A}|} - 1) \prod_{j=1}^{m}(2^{|X_j|} - 1) \tag{1}$$

where $|A|$ is the cardinal of the set A. Then, N_d is at most $min(S_n, S_v)$.

The previous equation has two terms, each of which represents the maximum possible lines within the table, the first one when the premise is true, the second one when the premise is false.

The first term represents the number of all possible combinations of values of the premise variable when the premise is true. This term is not a product because all the conclusion variables have a unique value: NA.

The second term of the equation is a product, the last factor of this product represents the number of all possible combinations of values of the different conclusion variables. The first factor represents all the possible combinations of values of the premise variable when the premise is false.

We define the *local growing factor* F_l as the ratio between the number of lines N_d of a daughter table and the number of lines of its mother table N_m: $F_l = N_d/N_m$. Its upper bound is given by:

$$F_l = \frac{N_d}{N_m} = \frac{min(S_n, S_v)}{N_m} \leq 2 \tag{2}$$

Conclusion: the size of the daughter table is at most twice as big as the mother table.

Globally

If the set of rules forms a tree, one can wonder if there is an increase in size depending on the depth of the tree. This idea is natural because, if for a single level the number of lines doubles, one may suppose that for two levels the number of lines will quadruple.

In fact, that is not the case. One must consider that if the number of lines doubles, half of these lines (in fact all the newly introduced lines) are filled with NA values. These lines, because of the NA semantic, can not refer to any line of another table, so, only some of the N_m initial lines will lead to values of the second table.

We define the *global growing factor* F_g as the ratio between the number of lines of a leaf table and the number of lines of the initial table. Its upper bound is given by:

$$F_g = \frac{N_d}{N} = \frac{min(S_n, S_v)}{N} \leq 2 \tag{3}$$

3.2 When All the Conclusions are Determined by Different Sets of Values

Locally

At first we will consider the case where we have y, the premise variable, and two disjoint sets of conclusion variables: $\{x_1, \ldots, x_m\}$ and $\{z_1, \ldots, z_t\}$. The domain of x_j and z_k is respectively $\mathcal{X}_j$ and $\mathcal{Z}_k$.

Let the domain $\mathcal{Y}$ of y be partitioned into three sets A_1, A_2 and A_3. A_1 is the set of values that makes x_j inapplicable, A_2 is the set of values that makes z_k inapplicable and $A_3 = \overline{A_1 \cup A_2}$.

According to $CQ1$ the size N_d of the daughter table is at most $S_n = 3 \times N_m$.

According to the domain of the variables the size N_d is at most :

$$S_v = (2^{|A_1|} - 1) \prod_{j=1}^{t} (2^{|Z_j|} - 1) + (2^{|A_2|} - 1) \prod_{k=1}^{m} (2^{|X_k|} - 1) +$$

$$(2^{|A_3|} - 1) \prod_{k=1}^{m} (2^{|X_k|} - 1) \prod_{j=1}^{t} (2^{|Z_j|} - 1) \tag{4}$$

So, N_d is at most $min(S_n, S_v)$ and the upper bound of F_l is

$$F_l = \frac{N_d}{N_m} = \frac{min(S_n, S_v)}{N_m} \leq 3 \tag{5}$$

More generally, if the domain of the variable y is partitioned into n parts, then $F_l \leq n$.

Globally

When the premise value is partitioned into 3 sets, the same kind of phenomenon as we observed in paragraph 3.1 occurs. Each table is divided into three parts, in the first part all the x_j have a value and all z_k are NA. In the second part all the x_j are NA and all z_k have a value. In the third part all the x_j have the same value as in the first part. So, if x_j is a premise variable, it will reference the same line of the daughter table in the first and in the third part, so we have $F_g \leq 3$. More generally, if the domain of the variable y is partitioned into n parts, then $F_g \leq n$.

4 Application

We consider a biological knowledge base which describes the male Phlebotom sandflys from French Guyana (Vignes (1991)). There are 73 species (each corresponding to a symbolic description) described by 53 categorical set-valued variables. There are 5 hierarchical dependencies corresponding to 5 rules. These rules are represented by 3 different connected graphs involving 8 variables. These variables take a discrete set of values.

The size of the different *secondary tables* was 32, 18 and 16 lines. In this particular example the local and global growing factors are identical for each *secondary table*, because there is only one mother table, the main table. The values of the different growing factors are 32/73, 18/73 and 16/73. As we expected, we have a reduction in memory requirement.

5 Final Comments and Conclusion

Until now we just indicated how the growth of memory requirement is bounded, using nominal variable and hierarchical rules. But in real applications, we observed a reduction rather than a growth. This reduction concerns the *secondary tables* and is due to the fact that with a limited number of variables (as in the *secondary tables*) some different SO can have the same description which will appear only once in the table. The greater the number of SO, the greater is the chance of obtaining a factorization is.

In the case of a logical dependencies a major change occurs. If the local growing factor is exactly the same, the global one is different, because the limitation due to the NA semantic will not appear. The size of each *secondary table* can grow according to its position within the table tree. However, if the tree is deep and the number of variables within a table is small then S_v is smaller than S_n and the growth will be bounded.

In the case of continuous variables, the number of lines in a *secondary table* cannot be bounded by S_v because the domain of each continuous variable is infinite. As a consequence, the factorization is less likely to occur. Nevertheless, as there is little chance of having continuous variables as premise, they can just induce a local growth.

To conclude we can say that in every case NSF transforms computation time from exponential to polynomial. If we have nominal variables, and only hierarchical rules the possible memory growth is bounded by a small constant factor and in most cases, we obtain a memory reduction. If we have logical rules, the memory growth is usually linear depending on the number of rules, but in the worst cases the memory growth can be exponential but still bounded by S_v. In both cases a factorization may occur and reduce the memory growth. If we use continuous variables there are no longer any bounds to the possible growth, and very little chance of obtaining a factorization. However, they may still be used without introducing an exponential growth in the memory requirement.

Acknowledgements. This paper is supported by grants from the joint project INRIA-France CNPq-Brasil CLADIS (Proc. 480190/00-1) and by grants from CNPq (Proc. 301387-92-3)

References

BOCK, H. -H. and DIDAY, E. (2000): Analysis of Symbolic Data. Springer, Berlin .

CODD, E. E. (1972): Further Normalization of the Data Relational Model. in R. Rustin (Ed.): Data Base Systems, Prentice-Hall, Englewood Cliffs, N.J., 33–64.

CSERNEL, M. (1998): On the Complexity of Computation with Symbolic Objects using Domain Knowledge, in Rizzi, Vicchi, Bock (Eds.): Advances in Data Science and Classification, Springer, Berlin, 403–408.

CSERNEL, M. and DE CARVALHO, F. A. T. (1999): Usual Operations with Symbolic Data under Normal Symbolic Form. *Applied Stochastic Models in Business and Industry, Vol. 15, 241–257.*

DE CARVALHO, F. A. T. (1998): Extension based Proximities between Constrained Boolean Symbolic Objects, in Hayashi, Ohsumi, Yajima, Tanaka, Bock, Baba (Eds.): Data Science, Classification and Related Methods, Springer, Berlin, 370–378.

VIGNES, R. (1991): Caractérisation Automatique de Groupes Biologiques. Thèse de Doctorat. Université Paris VI, Paris.

On the Roles of Galois Connections in Classification

F. Domenach[1], B. Leclerc[2]

[1]Centre d'Étude et de Recherche en Mathématiques, Statistiques et Économie Mathématique,
Université Paris 1 Panthéon - Sorbonne,
106 - 112 Bd de l'Hopital 75647 Paris Cedex 13, France,
e-mail: domenach@univ-paris1.fr

[2]Centre d'Analyse et de Mathématiques Sociales,
École des Hautes Études en Sciences Sociales,
54 bd Raspail, F-75270 Paris Cedex 06, France,
e-mail: leclerc@ehess.fr

Abstract: Galois connections (or residuated mappings) are of growing interest in various domains related with or relevant from Classification. Among their many uses, we select some topics related with modelization and aggregation of dissimilarities and conceptual classification. We partially revisit them in a common frame provided by a recent study about Galois connections between closure spaces.

1 Introduction

Since the creation of the notion of Galois connection by Öre (1944) and Birkhoff (see the 1967 edition of his book), Galois mappings (or equivalently, polarized, or residuated, mappings) were strongly associated to many fundamental mathematical notions: among others, closure operators, completion by cuts, and full implicational systems (see Caspard and Monjardet 2001).

In conceptual classification, they first provide the mathematical formalization of the classical extent/intent scheme of objects described by properties (see, e.g., Barbut and Monjardet 1970, Duquenne 1987, Ganter and Wille 1999, Wille 1982). They also constitute an important tool for the modelization, the comparison and the aggregation of classification methods and structures (Barthélemy, Leclerc and Monjardet 1986, Janowitz 1978, Janowitz and Wille 1995, Leclerc 1994).

This paper proposes, after a short presentation of the above domains (Section 2), a common frame for these uses of Galois connections in classification (Section 3). It allows us to think of an approach of conceptual classification, illustrated with a short example.

2 Galois Connections in Classification

2.1 Some Definitions

In this section, we recall some basic facts and properties about ordered sets, complete lattices, closure mappings and Galois connections. For more information on these topics, see Birkhoff (1967).

A *(partially) ordered set* P is a set endowed with a reflexive, antisymmetric and transitive binary relation, generally denoted by $\leq_P$, or simply by $\leq$ (by $\subseteq$ in the case of inclusion orders). The order $\leq$ is *linear* if, moreover, either $x \leq y$ or $y \leq x$ for all $x, y \in P$. In this paper, several types of such ordered sets will be considered: with the inclusion order, subsets of a fixed set E, intervals of a linear order, classes of a hierarchy, binary relations on a given set; with the usual order, sets of numerical values (linearly ordered) or products of such sets pointwise ordered ...

Let $(P, \leq)$ and $(Q, \leq)$ be two ordered sets. A mapping $f : P \to Q$ is said to be *antitone* if $x \leq x'$ implies $f(x') \leq f(x)$ and *extensive* if, for any $x \in P, x \leq f(x)$ holds.

A pair (f, g) is a *Galois connection* between two ordered sets P and Q if it satifies the following conditions:

- f is an antitone mapping from P to Q and g is an antitone mapping from Q to P;

- both composed mappings $\varphi = gf$ and $\psi = fg$ are extensive.

Then, the sets $f(P) = \psi(Q) = \Psi$ and $g(Q) = \varphi(P) = \Phi$ are in a one-to-one correspondence by the restrictions of f and g, with a dual order isomorphism: for $x, y \in g(Q), x \leq y \iff f(y) \leq f(x)$. This is one of the strong properties of a Galois connection: it extracts a common (under duality) ordered substructure from the ordered sets P and Q.

A lower (upper) bound of a subset A of P is an element $x \in P$ such that $x \leq a \ (a \leq x)$ for all $a \in A$. A *lattice* $(P, \vee, \wedge, \leq)$ is an ordered set $(P, \leq)$ such that any pair x, y of elements of P has both a least upper bound (join), denoted by $x \vee y$ and a greatest lower bound (meet), denoted by $x \wedge y$. In the sequel, both P and Q will be assumed to be *complete lattices*, where any subset A has a join $\bigvee A$ and a meet $\bigwedge A$. This is the case of a finite latice, or the lattice of the subsets of a given set, or the closed real interval $[0, 1]$, or products of such complete lattices.

2.2 A Basic Fact for Conceptual Classification

Galois connections satisfy strong requirements that, at a first glance, could be expected to be rarely satisfied. In fact, as practitioners know, they are encountered in many fields, frequently according to the following frame. Set $P = \mathcal{P}(S)$, where S is the finite set of objects under study, and consider a (complete) lattice Q of *descriptions*, together with a description $d(s) \in Q$ of each element $s \in S$. The order of Q corresponds with a generalization order, where $q \leq q'$ means that description q is more general than description q'. Then, it is said that $s \in S$ satisfies description q if $q \leq d(s)$.

Consider, for any class (subset) $C \subseteq S$, and description $q \in Q$, the mappings $f(C) = \bigwedge\{d(s) : s \in C\}$ and $g(q) = \{s \in S : q \leq d(s)\}$; so, $f(C)$ is the less general description satisfied by all the elements of C (the *intent* of C), while $g(q)$ is the class of all the elements of S satisfying description q (the *extent* of q). Then, it is not difficult to see that the pair (f, g) constitutes a Galois connection between $\mathcal{P}(S)$ and Q; it defines a set C of *conceptual classes* which are pairs (C, D) with $C = g(D) \subseteq S$ and $D = f(C) \in Q$. Moreover, according to the properties of Galois connections between two lattices, C itself is a lattice (frequently called the *Galois lattice* of the triple (S, Q, d), with $(C, D) \leq (C', D') \iff C \subseteq C' \iff D' \leq D$; the lattice operations of C are given by $(C, D) \wedge (C', D') = (C \wedge C', f(C \wedge C'))$ and $(C, D) \vee (C', D') = (g(D \wedge D'), D \wedge D'))$.

The lattice Q may be a set of numbers, or number intervals (with the inclusion order), or product of such lattices (see Bock and Diday 2000). In a classical instance, one considers a set A of binary attributes that elements of S may have or have not and set $Q = \mathcal{P}(A)$. Then, the data consist of a binary relation $I \subseteq S \times A$, with sIa if object s has attribute a; equivalently, for $s \in S, d(s) = sI = \{a \in A : sIa\}$. Then, $f(C) = \bigcap\{d(s) : s \in C\} = \{a \in A : sIa \text{ for all } s \in C\}$ and $g(A') = \{s \in S : sIa \text{ for all } a \in A'\}$. This situation is well-known as the basic frame of Formal Concept Analysis (Ganter and Wille 1999) and the lattice obtained by a Galois connection of this type is also called the *concept lattice* of the context (S, A, I).

2.3 Conceptual Classification

As described above, Galois lattices fulfill one of the aims of Classification: they provide classes (the extents) of objects sharing similar characters (the intents), a description by attributes is associated to each class. The obtained correspondence between classes and descriptions is a basic fact in the domain of Symbolic Classification.

Another purpose of Classification is the organization of data to make them more readable or to recover some unknown structure. For instance, hierarchical clustering methods provide a classification tree, sometimes (e.g. in phylogenetic reconstruction or cognitive psychology) an estimation of an unknown tree. The concept lattice does not correspond to such an objective since it preserves the whole information of the data. So, as frequently observed in the literature, it has a great sensitivity to noise and deviation from the model. Also, the number of concepts potentially grows exponentially with the data size, leading to problems of computational complexity.

For these reasons, several authors have proposed methods to prune the concept lattice, for instance by limiting its construction to a convenient filter (Mephu-Nguifo 1993). Another approach, more related with classification, consists of retaining only a (small) part of C, in such a way as the set of the corresponding extents satisfy a desired structure, e.g. constitute a hierarchical tree or a pyramid (Guénoche 1993, Brito 1995, Polaillon 1998), the last two references in the more general context of Diday's (1988) symbolic objects.

2.4 Residuation Models

An apparently different use of Galois connections in classification started with a paper of Janowitz (1978). In fact, Janowitz took the equivalent formalization of Residuation Theory (Blyth and Janowitz 1972). For sake of homogeneity, our presentation differs from his, and is based on similarity considerations instead of dissimilarity. Consider the closed real interval $[0, 1]$ and the lattice $\mathcal{R}_S$ of all the binary relations on a given finite set S, ordered with inclusion. Here, a *dendrogram* is a mapping $g : [0, 1] \rightarrow \mathcal{R}_S$ satisfying the properties (D1) for all $\lambda, \mu \in [0, 1], \lambda \leq \mu$ implies $g(\mu) \subseteq g(\lambda)$, and (D2) for any $x, y \in S$, the set $\{\lambda \in [0, 1] : (x, y) \in g(\lambda)\}$ has a maximum $f(x, y)$. The value $f(x, y)$ can be thought of as a degree of proximity from x to y.

A consequence of (D2) is that, in fact, for any binary relation $R \subseteq S^2$, the set $\{\lambda \in [0, 1] : R \subseteq g(\lambda)\}$ has a maximum. Especially, $g(0) = S^2$. Then, g is a Galois mapping and, extending the range of f to $\mathcal{R}_S$ by setting $f(R) = min\{f(x, y) : (x, y) \in R\}$, the pair (f, g) constitutes a Galois connection between the lattices $\mathcal{R}_S$ and $[0, 1]$.

In the above scheme, a restricted lattice of relations can be considered instead of $\mathcal{R}_S$. The case of the lattice $\mathcal{S}_S$ of all the symmetric binary relations on S (with the classical similarity coefficients as function f) corresponds to Janowitz's initial paper. In Leclerc (1991, 1994) the lattices of preorders on S (with a kind of valued preorders as similarity coefficients) and of equivalences on S (with dual ultrametric coefficients) are considered. The last case corresponds to standard dendrograms (called *valued trees* in Boorman and Olivier 1973) of the classification literature. Other examples are also explicited in Leclerc (1994), where generalized dendrograms account for other one-to-one correspondence between, on the one hand, types of dissimilarities, and, on the other hand, seriation or classification structures. Two points may be emphasized in this approach: (i) the elements of Q are not necessarily binary relations; examples are given where they are subsets of S of a special, e.g. convex, type; (ii) the interval $[0, 1]$ may be replaced with a description lattice Q as in Sections 2.2-3 above.

Janowitz's residuation (or, as in the presentation here, Galois) model was successfully used in clustering method formalization (Janowitz, e.g. 1978, Janowitz and Wille 1995), comparison of classification trees (Boorman and Olivier 1973, Barthélemy, Leclerc and Monjardet 1986) or consensus of classification trees, valued preorders, and other objects of similar type (Barthélemy, Leclerc and Monjardet 1986, Leclerc 1991, Leclerc and Monjardet 1995).

3 A Paradigm for Conceptual Classification

3.1 A Unifying Frame

Table 1 below summarizes the previously mentioned uses of Galois lattices in Classification; of course, it does not claim exhaustivity in the field.

Consider the lattice Q of descriptions. In Table 1, Situation 2 appears as a generalization of Situation 1. In fact, as pointed out in Ganter and Wille (1999), the FCA scheme accounts for various types of situations and lattices. First, any complete lattice Q can be represented by a family of subsets of a set A of elementary descriptions, provided A generates Q by the join operation. Each description $q \in Q$ is representable by the set $A_q = \{a \in A : a \leq q\}$, with $q = \bigvee A_q$. A subset A' of A is obtainable as A_q for some $q \in Q$ if and only if it satisfies the following condition (C):

$$(C) \text{ for any } a \in A, a \leq \bigvee A' \text{ implies } a \in A'$$

We then say that A' is *closed with regard to* Q (or *Q-closed*). Let $\mathcal{G} \subseteq \mathcal{P}(A)$ be the set of all subsets satisfying (C). The correspondence $q \leftrightarrow A_q$ between Q and $\mathcal{G}$ is one-to-one. For instance, consider the case where elements of Q are intervals of some linearly ordered set $(L, \leq_L)$, ordered by inclusion. Then, we can take points intervals as elementary descriptions, and the closure constraints particularize as $a, a' \in q$ and $a \leq_L b \leq_L a'$ imply $b \in q$.

Now the case $\mathcal{G} = \mathcal{P}(A)$ of Situation 1 just appears as the unconstrained one. When examining the models of Section 2.4, we find in Situations 5 and 6 (and others) lattices P of objects that are no longer power sets.

Model	Lattice P of objects	Lattice Q of descriptions	Obtained system of classes
1. Formal Concept Analysis	$\mathcal{P}(S)$	$\mathcal{P}(A)$	Any lattice of subsets (classes) of S
2. Galois lattice classification	$\mathcal{P}(S)$	Any lattice of descriptions	Any lattice of subsets of S
3. Conceptual Classification	$\mathcal{P}(S)$	Any lattice of descriptions	A structured set of classe (hierarchy, pyramid) extracted from the above lattice with associated descriptions
4. Dendrogram (Janowitz 1978)	S_S (symmetric binary relations)	$[0,1]$ or $\mathbb{R}^+$	An indexed chain of symmetric binary relations on S
5. Standard dendrogram	$\mathcal{E}_S$ (equivalence relations)	$[0,1]$ or $\mathbb{R}^+$	An indexed classification tree
6. Generalized dendrogram	$\mathcal{T}_S$ (a specified lattice of binary relations)	$[0,1]$ or $\mathbb{R}^+$	An indexed chain of relations in $\mathcal{T}_S$

Table 1: Uses of Galois lattices in classification

They may be represented again, as above, by the P-closed subsets of a set S of elementary objects, and the set $\mathcal{F}$ of all the P-closed subsets of S is in one-to-one correspondence with P. For instance, when $P = \mathcal{E}_S$, elementary objects are unordered pairs xy of elements of S, and the closure constraints particularize as : $x, y, z \in S$ and $xy, yz \in p$ imply $xz \in p$ (the transitivity property).

In practice, the definition of elementary descriptions is frequently less easy than for objects, but such descriptions always exist. At worst, they are all the elements of Q. Assume sets S, A of elementary objects and descriptions have been defined, and P and Q replaced with the equivalent sets $\mathcal{F}$ and $\mathcal{G}$. Then, a relation $I \subseteq S \times A$ is defined by $sIa \iff a \leq d(s)$ and we are in a situation of the Formal Concept Analysis type, with precisely the following constraints:

- for any $s \in S, sI = \{a \in A : sIa\} \in \mathcal{G}$;

- for any $a \in A, Ia = \{s \in S : sIa\} \in \mathcal{F}$;

We then say that I is *PQ-biclosed*. When relation I is given by its table, sets sI correspond to the lines of the table and sets Ia to the columns. As shown in Domenach and Leclerc (2001), biclosed relations are in one-to-one correspondence with Galois connections between P and Q. With such a relation, a conceptual class (C, D) has an extent corresponding to an element of P (and an intent to a description in Q).

3.2 Conceptual Classification as a Fitting Problem

Many classification problems may be expressed by a requirement that the obtained classes belong to a lattice of some type. For instance, a set of classes $\mathcal{H} \subseteq \mathcal{P}(S)$ is a *hierarchy* on S if $S \in \mathcal{H}$ and $H, H' \in \mathcal{H}$ imply $H \cap H' \in \{\emptyset, H, H'\}$. It follows that $\mathcal{H}^* = \mathcal{H} \cup \{\emptyset\}$ is a lattice of subsets of S. The same remark holds for, e.g., pyramids; for sake of brevity we focuse here on the hierarchy case.

Consider data consisting, as in Section 2.2, of a set S endowed with individual descriptions $d(s), s \in S$, elements of some lattice Q. According to Section 3.1, such data correspond to a binary relation I_0 with Q-closed lines sI_0 and unconstrained columns I_0a. Now the problem is to fit to I_0 a relation I with P-closed columns. Two cases may be distinguished: (i) The lattice P is given. Then, there exists a canonical solution $\beta(I_0)$, the minimum (for inclusion) biclosed relation containing I_0. This existence is pointed out in Domenach and Leclerc (2001), with algorithmic indications. Further research could address the properties of this approach, and also methods providing alternative solutions, possibly more close to the data.
(ii) Only the type of P is defined. This situation may be compared with hierarchical classification, where the aim is to obtain a system of classes of the hierarchy form, but of course not fixed at the beginning of the classification process. Here, much work seems to remain to do.

	a_1	a_2	a_3	b_1	b_2	c_1	c_2	d	e_1	e_2	f_1	f_2	g_1	g_2	h
A				1	1	1	1	1	1	1	1	1	1	1	
B	1				1		1	1	1	1		1	1	1	1
C	1	1		1		1			1	1			1		
D				1	1	1		1							1
E	1	1		1		1			1		1				1
F	1	1	1				1				1	1	1	1	1

Table 2: Relation I_0

We illustrate by a simple example, adapted, essentially by reduction, from real data. Six scanners are described, in a consumer review, according to eight criteria, with, respectively, 4, 3, 3, 2, 3, 3, 3, 2 (totally ordered) levels. A criterion with k levels is represented by $k-1$ columns in Table 2 (the omitted columns have only "1"entries). So, the number of columns is 15.

In order to put ourselves back in Situation (i) above, we first use the well-known Jaccard similarity index between the lines of Table 2 to determine the complete linkage hierarchy $\mathcal{H}$ of Figure 1. Then, we require columns to correspond to classes of $\mathcal{H}$. Table 3 gives $\beta(I_0)$, while an alternative solution I', closer to the original table, is obtained by deleting the bold "1"s in $\beta(I_0)$. Besides the empty and the full ones, Table I' leads to three conceptual classes, the extent of which being, respectively, A, ABF, and CE.

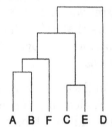

Figure 1: Hierarchy $\mathcal{H}$

	a_1	a_2	a_3	b_1	b_2	c_1	c_2	d	e_1	e_2	f_1	f_2	g_1	g_2	h
A	1	1		1	1	1	1		1	1	1	1	1	1	1
B	1	1		1	1	1			1	1	1	1	1	1	1
C	1	1		1	1	1		**1**	1	1	1		1		1
D				1	1	1		**1**							1
E	1	1		1	1	1		**1**	1	1	1		1		1
F	1	1	1	1	1	1			1	1	1	1	1	1	1

Table 3: Relation $\beta(I_0)$, and an alternative solution I'

4   Conclusion

All the uses of Galois connections recalled in Sections 2.2 to 2.4 are based on the strong hypothesis of a complete lattice structure for both set P of object and set Q of description. These hypotheses are not always satisfied but, as it appears in the literature of symbolic data analysis, enough realistic. Setting emphasis on the description of lattices P and Q in terms of closure constraint (the implications in the recent literature about lattices) allowed us to give a unified presentation of some parts of the previous literature and to propose an approach for further studies.

References

BARBUT, M. and MONJARDET, B. (1970): Ordre et classification, Algebre et combinatoire. Hachette, Paris.

BARTHÉLEMY, J.P. , LECLERC, B. and MONJARDET, B. (1986): On the use of ordered sets in problems of comparison and consensus of classifications. *Journal of Classification, Vol. 3, 187–224.*

BIRKHOFF, G. (1967): Lattice Theory (3rd ed.). Amer. Math.Soc, Providence, R.I.

BLYTH, T.S. and JANOWITZ, M.F. (1972): Residuation Theory. Pergamon Press, Oxford.

BOCK, H.H. and DIDAY, E. (2000): Analysis of symbolic data. Studies in Classification, Data analysis and Knowledge organization, Springer, Berlin.

BOORMAN, S.A. and OLIVIER, D.C. (1973): Metrics on spaces of finite trees. *Journal of Mathematical Psychology, Vol. 10, 26–59.*

BRITO, P. (1995): Symbolic objects: order structure and pyramidal clustering. *Annals of Operations Research, Vol. 55, 277–297.*

CASPARD, N. and MONJARDET, B. (2001): The lattice of closure systems, closure operators and implicational systems on a finite set; a survey. *Discrete Applied Math., to appear.*

DIDAY, E. (1988): The symbolic approach in clustering and related methods of data analysis: the basic choices, in Bock (Ed.), Classification and Related Methods od Data Analysis, North-Holland, Amsterdam, 673–683.

DOMENACH, F., and LECLERC, B. (2001): Biclosed binary relations and Galois connections. *Order, Vol. 18, 89–104.*

DUQUENNE, V. (1987): Contextual implications between attributes and some representation properties for finite lattices, in Ganter et al. (Eds.), Beiträge zur Begriffsanalyse, Wissenchaftverlag, Mannheim, 213–240.

GANTER, B. , and WILLE, R. (1999): Formal concept analysis. Springer-Verlag, Berlin.

GUÉNOCHE, A. (1993): Hiérarchies conceptuelles de données binaires. *Math. Inf. Sci. hum., Vol. 121, 23–34.*

JANOWITZ, M.F. (1978): An order theoretic model for cluster analysis. *SIAM J. Appl. Math., Vol. 37, 55–72.*

40

JANOWITZ, M.F. and WILLE, R. (1995): On the classification of monotone and equivariant cluster methods, in Cox, Hansen, Julesz (Eds.), Partitioning data sets, DIMACS Series in Discrete Mathematics and Theoretical Computer Science, Vol. 19, Amer. Math. Soc., Providence, RI, 117–141.

LECLERC, B. (1991): Aggregation of fuzzy preferences: a theoretic Arrow-like approach. *Fuzzy sets and systems, Vol. 43, 291–309.*

LECLERC, B. (1994): The residuation model for the ordinal construction of dissimilarities and other valued objects, in Van Cutsem (Ed.), Classification and Dissimilarity Analysis, Springer-Verlag, New York, 149–172.

LECLERC, B., and MONJARDET, B. (1995): Latticial theory of consensus, in Social choice, Welfare and Ethics, in Barnett, Moulin, Salles, Schofield (Eds.), Cambridge University Press, 145–159.

MEPHU NGUIFO, E. (1993): Une nouvelle approche basée sur le treillis de Galois pour l'apprentissage de concepts. *Math. Inf. Sci. hum., Vol. 124, 19–63.*

ÖRE, O. (1944): Galois connections. *Trans. Amer. Math. Soc., Vol. 55, 494–513.*

POLAILLON, G. (1998): Interpretation and reduction of Galois lattices of complex data, in Rizzi, Vichi, Bock (Eds.), Advances in Data Science and Classification, Studies in Classification, Data Analysis and Konwledge Organization, Springer-Verlag, Berlin, 433–440.

WILLE, R. (1982): Restructuring lattice theory: an approach based on hierarchies of concepts, in Rival (Ed.), Ordered Sets, D. Reidel, Dordrecht, 445–470.

Acknowledgement : The authors thank an anonymous referee for his careful reading and constructive remarks.

Mining Sets of Time Series: Description of Time Points

A. Dugarjapov, G. Lausen

University of Freiburg
Institut for Computer Science
Georges-Köhler-Allee, 51
79110 Freiburg, Germany

Abstract: Investigating time series with respect to statistical analysis and forecasting is a well established area. In this paper we study the problem of matching time series. This paper addresses the problem how to choose the attributes describing the time points with respect to certain matching problem. Furthemore, this paper introduce an algorithm, computing the domain error of a classifier of the model, which can be used as an evaluation function of the classifier with respect to the matching problem. The exeprimental results, presented in this paper, illustrate the usefulness of the evaluation.

1 Introduction

Investigating time series with respect to statistical analysis and forecasting is a well established area. In this paper we study the problem of matching time series. Consider a set of time series over some fixed time intervall $[t_0..t_k]$. Each time point of the intervall is described by a set of uniform tuples of attributes, where attributes correspond to certain time points in the intervall and each time series gives rise to one such tuple. The intervall is partitioned into subsets of time points. The set of tuples for each subset is used to build a classifier where the classes are given by the respective time points. A model of the set of time series if constituted by the corresponding set of classifiers; it can be used for a pointwise matching of some patterns of values against the set of time series. This approach of the modeling of a set of time series has been introduced in Savnik et al. (2000). It allows for matching of a sequence of real numbers with the set of time series, using a confidence-rated classification technique, such as C5.0 (Quinlan (1993)). The corresponding matching algorithm performs the *dating* of the sequence with respect to the given set of time series.

The dating of a sequence is considered as the assigning of a time point to the first item of the sequence. Actually, it is a generalization of the problem of *subsequence matching in time series databases* (Faloutsos at al. (1994)) to a case with *multivariate* time series. The matching algorithm was successfully used for dating of tree-ring time series (Savnik et al. (2000)) and is general enough to be effectively used for other domains including stock market data, sensor data, medical measurements and DNA sequences.

The matching algorithm compares the sequence, starting as a certain time point with a set of time series. As similarity measure is used the *collected probability*: the higher the collected probability, the higher the similarity between the set of time series and the sequence starting at the given time point. The matching algorithm computes the collected probabilities of some number of best matches, depending on certain tresholds. This approach to modeling a set of time series is different to other approaches. The most widely known approach to describe the characteristics of a set of time series is the creation of exactly one time series, representing the given set of time series. The time series can be created in different ways, e.g. the *longest common subsequence* (Sankoff, Kruskal (1983)), the *prototypical sequence* (Keogh, Pazzani(1998)) or the *master time series* (Holmes (1983)).

The structure of the paper is as follows. We start with a motivation of the current paper. The formal framework is given in the section 3. Section 4 presents the algorithms, which computes the error of a classifier. Some experimental results can be found in Section 5. Section 6 contains a comparison to other works and concludes the paper.

2 Motivation

The original algorithm of Savnik et al. (2000) has been improved in Lausen et al. (2000). This version allows to compute the collected probability for each possible match of a sequence with a model of a set of time series. The match with the highest collected probability is taken as the final dating. Since, we are able to create different models of a set of time series, we may search for a model, which is "good". As measure for the "goodness" of a model we use its *discriminatory power* (e.g. Sachs (1984)). That means, a model is "good", if the matching of a sequence with the model produce the best match with a high collected probability and all other matches have a much lower collected probability.

If we allow arbitrary partitions of the time interval and arbitrary subsets of features, describing time points, the number M of all possible models of a set of time series will be obviously exponential in the length of the time series: $M \in O(B_k \cdot (2^k)^k)$, where k is the length of the time series, B_k is the k-th Bell number and $(2^k)^k$ is the number of possibilities to combine each of $O(k)$ classifiers with once of the 2^k feature subsets. Consequently, exhaustive search for the best model is out of question.

The intention of this work is to find an evaluation function of a classifier which allows a reasonable choice of attributes, describing time points. So, we assume the partition of the time interval to be already choosen.

In the framework of this paper, each time point is described by a set of *windows* of values of neighboring time points. In other words, each time point is described by the values of time series, which appeared within a time window, whose central point is the described time point. The motivation is as follows: at first, we may assum, that the dependence between values of time series is local in the time and at second, we would like to be able to match short sequences.

Since we are looking for a model, which produce the matching with a high discriminatory power, we should care not only about the best match, but also about all other possible matches. That means, on the one hand, each classifer should classify a case, belonging to the class from its responsibility with a high confidence as possible, and on the other hand, it should classify a case, belonging to some other classes, with a low confidence as possible. The algorithm, presented in this paper, evaluate a given classifier with respect to both aspects. This evaluation function can be applied to several feature selection algorithms, respectively to algorithms, searching for an improved model.

The experiment, presented in this paper consists of two parts. The first part is the creation of the learning curves of the classifiers, using different window sizes. The second part is the matching of test time series with models, created with different window sizes.

3 Formal Framework

The following is based on Savnik et al. (2000). A time series $s = (v_1, \ldots, v_r)$ is a sequence of real numbers. Given a set of time series $D = \{s_1, \ldots, s_m\}$. D is called a *domain*. Each time series $s_i \in D$ is defined for one and the same time intervall $[t_0..t_k]$, that means, each time series $s_i \in D$ has a value for each time point $t_j \in [t_0..t_k]$. $s_i[t_j]$ denotes the value of time series s_i, appeared for the time point t_j.

The interval $[t_0..t_k]$ is partitioned into n non overlapping subsets $T_i \subset [t_0..t_k], i \in [1..n]$, each of which consits of at least two time points.

The values of a time series $s_i \in D$ in subsequent time points $t_j, \ldots, t_{j+k}$, $[t_j..t_{j+k}] \subset [t_0..t_k]$, is the projection of s_i on the given sequence of time points $s_i[t_j..t_{j+k}] = (s_i[t_j], \ldots, s_i[t_{j+k}])$. Let $w \geq 1$ and w odd, then the projection $s_i[t_j..t_{j+w-1}]$ is called a *window* of s_i and w is the size of the window. Furthermore the time point t_j is called the *central point* of the window $s_i[t_{j-\lfloor w/2 \rfloor}..t_{j+\lfloor w/2 \rfloor}]$ of the size w.

Each time point t_j is described by the values, which appeared in the domain time series within a window, whose central point is t_j. A pair $(s_i[t_{j-\lfloor w/2 \rfloor}..t_{j+\lfloor w/2 \rfloor}], t_j)$ can be seen as a training example.

Let w be a window size. The *dataset* S_i^w is the set of all windows of size w, whose central point belongs to T_i, with corresponding central point:

$S_i^w = \{(s_r[t_{j-\lfloor w/2\rfloor}..t_{j+\lfloor w/2\rfloor}], t_j) \mid r \in [1..m], t_j \in T_i\}$. The dataset can be used as the training set of a classifier. The classifier, created from the dataset S_i^w is denoted by C_i^w and can classify any sequences of real numbers of the length w to the time points from T_i. Let be U a sequence of real numbers of the length w. The result of the classification of U with C_i^w is the pair (t, p), where $t \in T_i$ is the *predicted central point* of U and $p \in (0, 1]$ the probability of the correctness of the prediction.

Let $W = (w_1, \ldots, w_n)$ be a sequence of window sizes. A sequence of classifiers $T = (C_1^{w_1}, \ldots, C_n^{w_n})$ is a *model* of the domain. Let $s = (u_1, \ldots, u_r)$ be a sequence of values. A match of s with the model T is an assumption, that s starts at the certain time point t. The time point t is called the *starting time point* of the match. Let C_i^w be a classifier of the model T, $U = (u_y, \ldots, u_{y+w-1})$ a window of s and (t_j, p) the result of the classification of U with $C_i^{w_i}$. The time point t_j is the predicted central point of the window U. Let $\hat{t} = t_j - \lfloor w/2\rfloor - (y + 1)$ be the time point, on which s must start, so that t_j is the central point of U. The classification is then said *consistent* with the match with starting time point $\hat{t}$.

The algorithm, introduced in Lausen et al. (2000), computes a *collected probability* for each possible match of the sequence with the model. The collected probability of a match is the sum of all probabilities of classifications, which are consistent with the match, divided by the number of all classifications, which could be consistent with the match. Consequently, a classification of a window with a classifier contributes those probability to the collected probability of the match, with which the classification is consistent. Let t^* be the starting point of s. All classification of windows of s, which are consistent with t^* are called *correct*, all other classification are called *incorrect*.

4 Domain Error

Let $D = \{s_1, \ldots, s_m\}$ be a domain, $W = (w_1, \ldots, w_n)$ a sequence of window sizes, $T = (C_1^{w_1}, \ldots, C_n^{w_n})$ a model and $C \in T$ a classifier. The question is, how "good" is C with respect to the matching algorithm. A classifier is "good", if the probabilities of it's correct classifications are high and the probabilities of it's incorrect classifications are low.

We consider a *domain error* $\epsilon(C)$ of a classifier C computed by the algorithm, presented in the figure 1. The algorithm classifies each possible window of size w, obtained from each domain time series. If the classifier is able to make a correct classification (line 5) and if it does (line 7), the algorithm accumulates the probability of the classification (line 8). If the classifier is not able to make a correct classification (line 10), the algorithm accumulates the complement of the probability of the classification. The resulting domain error is then the complement

Algorithm domain error

```
Input:   Domain D = {s₁,...,sₘ}, classifier Cᵢʷ with
         the label set Tᵢ and window size w
Output:  The domain error ε(Cᵢʷ)
Method:
```

1. $sum_1 = 0$; $sum_2 = 0$; $i_1 = 0$; $i_2 = 0$;
2. **foreach** ($s \in D$) **do**
3.   **foreach** (window $U = s[t_{j-\lfloor w/2 \rfloor}..t_{j+\lfloor w/2 \rfloor}]$) **do**
4.   ($t, prob$) = predictCentralPoint(U, C_i^w);
5.   **if** ($t_j \in T_i$) **then**
6. $i_1 = i_1 + 1$;
7.   **if** ($t_j = t$) **then**
8. $sum_1 = sum_1 + prob$;
9.   **fi**;
10.   **else**
11. $i_2 = i_2 + 1$;
12. $sum_2 = sum_2 + (1 - prob)$;
13.   **fi**;
14.   **od**;
15. **od**;
16. **return** $1 - (\frac{sum_1}{i_1} + \frac{sum_2}{i_2})/2$;

Figure 1: Algorithm domain error.

of the arithmetic mean of the weighted sums (line 16). The function predictCentralPoint(U, C_i^w) returns the pair ($t, prob$), where t is the class, to which the classifier C_i^w assigns the window U and p is the probability, that the prediction is correct.

5 Experimental Results

The data set used for the experiment consists of 50 tree-ring time series, obtained from Europen beech sample trees from two geographical areas in Germany (cf. experiment in Savnik et al. (2000)). A tree-ring time series is a sequence of values which represents the yearly increments of a tree trunk. The dating of tree-ring time series is an important problem of *Dendrochronolgy* (Cook, Kairiukstis(1990)). All time series are defined for the time interval [1880..1979]. The data set was randomly split into two parts: the domain, cosisting of 40 training time series, and 10 test time series. The time interval was partitioned into 10 equal subintervals, each of which was used to create a classifier of the model. As the classification technique was used the program C5.0 (Quinlan (1993)) with the boosting (e.g. Schapire (1999)).

In the first part of the experiment we build the learning curves of the classifiers (figure 2). A learning curve of a classifier is given by the plot of the domain error of the classifier using different window sizes.

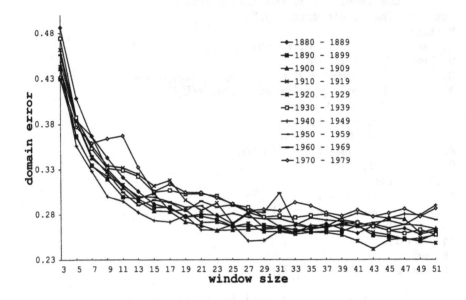

Figure 2: domain errors

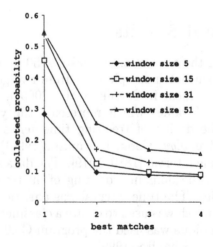

Figure 3: best matches

In the second part of the experiment we made the matching of the test time series with the models, which were build on different window sizes. The classifiers of a particular model were created with one and the same window size. We created four models with window sizes 5, 15, 31 and 51, respectively. The results of the matching are presented in the figure 3. Each test time series was matched with each model. We kept the four best matches of each matching. For each model and for each best match we calculated the average of the collected probability over all ten time series. So we have for each of the models four best matches.

The plots of the different models shows, that the model, created with the window size 5, produces the matching with relatively low discrimatory power as the difference between the probability of the best match and the probablities of the next best match is rather small. Moreover, the increasing of the window size beyond 15 does not result in a significant improvement of the discriminatory power. This results correspond very well with the structure of the learning curves of the classifiers.

6 Related Work

A great number of works has been done on *feature subset selection*. Blum, Langley (1997) and Dash, Liu (1997) can be seen as good surveys over different approaches and techniques of feature selection. The problem of the feature selection is the searching for a "good" set of attributes for a particular classifier. The goodnes of a set of attributes depends very strong on the purpose, for which the classifier is constructed. In our case it is the discriminatory power of the matching of a sequence with a set of classifiers. We proposed an evaluation function of a classifier, the domain error, which can be used in a *wrapper* feature selection algorithm (see e.g. Blum, Langley (1997)). The significant difference between the domain error and classical classifier evelution functions[1] is that the classifier is evaluated on training instances, which do not belong to a class from the classifiers label set. To the best of our knowledge, there are no works on this topic.

7 Conclusion

A technique for classifier evaluation, which allows a reasonable choice of attributes, describing time points was presented in this paper. The domain error, introduced in this paper, can be used as an evaluation function of a classifier. The experimental results shows, that the domain error can support the choice of the window size by the model creation.

---

[1] e.g. classifiers error rate(Dash, Liu (1997)) or estimated accuracy (John et al. (1994))

8 Aknowledgements

We thank Heinrich Spiecker and Hans-Peter Kahle for many insightful discussions and the opportunity to use real world data.

References

BLUM, A., LANGLEY, P. (1997): Selection of Relevant Features and Examples in Machine Learning. Artificial Intelligence, Vol. 97, 245–271.

COOK, E., KAIRIUKSTIS, L. (1990): Methods of Dendrochronology: Applications in the Environmental Siences. Kluwer Acad. Publishers.

DASH, M., LIU, H. (1997) : Feature Selection Methods for Classification. Intelligent Data Analysis : An International Journal, Vol.1, N.3.

FALOUTSOS, C., RANGANATHAN, M., MANOLOPOULOS, Y. (1994): Fast Subsequence Matching in Time-Series Databases. Proceedings of the 1994 ACM SIGMOD International Conference. ACM Press, 419-429

HOLMES, R. L. (1983): Computer-Assisted Quality Control in Tree-Ring Dating and Measurement. Tree-Ring Bulletin 43, 69–78.

JOHN, G. H., KOHAVI, R., PFLEGER, K. (1994): Irrelevant Features and the Subset Selection Problem. Proceedings of the 11th International Conference on Machine Learning 1994, Morgan Kaufmann, 121–129.

KEOGH, E. J., PAZZANI, M. J. (1998): An Enhanced Representation of Time Series Which Allows Fast and Accurate Classification, Clustering and Relevance Feedback. Proceedings of the 4th International Conference on Knowledge Discovery and Data Mining 1998, AAAI Press, 239–243.

LAUSEN, G., SAVNIK, I., DOUGARJAPOV, A. (2000): MSTS: A System for Mining Sets of Time Series. Principles of Data Mining and Knowledge Discovery. Proceedings of the 4th European Conference, PKDD 2000. Springer, 289–298.

QUINLAN, J. R. (1993): C4.5: Programs for Machine Learning. Morgan Kauffman.

SACHS, L. (1984): Applied statistics : a handbook of techniques. Springer.

SANKOFF, D. and KRUSKAL, J. B. (1983): Time Warps, String Edits and Macromolecules: The Theory and Practice of Sequence Comparison. Addison-Wesly.

SAVNIK, I., LAUSEN, G., KAHLE, H. P., SPIECKER, H., HEIN, S. (2000): Algorithm for Matching Sets of Time Series. Principles of Data

Mining and Knowledge Discovery. Proceedings of the 4th European Conference, PKDD 2000. Springer, 277–288.

SCHAPIRE, R. E. (1999): A Brief Introduction to Boosting. International Joint Conference on Artificial Intelligence, Proceedings of the 16th Conference, IJCAI 99. Morgan Kaufmann, 1401–1406.

Spatial Smoothing with Robust Priors in Functional MRI

L. Fahrmeir[1], C. Gössl[2], A. Hennerfeind[1]

[1]Department of Statistics, Ludwig-Maximilians-University Munich
Ludwigstr. 33, 80539 Munich, Germany

[2]Max-Planck-Institute of Psychiatry
Kraepelinstr. 2–10, 80804 Munich, Germany

Abstract: Functional magnetic resonance imaging (fMRI) has led to rapid advances in human brain mapping. The statistical assessment of brain areas that are activated by a particular stimulus requires elaborate models and efficient algorithms. Based on regression models of the observed MR signals on the presented stimulus, pixelwise and spatial techniques have been applied. Besides describing the stimulus dependence of the MR signal, the latter also include spatial correlations between neighboring areas in the brain. In a Bayesian context mostly Gaussian Markov random field priors are used for this purpose, possibly blurring the smoothed images. To preserve potentially existing edges, this article presents some robust alternatives to the Gaussian choice. Inference is based on MCMC techniques. The performance of the new approaches will be demonstrated for simulated as well as for real data.

1 Introduction

Functional magnetic resonance imaging (fMRI) is a non-invasive technique to examine sensory, motor and cognitive functions in a living human brain. In fMRI experiments, an external stimulus is presented which triggers in activated areas a local increase in blood oxygenation that can be visualized with specific MR sequences. In a time series of acquired MR images this effects an increase of the MR signal in comparison with a control or rest condition. This so called BOLD (blood oxygenation level dependent) effect allows for an exact localization of brain activation. In classical experiments the stimulus is presented in a so-called boxcar paradigm, i.e., a sequence of ON and OFF periods. For a visual stimulation, Figure 1 shows such a boxcar stimulus together with representative MR signal time courses. Conventional evaluation techniques are based on correlation, regression or time series analysis (see e.g. Lange, 1996, Friston et al., 1994). For each pixel, activation is assessed separately by comparing pixelwise MR signal time courses to the applied stimulus. In Gössl, Auer and Fahrmeir (2001) an extension of the pixelwise models has been presented that also considers spatial correlations between neighboring brain areas or pixels. A hierarchy of spatial

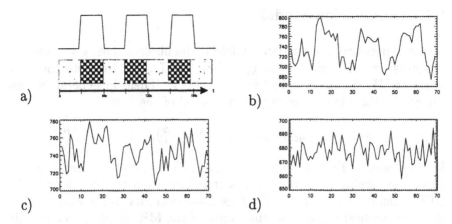

Figure 1: Visual fMRI : a) an 8 Hz flickering rectangular checkerboard (ON) is presented to the subject alternating every 30 s with an uniformly dark background and a fixation point (OFF); an experiment consists of 4 OFFs and 3 ONs; additionally, representative MR signal time courses from strongly (b), weak (c) and non-activated (d) pixels.

and spatio-temporal Bayesian models has been introduced that in the spatial dimension is mainly based on Gaussian intrinsic auto-regressions. However, Gaussian priors are well known for their heavy smoothing properties especially with regard to edges or areas of high curvature in the underlying surface, see, e.g., Geman and McClure (1987) and Künsch (1994) in image analysis, and Besag et al. (1991) in epidemiology. Robust priors avoiding these drawbacks have been suggested by these authors. In this article we transfer a class of robust priors to the analysis of fMRI experiments. Replacing the Gaussian auto-regression with robust alternatives, edge preserving robust spatial models for fMRI data will be derived. Estimation is done by MCMC techniques. The feasibility of this approach will be shown and results for simulated as well as for real data will be presented and discussed.

2 Bayesian Analysis of fMRI Data with Robust Spatial Smoothing

In principle, there are various ways for analyzing spatio-temporal fMRI data. We adopt the classical viewpoint from image analysis: Observation models are defined pixelwise for the time series of MR signals, and spatial or spatio-temporal correlation is introduced via appropriate priors in a second stage.

2.1 Observation Models

In our experience, a Gaussian model for the observations, given the parameters, is not critical. Let $y_i = (y_{it}, t = 1, \ldots, T)$ denote the time series of MR signals for pixel i, $i = 1, \ldots, I$. Then, in general, we assume a flexible time series regression model of the form

$$y_{it} = a_{it} + z_{it}b_{it} + \varepsilon_{it}, \quad \varepsilon_{it} \sim N(0, \sigma_i^2), \quad t = 1, \ldots, T$$

for each pixel i. Here, $a_i = (a_{it}, t = 1, \ldots, T)$ denotes the baseline trend. The time-varying covariate z_{it} is a transformation of the original ON-OFF-stimulus $x_t, t = 1, \ldots, T$. This transformation takes into account that the cerebral blood flow, the source of the MR signal, increases only approximately 6-8 s after the onset of the stimulus, and that flow responses do not occur suddenly, but more continuously and delayed. For more details see, e.g., Gössl, Auer and Fahrmeir (2001). The coefficients b_{it} are interpreted as - possibly time-varying - activation effects and $z_{it}b_{it}$ as the activation profile at time t.

Various forms of observation models are obtained by parametric or nonparametric specifications for the baseline and activation effects. In this article we focus on *parametric models*, which are defined by specifying the effects through

$$a_{it} = w_t' \alpha_i, \quad b_{it} = v_t' \beta_i, \tag{1}$$

where the design vectors w_t and v_t may include time-varying components like t, t^2 and sine/cosine terms, and α_i and β_i are time-constant parameters of fixed, low dimension. As a special case, with $v_t \equiv 1$, we obtain the classical parametric fMRI model

$$y_{it} = w_t' \alpha_i + z_{it} \beta_i + \varepsilon_{it}. \tag{2}$$

Such a model is appropriate if the assessment of activated areas, e.g. the visual cortex, is of primary interest.

2.2 Spatial Priors

To consider spatial dependencies between pixels in the models spatial priors are introduced for the parameters α and β. While Gaussian priors are appropriate for the smooth *temporal* variation of baseline trends and activation effects, they can blur edges or regions of high curvature between activated and non-activated *spatial* areas of the brain. Therefore, we consider a family of robust spatial priors with improved edge-preserving properties. They are based on (mixtures of) pairwise difference priors, including Gaussian priors as a special case.

We focus on a scalar component $\beta_i, i = 1, \ldots, I$, of the activation effect. Extensions to vectors are possible by assuming independent priors for its components or by multivariate versions of the following. The general form of the prior for the vector $\beta = (\beta_1, \ldots, \beta_i, \ldots, \beta_I)'$ is a pairwise interaction Markov random field (MRF)

$$p(\beta|\tau, w) \propto exp\{-\sum_{i \sim j} w_{ij}\Phi(\tau(\beta_i - \beta_j))\}, \tag{3}$$

where τ is a scale parameter, Φ is symmetric with $\Phi(u) = \Phi(-u)$, the summation is over all pairs of pixels $i \sim j$ that are neighbors, and the w_{ij}'s are corresponding weights. It is assumed that the full conditionals

$$p(\beta_i|\beta_{-i}, \tau, w) \propto \tau \cdot exp\{-\sum_{j \in \partial_i} w_{ij}\Phi(\tau(\beta_i - \beta_j))\}$$

are well defined, where $\partial_i = \{j : j \sim i\}$.

If the weights are specified deterministically, e.g. by setting them equal to one or by measuring the distance between neighboring sites, these are the well-known MRF priors described, e.g., in Besag et al. (1995). In the following we will also admit the w_{ij}'s to be random variables obeying a hyperprior. Then the prior $p(\beta|\tau)$ is a mixture of pairwise interaction priors. Such a prior gives additional flexibility when pixel i is near the border of an activated area, where some neighbors $j \in \partial_i$ have similar activation effects and others may be only weakly or not activated.

Starting with $\Phi(u) = \frac{1}{2}u^2$, the above general form becomes to

$$\beta_i|\beta_{-i}, \tau, w \sim N(\sum_{j \in \partial_i} \frac{w_{ij}\beta_j}{w_{i+}}, \frac{1}{\tau^2 w_{i+}}).$$

Setting $w_{ij} = 1$ for regular grids, this reduces to the *Gaussian priors* used in Gössl et al. (2001). In the present article, the w_{ij}'s are allowed to be *i.i.d.* random variables following a Gamma hyperprior

$$w_{ij} \sim GA(\frac{\nu}{2}, \frac{\nu}{2}). \tag{4}$$

The resulting mixture distribution $p(\beta_i|\beta_{-i}, \tau^2)$ is a *Student prior* with ν degrees of freedom, which is for $\nu = 1$ a *Cauchy prior*.

For $\Phi(u) = |u|$, we obtain the *Laplace prior*

$$p(\beta_i|\beta_{-i}, \tau, w) \propto \tau \cdot exp\{-\tau \sum_{j \in \partial_i} w_{ij}|\beta_i - \beta_j|\}. \tag{5}$$

With $w_{ij} = 1$, this is the L_1-based *Laplace prior*, considered to have improved edge-preserving properties compared to the Gaussian prior (see Besag et al. (1991)). It can be interpreted as a stochastic version of the median filter. Assuming again a Gamma hyperprior (4) for the w_{ij}'s, we define a *weighted Laplace prior*.

The Bayesian specification is completed by priors for the variances σ_i^2 and the precision parameter τ. We follow the common choice and assume (inverse) Gamma hyperpriors

$$\sigma_i^2 \sim IG(a, b), \quad \tau \sim GA(c, d).$$

As a standard option, we set $a = c = 1$ and $b = d = 0.005$, yielding highly dispersed hyperpriors.

In principle, the same spatial priors could be chosen for the baseline parameters α_i. However, because the focus is placed on the activation effect, we specify only separate independent diffuse priors $p(\alpha_i) \propto const$ or highly dispersed normal priors for each pixel $i = 1, \ldots, I$.

2.3 Bayesian Inference via MCMC

Gathering parameters of the observation model (1) in vectors $\alpha = (\alpha_1, \ldots, \alpha_I)$, $\beta = (\beta_1, \ldots, \beta_I)$, $\sigma^2 = (\sigma_1^2, \ldots, \sigma_I^2)$, prior parameters in $\tau = (\tau_1, \ldots, \tau_I)$, $w = (w_{ij}, i \sim j)$ and observations in $Y = (y_{it}, i = 1, \ldots, I, t = 1, \ldots, T)$, fully Bayesian inference is based on the posterior

$$p(\alpha, \beta, \sigma^2, \tau, w | Y) \propto L(Y | \alpha, \beta, \sigma^2) p(a) p(b | \tau, w) p(\sigma^2) p(\tau) p(w).$$

The likelihood $L(Y | \alpha, \beta, \sigma^2)$ is determined by the observation model, the other factors by the priors, together with conditional independence assumptions.

Inference is performed by standard MCMC simulation through repeated drawings from univariate or multivariate full conditionals. Details are given in Gössl (2001) and Hennerfeind (2000).

3 Simulation Studies

A number of simulation experiments were carried out to compare different spatial smoothers and to explore their properties. Hennerfeind (2000) investigated purely spatial smoothing based on Gauß, Laplace, Cauchy (=Gauß with weights) and weighted Laplace priors. It turned out that Cauchy and weighted Laplace priors were always superior in terms of MSE compared to the unweighted Laplace prior for surfaces with edges.

Therefore, we did not include the unweighted Laplace prior in the fol-

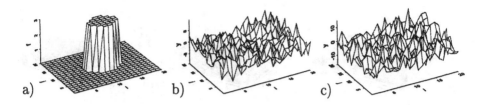

Figure 2: a) Activation area used for simulation study, b) observations for $t = 1$, c) observations for $t = 11$.

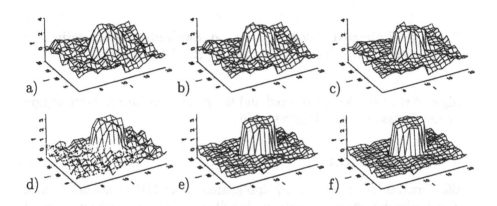

Figure 3: Estimated surfaces for $\sigma_i^2 = 25 + 2 \cdot N(0, 1)$ (a-c) and $\sigma_i^2 = 4 + 0.5 \cdot N(0, 1)$ (d-f): Gauß (a,d), Cauchy (b,e), and weighted Laplace (c,f).

lowing study. In this simulation experiment, spatio-temporal data were generated according to a structure resembling fMRI data. The surface $f(\cdot)$ in Figure 2a stylizes an activation area by a cylinder. Its height is the "activation effect". Outside of the cylinder, activation is zero. For each pixel i, data were generated by

$$y_{it} = f(i) \cdot z_{it} + \varepsilon_{it}, \quad \varepsilon_{it} \sim N(0, \sigma_i^2), \quad t = 1, \ldots, 70,$$

where z_{it} is 0-1-stimulus, with $z_{it} = 0$ for $t = 1, \ldots, 10$, $z_{it} = 1$ for $t = 11, \ldots, 20$, etc. To achieve a realistically low signal-to-noise ratio, we set $\sigma_i^2 = 25 + 2 \cdot N(0, 1)$. Figures 2b and 2c display the observations for $t = 1$ ($z_{it} = 0$) and $t = 11$ ($z_{it} = 1$). It is more or less impossible to recover the surface by eye. Figures 3a-3c show the estimated surfaces using a Gauß, Cauchy and weighted Laplace prior. The Cauchy and weighted Laplace prior seem to have improved edge preserving properties.

With a higher signal-to-noise ratio ($\sigma_i^2 = 4 + 0.5 \cdot N(0, 1)$), the true surface is recovered quite satisfactorily by the robust smoothers while

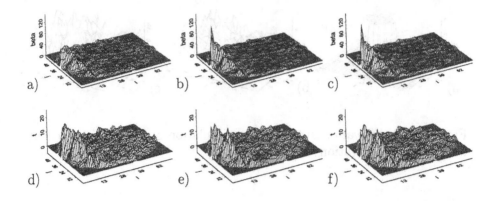

Figure 4: Estimated absolute (a-c) and standardized (d-f) activation effects: Gauß (a,d), Cauchy (b,e), and weighted Laplace (c,f)

edges of the cylinder are blurred and the plane is estimated more wiggly by the Gaussian prior (Figures 3d-3f).

4 Application

We illustrate our approach by application to an fMRI data set from a visual stimulation experiment as described in the introduction. Visual paradigms are known to elicit great activation amplitudes in the visual cortical areas, which are sharply separated from other functional areas.

We apply the parametric observation model (2) with time-constant activation effect β_i, where the transformed stimulus z_{it} was determined through a pilot estimate. The baseline trend was modelled parametrically by $a_{it} = \alpha_{i0} + \alpha_{i1} \cdot t + \alpha_{i2} \sin(\pi/16 \cdot t) + \alpha_{i3} \cos(\pi/25 \cdot t) + \alpha_{i4} \cos(\pi/40 \cdot t)$, with frequencies selected through a stepwise selection procedure.

All MCMC algorithms consisted of 23000 iterations, with the first 3000 being discarded as burn-in and every 20th iteration included in the final sample. Convergence was checked visually by sampling paths. Parameters of the (inverse) Gamma priors were set to $a = c = 1$, $b = d = 0.005$. For the neighborhoods in the spatial priors, the four nearest neighbors were chosen. Figures 4a-4c show posterior mean estimates $\{\hat{\beta}_i, i = 1, \ldots, I\}$ of the activation surface for the Gauß, Cauchy and weighted Laplace prior, respectively. Obviously, the Gauß prior oversmoothes peaks and steep slopes in the visual cortex, while it undersmoothes in non-activated areas, resulting in a comparably rough estimated surface. The results for the Cauchy and weighted Laplace priors clearly illustrate improved smoothing and edge-preserving properties.

However, this better performance in estimating activation surfaces is less clear when we consider standardized activation effects or t-values $t_i = \hat{\beta}_i / \hat{\sigma}_{\hat{\beta}_i}$, where $\hat{\sigma}_{\hat{\beta}_i}$ is the estimated standard deviation. So-called t-maps $\{t_i, i = 1, \ldots, I\}$ are commonly used to obtain activation maps by thresholding. Figures 4d-4f show no substantial gain in using robust priors for t-maps. A main reason for this are increased standard deviations $\hat{\sigma}_{\hat{\beta}_i}$ in areas with high curvature. As can be seen from posteriors not shown here, this is caused by a decrease of corresponding weights w_{ij}. As a remedy, incorporation of more informative prior knowledge on brain regions should prove useful in future research.

References

BESAG, J., GREEN, P., HIGDON, D. and MENGERSEN, K. (1995): Bayesian Computation and Stochastic Systems. *Statistical Science 10*, 3–66.

BESAG, J., YORK, J., and MOLLIE, A. (1991): Bayesian image restoration with two applications in spatial statistics. *Annals of the Institute of Statistical Mathematics*, 43-1, 1–59.

FRISTON, K. J., HOLMES, A. P., WORSLEY, K. J., POLINE, J.-B., WILLIAMS, S. C. R. and FRACKOWIAK, R. S. J. (1994): Analysis of functional MRI time-series. *Human Brain Mapping, 1, 153–171.*

GEMAN, S. and MCCLURE, D. E. (1987): Statistical methods for tomographic image reconstruction *Bulletin of the International Statistical Institute*, LII-4, 5-21.

GÖSSL, C. (2001): Bayesian Models in functional Magnetic Resonance Imaging: Approaches for Human Brain Mapping, Dissertation, Dept. of Statistics, Univ. of Munich, to appear in Shaker Verlag, Aachen.

GÖSSL, C., AUER, D. P., and FAHRMEIR, L. (2001): Bayesian spatio-temporal inference in functional magnetic resonance imaging. *Biometrics*, In Press.

GÖSSL, C., FAHRMEIR, L. and AUER, D. P. (2001): Bayesian Modeling of Hemodynamic Response Function in BOLD fMRI, *NeuroImage*, to appear.

HENNERFEIND, A. (2000): Bayesianische kantenerhaltende Glättung, Diplomarbeit, Dept. of Statistics, Univ. of Munich.

KÜNSCH, H. R. (1994): Robust priors for smoothing and image restoration. *Annals of the Institute of Statistical Mathematics, 46, 1, 1–19.*

LANGE, N. (1996): Statistical approaches to human brain mapping by functional magnetic resonance imaging. *Statistics in Medicine*, 15, 389–428.

Clustering in the Presence of Outliers

M.T. Gallegos

Fakultät für Mathematik und Informatik,
Universität Passau, D-94030 Passau, Germany

Abstract: Suppose that we are given a list of n $\mathbb{R}^d$-valued observations and a natural number $k \leq n$. We develop an estimator which simultaneously detects $n - k$ outliers and partitions the remaining k observations in N clusters. We also propose an algorithm that approximates this estimator. We study under which conditions this estimator is a maximum likelihood estimator.

1   Introduction

This paper deals at the same time with two well–known problems of multivariate statistics: cluster analysis and robust estimation of the parameters of multivariate distributions.

The task of the first one is partitioning of a list of objects (here n observations in $\mathbb{R}^d$) in N disjoint sublists or clusters. The elements of the N clusters are then used in order to estimate the parameters of the N underlying statistical populations. The aim of the second one is the detection of contaminations or spurious elements among a list of observations which are assumed to come from a (parametric) statistical population. Their previous discard make possible a clean estimation of the parameters of the generating statistical law.

If there is only *one* statistical population ($N = 1$) there exist many robust methods, cf. Barnett, Lewis (1994), Chapter 7. In order to *detect* contaminations in the case where there are *more than one* underlying statistical populations ($N \geq 2$) we extend one of these procedures: Rousseeuw's (1999) minimum covariance determinant (MCD). Here, one seeks a sublist k of the n observations whose covariance matrix has lowest determinant. The robust estimates of mean vector and covariance matrix are the sample mean and the sample covariance, respectively, of this sublist.

Concerning solutions for the clustering problem in absence of outliers many mathematical techniques have already been developed. Some of them are maximum likelihood methods for estimating the parameters, cf. Mardia et al.(1979), Chapter 13.2, or the EM algorithm studied in Csiszár and Tusnády (1984) or Peel and MacLachlan (2000). It is also possible to apply clustering algorithms; for an overview see Bock (1996). In this communication and in order to deal with the partitioning of a given number of observations in a given number of clusters and in the presence of outliers, we extend the so-called determinant criterion of cluster analysis, cf. Bock (1974), Chapter 4, §16(a). This method looks for a partition of the list of observations whose pooled within–groups sums of squares and products (SSP) matrix has least determinant.

In Section 2 we formalize the problem described above by specifying the way to follow in order to simultaneously detect contaminations and partitioning the remaining observation in a given number of clusters. The chosen procedure unites in a suitable way the MCD estimator and the determinant criterion already mentioned above. Extending the terminology and notation of Rousseeuw and Van Driessen (1999), we call any such solution a minimum covariance determinant clustering estimator and we denote it by MCDC. In Subsection 2.1 we give a probabilistic formulation of the problem. If the different statistical populations are normally distributed with equal and unknown covariance matrices, we show that the maximum likelihood estimate for the given statistical model coincides with our MCDC estimator.

Calculation of a MCDC estimator requires computation of a sublist of size k out of the n observations and the subsequent partitioning of it in N clusters. Hence, as in the case of the determinant criterion, computation of MCDC is infeasible for most real data lists and approximation algorithms are desirable. For MCD, Rousseeuw and Van Driessen (1999) have developed such an algorithm. In Section 3 we first generalize the building block of their algorithm obtaining the basic step of our procedure: the CC–step. With this tool at hand, we are able to generalize Rousseeuw–Van Driessen's algorithm.

The breakdown value of an estimator is an indicator of its robustness. Estimators with zero breakdown value are not robust. Section 4 deals with the computation of the breakdown value of the MCDC estimator when the size k of the sublist we are looking for is greater than or equal to $\left\lceil \frac{n+N(d+1)}{2} \right\rceil$. It turns out that it is positive, a fact which pleads for its robustness.

Omitted proofs of results presented here are given in Gallegos (2000).

1.1   General Notation

We denote by $\mathbb{N} = \{0, 1, 2, \cdots\}$ ($\mathbb{N}_> = \{1, 2, \cdots\}$) the set of (strictly positive) natural numbers. We write $\mathbf{A} > 0$ ($\mathbf{A} \geq 0$) in order to indicate that a matrix $\mathbf{A}$ is positive (semi–)definite. The symbol $N_d(\cdot; \boldsymbol{\mu}, \boldsymbol{\Sigma})$ denotes the d–dimensional multinormal distribution with mean vector $\boldsymbol{\mu} \in \mathbb{R}^d$ and covariance matrix $\boldsymbol{\Sigma} \in \mathbb{R}^{d \times d}$. It also denotes its Lebesgue density.

Let $\mathcal{C} = \{C_1, C_2, \ldots, C_N\}$ be a partition of a finite, nonempty sublist C of $\mathbb{R}^d$ in N clusters. (Some lists C_i may be empty). We will often identify C with the integral interval $1..N$. For all $j \in 1..N$ such that $\#C_j > 0$, let $\mathbf{m}_C(j)$ be the sample mean vector of the jth cluster and let $\mathbf{V}_C(j)$ be its sample covariance matrix. Let

$$\mathbf{W}_C := \sum_{j=1}^{N} \sum_{\mathbf{x} \in C_j} (\mathbf{x} - \mathbf{m}_C(j))(\mathbf{x} - \mathbf{m}_C(j))^t = \sum_{j=1}^{N} \#C_j \mathbf{V}_C(j)$$

be the pooled within–groups sums of squares and products (SSP) matrix of the partition C. Finally, let us denote by $\mathcal{P}(R)$ the set of all partitions of the list R in N clusters and by $\binom{S}{k}$ the set of all sublists of S with k elements.

2 The Problem

Let $n \in \mathbb{N}_>$, let $S = (\mathbf{x}_1, \mathbf{x}_2, \ldots, \mathbf{x}_n)$ be a sublist of $\mathbb{R}^d$ of n observations in general position, i.e., any $d+1$ elements in S are affinely independent. Let $k \in 1..n$, and let $N \in 1..k$. ($k \geq Nd + 1$). The problem (which we denote by (P)) is to find a *sublist* R of S, the list of *regular* observations, with k elements and a *partition* C of the set R in N clusters such that $\mathbf{W}_C$ has minimal determinant, i.e.,

(P) $\qquad\qquad\qquad$ min det $\mathbf{W}_C$,

where this minimum ranges over all partitions $C \in \bigcup_{D \in \binom{S}{k}} \mathcal{P}(D)$. Note that the condition $k \geq Nd + 1$ assures that at least one cluster has $d+1$ or more elements.

We call the estimator associated with (P) the *minimum covariance determinant clustering* estimator MCDC. Therefore, for $N = 1$, MCDC is Rousseeuw's MCD, and for $k = n$, MCDC is the usual determinant criterion of cluster analysis. The MCDC estimate of location is defined by the sample means of the N clusters, whereas the MCDC estimate of scatter is given by the SSP matrix divided by k.

2.1 A Probabilistic Formulation

This paragraph extends and adapts the usual statistical clustering setup, cf. Mardia et al.(1979), Section 13.2, and combines it with Mathar's (1981), Section 5.2, outlier model. Let $\Lambda := \{\ell : 1..n \to 0..N | \#\ell^{-1}(0) = n - k\}$. Moreover, let $(g_\nu)_{\nu \in \mathbb{R}^d}$ be a family of p.d.f.'s on $\mathbb{R}^d$. The parameter set of our statistical experiment is

$$\Theta := \Lambda \times (\mathbb{R}^d)^N \times \{\Sigma \in \mathbb{R}^{d \times d} | \Sigma > 0\} \times (\mathbb{R}^d)^n.$$

The first component of Θ stands for the unknown assignment of observations to clusters. The next two components stand for the unknown parameters of the N underlying statistical populations which generate the regular observations. Finally, the last component of Θ stands for the unknown parameters of the statistical laws generating the outliers. Let $X_1, X_2, \ldots, X_n : (\Omega, P) \to \mathbb{R}^d$ be n independent random variables such that the p.d.f of X_i, $i \in 1..n$, under the parameter $\theta = (\ell, \mu_1^N, \Sigma, \nu_1^n)$, $(\mu_1^N := (\mu_1, \ldots, \mu_N), \nu_1^n := (\nu_1, \ldots, \nu_n))$, is given by

$$f_{X_i, (\ell, \mu_1^N, \Sigma, \nu_1^n)}(\mathbf{x}) = \begin{cases} N_d(\mathbf{x}; \mu_{\ell(i)}, \Sigma), & \ell(i) \in 1..N, \\ g_{\nu_i}(\mathbf{x}), & \ell(i) = 0. \end{cases} \tag{1}$$

The list of observations S is a realization of these random variables. For an assignment $\ell \in \Lambda$, the sublist $\ell^{-1}(j)$, $j \in 1..N$, consists of all observations assigned to the jth cluster, whereas $\ell^{-1}(0)$ is the sublist of irregular observations. In the notation of the first paragraph of this section, we have $C_j = \ell^{-1}(j)$, $j \in 1..N$, and $S \backslash R = \ell^{-1}(0)$. Furthermore, we denote by $\mathcal{C}^\ell$ the partition of the list R induced by the assignment ℓ.

Clearly, these elements and assumptions define a statistical experiment whose likelihood function for the realization S is given by

$$L_S(\ell, \mu_1^N, \Sigma, \nu_1^n) = \left[\prod_{j=1}^N \prod_{i \in \ell^{-1}(j)} N_d(\mathbf{x}_i; \mu_j, \Sigma) \right] \left[\prod_{i \in \ell^{-1}(0)} g_{\nu_i}(\mathbf{x}_i) \right]. \tag{2}$$

With these preliminaries at hand we are now ready to specify some situations under which the MCDC is a maximum likelihood estimator. Loosely speaking, assume that for each identifying label which maximizes the part of the likelihood function corresponding to the choice of the regular observations and their subsequent assignment to a cluster also maximizes the part of the likelihood function corresponding to the choice of the outliers. Then the MLE of the assignment of the different observations to a cluster or to the set of irregular observations is the MCDC. To be more precise we have the following theoretical result. In another situation, the method used here and the Condition (3) appear already in Ritter and Gallegos (2002).

2.2 Proposition (Statistical Justification of MCDC)

Assume that,

$$\operatorname{argmax}_{\ell \in \Lambda} \prod_{j=1}^N \prod_{\mathbf{x} \in \ell^{-1}(j)} N_d(\mathbf{x}; \mathbf{m}_{\mathcal{C}^\ell}(j), \frac{1}{k} \mathbf{W}_{\mathcal{C}^\ell}) \tag{3}$$

$$\subseteq \operatorname{argmax}_{\ell \in \Lambda} \prod_{\mathbf{x}_i \in \ell^{-1}(0)} \max_{\nu_i} g_{\nu_i}(\mathbf{x}_i).$$

Then the MLE of ℓ (for the given statistical model) is the MCDC estimator. Moreover, if we denote it by C^* then the MLE of $(\boldsymbol{\mu}_1, \ldots, \boldsymbol{\mu}_N)$ is $(\mathbf{m}_{C^*}(1), \ldots, \mathbf{m}_{C^*}(N))$ and the MLE of the common covariance matrix $\boldsymbol{\Sigma}$ is $\frac{1}{k}\mathbf{W}_{C^*}$.

Proof. From (2) together with simple arguments on the computation of a maximum we have

$$\max_{(\ell,\boldsymbol{\mu}_1^N,\boldsymbol{\Sigma},\boldsymbol{\nu}_1^n)} L_S(\ell, \boldsymbol{\mu}_1^N, \boldsymbol{\Sigma}, \boldsymbol{\nu}_1^n)$$

$$= \max_\ell \left[\max_{\boldsymbol{\Sigma}} \prod_{j=1}^N \max_{\boldsymbol{\mu}_j} \prod_{\mathbf{x} \in \ell^{-1}(j)} N_d(\mathbf{x}; \boldsymbol{\mu}_j, \boldsymbol{\Sigma}) \prod_{\mathbf{x} \in \ell^{-1}(0)} \max_{\boldsymbol{\nu}_i} g_{\boldsymbol{\nu}_i}(\mathbf{x}) \right]. \quad (4)$$

Now, standard operations on normal distribution theory, cf. e.g. Mardia et al.(1979), pp. 103–105, show that for each $\ell \in \Lambda$ we have

$$\max_{\boldsymbol{\Sigma}} \prod_{j=1}^N \max_{\boldsymbol{\mu}_j} \prod_{\mathbf{x} \in \ell^{-1}(j)} N_d(\mathbf{x}; \boldsymbol{\mu}_j, \boldsymbol{\Sigma})$$

$$= \max_{\boldsymbol{\Sigma}} (\det 2\pi\boldsymbol{\Sigma})^{-\frac{k}{2}} \prod_{j=1}^N \exp -\frac{1}{2} \operatorname{trace} \left(\boldsymbol{\Sigma}^{-1} \# \ell^{-1}(j) \mathbf{V}_{C\ell}(j)\right)$$

$$= \max_{\boldsymbol{\Sigma}} (\det 2\pi\boldsymbol{\Sigma})^{-\frac{k}{2}} \exp -\frac{1}{2} \operatorname{trace}(\boldsymbol{\Sigma}^{-1}\mathbf{W}_{C\ell})$$

$$= K(\det 2\pi\mathbf{W}_{C\ell})^{-\frac{k}{2}} = \prod_{j=1}^N \prod_{\mathbf{x} \in \ell^{-1}(j)} N_d(\mathbf{x}; \mathbf{m}_{C\ell}(j), \frac{1}{k}\mathbf{W}_{C\ell}), \quad (5)$$

where K is a constant independent of the chosen ℓ and where the last but one equality follows as a direct application of Mardia et al. (1979), Theorem 4.2.1 on page 104. The claim follows now from (4) and (5) together with (3). $\square$

A simple situation for which the hypothesis (3) of the proposition above is satisfied is $g_{\boldsymbol{\nu}}(\cdot) = N_d(\cdot; \boldsymbol{\nu}, \mathbf{I}_d)$. Indeed, a simple argument shows that for each $\mathbf{x} \in \mathbb{R}^d$ we have

$$\max_{\boldsymbol{\nu} \in \mathbb{R}^d} N_d(\mathbf{x}; \boldsymbol{\nu}, \mathbf{I}_d) = N_d(\mathbf{x}; \mathbf{x}, \mathbf{I}_d) = (2\pi)^{-\frac{d}{2}},$$

i.e., a constant independent of the chosen $\mathbf{x}$. Therefore, for each $\ell \in \Lambda$ we have

$$\prod_{\mathbf{x} \in \ell^{-1}(0)} \max_{\boldsymbol{\nu}_i} g_{\boldsymbol{\nu}_i}(\mathbf{x}) = (2\pi)^{-\frac{d(n-k)}{2}},$$

a constant independent of the chosen ℓ. This last assertion implies that

$$\text{argmax}_{\ell \in \Lambda} \prod_{\mathbf{x}_i \in \ell^{-1}(0)} \max_{\nu_i} g_{\nu_i}(\mathbf{x}_i) = \Lambda.$$

Hence, for the situation considered the Condition (3) is satisfied.

3   The CC–Step as an Extension of the C–Step

In this section, we give an algorithm which computes an approximation to MCDC. The basic step of our procedure is the CC–step. In order to write it in algorithmic terms, we will first assume that there is just one point minimizing a given objective function and there is just one permutation ordering a given set of numbers in an increasing fashion, i.e., there are no ties. Moreover, for all $\mathbf{x}_i \in S$ and all $j \in 1..N$, define the (squared) distances

$$d_C(j,i)^2 := (\mathbf{x}_i - \mathbf{m}_C(j))^t \mathbf{W}_C^{-1}(\mathbf{x}_i - \mathbf{m}_C(j)). \tag{6}$$

Note that since S is in general position and $k \geq Nd+1$, $\mathbf{W}_C > 0$.

3.1   The CC–Step

// Input: The mean vectors $\mathbf{m}_C$ and the SSP matrix $\mathbf{W}_C$ of a
        configuration C.
// Output: A configuration C_{new} such that $\det \mathbf{W}_{C_{\text{new}}} \leq \det \mathbf{W}_C$.

(i) Compute the (squared) distances $d_C(j,i)^2$, $j \in 1..N$, $i \in 1..n$, defined in (6).

(ii) For each $\mathbf{x}_i \in S$, determine the nearest cluster c_i to it,
     i.e., $c_i := \text{argmin}_{j \in 1..N} d_C(j,i)^2$.

(iii) Determine the permutation $\pi : 1..n \to 1..n$ that satisfies

$$d_C(c_{\pi(1)}, \pi(1))^2 \leq d_C(c_{\pi(2)}, \pi(2))^2 \leq \ldots \leq d_C(c_{\pi(n)}, \pi(n))^2. \tag{7}$$

(iv) Put $R_{\text{new}} = \{\pi(1), \pi(2), \ldots, \pi(k)\}$ and, for each $j \in 1..N$, put $C_{\text{new},j} = \{i \in 1..k | c_{\pi(i)} = j\}$.

    Finally, let $C_{\text{new}} := \{C_{\text{new},1}, \ldots, C_{\text{new},N}\}$.

(v) Return C_{new}.

A first remark concerning the computation of the sample means of the new clusters: if $C_{\text{new},j}$ is an empty cluster then put $\mathbf{m}_{C_{\text{new}}}(j) := \mathbf{m}_C(j)$. A second remark concerning (possible) "ties": if in 3.1(ii) there exists an observation $\mathbf{x}_i \in S$ such that the set $\text{argmin}_{j \in 1..N}\, d_C(j,i)$ has two or more elements and/or in 3.1(iii) there exist more than one permutation satisfying (7) then from all equivalent configurations (defined according to 3.1(iv)) choose a partition C_{new} with least SSP determinant.

The following result asserts that a CC–step yields indeed a partition with lower or equal determinant of the corresponding SSP.

3.2 Theorem

We have $\det \mathbf{W}_{C_{\text{new}}} \leq \det \mathbf{W}_C$ with equality if and only if $\mathbf{m}_{C_{\text{new}}} = \mathbf{m}_C$ and $\mathbf{W}_{C_{\text{new}}} = \mathbf{W}_C$.

Now, starting from an initial partition C_0 and iterating the CC-step, we obtain a sequence of partitions $(C_n)_{n \geq 0}$ such that $\det \mathbf{W}_{C_{n+1}} \leq \det \mathbf{W}_{C_n}$ for all $n \in \mathbb{N}$. Since the sequence $(\det \mathbf{W}_{C_n})_{n \in \mathbb{N}}$ converges and there are only a finite number of possible partitions, this iterative process must terminate in a finite number of steps with $\det \mathbf{W}_{C_{n+1}} = \det \mathbf{W}_{C_n} (> 0)$. By Theorem 3.2, we have $\mathbf{m}_{C_{n+1}} = \mathbf{m}_{C_n}$ and $\mathbf{W}_{C_{n+1}} = \mathbf{W}_{C_n}$ and, therefore, $d_{C_n}(j,i) = d_{C_{n+1}}(j,i)$, $i \in 1..n$, $j \in 1..N$. Thus, the next CC–step yields the same partition C_n. This is our approximation to the MCDC.

4   On the Breakdown Value of MCDC

Results concerning the breakdown value of the MCD estimator ($N = 1$) are given in Lopuhaä and Rousseeuw (1991), Theorem 3.1. In this section we restrict matters to the case $N \geq 2$. Loosely speaking, the finite–sample version of the breakdown value of an estimator measures the minimum fraction of outliers that can *completely* spoil the estimate. In this communication we are dealing with estimation of the optimal partition of a subset of k elements out of a set of n observations in N clusters. Hence, corruption of the estimate is reflected by an arbitrarily large or small eigenvalue of the SSP matrix. Note that if the norm of the sample mean (of such an estimate) becomes arbitrarily large we are not (necessarily) dealing with a spoiled estimate: all the observations in one cluster may go to infinity in the same direction. Let $\mathbf{W}(S) = \sum_{j=1}^{N} n_j \mathbf{V}_j(S)$ be the sample SSP matrix of a clustering estimate for the set of observations S (n_j and $\mathbf{V}_j(S)$ are the cardinality and the scatter matrix of the jth cluster, respectively). The breakdown value of $\mathbf{W}(S)$, cf. Lopuhaä and Rousseeuw (1991), Section 2, is defined by

$$\epsilon(\mathbf{W}(S)) = \min_{1 \leq m \leq n} \left\{ \frac{m}{n} \,\Big|\, \sup_O d_M(\mathbf{W}(S), \mathbf{W}(O)) = \infty \right\},$$

where the supremum is taken over all possible sets O of n observations obtained by replacing m points of S with arbitrary points in $\mathbb{R}^d$ and where $d_M(\mathbf{A}, \mathbf{B}) := \max\{|\lambda_1(\mathbf{A}) - \lambda_1(\mathbf{B})|, |\lambda_d(\mathbf{A})^{-1} - \lambda_d(\mathbf{B})^{-1}|\}$, where $\lambda_1(\mathbf{A})$ and $\lambda_d(\mathbf{A})$ are the largest and smallest eigenvalues of $\mathbf{A}$, respectively.

Let us denote by $\mathbf{W}_{C^*}(S)$ the MCDC estimator of the SSP matrix for the set S of observations and for the parameters n, k and N.

4.1 Theorem

Assume that S is in general position. Let $N \geq 2$, let $n \geq N(d+1)$, and let $k \geq \lceil \frac{n+N(d+1)}{2} \rceil$. The breakdown value of the SSP matrix of the MCDC is $(n - k + N)/n$.

Proof. Note first that, since $k \geq Nd+1$ we have $\mathbf{W}_{C^*}(S) > 0$. We show first that $\epsilon(\mathbf{W}_{C^*}(S))$ is at most $(n - k + N)/n$. Replace $n - k + N$ observations of S with arbitrary points in $\mathbb{R}^d$ and denote by O this new set. Then each subset of O with k elements contains at least N replacements and at least $2k - n - N$ original observations. Therefore, in each configuration consisting of k elements in O we can find a cluster, say the first one, with two or more elements, and these two elements are either both replacements or one is a replacement and the other one is an original observation. This fact is particularly true for the MCDC configuration. Therefore, the largest eigenvalue $\lambda_{1,1}^*(O)$ of $\mathbf{V}_{C^*,1}(O)$ becomes arbitrarily large as we send the replacements of the first MCDC cluster to infinity in a suitable way. The inequalities trace $\mathbf{W}_{C^*}(O) \geq n_1$ trace $\mathbf{V}_{C^*,1}(O) \geq \lambda_{1,1}^*(O)$ together with this last fact imply that the largest eigenvalue of $\mathbf{W}_{C^*}(O)$ becomes also arbitrarily large. This is the proof of our first claim.

We now show that $\epsilon(\mathbf{W}_{C^*}(S))$ is at least $(n - k + N)/n$. Replace $n - k + N - 1$ or less original observations by arbitrary points in $\mathbb{R}^d$ and denote this corrupted set by O. Then each subset of O with k points contains at least $2k - n - N + 1$ original observations. By assumption, we have $2k \geq Nd + n + N$ or, equivalently, $2k - n - N + 1 \geq Nd + 1$. Therefore, at least one cluster, say the first one, contains $d + 1$ or more original observations. By hypothesis, we have $\mathbf{V}_{C^*,1}(O) > 0$ and, therefore, $\mathbf{W}_{C^*}(O) > 0$. Furthermore, there exists a configuration with k observations of O such that its first cluster contains (all) $k - N + 1$ original observations and the remaining $N - 1$ clusters are (possibly replaced) singletons. Plainly, the determinant of the SSP matrix $\mathbf{W}$ of this configuration is finite. By definition of MCDC, we have $0 < \det \mathbf{W}_{C^*}(O) \leq \det \mathbf{W} < \infty$. Therefore the largest eigenvalue of $\mathbf{W}_{C^*}(O)$ is finite, whereas the smallest eigenvalue of $\mathbf{W}_{C^*}(O)$ is strictly greater than 0. This is our second claim. $\qquad\square$

References

BARNETT, V. and LEWIS, T. (1994): Outliers in statistical data. 3rd Edition, Wiley, Chichester, UK.

BOCK, H. H. (1974): Automatische Klassifikation. Goettingen Vandenhoeck & Ruprecht.

BOCK, H. H. (1996): Probabilistic models in cluster analysis. *Computational Statistics and Data Analysis 23, 5–28.*

CZISZÁR, I. and TUSNÁDY, G. (1984): Information geometry and alternating minimization procedures. *Statistics and Decision, Supplement Issue,1, 203–237*

GALLEGOS M. T. (2000): A Robust Method for Clustering Analysis. Technical Report MIP–0013 October 2000. Fakultät für Mathematik und Informatik, Universität Passau.

LOPUHAÄ, H. P. and ROUSSEEUW, P. J. (1991): Breakdown Points of Affine Equivariant Estimators of Multivariate Location and Covariances Matrices. *The Annals of Statistics, Vol. 19, 229–248.*

MARDIA, K. V., KENT, J. T. and BIBBY, J. M. (1979): Multivariate Analysis. Academic Press, London, New York, Toronto, Sydney, San Francisco.

MATHAR, R. (1981): Ausreißer bei ein– und mehrdimensionalen Wahrscheinlichkeitsverteilungen. Dissertation, Mathematisch–Naturwissenschaftliche Fakultät der Rheinisch–Westfälichen Technischen Hochschule, Aachen.

PEEL, D. and MACLACHLAN, G. (2000): Finite mixture models. Wiley, New York.

RITTER, G. and GALLEGOS, M.T. (2002): Bayesian Object Identification: Variants. To appear in *J. Multivariate Analysis.*

ROUSSEEUW, P. J. and VAN DRIESSEN, K. (1999): A Fast algorithm for the Minimum Covariance Determinant Estimator. *Technometrics, Vol. 41, 212–223.*

Two Models of Random Intersection Graphs for Classification

E. Godehardt[1], J. Jaworski[2]

[1]Clinic of Thoracic and Cardiovascular Surgery,
Heinrich Heine University, D-40225 Düsseldorf, Germany

[2]Faculty of Mathematics and Computer Science,
Adam Mickiewicz University, PL-60769 Poznań, Poland

Abstract: Graph concepts generally are useful for defining and detecting clusters. We consider basic properties of random intersection graphs generated by a random bipartite graph $\mathcal{BG}_{n,m}\left(n, \mathcal{P}_{(m)}\right)$ on $n+m$ vertices. In particular, we focus on the distribution of the number of isolated vertices, and on the distribution of the vertex degrees. These results are applied to study the asymptotic properties of such random intersection graphs for the special case that the distribution $\mathcal{P}_{(m)}$ is degenerated. The application of this model to find clusters and to test their randomness especially for non-metric data is discussed.

1 Introduction

Consider a set $\mathcal{V}$ of objects and a set $\mathcal{W}$ of their possible properties and let $\mathcal{BG} = \mathcal{BG}(\mathcal{V} \cup \mathcal{W}, \mathcal{E})$ be a bipartite graph with the 2-partition $(\mathcal{V}, \mathcal{W})$ of the vertex set and with the edge set $\mathcal{E}$ of choices; edges go only from elements of $\mathcal{V}$ to elements of $\mathcal{W}$ and the edge between $v \in \mathcal{V}$ and $w \in \mathcal{W}$ means that an object v has a property w. Each graph $\mathcal{BG}$ generates two intersection graphs. The first one with the vertex set $\mathcal{V}$ has two vertices joined by an edge if and only if the sets of neighbors of these vertices in $\mathcal{BG}$ have a non-empty intersection (or, more generally, an intersection consisting of at least s elements). The second intersection graph generated by $\mathcal{BG}$ is defined on the vertex set $\mathcal{W}$ of properties analogously.

We will first give the definition and study some basic properties of a random bipartite graph $\mathcal{BG}_{n,m}\left(n, \mathcal{P}_{(m)}\right)$ on a vertex set $\mathcal{V} \cup \mathcal{W}$, $|\mathcal{V}| = n$, $|\mathcal{W}| = m$, where each vertex from $\mathcal{V}$ "chooses" its degree according to a discrete distribution $\mathcal{P}_{(m)} = (P_0, P_1, \ldots, P_m)$, and then uniformly its set of neighbors from $\mathcal{W}$ independently of all other vertices from $\mathcal{V}$.

The main purpose of the paper is to study some properties of two models of random intersection graphs generated by this random bipartite graph for the case $s = 1$. The first model $\mathcal{IG}_n^{active}\left(n, \mathcal{P}_{(m)}\right)$, which we will call the "active" one, is that with the set $\mathcal{V}$ of active vertices, i.e., with the vertices which choose their neighbors in the original bipartite graph. The second intersection graph model can be treated as the "passive" one (since its vertices were chosen in the bipartite graph) and therefore will be denoted by $\mathcal{IG}_m^{passive}\left(n, \mathcal{P}_{(m)}\right)$.

Finally, we show how a mathematical definition of the term "cluster" can be given using this model of bipartite graphs. The corresponding probability models can be applied to test the validity of such groups detected by some numerical cluster uncovering algorithm. For more information on how graph-theoretical concepts can be used in defining cluster models, revealing hidden clusters in a data set, and testing the randomness of such clusters, we refer the reader to Bock (1996), Godehardt (1990), Godehardt and Jaworski (1996), and Jaworski and Palka (1999).

2 Definitions, Basic Properties of $\mathcal{BG}_{n,m}\left(n, \mathcal{P}_{(m)}\right)$

We will start with the formal definition of a random bipartite graph which serves to introduce the two models of active and passive intersection graphs. This bipartite model is an analogue of a general model of random digraphs which was introduced in Jaworski and Smit (1987). Let $\mathcal{P}_{(m)} = (P_0, P_1, \ldots, P_m)$ be a discrete probability distribution, i.e., an $(m + 1)$-tuple of non-negative real numbers which satisfy $P_0 + P_1 + \cdots + P_m = 1$. Denote by $\mathcal{BG}_{n,m}\left(n, \mathcal{P}_{(m)}\right)$ a random bipartite graph on a vertex set $\mathcal{V} \cup \mathcal{W} = \{v_1, v_2, \ldots, v_n\} \cup \{w_1, w_2, \ldots, w_m\}$, such that

1. each vertex $v \in \mathcal{V}$ chooses first its degree $X^*(v)$ and then its set of neighbors $\Gamma(v) \subset \mathcal{W}$, independently of all other vertices from $\mathcal{V}$;

2. the vertex v chooses its degree according to the distribution $\mathcal{P}_{(m)}$,

$$\Pr\{X^*(v) = k\} = P_k, \quad k = 0, 1, \ldots, m;$$

3. the vertex v with degree k chooses its set of neighbors $\Gamma(v)$ uniformly from all k-element subsets of $\mathcal{W}$, i.e., if we denote by p_k the probability that a subset $\mathcal{S} \subseteq \mathcal{W}$, with $|\mathcal{S}| = k$, coincides with the set of neighbors of a vertex v, then

$$p_k = \Pr\left\{\Gamma(v) = \mathcal{S}\right\} = \Pr\{|\Gamma(v)| = k\} \bigg/ \binom{m}{k} = P_k \bigg/ \binom{m}{k}.$$

Many properties of $\mathcal{BG}_{n,m}\left(n, \mathcal{P}_{(m)}\right)$ can be proved in the same manner as for the general digraph model (see Jaworski and Palka (2001), Jaworski and Smit (1987)). In the case that $\mathcal{P}_{(m)}$ is a binomial distribution $\mathcal{B}(m, p)$, i.e.,

$$\mathcal{P}_{(m)} = \left(p^0 q^m, \ldots, \binom{m}{k} p^k q^{m-k}, \ldots, p^m q^0\right),$$

the model $\mathcal{BG}_{n,m}\left(n, \mathcal{B}(m, p)\right)$ is equivalent to the well known random bipartite graph $\mathcal{G}_{n,m,p}$ on $n + m$ labeled vertices where each of all $n\,m$

possible edges between the sets $\mathcal{V}$ and $\mathcal{W}$ appears independently with a given probability p (see Janson, Łuczak and Ruciński (2001)).

A common and very useful property of classical models of random graphs is "monotonicity". Using the standard terminology of stochastic ordering, we say that a random variable Y is smaller than a random variable X in the usual stochastic order, and we will write $Y \preceq X$, if and only if $\Pr\{Y > u\} \leq \Pr\{X > u\}$ for all $u \in (-\infty, \infty)$. In the particular case of discrete random variables X and Y with

$$\Pr\{X = k\} = P_k \quad \text{and} \quad \Pr\{Y = k\} = Q_k,$$

for $k = 0, 1, 2, \ldots, m$, the condition above is equivalent to

$$\sum_{k=0}^{u} P_k \leq \sum_{k=0}^{u} Q_k$$

for all $u = 0, 1, 2, \ldots, m$. In such a case we will also write $\mathcal{Q}_{(m)} \preceq \mathcal{P}_{(m)}$ for distributions $\mathcal{P}_{(m)} = (P_0, P_1, \ldots, P_m)$ and $\mathcal{Q}_{(m)} = (Q_0, Q_1, \ldots, Q_m)$. A graph property $\mathcal{A}$ is called a "monotone increasing property" if for a graph $\mathcal{G}$ having $\mathcal{A}$, every spanning supergraph of $\mathcal{G}$ (every graph with the same vertex set and an edge set which contains that of $\mathcal{G}$) also has this property.

Our first result confirms that $\mathcal{BG}_{n,m}\left(n, \mathcal{P}_{(m)}\right)$ obeys the monotonicity condition. Since the proof follows the same lines as for the general model of random digraphs (see Jaworski and Palka (2001)), we omit it here.

Property 2.1 *If $\mathcal{Q}_{(m)} \preceq \mathcal{P}_{(m)}$ then for any monotone increasing graph property $\mathcal{A}$, it holds that*

$$\Pr\left\{\mathcal{BG}_{n,m}\left(n, \mathcal{Q}_{(m)}\right) \text{ has } \mathcal{A}\right\} \leq \Pr\left\{\mathcal{BG}_{n,m}\left(n, \mathcal{P}_{(m)}\right) \text{ has } \mathcal{A}\right\}.$$

The degree of a given vertex $v \in \mathcal{V}$ in $\mathcal{BG}_{n,m}\left(n, \mathcal{P}_{(m)}\right)$ was defined as the random variable $X^*(v)$ with the distribution $\mathcal{P}_{(m)}$. It follows from the definition of this random bipartite graph that we have n independent, identically distributed random variables $X^*(v_1), X^*(v_2), \ldots, X^*(v_n)$, all with the distribution $\mathcal{P}_{(m)}$. Therefore we can write shortly X^* instead of $X^*(v_i)$. As usual, $(n)_k = n(n-1)\cdots(n-k+1)$ and $\mathrm{E}_k(X) = \mathrm{E}((X)_k)$ denotes the k-th factorial moment of a random variable X. The next property expresses, by appropriate factorial moments of X^*, the probability that the set of neighbors of a vertex $v \in \mathcal{V}$ contains a given subset of vertices from $\mathcal{W}$.

Property 2.2 *For a given v, $v \in V$, let $U \subseteq W$ and $|U| = t \geq 1$. Then*

$$\Pr\{\Gamma(v) \supseteq U\} = \frac{1}{(m)_t} E_t(X^*).$$

Proof. Assume that we are given a vertex $v \in V$ and a t-element subset $U \subseteq W$. Let $\mathcal{L}$ be an l-element subset of $W \setminus U$. Then

$$\Pr\{\Gamma(v) \supseteq U\} = \sum_{l=0}^{m-t} \sum_{\substack{\mathcal{L} \subseteq W \setminus U \\ |\mathcal{L}|=l}} \Pr\{\Gamma(v) = U \cup \mathcal{L}\}.$$

Thus, by the definition of our model we have

$$\Pr\{\Gamma(v) \supseteq U\} = \sum_{l=0}^{m-t} \binom{m-t}{l} p_{t+l} = \sum_{k=t}^{m} \binom{m-t}{k-t} p_k = \frac{1}{(m)_t} \sum_{k=t}^{m} (k)_t \, P_k,$$

which implies the result.   ●

In particular, for $t = 1$ the above property defines an edge occurrence probability in a bipartite graph $BG_{n,m}(n, P_{(m)})$. Denoting this probability by p^* we have

$$p^* = \Pr\{w \in \Gamma(v)\} = \frac{1}{m} E(X^*).$$

Let $Y^* = Y^*(w)$ be the degree of a given vertex w, $w \in W$, of a random bipartite graph $BG_{n,m}(n, P_{(m)})$. Clearly, the probability distribution of $Y^*(w)$ does not depend on the specific choice of $w \in W$. Moreover, we can show the following result.

Property 2.3 *For any vertex w, $w \in W$, the random variable $Y^*(w)$ follows a binomial distribution $\mathcal{B}(n, p^*)$ with p^* given above.*

Proof. For a given vertex $w \in W$, let $\Gamma(w)$ be the set of neighbors of this vertex w. Then, due to the homogeneous structure of $BG_{n,m}(n, P_{(m)})$ (which is a consequence of its definition), we obtain

$$\Pr\{Y^*(w) = k\} = \Pr\{|\Gamma(w)| = k\}$$
$$= \binom{n}{k} \Pr\{w \in \Gamma(v_1), \dots, w \in \Gamma(v_k), w \notin \Gamma(v_{k+1}), \dots, w \notin \Gamma(v_n)\}$$
$$= \binom{n}{k} \left(\Pr\{w \in \Gamma(v_1)\}\right)^k \left(\Pr\{w \notin \Gamma(v_{k+1})\}\right)^{n-k}$$

and the result follows from Property 2.2. •

In contrast to the degrees $X^*(v_i)$ of vertices from $\mathcal{V}$, the degrees of vertices from $\mathcal{W}$, i.e., the random variables $Y^*(w_j)$, $j = 1, 2, \ldots, m$, are, in general, not independent. As a matter of fact these variables are independent only if X^* is binomially distributed.

Property 2.4 *The random variables* $Y^*(w_1), Y^*(w_2), \ldots, Y^*(w_m)$ *are independent if and only if the degrees* $X^*(v)$ *of all vertices* v *from* $\mathcal{V}$ *have the same binomial distribution.*

Proof. Let $Y^*(w_1), Y^*(w_2), \ldots, Y^*(w_m)$ be independent random variables. Since, by Property 2.3, they have a binomial distribution $\mathcal{B}(n, p^*)$ with $p^* = \mathbb{E}(X^*)/m$, our model is equivalent to $\mathcal{BG}_{m,n}(m, \mathcal{B}(n, p^*))$. This, by Property 2.3 again, means that the degrees of vertices from $\mathcal{V}$ follow a binomial distribution with parameters m and p^*.

Conversely, let $\mathcal{P}_{(m)}$ be a binomial distribution $\mathcal{B}(m, p^*)$. Then for each vertex v, the appearence of edges going from v to its neighbors defines independent events. Thus the definition of $\mathcal{BG}_{n,m}(n, \mathcal{P}_{(m)})$ implies that all edges appear independently and with probability p^* in such a bipartite graph. This ensures the independence of degrees of vertices from $\mathcal{W}$ in the random bipartite graph $\mathcal{BG}_{n,m}(n, \mathcal{P}_{(m)})$. In both cases we have $\mathcal{BG}_{n,m}(n, \mathcal{B}(m, p^*)) = \mathcal{BG}_{m,n}(m, \mathcal{B}(n, p^*)) = \mathcal{G}_{n,m,p^*}$. •

With the last three properties we give some formulas which serve as the first step in the derivation of probability distributions for many characteristics of the two models of random intersection graphs generated by $\mathcal{BG}_{n,m}(n, \mathcal{P}_{(m)})$. Denote by $\Gamma(\mathcal{A})$ the set of neighbors of vertices from the subset $\mathcal{A}$ of $\mathcal{V}$ in the random bipartite graph $\mathcal{BG}_{n,m}(n, \mathcal{P}_{(m)})$.

Property 2.5 *For any* $\mathcal{S}, \mathcal{S} \subseteq \mathcal{W}, |\mathcal{S}| = k \geq 1$, *and any* $\mathcal{A}, \mathcal{A} \subseteq \mathcal{V}$ *and* $|\mathcal{A}| = t \geq 1$, *we have*

$$\Pr\left\{\Gamma(\mathcal{A}) \subseteq \mathcal{W} \setminus \mathcal{S}\right\} = \left[\frac{\mathbb{E}_k(m - X^*)}{(m)_k}\right]^t.$$

Proof. For a given probability distribution $\mathcal{P}_{(m)} = (P_0, P_1, \ldots, P_m)$ of a random variable X^*, let denote by $\mathcal{P}'_{(m)}$ the distribution of the random variable $Z^* = m - X^*$, i.e.,

$$(P'_0, P'_1, \ldots, P'_m) = (P_m, P_{m-1}, \ldots, P_0).$$

In terms of our definition of the random bipartite graph, such a distribution corresponds to the complement of the graph $\mathcal{BG}_{n,m}(n, \mathcal{P}_{(m)})$ to the complete bipartite graph $\mathcal{K}_{n,m}$. Therefore it is easy to check the following equivalence

$$\Gamma_{\mathcal{P}_{(m)}}(\mathcal{A}) \subseteq \mathcal{W} \setminus \mathcal{S} \quad \text{if and only if} \quad \mathcal{S} \subseteq \Gamma_{\mathcal{P}'_{(m)}}(v) \text{ for all } v \in \mathcal{A}.$$

Hence, the independent choices of neighbors and Property 2.2 imply the result. $\bullet$

In the following property we derive the probability that two vertices from $\mathcal{V}$ share exactly s neighbors in $\mathcal{W}$.

Property 2.6 *For any two vertices v and z from $\mathcal{V}$ we have*

$$\Pr\left\{|\Gamma(v) \cap \Gamma(z)| = s\right\} = \sum_{k=s}^{m} \sum_{j=s}^{m-k+s} \frac{\binom{j}{s}\binom{m-j}{k-s}}{\binom{m}{k}} P_k P_j.$$

In particular we have

$$\Pr\left\{|\Gamma(v) \cap \Gamma(z)| > 0\right\} = 1 - \sum_{k=0}^{m} P_k \frac{E_k(m - X^*)}{(m)_k}$$

as the probability that two vertices v and z from $\mathcal{V}$ share at least one neighbor in $\mathcal{W}$.

Proof. For $v \in \mathcal{V}$ and $z \in \mathcal{V}$,

$$\Pr\left\{|\Gamma(v) \cap \Gamma(z)| = s\right\}$$

$$= \sum_{S} \sum_{A} \sum_{B} \Pr\left\{\Gamma(v) = S \cup A \text{ and } \Gamma(z) = S \cup B\right\},$$

$$= \sum_{k=s}^{m} \sum_{j=s}^{m-k+s} \binom{m}{s}\binom{m-s}{k-s}\binom{m-k}{j-s} p_k\, p_j,$$

where $S \subseteq \mathcal{W}$, $|S| = s$, $A \subseteq \mathcal{W} \setminus S$, and $B \subseteq \mathcal{W} \setminus (S \cup A)$. This immediately implies the first assertion. Since

$$\Pr\left\{|\Gamma(v) \cap \Gamma(z)| > 0\right\} = 1 - \Pr\left\{|\Gamma(v) \cap \Gamma(z)| = 0\right\}$$

$$= 1 - \sum_{k=0}^{m} P_k \sum_{j=0}^{m-k} \frac{(m-k)_j}{(m)_j} P_j,$$

the second formula follows as well. $\bullet$

We will also need the respective probability that two vertices from $\mathcal{W}$ share exactly s neighbors in $\mathcal{V}$.

Property 2.7 *For any two vertices w and u from $\mathcal{W}$ we have*

$$\Pr\Big\{|\Gamma(w) \cap \Gamma(u)| = s\Big\} = \binom{n}{s}\left[\frac{E_2(X^*)}{(m)_2}\right]^s\left[1 - \frac{E_2(X^*)}{(m)_2}\right]^{n-s}.$$

In particular we have

$$\Pr\Big\{|\Gamma(w) \cap \Gamma(u)| > 0\Big\} = 1 - \left[1 - \frac{E_2(X^*)}{(m)_2}\right]^n$$

as the probability that two vertices w and u from $\mathcal{W}$ share at least one neighbor in $\mathcal{V}$.

Proof. Let $w, u \in \mathcal{W}$. Similarly as in the previous proof, we have

$$\Pr\Big\{|\Gamma(w) \cap \Gamma(u)| = s\Big\}$$

$$= \sum_{S}\sum_{A}\sum_{B}\Pr\Big\{\Gamma(w) = S \cup A \text{ and } \Gamma(u) = S \cup B\Big\},$$

where $S \subseteq \mathcal{V}$, $|S| = s$, $A \subseteq \mathcal{V} \setminus S$, and $B \subseteq \mathcal{V} \setminus (S \cup A)$. Moreover the event $\{\{\Gamma(w) = S \cup A\} \cap \{\Gamma(u) = S \cup B\}\}$ is equivalent to

$$\prod_{v \in S}\Big\{\{w, u\} \subseteq \Gamma(v)\Big\} \cap \prod_{v \in A}\Big\{\Gamma(v) \cap \{w, u\} = \{w\}\Big\}$$

$$\cap \prod_{v \in B}\Big\{\Gamma(v) \cap \{w, u\} = \{u\}\Big\} \cap \prod_{v \in \mathcal{V} \setminus (S \cup A \cup B)}\Big\{\Gamma(v) \cap \{w, u\} = \emptyset\Big\}.$$

Hence, we can write

$$\Pr\Big\{|\Gamma(w) \cap \Gamma(u)| = s\Big\} = \sum_{k=s}^{n}\sum_{j=s}^{n-k+s}\binom{n}{s}\binom{n-s}{k-s}\binom{n-k}{j-s}$$

$$\times \left[\sum_{i=0}^{m-2}\frac{\binom{m-2}{i}}{\binom{m}{i+2}}P_{i+2}\right]^s\left[\sum_{i=0}^{m-2}\frac{\binom{m-2}{i}}{\binom{m}{i+1}}P_{i+1}\right]^{k+j-2s}\left[\sum_{i=0}^{m-2}\frac{\binom{m-2}{i}}{\binom{m}{i}}P_i\right]^{n-k-j+s}$$

$$= \sum_{k=s}^{n}\sum_{j=s}^{n-k+s}\binom{n}{s}\binom{n-s}{k-s}\binom{n-k}{j-s}$$

$$\times \left[\frac{E_2(X^*)}{(m)_2}\right]^s\left[\frac{E(X^*(m - X^*))}{(m)_2}\right]^{k+j-2s}\left[\frac{E_2(m - X^*)}{(m)_2}\right]^{n-k-j+s},$$

which implies the result. ●

3 The Active Intersection Graph

A random graph $\mathcal{BG}_{n,m}\left(n, \mathcal{P}_{(m)}\right)$ generates two models of random intersection graphs. The first one, the *active random intersection graph* $\mathcal{IG}_n^{active}\left(\mathcal{BG}_{n,m}\left(n, \mathcal{P}_{(m)}\right)\right) = \mathcal{IG}_n^{active}\left(n, \mathcal{P}_{(m)}\right)$ is a random graph with the vertex set $\mathcal{V}$, $|\mathcal{V}| = n$, such that any pair v_i and v_j of vertices from $\mathcal{V}$, is joined by an edge $v_i v_j$ if and only if the sets of their neighbors (subsets of $\mathcal{W}$) in the original bipartite graph intersects, i.e., if $\Gamma(v_i) \cap \Gamma(v_j) \neq \emptyset$. Property 2.6 implies that

$$\Pr\left\{v_i v_j \text{ is an edge of } \mathcal{IG}_n^{active}\left(n, \mathcal{P}_{(m)}\right) \right\} = 1 - \sum_{k=0}^{m} P_k \frac{\mathrm{E}_k(m - X^*)}{(m)_k}.$$

Moreover, the vertex-disjoint edges of $\mathcal{IG}_n^{active}\left(n, \mathcal{P}_{(m)}\right)$ appear independently of each other.

Remark. If $\mathcal{P}_{(m)}$ is a binomial distribution, then $\mathcal{IG}_n^{active}\left(n, \mathcal{P}_{(m)}\right)$ is the model introduced in Karoński, Scheinerman and Singer-Cohen (1999).

The model for the active intersection graphs as given above can be generalized by replacing the respective condition with $|\Gamma(v_i) \cap \Gamma(v_j)| \geq s$ for $s \geq 1$, i.e., two vertices should have at least s neighbors in common to be joined by an edge in $\mathcal{IG}_n^{active}\left(n, \mathcal{P}_{(m)}\right)$. We also can use Property 2.6 to derive the probability of an edge in these generalizations of our active random intersection graph model. In this paper, however, we restrict ourselves to the case $s = 1$.

In the first theorem, we will give the formula for the k-th factorial moments of the number of isolated vertices in the active model of random intersection graphs.

Theorem 3.1 *Let* $C^a = C^a\left(\mathcal{IG}_n^{active}\right)$ *denote the number of isolated vertices in* $\mathcal{IG}_n^{active}\left(n, \mathcal{P}_{(m)}\right)$. *Then*

$$\mathrm{E}_k(C^a) = (n)_k \sum_{j=0}^{m} \sum_{(j_1,\ldots,j_k)} \frac{(m)_j}{(m)_{j_1} \cdots (m)_{j_k}} P_{j_1} \cdots P_{j_k} \left[\frac{\mathrm{E}_j(m - X^*)}{(m)_j}\right]^{n-k},$$

where the second sum is over all k-tuples $(j_1, j_2, \ldots, j_k)$, with $0 \leq j_i \leq m$ for $i = 1, 2, \ldots, k$, and $j_1 + j_2 + \cdots + j_k = j \leq m$, assuming additionally $(m)_0 = 1$, and $\mathrm{E}_0(\cdot) = 1$.

Proof. Let $\mathcal{S} = \{v_1, v_2, \ldots, v_k\} \subseteq \mathcal{V}$. Then

$$\Pr\Big\{\text{All vertices in } \mathcal{S} \text{ are isolated in } \mathcal{IG}_n^{active}\left(n, \mathcal{P}_{(m)}\right)\Big\}$$

$$= \Pr\Big\{\Gamma\left(v_1\right), \Gamma\left(v_2\right), \ldots, \Gamma\left(v_k\right) \text{ are pairwise disjoint}$$

$$\text{and } \Gamma\left(\mathcal{V} \setminus \mathcal{S}\right) \subseteq \mathcal{W} \setminus \bigcup_{i=1}^{k} \Gamma\left(v_i\right)\Big\}.$$

Now Property 2.5 and arguments similar to those in the proof of Property 2.6 imply

$$\Pr\Big\{\text{All vertices in } \mathcal{S} \text{ are isolated in } \mathcal{IG}_n^{active}\left(n, \mathcal{P}_{(m)}\right)\Big\}$$

$$= \sum_{j=0}^{m} \sum_{(j_1, j_2, \ldots, j_k)} \frac{(m)_j}{(m)_{j_1}(m)_{j_2}\cdots(m)_{j_k}} P_{j_1} P_{j_2} \cdots P_{j_k} \left[\frac{\mathrm{E}_j(m-X^*)}{(m)_j}\right]^{n-k},$$

where the second sum is over all k-tuples $(j_1, j_2, \ldots, j_k)$ as defined in the theorem. This implies the assertion immediately. •

In particular, for $k = 1$, it follows from the proof of Theorem 3.1 that

$$\Pr\Big\{v \text{ is isolated in } \mathcal{IG}_n^{active}\left(n, \mathcal{P}_{(m)}\right)\Big\} = \sum_{j=0}^{m} P_j \left[\frac{\mathrm{E}_j(m - X^*)}{(m)_j}\right]^{n-1}.$$

In the second theorem, we generalize this result by giving the degree distribution of an arbitrary vertex in the active model.

Theorem 3.2 *The degree distribution of an arbitrary vertex v in the active model is given by*

$$\Pr\Big\{v \text{ has degree } i \text{ in } \mathcal{IG}_n^{active}\left(n, \mathcal{P}_{(m)}\right)\Big\}$$

$$= \binom{n-1}{i} \sum_{k=1}^{m} P_k \left(1 - \frac{\mathrm{E}_k(m - X^*)}{(m)_k}\right)^{i} \left(\frac{\mathrm{E}_k(m - X^*)}{(m)_k}\right)^{n-1-i}.$$

Proof. Let $\emptyset \neq \mathcal{S} = \{v_1, v_2, \ldots, v_i\} \subseteq \mathcal{V}$ be the set of neighbors of a given vertex v in $\mathcal{IG}_n^{active}\left(n, \mathcal{P}_{(m)}\right)$. Given the set of neighbors of the vertex v in $\mathcal{BG}_{n,m}\left(n, \mathcal{P}_{(m)}\right)$, the independence of choices implies that the events $\{vv_j \text{ is an edge of } \mathcal{IG}_n^{active}\left(n, \mathcal{P}_{(m)}\right)\}$ as well as the events $\{vu \text{ is not an edge of } \mathcal{IG}_n^{active}\left(n, \mathcal{P}_{(m)}\right)\}$ for all $j = 1, 2, \ldots, i$ and all

$u \in \mathcal{V} \setminus \mathcal{S}$ are independent. Hence using the reasonings similar to that in the proof of Property 2.6 we get

$$\Pr\left\{ \mathcal{S} \text{ is the set of all neighbors of } v \text{ in } \mathcal{IG}_n^{active}\left(n, \mathcal{P}_{(m)}\right) \right\}$$
$$= \sum_{k=1}^{m} P_k \left(1 - \frac{E_k(m - X^*)}{(m)_k} \right)^i \left(\frac{E_k(m - X^*)}{(m)_k} \right)^{n-1-i}.$$

This leads to the assertion of theorem. $\quad\bullet$

For an application of the general model discussed above, we will focus on the special case where $\mathcal{P}_{(m)} = (P_d)$ is the degenerate distribution, i.e., where each active vertex from $\mathcal{V}$ in the original bipartite graph chooses exactly d neighbors from $\mathcal{W}$. The following two corollaries show that active random intersection graphs $\mathcal{IG}_n^{active}\left(n, (P_d)\right)$ for large n and appropriately chosen m and d, then behave almost like the classical $\mathcal{G}_{n,p}$-model. This holds, since the edges "appear" almost independently (see the proof of Theorem 3.2).

Let us consider first the number of isolated vertices.

Corollary 3.1 *Let $m = m(n)$, $d = d(n)$, and let $nd^2/m - \log n \to c$ as $n \to \infty$ for any constant c. Then for sequences $\left(\mathcal{IG}_n^{active}\left(n, (P_d)\right)\right)_{n \to \infty}$ of active random intersection graphs, the distribution of the number C^a of isolated vertices tends to a Poisson distribution with parameter e^{-c}.*

Proof. Using Theorem 3.1, we can show that

$$E_k(C^a) = (n)_k \frac{(m)_{kd}}{[(m)_d]^k} \left[\frac{(m - d)_{kd}}{(m)_{kd}} \right]^{n-k}.$$

This implies immediately

$$E_k(C^a) = [E(C^a)]^k \left(1 + o(1) \right) = \lambda^k + o(1)$$

with $\lambda = e^{-c}$, whenever $nd^2/m - \log n \to c$ holds, which proves the assertion. $\quad\bullet$

Theorem 3.2 implies directly the following result for the degree of any vertex v, when $\mathcal{P}_{(m)} = (P_d)$.

Corollary 3.2 *For the degenerate distribution $\mathcal{P}_{(m)} = (P_d)$, the degree distribution of any vertex v in an active random intersection graph $\mathcal{IG}_n^{active}\left(n, (P_d)\right)$ follows a binomial distribution with parameters $n - 1$ and p_a where*

$$p_a = \Pr\left\{ v_i v_j \text{ is an edge of } \mathcal{IG}_n^{active}\left(n, (P_d)\right) \right\} = 1 - \frac{(m - d)_d}{(m)_d}.$$

In particular, if $m = m(n)$, $d = d(n)$, and $nd^2/m \to \lambda$ as $n \to \infty$ for any constant λ, then for sequences $\left(\mathcal{IG}_n^{active}\left(n, (P_d)\right)\right)_{n\to\infty}$, the degree distribution tends to a Poisson distribution with parameter λ.

4 The Passive Intersection Graph

Using $\mathcal{BG}_{n,m}\left(n, \mathcal{P}_{(m)}\right)$, we introduce the *passive random intersection graph* $\mathcal{IG}_m^{passive}\left(\mathcal{BG}_{n,m}\left(n, \mathcal{P}_{(m)}\right)\right) = \mathcal{IG}_m^{passive}\left(n, \mathcal{P}_{(m)}\right)$ as a random graph with the vertex set $\mathcal{W}$, $|\mathcal{W}| = m$, such that any pair w_k and w_l of vertices is joined by an edge $w_k w_l$ if and only if the sets of their neighbors in the original bipartite graph (now subsets of $\mathcal{V}$) intersect.

Property 2.7 gives the probability of an edge in $\mathcal{IG}_m^{passive}\left(n, \mathcal{P}_{(m)}\right)$,

$$\Pr\left\{w_k w_l \text{ is an edge of } \mathcal{IG}_m^{passive}\left(n, \mathcal{P}_{(m)}\right)\right\} = 1 - \left[1 - \frac{E_2(X^*)}{(m)_2}\right]^n .$$

The edges of $\mathcal{IG}_m^{passive}\left(n, \mathcal{P}_{(m)}\right)$ do not appear independently of each other, even those with no vertex in common.

Remark. We already noticed that for the binomial distribution, we have $\mathcal{BG}_{n,m}\left(n, \mathcal{B}(m, p)\right) = \mathcal{BG}_{m,n}\left(m, \mathcal{B}(n, p)\right)$. Hence, the intersection graph $\mathcal{IG}_m^{passive}\left(n, \mathcal{B}(m, p)\right) = \mathcal{IG}_m^{active}\left(m, \mathcal{B}(n, p)\right)$ is again the model which was introduced in Karoński, Scheinerman and Singer-Cohen (1999).

Like the previous model, this one also can be generalized, now by the condition $|\Gamma(w_k) \cap \Gamma(w_l)| \geq s$ for $s \geq 1$, on the sets of neighbors of w_k and w_l. We also can use Property 2.7 to derive formulas for the probability of an edge in these generalizations of the passive random intersection graph model.

In the following theorem, we will give the formula for the k-th factorial moments of the number of isolated vertices in the passive model of random intersection graphs.

Theorem 4.1 Let $C^p = C^p\left(\mathcal{IG}_m^{passive}\right)$ denote the number of isolated vertices in $\mathcal{IG}_m^{passive}\left(n, \mathcal{P}_{(m)}\right)$. Then

$$E_k(C^p) = (m)_k \left(\frac{k}{m} P_1 + \frac{E_k(m - X^*)}{(m)_k}\right)^n .$$

Proof. Let $\mathcal{U} = \{u_1, u_2, \ldots, u_k\} \subseteq \mathcal{W}$. Then

$$\Pr\left\{\text{All vertices in } \mathcal{U} \text{ are isolated in } \mathcal{IG}_m^{passive}\left(n, \mathcal{P}_{(m)}\right)\right\}$$

$$= \Pr\left\{\bigcap_{v \in \mathcal{V}}\left\{\bigcup_{u \in \mathcal{U}}\{\Gamma(v) = \{u\}\} \cup \{\Gamma(v) \subseteq \mathcal{W} \setminus \mathcal{U}\}\right\}\right\}.$$

Hence, by Property 2.5 and the definition of $\mathcal{BG}_{n,m}\left(n, \mathcal{P}_{(m)}\right)$, we obtain

$$\Pr\left\{\text{All vertices in } \mathcal{U} \text{ are isolated in } \mathcal{IG}_m^{passive}\left(n, \mathcal{P}_{(m)}\right)\right\}$$
$$= \left(\frac{k}{m} P_1 + \frac{E_k(m - X^*)}{(m)_k}\right)^n.$$

This implies the assertion immediately. ●

For $k = 1$, the last formula in the proof of Theorem 4.1 implies

$$\Pr\left\{w \text{ has degree 0 in } \mathcal{IG}_m^{passive}\left(n, \mathcal{P}_{(m)}\right)\right\} = \left(\frac{P_1}{m} + \frac{E(m - X^*)}{m}\right)^n.$$

We can generalize this result by giving the distribution of the degree of an arbitrary vertex in the passive model as follows.

Theorem 4.2 *The degree distribution of an arbitrary vertex w in the passive model is given by*

$$\Pr\left\{w \text{ has degree } i \text{ in } \mathcal{IG}_m^{passive}\left(n, \mathcal{P}_{(m)}\right)\right\}$$
$$= \binom{m-1}{i} \sum_{s=0}^{i} \binom{i}{s}(-1)^s \left(1 - \frac{EX^*}{m} + \frac{E\left(X^*(m - X^*)_{m-i+s-1}\right)}{(m)_{m-i+s}}\right)^n.$$

Proof. To find the degree distribution for the passive model let us first derive the probability that a given vertex $w \in W$ is not connected by an edge with any vertex from the set $\mathcal{U} \subseteq W$, $|\mathcal{U}| = k$, in $\mathcal{IG}_m^{passive}\left(n, \mathcal{P}_{(m)}\right)$. Conditioning on the subset of $\mathcal{V}$ which consists of all vertices which choose vertex w, we obtain after some tedious but standard calculations

$$\Pr\left\{\text{There is no edge between } w \text{ and vertices of } \mathcal{U} \subseteq W \setminus \{w\}\right\}$$
$$= \sum_{j=0}^{n} \binom{n}{j} \Pr\left\{\Gamma(v) \cap (\mathcal{U} \cup \{w\}) = \{w\}\right\}^j \Pr\left\{\Gamma(v) \subseteq W \setminus \{w\}\right\}^{n-j}$$
$$= \Pr\left\{\Gamma(v) \cap (\mathcal{U} \cup \{w\}) = \{w\}\right\} + \Pr\left\{\Gamma(v) \subseteq W \setminus \{w\}\right\}^n$$
$$= \left(\frac{E(m - X^*)}{m} + \frac{E_k(m - X^*)}{(m)_k} - \frac{E_{k+1}(m - X^*)}{(m)_{k+1}}\right)^n,$$

where v is a given element from $\mathcal{V}$. The inclusion-exclusion formula immediately implies the assertion. ●

As in the previous section, let us give some consequences of the theorems for the special case of the distribution $\mathcal{P}_{(m)}$, the degenerate distribution.

Corollary 4.1 *Let* $m = m(n)$, $d = d(n)$, *and let* $nd/m - \log m \to c$ *as* $n \to \infty$ *for any constant* c. *For sequences* $\left(\mathcal{IG}_m^{passive}(n, (P_d))\right)_{n \to \infty}$ *of passive random intersection graphs, the distributions of the numbers* C^p *of isolated vertices tend to a Poisson distribution with parameter* e^{-c}.

Proof. From Theorem 4.1,

$$\mathrm{E}_k(C^p) = (m)_k \left[(m - d)_k / (m)_k\right]^n$$

follows. Thus, using standard approximations, one can show that

$$\mathrm{E}_k(C^p) = [\mathrm{E}(C^p)]^k \left(1 + O\left(\frac{k^2 dn}{m^2}\right)\right) = \lambda^k + o(1)$$

with $\lambda = e^{-c}$, whenever $nd/m - \log m \to c$ holds. $\qquad\bullet$

From Corollaries 3.1 and 4.1, we see that in the case of a degenerate distribution, $\mathcal{P}_{(m)} = (P_d)$, both intersection graph models $\mathcal{IG}_n^{active}(n, (P_d))$ and $\mathcal{IG}_m^{passive}(n, (P_d))$ differ much. This difference becomes even more evident, when the degree distribution is studied. As we already mentioned, the degree distribution of any vertex in this active random intersection graph is binomially distributed. This is not the case for the passive model. From Theorem 4.2, it follows that

$$\mathrm{Pr}\left\{w \text{ has degree 0 in } \mathcal{IG}_m^{passive}(n, (P_d))\right\} = \left(1 - \frac{d}{n}\right)^n,$$

$$\mathrm{Pr}\left\{w \text{ has degree between 1 and } d - 2 \text{ in } \mathcal{IG}_m^{passive}(n, (P_d))\right\} = 0,$$

and for $i \geq d - 1$

$$\mathrm{Pr}\left\{w \text{ has degree } i \text{ in } \mathcal{IG}_m^{passive}(n, (P_d))\right\}$$

$$= \binom{m-1}{i} \sum_{s=0}^{i-d+1} \binom{i}{s} (-1)^s \left(1 - \frac{d}{m}\left(1 - \frac{(i-s)_{d-1}}{(m-1)_{d-1}}\right)\right)^n.$$

Moreover, it is easy to see that in the passive model with degenerate distribution, a vertex either is isolated or it has to belong to a complete subgraph on d vertices.

5 Application as a Model for Classification

Consider a set $\mathcal{V}$ of n objects and a set $\mathcal{W}$ of m properties. Each object chooses some items. In ancient graves, e.g., we find with every buried person a number of gifts which were chosen out of a "list", and archaeologists want to classify the buried individuals into different groups (they want, e.g., to check whether the persons belong to different clans). This could be done by grouping two persons together if in their graves, at least s of the gifts are the same. In an example from medicine, we have n patients, and each of them chooses some properties from a list of m symptomes, for example high blood pressure, head-ache, and others. To define clusters of similar patients, we can group persons together in the same group if and only if they share at least s symptomes.

This classification concept can be based on the model of the bipartite graph $\mathcal{BG}(\mathcal{V} \cup \mathcal{W}, \mathcal{E})$ as can be seen from the definition of the model. It provides the user with a procedure to detect clusters, especially in sets of non-metric data. We define the components of the active intersection graph constructed from the data, as clusters, similarly as the single-linkage clusters in metric data sets are the components of a coincidence graph (see Godehardt and Jaworski (1996)). If we want to check a set of non-metric data for a structure between the properties, then we can define clusters of properties as subgraphs of the intersection graph on $\mathcal{W}$.

If we are heading for "real" clusters in a data set, then we have to test the validity of the groups detected by some numerical cluster uncovering algorithm. The corresponding probability models for active or passive intersection graphs can be used to derive test statistics to test the hypothesis of randomness of the outlined clusters for both, clusters of objects and clusters of properties.

6   Concluding Remarks

In this paper, we studied the active and passive intersection graphs under the assumption that an edge arises if the respective sets of neighbors have a non-empty intersection in the original bipartite graph. In many applications, however, especially in cluster analysis, it is more reasonable to replace this condition by a stronger one, namely that the sets of neighbors have at least s elements in common. For $s = 1$, the components of the active intersection graphs based archaeological or medical data would put persons into the same cluster if they share only one property.

In applying the general results, we restricted ourselves to the degenerate distribution (P_d). But again, from an application point of view, many other distributions should also be taken into consideration. At present, the authors are continuing their investigation using the beta-binomial distribution.

Acknowledgement   This work partially had been supported by grants *436 POL 17/11/99* and *436 POL 17/14/00* from the German Research Foundation *Deutsche Forschungsgesellschaft, DFG).*

References

BOCK, H.H. (1996): Probabilistic models in cluster analysis. *Computational Statistics and Data Analysis, Vol. 23, 5–28.*

GODEHARDT, E. (1990): Graphs as structural models: The application of graphs and multigraphs in cluster analysis. Vieweg, Braunschweig.

GODEHARDT, E. and JAWORSKI, J. (1996): On the connectivity of a random interval graph. *Random Structures and Algorithms, 9, 137–161.*

JANSON, S., ŁUCZAK, T. and RUCIŃSKI, A. (2001): *Random graphs.* Wiley, New York.

JAWORSKI, J. and PALKA, Z. (1999): Match-graphs of random digraphs. In: W. Gaul, H. Locarek–Junge, (eds.): *Classification in the information age (Proceedings 22nd Annual Conference of the Gesellschaft für Klassifikation e.V., Dresden, March 4–6, 1999).* Springer, Berlin – Heidelberg – New York, 310–319.

JAWORSKI, J. and PALKA, Z. (2001): Remarks on a general model of a random digraph. *Ars Combinatoria (to appear).*

JAWORSKI, J. and SMIT, I. (1987): On a random digraph. *Annals of Discrete Mathematics, 33, 111–127.*

KAROŃSKI, M., SCHEINERMAN, E. R. and SINGER-COHEN, K. B. (1999): On random intersection graphs: The subgraph problem. *Combinatorics, Probability and Computing, 8, 131–159.*

Classification Techniques based on Methods of Computational Intelligence

A. Grauel[1], I. Renners, E. Saavedra

University of Paderborn
Campus Soest, Department of Mathematics
Steingraben 21, D-59494 Soest, Germany
[1]Phone: +49-2921-378173, Fax: +49-2921-378180
E-mail: grauel@ibm5.uni-paderborn.de

Abstract: The main focus of this contribution is to present a general methodology for the structure optimization of fuzzy classifiers. This approach does not depend on a special type of membership function either it is restricted to small or medium sized input dimension. On a well-known classification problem the algorithm performs an input selection over 9 observed characteristics yielding in a statement which attributes are important with respect to the diagnosis of malignant or benign type of cancer. Results achieved by using different types of basis functions are presented.

1   Introduction

In the next section we introduce the basic model we used for classification. By interpreting the basic model as fuzzy system the rule-conclusions defining each model output are determined by a fuzzy classification algorithm presented in section 3. This is followed by a short introduction to Genetic Algorithms (GAs) and Evolutionary Strategies (ESs) because we used a mixed Evolutionary Algorithm to optimize the structure of the basic model. By structure optimization we mean the selection of relevant input features out of a bulk of possible inputs and the finding of a set of rules composed of univariate functions which are optimal adapted in position and shape. Finally classification rates of the different structures are presented.

2   The Basic Model Used for Classification

The used classification model is founded on the superstructure of a Mamdani fuzzy system (Fig. 1). Each processing node is a n-variate function composed of n univariate functions. Unmistakable the similarity to (radial) Basis Function Networks (BFNs) and Takagi-Sugeno-Kang (TSK) fuzzy systems can be seen. The only basic difference of a Mamdani based fuzzy system to a BFN or TSK superstructure is the linear combination of the outputs of the processing nodes of the latter ones. By interpreting this superstructure as a Mamdani based fuzzy system and therefore considering classification in a fuzzy environment [(Gr02)], the inputs are represented by fuzzy labels (or linguistic variables) and the system output

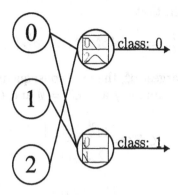

Figure 1: Exemplary superstructure of a Mamdani fuzzy system with three inputs (0-2) and two processing nodes (rules). Both rules consist of two premises (univariate functions).

is defined by fuzzy rules. Let $[a_n, b_n]^N$ be an N-dimensional continuous domain (in the form of a hyper-rectangle) of J real-valued vectors

$$\mathbf{x}_j = (x_{j1}, x_{j2}, \ldots, x_{jN}) \text{ with } x_{jn} \in [a_n, b_n] \tag{1}$$

given as points for a pattern classification into one of M classes. Each fuzzy classification rule k (a processing node in Fig. 1) has the following form:

IF x_{j1} is A_1^k **AND** ... **AND** x_{jN} is A_N^k **THEN** $\mathbf{x}_j$ belongs to class C^k,

where A_n^k are the linguistic terms or premises (univariate functions) and $C^k \in \{1, 2, \ldots, M\}$ is the class assigned.

3 Calculating the Rule Conclusion

The algorithm which computes the rule-conclusion is the same as introduced by [(IsNoYa93)]. To compute the activation $\alpha^k(\mathbf{x}_j)$ of rule k by point $\mathbf{x}_j$ use

$$\alpha^k(\mathbf{x}_j) = \mu_1^k(x_{j1}) + \mu_2^k(x_{j2}) + \cdots + \mu_N^k(x_{jN}) \tag{2}$$

with the degree of membership $\mu_n^k(x_{jn})$ of the vector component x_{jn} to the linguistic term A_n^k. The algebraic sum as a s-norm (for the effect of using different norms see [(GrReLu00)]) represents the AND-conjunction of the premises. Then let

$$\beta_m^k = \sum_{x_j \in C_m} \alpha^k(\mathbf{x}_j) \text{ with } m = 1, 2, \ldots, M. \tag{3}$$

Assign the class C^k such that

$$\beta^k_{C^k} = \max\{\beta^k_1, \beta^k_2, \ldots, \beta^k_M\}. \tag{4}$$

If there is no exclusive largest β^k_m then the consequent C^k of the fuzzy rule k cannot be determined uniquely and $C^k = \emptyset$ and $CF^k = 0$. Otherwise determine CF^k by

$$CF^k = \frac{|\beta^k_{C^k} - \beta|}{\sum_{m=1}^M \beta^k_m}, \tag{5}$$

where

$$\beta = \frac{\sum_{m \neq C^k} \beta^k_m}{M - 1}. \tag{6}$$

To classify of a new point $\mathbf{x}_j$ the rule with the highest product of degree of membership and degree of certainty determines the class assigned $C^{\tilde{k}}$:

$$\alpha^{\tilde{k}}(\mathbf{x}_j) \cdot CF^{\tilde{k}} = \max_k\{\alpha^k(\mathbf{x}_j) \cdot CF^k\}. \tag{7}$$

The classification is invalid if several fuzzy rules k supply the same membership $\alpha^k(\mathbf{x}_j)$ but assign the point $\mathbf{x}_j$ to different classes C^k, or if no rule is activated by the input point $\mathbf{x}_j$, i.e. if $\alpha^k(\mathbf{x}_j) = 0 \ \forall k$.

4 Evolutionary Algorithms

Evolutionary algorithms are inspired by the principles of natural selection which adapt the genes of a species towards an optimal configuration concerning to their environment. Like the interaction of a biological individual with its environment, the interaction of a mathematical formalized chromosome with a fitness function provides a numerical valued measure of fitness. This fitness represents the quality of an individual and is used to bias the selection by preserving better individuals a higher probability to reproduce. Offspring share different chromosome parts of their parents through recombination and diversity is sustained by mutation. Thus all evolutionary algorithms share the same general proceeding of

- Step 0: initialize a start population of solutions
- Step 1: create new solutions (offsprings) by recombination
- Step 2: mutate each solution
- Step 3: deterministic selection by fitness
- Step 4: goto step 1 until stop criteria
- Step 5: present the best solution in population

4.1 Genetic Algorithms

GAs were developed for optimizing binary (or at most integer) based solutions. The main operator in GAs is crossover which directs the search to new and better solutions areas. This simplest crossover schema determines by randomly chosen Crossover Points (CPs) which parts of the parent chromosomes are passed on the child (Fig. 2a). By using only

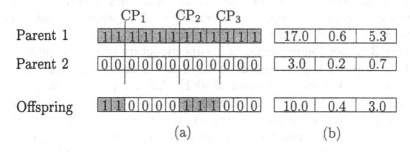

Figure 2: Discrete 3-point crossover (a) and intermediate crossover (b).

crossover a gene with identical parameter values in all individuals will never change. Therefore mutation is important to avoid stagnation of the search process in a subspace of the original search space. Mutation is implemented by inverting each gene with a priori given mutation probability.

4.2 Evolutionary Strategies

ESs were mainly developed to optimize real decision variables [(Sc95)]. Every ES-individual is equivalent to a vector of real numbers and contains values for all decision variables $x_j \in \mathbf{R}(j = 1, 2, \ldots, n)$ for the stated problem. Furthermore each individual contains $n_\sigma (1 \leq n_\sigma \leq n)$ standard deviations $\sigma_k \in \mathbf{R}^+(k = 1, 2, \ldots, n_\sigma)$ which are called (average) *rate of mutations*. These σ_k are strategy parameters which are adjusted self-adaptively during the optimization process. The decision variables of an offspring are inherited by one of the parents (same as in GAs), whereas the strategy parameters are inherited by intermediate crossover (see Fig. 2b). Mutation is the main operator in ES and is done (by working with only one standard deviation) by changing the values of σ_j and x_j by two different methods. First the values σ_j are recalculated [(Ni97)] by $\sigma'_j = \sigma_j \cdot exp(\tau_1 \cdot N(0,1) + \tau_2 \cdot N_h(0,1)$, where $N(0,1)$ and $N_h(0,1)$ are normal distributed random numbers. Then every decision variable (x_j) is changed by $x'_j = \sigma'_j \cdot N_h(0,1)$.

4.3 The Used Encoding

The complete Mamdani fuzzy system genotype consists of genes of four different types. We have defined promotor (p)-genes which (de)activate gene-sequences, integer (i)-genes which encode integer values, double (d)-genes and strategy parameter (s)-genes which encode real values. Hence all (d)-genes are processed by evolutionary strategies each (d)-gene is assigned a (s)-gene encoding the concerning σ. The promotor and integer genes are processed by genetic algorithm operations.

Picture 3 shows the encoding of a superstructure of two possible rules with three possible premises. Each (d)-gene encoding position, width or balance is scaled on an interval of $[0; 1]$. Thus a position value of 0.1 means that the center (not the peak point) of the concerning premise (univariate function) is on the very left side of the concerning input. A balance value of 0.5 means that the peak point of the concerning premise is places on the center of the premises definition interval. Thus the balance value has no effect on (per definition) symmetric functions (e.g. symmetric Gaussians). Implemented types of functions are triangular, trapezoidal, cubic b-spline, symmetric Gaussian and asymmetric Gaussian.

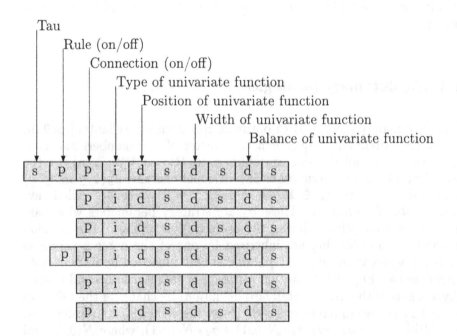

Figure 3: Encoding of two rules each with three premises.

4.4 The Used Evolutionary Parameters

Tab. 2 list the main parameters used by the evolutionary algorithm to achieve the results as shown in section 6. As a first instance an individual is rated better than another if it classifies more patterns correctly. If both individuals share the same number of correct classified patterns, the individual with the lower complexity is fitter.

parameter	value
population size	200
tournament selection	10 individuals (elite)
stop criteria	75th generation
encoded rules	4
used rules	2
established connections	1-10
1st fitness	proportional to correct classified patterns
2nd fitness	proportional to model complexity
crossover points	random but with higher crossover attraction on promotors
τ_1 (ES)	0.01 - 0.5 (self-adaptive)
τ_2 (ES)	0.01 - 0.5 (self-adaptive)
σ_n (ES)	0.000001 - 1.0 (self-adaptive)
mutation rate (GA)	0.01

Table 1: Main parameters for the evolutionary algorithm.

5 The Data

The Wisconsin breast cancer dataset was compiled by Dr. William H. Wolberg from the University of Wisconsin Hospitals, Madison [(WoMa90a; WoMa90b)]. It contains $J = 699$ cases (458 benign, 241 malignant), where 16 cases have missing values, separated in $M = 2$ classes (sample diagnosis benign or malignant) and characterized by $N = 9$ ordinal discrete features: Clump Thickness (0), Uniformity of Cell Size (1), Uniformity of Cell Shape (2), Marginal Adhesion (3), Single Epithelial Cell Size (4), Bare Nuclei (5), Bland Chromatin (6), Normal Nucleoli (7) and Mitoses (8). The discrete features are visually assessed characteristics on a scale from 1 to 10. 16 cases containing missing values were removed from the data set, leading to $J = 683$ cases used.

6   Results and Discussion

Five optimized structures were computed by restricting the evolutionary algorithm to use a specific type of function. One optimized structure was calculated without this constraint. In all cases we restricted the evolutionary algorithm to compute only two rules. All function types result in a very similar output performance. Performing no crossover in the evolutionary process yields as expected in slower convergence and worser classification results (3-5 patterns less correct classified).

The interpretability of the resulting fuzzy systems is always high because the complexity is a fitness factor. Thus in general the found models fulfill the requirements that a rule should use as few as possible variables and that no redundant rules occur in the model. By using the above described algorithm to determine the conclusion a confidence value of each rule is calculated and furthermore no contradictory rules (rules with identical antecedents but different consequences) can be produced. The performance is comparable to the free available software tool NEFCLASS [(NaKlKr97)] but our algorithm can also be used to select input features out of a high-dimensional space.

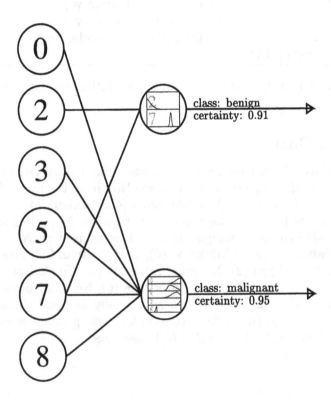

Figure 4: Structure of the model using asymmetric Gaussian functions.

allowed function type	correct classified patterns
triangular	669 (97.95%)
trapezoidal	671 (98.24%)
cubic bspline (symmetric)	670 (98.09%)
Gaussian (symmetric)	669 (97.95%)
Gaussian (asymmetric)	671 (98.24%)
all	672 (98.39%)

Table 2: Classification rates of the different structures.

References

Grauel A.: Taschenbuch der Mathematik, Fuzzy Logic, Chapter 5.9 (Eds. Bronstein, Semendjajew, Musiol, Mühlig) Physica-Verlag, Ed. Springer-Verlag, Heidelberg-Berlin, Germany, (2002) 372–391.

Grauel A., Renners I., Ludwig L.A.: Optimizing Fuzzy Classifiers By Evolutionary Algorithms. Proc. KES2000, 4th Int. Conf. on Knowledge Based Intelligent Engineering System & Allied Technologies, Brighton, Great Britain, (2000) 353–356.

Ishibuchi H., Nozaki K., Yamamoto N.: Selecting fuzzy rules by genetic algorithm for Classification. Proc. IEEE Int. Conf. Fuzzy Syst., Vol. II, San Francisco (1993) 1119–1124.

Schwefel H.P.: Evolution and Optimum Seeking, New York: Wiley&Sons, (1995).

Nissen V.: Einführung in Evolutionäre Algorithmen, Vieweg Verlag, Wiesbaden (1997) p.159.

Wolberg, W.H., Mangasarin, O.L.: Cancer Diagnosis via linear programming. SIAM News 23 (5),(1990) 1–18.

Wolberg, W.H., Mangasarin, O.L.: Multisurface method of pattern separation for medical diagnosis applied to breast cytology. Proc. National Academy of Sciences U.S.A. 87 (1990) 9193–9196.

Nauck D., Klawonn F., Kruse R.: Foundations of Neuro-Fuzzy Systems. Wiley, Chinchester (1997).

Constrained Correspondence Analysis for Seriation in Archaeology Applied to Sagalassos Ceramic Tablewares

P.J.F. Groenen[1], J. Poblome[2]

[1]Data Theory Group, Department of Education,
Faculty of Social and Behavioral Sciences, Leiden University,
P.O. Box 9555, 2300 RB Leiden, The Netherlands

[2]Department of Archaeology, Catholic University Leuven,
Sagalassos Archaeological Research Project,
Blijde Inkomststraat 21, 3000 Leuven, Belgium

Abstract: Correspondence analysis is a well known technique for seriation of archaeological artefactual assemblages. One problem with the seriation solution is that no explicit time frames are obtained, only a relative ordering. However, in some cases additional information is available allowing absolute dating of some of the deposits. Such explicit dating information may be obtained from associated categories of finds, such as coin series. Additional information may be available that logically restricts the order of the seriation. For example, in case of a superposed stratigraphical sequence, the lower stratum is associated with events which took place earlier than the upper layer and consequently this ordering should be replicated in the seriation.

In this paper, we propose a constrained form of correspondence analysis that takes such restrictions into account. Using these constraints we are able to assign explicit dates to a seriated solution. This new method of seriation is applied to a series of ceramic assemblages consisting of the locally produced tableware from Sagalassos (SW Turkey). These tableware assemblages have already been seriated and dated independently via empirical archaeological techniques (Poblome, 1999). The coarse classification of these assemblages are used as constraints in a confirmatory analysis of constrained correspondence analysis.

1 Introduction

Correspondence analysis (CA) is a well known technique for seriation or relative sequencing of archaeological finds (Shennan, 1988; Cool and Baxter, 1999, with further references). In many cases, extra information on the artefact assemblages is available, which, however, is mostly ignored in conventional CA. In this paper, we extend CA to make use of such additional information. In particular, we constrain CA to allow for inequalities between the ordering of assemblages, for equality of two or more assemblages, and to use available chronological information. From given dates of some assemblages, the chronological position can be derived for those assemblages for which no date could be reconstructed. This methodology is applied to Sagalassos pottery data.

Ancient Sagalassos (Pisidia, SW Turkey) has been the focus of an inter-disciplinary archaeological research project coordinated by the Catholic University of Leuven since 1985 (Waelkens and the Sagalassos team, 1997). In 1987, a potters' quarter was located to the east of the town and the study of the production organization of this mass-production center has substantially improved our understanding of the functioning of the settlement and the everyday life of its inhabitants. Reconstructing the chronological evolution of the local tableware, called *Sagalassos red slip ware*, has proven to be difficult, as no well dated imported pottery types were available. Therefore, an independent chronological framework had to be established. By applying a newly developed archeological method-ology (Poblome, 1999), anchor assemblages were determined from which the chronological position of the other assemblages could be extrapo-lated. However, each new excavation season at Sagalassos adds new as-semblages, invoking a continuous updating of the seriated sequence and a constant evaluation of the anchor assemblages. Our proposed model of constrained CA may provide a solution to both issues.

2 Constrained Correspondence Analysis

In this section, we discuss how the constrained correspondence analysis solution is obtained. In our approach, we consider correspondence anal-ysis as a least-squares minimization problem, where the $n_r \times n_c$ observed contingency matrix $\mathbf{F}$ of n_r assemblages by n_c pottery types is approx-imated. In this section, we use conventional matrix notation as can be found, for example, in Greenacre (1984) and Gifi (1990). Let $\mathbf{D}_r$ and $\mathbf{D}_c$ be a diagonal matrices with the row and column sums of $\mathbf{F}$, and let n be the sum of all elements in $\mathbf{F}$. Also, let $\mathbf{E}$ be the matrix of estimates of the independence model, that is, $\mathbf{E} = n^{-1}\mathbf{D}_r\mathbf{1}\mathbf{1}'\mathbf{D}_c$. Then, correspondence analysis can be seen as the minimization of

$$L(\mathbf{r}, \mathbf{c}) = \|\mathbf{D}_r^{-1/2}(\mathbf{F} - \mathbf{E} - n^{-1}\mathbf{D}_r\mathbf{r}\mathbf{c}'\mathbf{D}_c)\mathbf{D}_c^{-1/2}\|^2 \qquad (1)$$

over the assemblage coordinates vector $\mathbf{r}$ and the pottery type coordi-nates vector $\mathbf{c}$. Note that in (1) the norm notation $\|\mathbf{A}\|^2$ simply denotes the sum of all squared elements of $\mathbf{A}$, i.e., $\|\mathbf{A}\|^2 = \sum_{i=1}^{n_r} \sum_{j=1}^{n_c} a_{ij}^2$. It is well known that an analytic solution can be obtained by using a singular value decomposition of $\mathbf{D}_r^{-1/2}(\mathbf{F} - \mathbf{E})\mathbf{D}_c^{-1/2}$. However, we use an alter-nating least squares algorithm to minimize (1) because the additional constraints proposed below will need to be incorporated.

The alternating least squares approach to correspondence analysis con-sists of minimizing (1) using the following steps. First, find initial coor-dinates for $\mathbf{r}$ and $\mathbf{c}$, possibly by using random numbers. Then, alternate the next two steps until no change in coordinates occurs: (a) update $\mathbf{c}$ for fixed $\mathbf{r}$ and (b) update $\mathbf{r}$ for fixed $\mathbf{c}$. Because in (1) the vectors $\mathbf{r}$ and c are multiplied, the coordinates are unique upon a multiplication factor. To avoid this nonuniqueness, we constrain c without loss of generality to have sum of squares one.

In constrained correspondence analysis, we have additional constraints on r. Thus, the optimization task in Step (b) is to minimize (1) over r subject to the restrictions. Let us consider an example for the moment. Suppose we have six assemblages (A_1 to A_6) and archaeological findings indicate that the year of A_1 is $y_1 = 100$ AD, A_4 is $y_1 = 425$ AD, and A_6 is $y_1 = 600$ AD. Then, the linear constraints on the correspondence analysis coordinates r_1, r_4, and r_6 are $r_1 = a + by_1, r_4 = a + by_4$, and $r_6 = a + by_6$, where a and b need to be estimated. Let us further assume that the year of A_2 must be equal to that of the year of A_3. This equality constraint indicates that $r_2 = r_3$. Both type of constraints can be imposed by restricting r to be a linear combination of the columns of matrix H derived from these two types of constraints, i.e., $r = Hb$, where b are the parameters to be estimated. In our example, matrix H equals

$$H = \begin{bmatrix} 1 & 100 & 0 & 0 \\ 0 & 0 & 1 & 0 \\ 0 & 0 & 1 & 0 \\ 1 & 425 & 0 & 0 \\ 0 & 0 & 0 & 1 \\ 1 & 600 & 0 & 0 \end{bmatrix},$$

so that the restriction $r = Hb$ implies

$$
\begin{aligned}
r_1 &= b_1 + 100b_2 \\
r_2 &= b_3 \\
r_3 &= b_3 \\
r_4 &= b_1 + 425b_2 \\
r_5 &= b_4 \\
r_6 &= b_1 + 600b_2.
\end{aligned}
$$

We shall use a slightly different way to impose the same linear restriction on r. Let H_0 be the null-space of H so that $H'H_0 = 0$. To state that r must be a linear combination of the columns of H is the same as requiring that r is not allowed to be in the null-space of H, that $H_0'r = 0$. An advantage of the latter formulation of the restrictions is that it is expressed in terms of r without introducing new parameters b as in the expression $r = Hb$.

We now turn to the order restrictions on the coordinates of r. Let us assume that assemblage A_5 must be older than A_6, and that A_2 must be younger than A_1 but older than A_4. Note that the latter inequality restrictions also imply that A_3 must be younger than A_1 but older than A_4 due to the equality restrictions of assemblages A_2 and A_3. The ordering restrictions on the assemblages mean that $r_5 \leq r_6, r_1 \leq r_2$, and $r_2 \leq r_4$. In matrix algebra, these inequalities can be written as $\mathbf{Gr} \geq \mathbf{0}$, where

$$\mathbf{G} = \begin{bmatrix} -1 & 1 & 0 & 0 & 0 & 0 \\ 0 & -1 & 0 & 1 & 0 & 0 \\ 0 & 0 & 0 & 0 & -1 & 1 \end{bmatrix}.$$

Considering all constraints simultaneously, the optimization task in Step (b) is to minimize (1) over $\mathbf{r}$ subject to the restrictions $\mathbf{H}_0'\mathbf{r} = \mathbf{0}$ and $\mathbf{Gr} \geq \mathbf{0}$. Rewriting (1) gives

$$L(\mathbf{r}, \mathbf{c}) = \|n^{-1}\mathbf{D}_r^{-1/2}\mathbf{r} - \mathbf{t}\|^2 - \|\mathbf{t}\|^2 + \|\mathbf{D}_r^{-1/2}(\mathbf{F} - \mathbf{E})\mathbf{D}_c^{-1/2}\|^2 \quad (2)$$

where $\mathbf{t} = \mathbf{D}_r^{-1/2}(\mathbf{F} - \mathbf{E})\mathbf{c}$. For fixed $\mathbf{c}$, (2) is quadratic in $\mathbf{r}$. This type of problem is called the least-squares problem with linear equality and inequality constraints and a solution can be obtained by transforming it to a nonnegative least-squares problem (see, Lawson and Hanson, 1974).

A scheme of the alternating least squares constrained correspondence analysis (ALS-CCA) algorithm is:

1. Choose initial $\mathbf{c}_0$ (with $\mathbf{c}_0'\mathbf{c}_0 = 1$) and $\mathbf{r}_0$ satisfying the constraint $\mathbf{H}_0'\mathbf{r} = \mathbf{0}$ and $\mathbf{Gr} \geq \mathbf{0}$. Set iteration counter $k = 0$ and set a convergence criterion ϵ to a small positive value, e.g., $\epsilon = 10^{-6}$.

2. $k := k + 1$.

3. Update $\mathbf{c}$:
   Set $\mathbf{c} = n^{-1}\mathbf{D}_c^{-1/2}(\mathbf{F} - \mathbf{E})'\mathbf{r}_{k-1}$ and compute $\mathbf{c}_k = \mathbf{c}/(\mathbf{c}'\mathbf{c})^{1/2}$.

4. Update $\mathbf{r}$:
   Minimize $\|n^{-1}\mathbf{D}_r^{-1/2}\mathbf{r} - \mathbf{t}\|^2$ over $\mathbf{r}$ subject to $\mathbf{H}_0'\mathbf{r} = \mathbf{0}$ and $\mathbf{Gr} \geq \mathbf{0}$ by using Lawson and Hanson (1974, see pages 168-169) and set the result into $\mathbf{r}_k$.

5. If $L(\mathbf{r}_{k-1}, \mathbf{c}_{k-1}) - L(\mathbf{r}_k, \mathbf{c}_k) > \epsilon$ then go to step 2, otherwise stop.

3   Reconstructing the Dates

The seriation results can be used to reconstruct the dates. For some of the assemblages, absolute dates could be determined and these are used in the constrained correspondence analysis to separate them on the seriation axis $\mathbf{r}$ proportionally to the dates. The other assemblages also obtain a coordinate on the seriation axis of $\mathbf{r}$. Therefore, it is possible to inter- and extrapolate the dates for the assemblages with unknown a

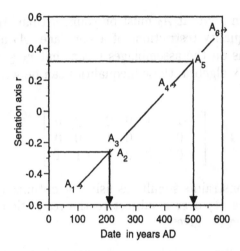

Figure 1: Relation between seriation axis **r** (vertically) found by constrained correspondence analysis and the derived years.

priori dates. This process is illustrated in Figure 1. The assemblages with known dates $(A_1, A_4,$ and $A_6)$ define the straight line between their seriation coordinate r_i and their date. For the assemblages with unknown dates $(A_2, A_3,$ and $A_5)$ their seriation coordinate r_i is projected on the line and from the line on the date axis. In this way, the date can be derived for an assemblage with an unknown *a priori* date.

4 Constrained CA Applied to Sagalassos Red Slip Ware Data

In the present analysis, we use a selection of 34,392 shards of the Sagalassos pottery data, from 27 assemblages and classified into 85 pottery types and variants. Due to the large difference in marginal frequencies of the pottery types (ranging from 1 to 3,384), we analyzed pottery proportions and not frequency counts. For four of the assemblages the dates were known in advance: assemblage 1 was fixed to date 1 AD, 4 to 100 AD, 22 to 410 AD, and 27 to 650 AD. Two pairs of assemblages were fixed to equal dates, i.e., assemblages 6 and 7, and assemblages 24 and 25. The assemblages were ordered by the archaeologist *a priori* into nine phases. This ordering can be found in Table 1 in the column 'Phase date'. The seriation obtained by constrained correspondence analysis together with the reconstructed dates are also found in Table 1.

Figure 2 shows the frequencies of the assemblages and the pottery types using the order found by constrained correspondence analysis of **r** and **c**. In general, three main phases have been distinguished in the evolution of Sagalassos red slip ware: early Roman (1 st-2nd century AD), late Ro-

Table 1: Seriation results by constrained correspondence analysis.

Recon-structed Year	Assemblage Nr	Assemblage Label	Phase Nr	Phase Date	r_i
1	1	TSW2	1	0-50	-10.9
15	2	NoN 5-8	1	0-50	-10.4
46	3	L 10-16N	1	0-50	-9.0
100	4	L 9-18S	2	50-100	-6.7
122	8	LW 18-20C	3	100-150	-5.8
134	6	EoN 11-18	3	100-150	-5.3
134	7	NoN 2-4	3	100-150	-5.3
136	10	RB-R3, B	3	100-150	-5.2
138	5	L 8-9N	3	100-150	-5.1
153	9	RB-R3, A	3	100-150	-4.5
153	11	L 3-4N	4	150-200	-4.5
153	12	L 5-7N	4	150-200	-4.5
153	13	EoN 4-8	4	150-200	-4.5
153	14	F	5	200-300	-4.5
182	15	TSW4 4-6	5	200-300	-3.2
293	16	Lib	6	300-350	1.5
320	17	LW 16-17C	6	300-350	2.7
410	18	LE 4-6	7	350-450	6.6
410	19	LW 9-14C	7	350-450	6.6
410	22	H Floor	8	450-575	6.6
490	23	H Fill	8	450-575	10.0
592	26	B3 D1 pre	8	450-575	14.3
601	20	Nymph	8	450-575	14.7
608	21	WDT	8	450-575	15.0
617	24	Inn Corr S, 7	8	450-575	15.4
617	25	B3 D1 post	8	450-575	15.4
650	27	LA	9	575-650	16.8

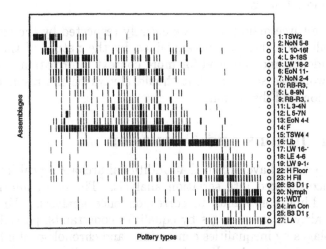

Figure 2: Battleship plot showing frequencies of the seriated assemblages and pottery types.

man (4th-5th century AD) and early Byzantine (6 th-7th century AD). Clearly, ceramic production continued in the third century AD, but the typological constitution of the contemporary ceramic assemblage seems as yet difficult to distinguish from both previous centuries. The material has, moreover, never been found as part of a stratigraphical sequence. The assemblages of phases 1 to 4 are well placed. The distance between phases 1 and 2 is respected, as well as the superposed assemblages 2 and 7, 3, 5, 11 and 12, 10 and 9, and 6 and 13. Five assemblages are attributed to 153 AD. As the typological content of these assemblages is very similar, this sequencing comes as no surprise. At this point, the assemblages should be integrated into the wider context of the evolution of the ancient town, in order to differentiate between them based purely on archaeological interpretation. The position of assemblages 14 and 15 may be regarded as an anomaly, but until more stratigraphical evidence becomes available this particular problem cannot be resolved. The assemblages of phases 6 and 7 are well placed and respect the stratigraphical superposition.

As far as the early Byzantine period is concerned, the actual chronological distance between phases 8 and 9 is recognized. Phase 8 is as such the longest of the entire sequence. Although there are some discrepancies with the actual dates, they may have resulted from our deliberate choice not to add additional constraints for the 518 AD and 528 AD earthquakes in Sagalassos. As the interpretation of Phase 8 has still not been completed, we considered adding this fixed point premature. However, the proposed sequence entirely respects the stratigraphical sequencing, with large enough distances between assemblages 22 and 23 and 25 and 26. The position of assemblages 20 and 21 is also acceptable, irrespective of the proposed dates.

In sum, the three main phases are entirely respected by the constrained CA and all anomalies can be explained by the nature of the archaeological evidence and the need for more context information. Our model provides workable solutions, but should always be the result of a close collaboration and discussion.

5 Conclusions and Discussion

In this paper, we have shown that additional archaeological information can be incorporated into seriation analysis. The extra information is used by adding constraints to correspondence analysis. We have imposed equality of assemblages by equalities constraints, partial ordering of assemblages by inequalities constraints, and chronological information by linear constraints. An algorithm for constrained correspondence analysis is proposed. As an important consequence, our seriation results can be used to reconstruct the chronological position of for assemblages with unknown dates. However, the reconstructed dates have to be interpreted with care because it is based on inter and extrapolation. Therefore, the quality of the reconstructed dates is dependent on the range of the known dates, and the fit of the solution.

When correspondence analysis is used for seriation, only the first correspondence analysis axis is used. However, it may be so that the time axis is not the first dimension, but comes up in, say, the second dimension. In such cases, our methodology is still applicable, and a constrained correspondence analysis algorithm can be obtained by sequential fitting of dimensions.

References

COOL, H.E.M. and BAXTER, M.J. (1999): Peeling the onion: An approach to comparing vessel glass assemblages. *Journal of Roman Archaeology, 12,* 72—100.

GREENACRE, M.J. (1984): Theory and applications of correspondence analysis. Academic Press, New York.

GIFI, A. (1990): Nonlinear multivariate analysis. Wiley, Chichester.

LAWSON, C.L. and HANSON, R.J. (1974): Solving least squares problems. Englewood Cliffs, NJ: Prentice Hall.

POBLOME, J. (1999): Sagalassos red slip ware: Typology and chronology. *Studies in Eastern Mediterranean Archaeology, 2.* Brepols, Turnhout.

SHENNAN, S. (1988): Quantifying archaeology. Edinburgh University Press, Edinburgh.

WAELKENS, M. and THE SAGALASSOS TEAM (1997): Interdisciplinarity in classical archaeology. A case study: the Sagalassos Archaeological Research Project. In: M. Waelkens and J. Poblome (Eds.), Sagalassos IV, *Acta Archaeologica Lovaniensia Monographiae, 9,* Leuven, 225-252.

The Moment Preservation Method of Cluster Analysis

B. Harris

Department of Statistics,
University of Wisconsin, Madison, Wisconsin 53706 1693, U.S.A.

Abstract: A technique for cluster analysis, known as the moment preservation method is found primarily in the engineering literature. The method appears to be motivated by mixture models. A brief description of this technique is as follows. If q clusters are to be determined, then $2q - 1$ linearly independent functions of the data are calculated. (In many of the applications that I have encountered, these are the first $2q - 1$ sample moments.) For univariate data, the values of these functions are used to determine thresholds. The thresholds are chosen so that these moments are preserved. For multivariate data, the thresholds are replaced by their natural analogue, linear manifolds. The theoretical properties of this process can be determined from a substantial body of mathematical analysis, known as the "Reduced Moment Problem". These properties facilitate determining the solutions sets and their properties, including the determination of optimal solutions.

1 Introduction and Summary

A technique for determining clusters, known as the moment preservation technique for cluster analysis has appeared in the engineering literature; in particular, see Tabatai and Mitchell (1984), Tsai (1985), Pei and Cheng (1986), Delp and Mitchell (1979), Tabatai (1981), Lin and Tsai (1994). The proponents of this method claim that it is highly accurate and easily computable. The authors of the above papers appeared to be unaware that there is a substantial mathematical literature applicable to their procedure and that these results can be readily extended to more general situations.

2 The Moment Preservation Method

The present form of the moment preservation method will now be described for univariate data. Let $x_1, x_2, \ldots, x_n$ be the realizations of a random sample from a continuous distribution. The number of clusters to be obtained needs to be specified in advance. Assume that q ($q < n$) clusters are desired. Then the first $2q - 1$ non-central sample moments are calculated. The unique discrete probability distribution with positive probability at q points, which possesses these moments, is determined. Thus, to obtain two clusters, the first three sample moments are calculated, which determines the two point distribution, $(p_1, y_1 : 1 - p_1, y_2, 0 < p_1 < 1, y_1 < y_2)$. In general, the solution is given by $(p_i, y_i, i = 1, 2, \ldots, q : p_i \geq 0, \sum_{i=1}^{q} p_i = 1, y_1, y_2, \ldots, y_q)$; y_i is the center of the ith cluster. The data points x_i, $i = 1, 2, \ldots, n$, are assigned to the q clusters as follows. Arrange the n data points in increasing order. The observations whose rank does not exceed $n\,p_1$ are placed in the first cluster. From the remaining observations, those whose rank does not exceed $n(p_1 + p_2)$ are placed in the second cluster. This pocess is continued until all observations have been assigned to a cluster. The process of obtaining the q-point probability distribution, referred to above, is computationally identical to solving the "classical" reduced moment problem.

3 The Reduced Moment Problem and Its Solutions

Let $M_k = \{\mu_1, \mu_2, \ldots, \mu_k\}$ be a given realizable sequence of the first k non-central moments. Realizable means that there exists a cumulative distribution function $F(x)$ such that

$$\int_{-\infty}^{\infty} x^j \, dF(x) = \mu_j, \qquad j = 1, 2, \ldots, k.$$

$F(x)$ is unknown. The problem of determining the set of possible probability distributions with these moments is called the (classical) *reduced moment problem*. There are three cases.

– The Hausdorff moment problem. The carrier set of $F(x)$ is a subset of $[0, 1]$.

– The Stieltjes moment problem. The carrier set is a subset of $[0, \infty)$.

– The Hamburger moment problem. The carrier set is a subset of $(-\infty, \infty)$.

There is an extensive body of literature on this subject as well as on various modifications and extensions of it. Space limitations prevent even a cursory discussion of this subject, nevertheless, for interested readers, the following references may be of interest: Shohat and Tamarkin (1943), Karlin and Shapley (1953), Ahiezer and Krein (1938).

To characterize the solution set, we will use a geometric approach, based on the theory of convex functions and convex sets. The following basic results are relevant. Let $\mathcal{X}$ denote a subset of the real line (such as $\{0, 1\}$, $[0, \infty)$, and so forth).

1. The set of probability distributions on $\mathcal{X}$ is a convex set.

2. The set of discrete probability distributions on $\mathcal{X}$ is a convex set.

3.   The set of probability distributions on $\mathcal{X}$ with finite support is a convex set.

In each of the above situations, the set of extreme points is the set of degenerate (one-point) distributions. In particular, the set of probability distributions on $\mathcal{X}$ is the weak closure of the set of finite convex combinations of the degenerate distributions. That is, all distributions are "limits" of finite convex combinations of the degenerate distributions.

4.   Let $\mathcal{M}_k$ be the set of realizable moment sequences of length k for probability distributions on $\mathcal{X}$. Then $\mathcal{M}_k$ is a convex set in $\mathcal{E}_k$ (k-dimensional Euclidean space).

5.   The set of probability distributions $\mathcal{F}(M_k)$ on $\mathcal{X}$ with a given realizable moment sequence M_k is a convex set.

The extreme points of $\mathcal{M}_k$ are the vectors of the form $(a, a^2, \ldots, a^k)$, that is, the reduced moment sequences of the degenerate distributions. Thus, for example, for $k = 2$, $\mathcal{X} = [0, 1]$, the set of realizable moments is the set bounded by the line from $(0, 0)$ to $(1, 1)$ and the curve $\{(x, x^2), 0 \leq x \leq 1\}$. A useful characterization of moment sequences in terms of the extreme points is provided by a theorem of Carathéodory, which we will now state and subsequently, we will show how this theorem can be utilized to select solutions of the reduced moment problem with specific properties.

Carathéodory's Theorem. *Let C be a compact convex set in $\mathcal{E}_k$. Let $u \in C$. Then u has a representation as a convex combination of at most $k + 1$ extreme points of C. If u is a boundary point of C, then u has a representation as a convex combination of at most k extreme points of C.*

To see how Carathéodory's Theorem can be exploited, we consider the specific case $k = 2$ and $\mathcal{X} = [0, 1]$. If a point u is on the boundary, there are two possibilities. A point on the "lower boundary" is the image of a degenerate distribution, which has probability one at some value in $[0, 1]$. A point u on the upper boundary is a convex combination of degenerate distributions at 0 and 1. Now assume that u is in the interior of C. Then u can be represented as a convex combination of two points on the lower boundary or one point on the lower boundary and one point on the upper boundary or three points on the lower boundary. The coefficients of the convex combination can be readily determined by comparing the distances of u to each of the extreme points in the convex combination.

4 Some Extensions and Generalizations

As far as the mathematical theory is concerned, there is no reason to restrict to the first k monomials; $x, x^2, \ldots, x^k$, which determine the first k non-central moments. In principle any set of k linearly independent functions could be used. However, the computations necessary for determining the solutions are more complicated if the functions other than x are not strictly convex.

Some classes of variational problems can be treated using the same theory. For background material, see Rustagi (1976), Harris (1959), Harris (1962).

Let $g(x)$ be a real valued function, which is integrable on $\mathcal{F}(M_k)$. Then we wish to maximize (minimize) $\int g(x)\, dF(x)$, $F \in \mathcal{F}(M_k)$. This is of interest in the clustering problem, which motivated this study, in the following way. If $g(x)$ is some measure of merit of the clustering procedure, such as the probability of misclassification, then we would wish to choose the best solution from among the solutions of the reduced moment problem. The procedure requires considering the "reduced moment set" generated by $\{g(x), x, x^2, \ldots, x^k\}$. It is not possible to describe the possible procedures in the available space. The methodology uses the supporting hyperplane theorem for convex sets and (or) the Hahn-Banach extension theorem. However, the following brief remarks can be provided. In general, extremal solutions tend to be generated by distributions with small carrier sets. The computations are easily carried out if $g(x)$ is a well-behaved function. Depending on the particular probem, such include completely monotonic functions, absolutely monotonic functions, or strictly convex functions.

In the preceding material, only the univariate case was discussed. The multivariate case is feasible, but substantially more complicated. There are some problems with identifiability in multidimensional situations. One alternative that has been utilized is to subject the data to principal component analysis. This permits the reduction of the multivariate problem to a sequence of one dimensional problems. Treating the first principal component as a one dimensional problem, carry out the moment preservation method for projections of the data onto the first principal component. Then project the residuals onto the second principal component and so forth. There is also some work in progress in using binary decision trees for such problems, but this is in an early stage.

5 A Numerical Illustration

We consider two problems using the same data. For simplicity, the data is artificial, so that the solutions are patrticularly simple. For the first, we determine two clusters using the first three sample moments. This is precisely the engineering application that motivated this study. The second problem will utilize two moments and characterize the set of distributions that have these moments. We will obtain a parametric reprersentation for the elements of this set. If we choose a specific parameter, then we get the unique solution that is obtained as the solution to the first problem.

Denote the sample moments by m_1, m_2, m_3. Then the solution is given by

$$a = \frac{m_1 m_3 - m_2^2}{m_2 - m_1^2}, \qquad b = \frac{m_1 m_2 - m_3}{m_2 - m_1^2};$$

$$y_1 = \frac{1}{2}\left\{-b - (b^2 - 4a)^{1/2}\right\}, \qquad y_2 = \frac{1}{2}\left\{-b + (b^2 - 4a)^{1/2}\right\};$$

$$p_1 = \frac{y_2 - m_1}{y_2 - y_1}, \qquad p_2 = 1 - p_1.$$

Specifically, let $m_1 = .5$, $m_2 = .375$, $m_3 = .3125$. We restrict to distributions on $[0, 1]$. Proceeding according to the above recipe, we get $a = .125$, $b = -1$. Then $y_1 = .1464$, $y_2 = .8536$. Finally, we obtain $p_1 = p_2 = .5$. It is easily seen that these satisfy the moment conditions. Assume that n, the sample size is an even number. This generates the following two clusters as follows, rank the data, split the data at the sample median and put the $n/2$ smallest observations in the first cluster and the remainder in the second cluster.

Now we characterize all two point distributions on $[0, 1]$ with mean .5 and second moment .375. The point $(.5, .375)$ is in the interior of $\mathcal{M}_2$. The totality of two point representations can be constructed as follows. Choose any $x \in [0.1]$. Draw the line segment from (x, x^2) through $(.5, .375)$ and continue until it reaches the boundary of $\mathcal{M}_2$. The set of two point distributions are characterized by the set of such lines which reach the lower boundary. It is easily seen that every solution has $y_2 \in [.75, 1]$ and symmetrically $y_1 \in [0, .25]$. Also, at the extremal solutions, $(0, .75)$, $p_1 = 1/3$ and $p_2 = 2/3$; $(.25, 1)$, $p_1 = 2/3$ and $p_2 = 1/3$. For every x in this interval, p_1 and p_2 are easily calculated. A simple parametric representation is possible which can be used to solve variational problems.

6   Concluding Remarks

This paper provides a very brief description of the moment preservation method of cluster analysis and some of the ramifications of this technique. Space limitations prevented an extensive discussion. An extensive discussion, which will include mathematical details and proofs will be prepared in the near future.

References

AHIEZER, N. I. and KREIN, M. G. (1938): *Some Problems in the Theory of Moments (in Russian).* Izdateltstvo Ukraini, Kharkov.

DELP, E. J. and MITCHELL, O. R. (1979): Image Compression Using Block Truncation Coding. *IEEE Transactions on Communication, Vol. 27, 1335–1341.*

HARRIS, B. (1959): Determining Bounds on Integrals with Applications to Cataloging Problems, *Annals of Mathematical Statistics, Vol. 30, 521–548.*

HARRIS, B. (1962): Determining Bounds on Expected Values of Certain Functions. *Annals of Mathematical Statistics, Vol. 33, 1454–1457.*

KARLIN, S. and SHAPLEY, L. (1953): *The Geometry of Moment Spaces.* American Mathematical Society, Providence, Rhode Island.

LIN, J.-C. and TSAI, W. H. (1994): Feature Preserving Clustering of 2D Data for Two-Class Problems Using Analytic Formulas: An Automatic and Fast Approach, *IEEE Transactions on Pattern Analysis and Machine Intelligence, Vol. 16, 554–560.*

PEI, S.-C. and CHENG, C.-M. (1996): A Fast Two-Class Classifier for 2D Data Using Complex-Moment-Preserving Principle. *Pattern Recognition, Vol. 29, 519–531.*

RUSTAGI, J. S. (1976): *Variational Methods in Statistics.* Academic Press, New York.

SHOHAT, J. A. and TAMARKIN, J. D. (1943): *The Problem of Moments.* American Mathematical Society, Providence, Rhode Island.

TABATAI, A. J. (1981): *Edge Location and Data Compression for Digital Imagery.* PhD. Dissertation, School of Electrical Engineeering, Purdue University, Lafayette, Indiana, U.S.A.

TABATAI, A. J. and MITCHELL, O. R. (1984): Edge Location to Subpixel Values in Digital Imagery. *IEEE Transactions on Pattern Analysis and Machine Intelligence, Vol. 6, 188–201.*

TSAI, W. H. (1985): Moment Preserving Thresholding: A New Approach, *Comput. Vis. Graphics Image Proc., Vol. 29, 377–393.*

On The General Distance Measure

K. Jajuga, M. Walesiak, A. Bak

Wroclaw University of Economics,
Komandorska 118/120,
53-345 Wroclaw, Poland

Abstract: In Walesiak [1993], pp. 44-45 the distance measure was proposed, which can be used for the ordinal data. In the paper the proposal of the general distance measure is given. This measure can be used for data measured in ratio, interval and ordinal scale. The proposal is based on the idea of the generalised correlation coefficient.

1 Introduction

The construction of the particular dependence (e.g. correlation) and distance measure depends on the measurement scale of variables. In the measurement theory four basic scales are distinguished (see e.g. Stevens [1959]): nominal, ordinal, interval and ratio scale. Among them, the nominal scale is considered as the weakest, followed by the ordinal, the interval, and the ratio scale, which is the strongest one. The systematic of scales is based on the transformations that retain the relations of respective scale. These results are well-known and given for example in the paper by Jajuga and Walesiak [2000], p. 106.

2 The Generalised Correlation Coefficient

Consider two variables, say the j-th and the h-th one. A generalised correlation coefficient is given by the following equation (see Kendall and Buckland [1986], p. 266; Kendall [1955], p. 19):

$$\Gamma_{jh} = \frac{\sum_{i=2}^{n} \sum_{k=1}^{i-1} a_{ikj} b_{ikh}}{\left[\sum_{i=2}^{n} \sum_{k=1}^{i-1} a_{ikj}^2 \sum_{i=2}^{n} \sum_{k=1}^{i-1} b_{ikh}^2 \right]^{\frac{1}{2}}}, \tag{1}$$

where: $i, k = 1, ..., n$ – the number of objects,
$j, h = 1, ..., m$ – the number of variables.

Let us take the vectors of observations $(x_{1j}, \ldots, x_{nj})$, $(x_{1h}, \ldots, x_{nh})$ on the variables measured on ratio and (or) interval scale. Suppose that a_{ikj}, b_{ikh} are given as:

$$a_{ikj} = (x_{ij} - x_{kj}),$$
$$b_{ikh} = (x_{ih} - x_{kh}). \tag{2}$$

Then Γ_{jh} becomes Pearson's product-moment correlation coefficient (where x_{ij}, x_{kj} (x_{ih}, x_{kh}) denote i-th, k-th observation on j-th (h-th) variable). The proof is given in Kendall [1955], p. 21.

Let us now take the vectors of observations $(x_{1j}, \ldots, x_{nj}), (x_{1h}, \ldots, x_{nh})$ on the variables measured on ordinal scale. Suppose that a_{ikj}, b_{ikh} are given as:

$$a_{ikj}(b_{ikh}) = \begin{cases} 1 & \text{if} \quad x_{ij} > x_{kj}(x_{ih} > x_{kh}) \\ 0 & \text{if} \quad x_{ij} = x_{kj}(x_{ih} = x_{kh}) \\ -1 & \text{if} \quad x_{ij} < x_{kj}(x_{ih} < x_{kh}) \end{cases} . \tag{3}$$

Then Γ_{jh} becomes Kendall's tau correlation coefficient (Kendall [1955], pp. 19-20). Similarly as Pearson's coefficient, Kendall's tau correlation coefficient takes the values from the interval $[-1; 1]$. The value equal to 1 indicates the perfect consistency between two orders and the value equal to -1 indicates the perfect inconsistency (one order is the inverse of the other one).

In fact, in the Kendall's work in the formula (3) the equality was not considered. We took the more general approach. The value of Kendall's tau coefficient calculated by means of (1) and (3) for raw data is exactly the same as the value of Kendall's tau coefficient calculated by means of the formula (3.3) given in Kendall [1955], p. 35 only for the data for which the ranks were calculated. On the other hand, the application of the formulas (1) and (3) gives the same result for raw data and for the data for which the ranks were calculated. If we use formula by Kendall (formula 3.3 given in Kendall [1955], p. 35) then the observations must be given ranks.

3   The General Distance Measure

Some multivariate statistical methods (for example classification methods, multidimensional scaling methods, ordering methods) are based on the formal notion of the distance between objects (observations). One usually imposes three constraints for the function $d : A \times A \to R$ (A – set of objects, R – set of real numbers) in order to be a distance measure. This function has to be:

- Non-negative: $d_{ik} \geq 0$ for $i, k = 1, \ldots, n$;
- Reflexive: $d_{ik} = 0 \Leftrightarrow i = k$ for $i, k = 1, \ldots, n$;
- Symmetric: $d_{ik} = d_{ki}$ for $i, k = 1, \ldots, n$.

It is easy to notice that the generalised correlation coefficient (including Pearson's and Kendall's coefficient) does not meet the constraints of non-negativity and reflexivity. The constraint of non-negative value can be satisfied by using the transformation $d_{ik} = (1 - \Gamma_{ik})/2$ (the values fall into interval $[0; 1]$). However the constraint of reflexivity is still not fulfilled.

We propose here a general distance measure, which meets all three constraints. It is based on the idea of the generalised correlation coefficient. The general distance measure is given by the following equation (see Walesiak [2000]):

$$d_{ik} = \frac{1 - s_{ik}}{2} = \frac{1}{2} - \frac{\sum\limits_{j=1}^{m} a_{ikj}b_{kij} + \sum\limits_{j=1}^{m}\sum\limits_{\substack{l=1 \\ l \neq i,k}}^{n} a_{ilj}b_{klj}}{\left[\sum\limits_{j=1}^{m}\sum\limits_{l=1}^{n} a_{ilj}^2 \sum\limits_{j=1}^{m}\sum\limits_{l=1}^{n} b_{klj}^2\right]^{\frac{1}{2}}}, \tag{4}$$

where: $d_{ik}(s_{ik})$ – distance (similarity) measure,
$i, k, l = 1, \ldots, n$ – the number of objects,
$j = 1, \ldots, m$ – the number of variables,
$x_{ij}(x_{kj}, x_{lj})$ – i-th (k-th, l-th) observation on the j-th variable.

For the variables measured on ratio and (or) interval scale we take a_{ipj}, b_{krj} given as:

$$a_{ipj} = x_{ij} - x_{pj} \text{ for } p = k, l$$
$$b_{krj} = x_{kj} - x_{rj} \text{ for } r = i, l. \tag{5}$$

Now let us consider the ordinal scale. The only feasible empirical operation on the ordinal scale is counting (the number of the relations: "equal to", "higher than", "lower than"). Therefore in the distance measure we use the relations between the particular object and the other objects.

For the variables measured on ordinal scale we take a_{ipj}, b_{krj} given as (Walesiak [1993], pp. 44-45):

$$a_{ipj}(b_{krj}) = \begin{cases} 1 & \text{if } x_{ij} > x_{pj}(x_{kj} > x_{rj}) \\ 0 & \text{if } x_{ij} = x_{pj}(x_{kj} = x_{rj}) \\ -1 & \text{if } x_{ij} < x_{pj}(x_{kj} < x_{rj}) \end{cases} \text{ for } p = k, l; r = i, l. \tag{6}$$

Therefore in the denominator of the formula (4) the first factor is the number of the relations "higher than" and "lower than" for object i and the second factor is the number of relations "higher than" and "lower than" for object k.

The generalised correlation coefficient is used for the variables, and general distance measure (GDM) for the cases (objects). In the formula for GDM we used only the idea of the generalised correlation coefficient. The references for the construction of measure (4) with the use of (5) and (6) are respectively Pearson's correlation coefficient (for the variables measured on the interval and ratio scale) and Kendall's tau coefficient (for the variables measured on the ordinal scale). The construction of GDM is based on the relations between two analysed objects and the other objects. This approach is not necessary in the case of the variables measured on the interval and ratio scale, however it is necessary in the case of the variables measured on the ordinal scale. In the case of the ordinal scale the number of the relations: "equal to", "higher than", "lower than" is important, therefore in the construction of the measure the information on the relations between the object and the other objects should be taken into account. The similar method was used in the case of the interval and ratio scale, due to the similarity of the measure (4) to the measure (1).

The measure given as (4) with the use of (5) is applied as the distance measure for the variables measured on the interval and (or) ratio scale. When the formula (6) instead of (5) is used, we get the distance measure for the variables measured on the ordinal scale. Therefore, the distance measure given by (4) cannot be used directly when the variables are measured on different scales. Using (4) and (6) can partially solve this problem, however due to the transformation of data measured on interval and (or) ratio scale into ordinal scale, we loose the information.

4 The Properties of the General Distance Measure

The proposed general distance measure d_{ik} has the following properties:

- it can be applied when the variables are measured on the ordinal, interval and ratio scale,

- it takes values from the $[0; 1]$ interval. Value 0 indicates that for the compared objects i, k between corresponding observations of variables, only relations "equal to" take place. If the formula (6) is used, the value 1 indicates that for the compared objects i, k between corresponding observations on ordinal variables, relations "greater than" take place (or relations "greater than" and "equal to") and they are held for other objects (i.e. objects numbered $l = 1, ..., n$ where $l \neq i, k$),

- it satisfies the conditions: $d_{ik} \geq 0, d_{ii} = 0, d_{ik} = d_{ki}$ (for all $i, k = 1, \ldots, n$),

- the empirical analysis proves that distance sometimes does not satisfy the triangle inequality,

- it needs at least one pair of non-identical objects in order to avoid zero in the denominator,

- the transformation of data by any strictly increasing function (formula (6)) or by any linear function (formula (5)) does not change the value of d_{ik}.

The distance measure (4) takes care of variables equally weighted. If the weights are not equal then the general distance measure is defined as (see Walesiak [1999]):

$$
d_{ik} = \frac{1}{2} - \frac{\sum\limits_{j=1}^{m} w_j a_{ikj} b_{kij} + \sum\limits_{j=1}^{m} \sum\limits_{\substack{l=1 \\ l \neq i,k}}^{n} w_j a_{ilj} b_{klj}}{\left[\sum\limits_{j=1}^{m} \sum\limits_{l=1}^{n} w_j a_{ilj}^2 \sum\limits_{j=1}^{m} \sum\limits_{l=1}^{n} w_j b_{klj}^2 \right]^{\frac{1}{2}}}, \tag{7}
$$

and the weights w_j $(j = 1, \ldots, m)$ satisfy conditions $w_j \in (0; m)$, $\sum_{j=1}^{m} w_j = m$.

Three major methods of variable weighting have been developed: an *a priori* method based on the opinions of experts, the procedures based on information included in the data and the combination of these two methods. Gordon [1999], pp. 30-33 and Milligan [1989], pp. 318-325 discuss the problem of variable weighting in multivariate statistical analysis.

We performed simulation study in which the data sets consists of 50 bivariate normal observations representing 4 separated classes. Here the procedures RNMNGN and RNMNPR were used. They generate the multivariate normal data with given mean vectors and covariance matrices (Brandt [1998], pp. 111-112).

For these data sets the distance matrices were determined by using the distances GDM1 (for the variables measured on the ordinal scale), GDM2 (for the variables measured on the interval scale or the ratio scale), L1 (Manhattan distance), L2 (Euclidean distance) and LN (Chebychev distance). Then the objects were classified by means of four hierarchical methods: average linkage (between groups), average linkage (within groups), nearest neighbour, furthest neighbour. Then it was checked which distances and classification methods lead to the identification of natural clusters. For 12 different data structures and 4 classification methods the best results were obtained in the case when the distances GDM2 and L2 were used.

5 Summary

In the paper the general distance measure was proposed. This measure is given by (4) and (5) in the case of the variables measured on the ratio and interval scales and by (4) and (6) in the case of the variables measured on the ordinal scale. The measure is based on the idea of the generalised correlation coefficient. The properties and the results of the simulation studies are also presented. In addition, the computer program GDM in the C++ language, working under Windows 95/98, was written.

Acknowledgements: The research presented in the paper was partly supported by the project KBN 5 H02B 030 21.

References

BRANDT, S. (1998): Analiza danych. Metody statystyczne i obliczeniowe, PWN, Warszawa [Brandt, S. (1997): Statistical and Computational Methods in Data Analysis, Springer-Verlag, New York].

GORDON, A. D. (1999): Classification. Chapman & Hall, London.

JAJUGA, K. and WALESIAK, M. (2000): Standardisation of Data Set Under Different Measurement Scales. In: Decker, R. and Gaul, W. (Eds.): Classification and Information Processing at the Turn of the Millennium. Springer-Verlag, Berlin, Heidelberg, 105-112.

KENDALL, M. G. (1955): Rank Correlation Methods. Griffin, London.

KENDALL, M. G. and BUCKLAND, W. R. (1986): Slownik terminów statystycznych (A Dictionary of Statistical Terms). PWE, Warszawa.

MILLIGAN, G. W. (1989): A Validation Study of a Variable Weighting Algorithm for Cluster Analysis. *Journal of Classification, No. 1, 53-71.*

STEVENS, S. S. (1959): Measurement, Psychophysics and Utility. In: Churchman, C.W. and Ratooch, P. (Eds.): Measurement. Definitions and Theories. Wiley, New York, 18-63.

WALESIAK, M. (1993): Statystyczna analiza wielowymiarowa w badaniach marketingowych [Multivariate Statistical Analysis in Marketing Research]. Wroclaw University of Economics, Research Papers no. 654.

WALESIAK, M. (1999): Distance Measure for Ordinal Data. *Argumenta Oeconomica. No 2 (8), 167-173.*

WALESIAK, M. (2000): Propozycja uogólnionej miary odleglosci w statystycznej analizie wielowymiarowej [The Proposal of the Generalised Distance Measure in Multivariate Statistical Analysis]. In: Paradysz, J. (Ed.): Statystyka regionalna w sluzbie samorzadu lokalnego i biznesu. Wydawnictwo AE, Poznan (in press).

Tests for One-dimensional Nearest Neighbors Clusters

J. Krauth

Department of Psychology,
University of Düsseldorf, D-40225 Düsseldorf, Germany

Abstract: A finite population of points is generated on the real line by a Poisson process. From each point an arrow is drawn to its nearest neighbor (NN). In this way clusters of points are generated which are connected by arrows. Enns et al. (1999) studied the distribution of the number (M) of NN clusters. Because too small or too large values of M may indicate deviations from the null hypothesis of no clustering, tests for the detection of clusters can be based on the distribution of M. Such tests are of interest in epidemiology if one wants to know whether the times of occurrence of a certain disease cluster. We observe that the problem formulated by Enns et al. (1999) is equivalent to the problem of deriving the distribution of the number of peaks in randomization problems which was studied by David and Barton (1962) and others. Because the distribution of M has to be calculated by means of a recurrence equation, which makes the performance of the tests mentioned above difficult, we propose some bounds for the P-values and compare these with the exact distribution and a normal approximation.

1 Introduction

We consider a population consisting of n^* points, $X_i, 1 \leq i \leq n^*$, which are generated on the real line by a Poisson process. Denote by $X_{(i)}$ the corresponding order statistics and by $Y_i = X_{(i+1)} - X_{(i)}, 1 \leq i \leq n^* - 1$, the spacings between adjacent points. The Y_i are identically distributed and the distribution of the vector of ranks of the Y_i is discrete uniform. From each point an arrow is drawn to its nearest neighbor (NN) which exists uniquely with probability one. In this way clusters of points are generated which are connected by arrows.

Several results for this kind of problem were derived by Enns et al. (1999). In particular, the authors considered the probability of occurrence of exactly m clusters in a population of n^* points ($P_{n^*}(m)$), where $1 \leq m \leq \lfloor n^*/2 \rfloor$. For this probability two different recurrence equations were derived and by means of these explicit expressions for $P_{n^*}(1)$, $P_{n^*}(2)$, and $P_{n^*}(3)$. Further, the corresponding generating functions were obtained and by means of them the mean and variance of the number (M) of clusters. Finally, an explicit expression is given for the probability of the maximum number of clusters. For selected values of n^* and m the values of $P_{n^*}(m)$ are listed in two tables.

The results are of interest if clusters in epidemiological time-series are to be detected as this has been done, e.g., in Nagarwalla (1996), Krauth (1998). If the distribution of the number of clusters is known, two different tests for testing the null hypothesis of no clustering could be conceived. One alternative hypothesis might be that more clusters occur than are due to chance, and the other alternative hypothesis might be that fewer clusters occur. In other words, high as well as low values of M may indicate the presence of clustering.

However, it seems difficult to perform these tests, not only because calculating the distribution of M by means of the recurrence equations requires the use of corresponding computer programs but also because the numerators and denominators of $P_{n^*}(m)$ become very large even for moderate values of n^*. Therefore, the need for approximations or exact bounds for the distributions of M arises. This problem is studied here.

It is interesting that neither in Enns et al. (1999) nor in former articles cited by these authors it was observed that the problem of computing the distribution of M is equivalent to another problem which has been studied by other authors in another context. If we consider the ranks $R_1, ..., R_n$ of the spacings $Y_1, ..., Y_n$ between $X_1, ..., X_{n^*}$, where $n = n^* - 1$, we observe that two NN clusters are always separated by a peak of the ranks, where a peak R_i is defined as a rank for which $R_{i-1} < R_i > R_{i+1}$, $2 \le i \le n - 1$, holds. Denoting the number of peaks by S, the number of NN clusters is given by $M = S + 1$, $0 \le S \le \lfloor (n-1)/2 \rfloor$. Under the null hypothesis of no clustering the distribution of $(R_1, ..., R_n)$ is discrete uniform and we can derive the distribution of S and by this the distribution of M.

The distribution of S was studied in David and Barton (1962, pp. 162 - 164). These authors derived one of the recurrence equations given in Enns et al. (1999), the corresponding generating functions, the first four central cumulants and gave also a table of the distribution of S for $2 \le n \le 10$. This table was extended to cover the range of $2 \le n \le 15$ in David et al. (1966, p. 261). In David et al. (1966, p. 53) a normal approximation is given for the distribution of S. Further results concerning the distribution of S are derived in Warren and Seneta (1996). These authors consider, in particular, the rate of convergence of the distribution of S to the normal distribution and cite papers of Bienaymé and André from the nineteenth century where similar problems were discussed.

Mean (μ_n) and variance (σ_n^2) of S is given by

$$
\begin{aligned}
\mu_n &= (n-2)/3 \quad \text{for} \quad n \geq 2, \\
\sigma_2^2 &= 0, \\
\sigma_3^2 &= 2/9, \\
\sigma_n^2 &= 2(n+1)/45 \quad \text{for} \quad n \geq 4,
\end{aligned}
$$

and normal approximations including a correction for continuity are obtained as

$$
P_n(S \leq s) \approx \Phi\left(\frac{s + 0.5 - \mu_n}{\sigma_n}\right).
$$

As mentioned above, Enns et al. (1999) derived an explicit expression for the probability of the maximum number of clusters. However, they seem not to have been aware of the fact that the numerator of this probability is a so-called tangent number (T_n), i.e. we have

$$
P_n(S = \lfloor (n-1)/2 \rfloor) = \begin{cases} T_n/n! & \text{for} \quad n \text{ odd,} \\ T_{n+1}/n! & \text{for} \quad n \text{ even,} \end{cases}
$$

$$
P_n(M = \lfloor n^*/2 \rfloor) = \begin{cases} T_{n^*}/(n^* - 1)! & \text{for} \quad n^* \text{ odd,} \\ T_{n^*-1}/(n^* - 1)! & \text{for} \quad n^* \text{ even.} \end{cases}
$$

In Knuth and Buckholz (1967) properties of tangent numbers are derived including a recurrence equation, and a table of T_n for $n \leq 120$ is given. The name 'tangent numbers' stems from the fact that these numbers are the coefficients in a power series expansion of the tangent.

Our aim was to derive simple upper bounds for $P_n(S \leq s)$ and $P_n(S \geq s)$ in order to be able to perform the tests for the existence of NN clusters mentioned above.

2 Bounds for the P-values of S

The probability $P_n(S \geq s)$ is given by $|S_n \geq s|/n!$, where $|S_n \geq s|$ denotes the number of permutations of n ranks with s or more peaks. A lower bound of $|S_n \geq s|$ is given by

$$
|S_n \geq s|_L = \frac{(n-s-2)!(n-s-1)!}{(n-2s-1)!}[(n-s)(n-s-1) + (n-2s-1)
$$

$$
\times s(s+1) + s(n-1) + 2\binom{s}{3}] \quad \text{for} \quad 3 \leq s < \lfloor (n-1)/2 \rfloor,
$$

while in addition the exact results (based on Enns et al. (1999))

$$P_n(S \geq s) = \begin{cases} 1 \text{ for } s = 0, \\ 1 - (2^{n-1}/n!) \text{ for } s = 1, \\ 1 - ((2^{n-1} + 2^{2n-3} - n2^{n-2})/n!) \text{ for } s = 2, \\ T_n/n! \text{ for } s = \lfloor (n-1)/2 \rfloor \text{ and } n \text{ odd}, \\ T_{n+1}/n! \text{ for } s = \lfloor (n-1)/2 \rfloor \text{ and } n \text{ even}, \\ 0 \text{ for } s > \lfloor (n-1)/2 \rfloor \end{cases}$$

are valid. The lower bound is derived by enumerating a subset of the set of all permutations where at least s peaks occur. We consider first the situation where the s largest ranks $n, n-1, ..., n-s+1$ constitute peaks. Here we have $s!$ permutations of these s peaks and $\binom{n-s}{s+1}$ ways to select $(s+1)$ ranks to put on both sides and between the s peaks. These ranks can be permuted in $(s+1)!$ ways, while there are $(n-2s-1)!$ ways for the remaining $(n-2s-1)$ ranks. There are $\binom{n-s-2}{i}$ ways to split a given permutation of the remaining ranks into $(i+1)$ substrings which can be distributed in $\binom{s+1}{i+1}$ ways between and on both sides of the peaks, for $i = 0, 1, ..., n-2s-2$. Using formula (3.20) in Gould (1972, p. 24) we get

$$\sum_{i=0}^{n-2s-2} \binom{n-2s-2}{i}\binom{s+1}{i+1} = \binom{n-s-1}{n-2s-1}.$$

Multiplying the five terms we get

$$s!\binom{n-s}{s+1}(s+1)!(n-2s-1)!\binom{n-s-1}{n-2s-1}$$
$$= \frac{(n-s-2)!(n-s-1)!}{(n-2s-1)!}(n-s)(n-s-1)$$

which corresponds to the first term in $|S_n \geq s|_L$ above.

Next, we consider the $(s+1)$ largest ranks. We delete the rank $(n-k)$ for $k = 0, 1, ..., s-1$ and require that the remaining s ranks constitute peaks but not the rank $(n-k)$. By similar arguments as above the three last terms in $|S_n \geq s|_L$ are derived. The second term arises if the rank $(n-k)$ is posed at one of the two ends or right or left of a larger rank. The third term results if $(n-k)$ is posed at one of the two ends and a larger rank directly beside it, while the fourth term covers the case where two larger ranks are the direct neighbors of $(n-k)$.

In David and Barton (1962, p. 163), Warren and Seneta (1996, Formula (7)) and Enns et al.(1999, Formula (4.1)) a recurrence equation is derived which is given in our notation by

$$|S_n = s| = (2s + 2)|S_{n-1} = s| + (n - 2s)|S_{n-1} = s - 1|$$

with

$$|S_1 = s| = 1 \quad \text{for} \quad s = 0,$$
$$|S_1 = s| = 0 \quad \text{for} \quad s = 1,$$
$$|S_n = s| = 0 \quad \text{for} \quad s < 0 \text{ and } s > \lfloor(n - 1)/2\rfloor.$$

We present here an elementary proof of this relation which is simpler than the proofs given by the authors above.

Again, we denote by $|S_{n-1} = s|$ the number of permutations of the $(n-1)$ ranks $1, ..., n - 1$ with $(n - 2)$ spacings between them and s peaks. By adding the rank n to a given permutation with $(n - 1)$ ranks there are n ways of getting a new permutation with $(n - 1)$ spacings. If the rank n is set at one of the two ends, the number s of peaks is not altered. This yields 2 possible ways. If the rank n is set between two other ranks, one of which constitutes a peak, again the number s of peaks is not altered. For this, there exist $(2s)$ possible ways. Finally, if the rank n is set between two ranks, none of which represents a peak, the number of peaks is increased from s to $(s + 1)$. For this, there are $(n - 2 - 2s)$ ways. Thus we find

$$|S_n = s| = (2 + 2s)|S_{n-1} = s| + (n - 2 - 2(s - 1))|S_{n-1} = s - 1|$$

as stated above.

If we define

$$H_n(s) = \begin{cases} \sum_{i=0}^{s} |S_n = i| = |S_n \leq s|, s = 0, ..., \lfloor(n - 1)/2\rfloor \\ 0 \text{ for } s < 0 \\ n! \text{ for } s \geq \lfloor(n - 1)/2\rfloor \end{cases}$$

we find by means of the recurrence relation above

$$H_n(s) = (2s + 2)H_{n-1}(s) + (n - 2s - 2)H_{n-1}(s - 1).$$

By applying this recurrence equation k times in succession we obtain

$$H_n(s) = \sum_{l=0}^{k} \sum_{i_1=0}^{k-l} \sum_{i_2=0}^{k-l-i_1} \sum_{i_3=0}^{k-l-i_1-i_2} \cdots \sum_{i_l=0}^{k-l-i_1-i_2 \ldots -i_{l-1}}$$

$$\left(\prod_{t-1}^{l} (2s+4-2t)^{i_t} (n-2s-3+t-i_1-\ldots-i_t) \right)$$

$$\times (2s+2-2l)^{k-l-i_1-\ldots-i_l} H_{n-k}(s-l)$$

$$= (2s+2)^k H_{n-k}(s) + \sum_{i=0}^{k-1} (2s+2)^i (n-2s-2-i)(2s)^{k-1-i}$$

$$\times H_{n-k}(s-1) + \ldots + \prod_{t=1}^{k} (n-2s-3+t) H_{n-k}(s-k)$$

$$= (2s+2)^k H_{n-k}(s)$$

$$+ 2^{k-1}[(n-s-1-k)(s+1)^k - (n-s-1)s^k] H_{n-k}(s-1)$$

$$+ \ldots + \prod_{t=1}^{k} (n-2s-3+t) H_{n-k}(s-k).$$

The coefficient corresponding to $H_{n-k}(s-l)$ is nonnegative for $i_1 + \ldots + i_t \leq n - 2s - 3 + t$ for $t = 1, \ldots, l$. Observing $i_1 + \ldots + i_t \leq k - l$ and $n - 2s - 3 + t \geq n - 2s - 2$ we have nonnegativity for $k \leq n - 2s - 2 + l$ for $l = 0, 1, \ldots, k$. Because the coefficient corresponding to $l = 0$, which is given by $(2s+2)^k$, is always nonnegative, the other coefficients are nonnegative for $k \leq n - 2s - 1$ for $l = 1, \ldots, k$. Considering only the first two terms and choosing $k = n - 2s - 1$, we get for $H_n(s)$ the lower bound

$$|S_n \leq s|_L = (2s+2)^{n-2s-1} H_{2s+1}(s)$$

$$+ 2^{n-2s-2}[s(s+1)^{n-2s-1} - (n-s-1)s^{n-2s-1}]$$

$$\times H_{2s+1}(s-1)$$

$$= (2s+2)^{n-2s-1}(2s+1)!$$

$$+ 2^{n-2s-2}[s(s+1)^{n-2s-1} - (n-s-1)s^{n-2s-1}]$$

$$\times [(2s+1)! - T_{2s+1}] \quad \text{for } 1 \leq s < \lfloor (n-1)/2 \rfloor,$$

while in addition the exact results

$$P_n(S \leq s) = \begin{cases} 0 & \text{for } s < 0, \\ 2^{n-1}/n! & \text{for } s = 0, \\ 1 & \text{for } s = \lfloor (n-1)/2 \rfloor \end{cases}$$

are valid.

Here, we used

$$
\begin{aligned}
H_{2s+1}(s) &= (2s+1)!, \\
H_{2s+1}(s-1) &= H_{2s+1}(s) - |S_{2s+1} = s| = (2s+1)! - T_{2s+1}
\end{aligned}
$$

with the tangent number T_{2s+1}. The identity $|S_n = s| = T_{2s+1}$ was proved by Enns et al. (1999, Formula (5.3)), though the authors were not aware of this. From Formula (11) in Knuth and Buckholtz (1967) we derive

$$
T_{2s+1} = (-1)^s \frac{2^{s+2}(2^{2s+2}-1)}{2s+2} B_{2s+2},
$$

where B_{2s+2} denotes the corresponding Bernoulli number. The computation of the Bernoulli numbers is facilitated by observing the inequalities

$$
\frac{2(2s+2)!}{(2\pi)^{2s+2}} < (-1)^{s+2} B_{2s+2} < \frac{2(2s+2)!}{(2\pi)^{2s+2}} \frac{1}{1 - 2^{-(2s+1)}}, \quad s = 0,1,2,\dots
$$

as given in Abramowitz and Stegun (1972, Formula (23.1.15) on p. 805). From these we get the corresponding inequalities

$$
\frac{2(2s+1)!(2^{2s+2}-1)}{\pi^{2s+2}} < T_{2s+1} < \frac{2(2s+1)!(2^{2s+2}-1)}{\pi^{2s+2}} \frac{2^{2s+1}}{2^{2s+1}-1}
$$

for the tangent numbers. Because the tangent numbers are integers, the exact value of T_{2s+1} can be derived from these inequalities for small values of s. In any case, the mean of the two bounds is a good approximation of T_{2s+1}.

Using the lower bounds $|S_n \geq s|_L$ and $|S_n \leq s|_L$ derived above we get

$$
\frac{|S_n \leq s|_L}{n!} \leq P_n(S \leq s) \leq 1 - \frac{|S_n \geq s+1|_L}{n!},
$$

$$
\frac{|S_n \geq s|_L}{n!} \leq P_n(S \geq s) \leq 1 - \frac{|S_n \leq s-1|_L}{n!},
$$

and from this

$$
\frac{|S_{n^*-1} \leq m-1|_L}{(n^*-1)!} \leq P_{n^*}(M \leq m) \leq 1 - \frac{|S_{n^*-1} \geq m|_L}{(n^*-1)!},
$$

$$
\frac{|S_{n^*-1} \geq m-1|_L}{(n^*-1)!} \leq P_{n^*}(M \geq m) \leq 1 - \frac{|S_{n^*-1} \leq m-2|_L}{(n^*-1)!}.
$$

3 Numerical Comparisons

In Table 1 we compare the exact probabilities $P_{im}^{n^*}$, where $P_{1m}^{n^*} = P_{n^*}(M \leq m)$ and $P_{2m}^{n^*} = P_{n^*}(M \geq m)$, for $n^* = 10, 20$, and 30 with the normal approximations $N_{im}^{n^*}$, the lower bounds $L_{im}^{n^*}$ and the upper bounds $U_{im}^{n^*}$, for $i = 1, 2$. Obviously, the bounds give good results only for small values of n^*, while the normal approximation is of an astonishing accuracy even for small and medium values of n^*.

m	L_{1m}^{10}	P_{1m}^{10}	U_{1m}^{10}	N_{1m}^{10}	L_{2m}^{10}	P_{2m}^{10}	U_{2m}^{10}	N_{2m}^{10}
1	.001	.001	.001	.003	1.000	1.000	1.000	1.000
2	.088	.088	.187	.106	.999	.999	.999	.997
3	.580	.600	.683	.599	.912	.912	.912	.894
4	.968	.978	.978	.960	.317	.400	.420	.401
5	1.000	1.000	1.000	.999	.022	.022	.032	.040

m	L_{1m}^{20}	P_{1m}^{20}	U_{1m}^{20}	N_{1m}^{20}	L_{2m}^{20}	P_{2m}^{20}	U_{2m}^{20}	N_{2m}^{20}
4	.006	.008	.397	.011	.861	1.000	1.000	1.000
5	.066	.100	.676	.108	.603	.992	.994	.989
6	.279	.424	.879	.430	.324	.900	.934	.892
7	.624	.820	.973	.812	.121	.576	.721	.570
8	.899	.982	1.000	.974	.027	.180	.376	.188
9	.994	1.000	1.000	.999	.003	.018	.101	.026

m	L_{1m}^{30}	P_{1m}^{30}	U_{1m}^{30}	N_{1m}^{30}	L_{2m}^{30}	P_{2m}^{30}	U_{2m}^{30}	N_{2m}^{30}
6	.000	.001	.570	.001	.643	1.000	1.000	1.000
7	.005	.013	.756	.015	.430	.999	1.000	.999
8	.031	.092	.887	.097	.244	.987	.995	.985
9	.119	.327	.959	.333	.113	.908	.969	.903
10	.308	.669	.989	.667	.041	.673	.881	.667
11	.571	.910	.998	.903	.011	.331	.692	.333
12	.814	.989	1.000	.985	.002	.090	.429	.097
13	.955	1.000	1.000	.999	.000	.011	.186	015
14	.998	1.000	1.000	1.000	.000	.001	.045	.001

Table 1:
Exact probabilities (P), lower bounds (L), upper bounds (U), and normal approximations (N)

By enumerating a larger subset of permutations we might improve the lower bound $|S_n \geq s|_L$, and by considering more terms in $H_n(s)$ we might improve $|S_n \leq s|_L$. However, it seems preferable to use for small values of n^* the table given in David et al. (1966, p. 261) or the recurrence equation, while for larger values of n^* the normal approximation will give reliable results.

References

ABRAMOWITZ, M. and STEGUN, I. A. (1972): Handbook of Mathematical Functions with Formulas, Graphs, and Mathematical Tables. Dover Publications, New York

DAVID, F. N. and BARTON, D. E. (1962): Combinatorial Chance. Charles Griffin, London

DAVID, F. N., KENDALL, M. G. and BARTON, D. E. (1966): Symmetric Function and Allied Tables. University Press, Cambridge

ENNS, E. G., EHLERS, P. F. and MISI, T. (1999): A Cluster Problem as Defined by Nearest Neighbours. *Canadian Journal of Statistics, 27, 843-851.*

GOULD, H. W. (1972): Combinatorial Identities. A Standardized Set of Tables Listing 500 Binomial Coefficient Summations. Morgantown

KNUTH, D. E. and BUCKHOLTZ, T. J. (1967): Computation of Tangent, Euler, and Bernoulli Numbers. *Mathematics of Computation, 21, 663-688.*

KRAUTH, J. (1998): Upper Bounds for the P-Values of a Scan Statistic with a Variable Window, in Balderjahn, Mathar, and Schader (Eds.): Classification, Data Analysis, and Data Highways. Springer, Berlin, Heidelberg, New York, 155-163

NAGARWALLA, N. (1996): A Scan Statistic with Variable Window. *Statistics in Medicine, 15, 799 - 810.*

WARREN, D. and SENETA, E. (1996): Peaks and Eulerian Numbers in a Random Sequence. *Journal of Applied Probability, 33, 101-114.*

Integration of Cluster Analysis and Visualization Techniques for Visual Data Analysis

M. Kreuseler, T. Nocke, H. Schumann

Institute of Computer Graphics
University of Rostock, D-18059 Rostock, Germany

Abstract: This Paper investigates the combination of numerical and visual exploration techniques focused on cluster analysis of multi-dimensional data. We describe our new developed visualization approaches and selected clustering techniques along with major concepts of the integration and parameterization of these methods. The resulting frameworks and its major features will be discussed.

1 Introduction

The analysis of complex heterogenous data requires sophisticated exploration methods. Especially complex data mining processes which apply many different analysis techniques can benefit from visual data processing and new visualization paradigms. Additionally, visualization provides a natural method of integrating multiple data sets and has been proven to be reliable and effective across a number of application domains. Still visual methods can not replace analytic non visual mining algorithms. Rather it is useful to combine multiple methods during data exploration processes (Westphal, Blaxton (1998)).

The new area of visual data mining focuses on this combination of visual and non-visual techniques as well as on integrating the user in the exploration process. Ankerst (2001) classifies current visual data mining approaches into three categories. Methods of the first group apply visualization techniques independent of data mining algorithms. The second group uses visualization in order to represent patterns and results from mining algorithms graphically. The third category tightly integrates both mining and visualization algorithms in such a way that intermediate steps of the mining algorithms can be visualized. Furthermore, this tight integration allows users to control and steer the mining process directly based on the given visual feedback.

The focus of our research is to support each of these groups. In this context the goal is to create computer-supported interactive visual representations of abstract raw data to amplify cognition (Card et al. (1999)) and to solve a variety of exploration tasks. In order to achieve this and to support the selection and parameterization of suitable exploration techniques, new concepts for obtaining and handling meta-data have to be introduced. These concepts have to be general and flexible in order to be appliable for all 3 groups of visual data mining approaches (cf. classification of visual data mining above). Furthermore it is necessary to reduce the active size of large data volumes to processible levels without losing relevant information.

Summarizing the discussion above, the combination of non-visual and visual exploration techniques along with applying meta-data concepts to control the exploration process, seems to be an promising approach to support complex exploration scenarios.

The research in our paper is focused on the integration of different clustering techniques with our new developed visualization paradigms and meta-data concepts. We suggest a flexible framework which is scalable with respect to the characteristics of the data, the exploration tasks and user profiles.

We describe selected clustering techniques and introduce our new visualization methods in (section 2). Our framework which integrates the techniques and concepts mentioned above is discussed in (section 3). Section 3.1 covers the configuration and parameterization of the techniques in our scalable framework based on influencing factors such as characteristics of the data and exploration tasks. Basic concepts to specify and obtain meta-data are introduced in (section 3.2). Finally we discuss our future work regarding the selection and parametrization of suitable techniques in section (4).

2 Cluster and Visualization Techniques

2.1 Techniques for Clustering Data

Based on the literature referring to the classification of data (Bock (1974), Backhaus et al. (1996)), we identified 3 sub-processes for the application of clustering techniques in the field of visualization: standardization(1), determining similarities, distances, heterogenities or homogenities(2) and grouping of the data (3).

Standardization
Using proper standardization algorithms is crucial for the applicability of certain similarity measures and for achieving valuable clustering results.

Standard methods for variables of metric scale type are used for data standardization. Basically we apply data normalization (interval 0-1 normalization, mean value 0 - variance 1 - scaling), elimination of outliers (based on proximity matrix we eliminate those data records which are very dissimilar compared to the majority of the data records), treatment of identical data records and weighting or elimination of variables.

Similarity and distance measures
We provide several different similarity measures in order to adapt the

clustering process according to analysis tasks and data characteristics. Basically standard measures are applied. These are the m-coefficient for binary variables, the generalized m-coefficient for nominal data and L_r-distances, the Mahalanobis distance and the correlation coefficient for metric data.

Furthermore hybrid measures have been integrated in order to handle data records with mixed scale types. Therefore similarities are calculated separately for those variables that have the same scale type. Then the single similarity values are composed for obtaining the similarity between two data records.

Cluster analysis techniques

Two methods for automatic clustering are utilized within our framework: hierarchical agglomerative clustering with dynamic derivation of hierarchy trees and Self-organizing maps.

Self-organizing maps (SOM) as introduced by Kohonen (1995) provide an effective mechanism for organizing unstructured data by extracting groups of similar data objects. SOMs can be described as nonlinear projection from n-dimensional input space onto two-dimensional display space. After the training process neighboring locations in the display space correspond to neighboring locations in the data space. Thus SOMs provide a useful topological arrangement of data vectors by grouping similar data objects.

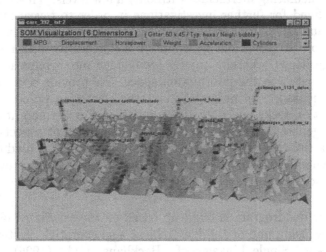

Figure 1: Clustering of data objects based on self-organizing maps

Figure 1 illustrates the use of SOMs for clustering unorganized data. The picture was generated from a car data set with 6 variables. Each peak in the map displays a cluster of similar data records. The number of records within a single cluster is mapped to the height of the peak.

Color is used for displaying similarities between adjacent clusters where bright intensities denote a higher degree of similarity.

Moreover we introduce cylinder icons for visualizing cluster properties, i.e. a small opaque cylinder is used for displaying the concrete value for each single variable of the map vectors. The height of the outer transparent cylinder corresponds to the maximum data value of the related dimension. Color is used to distinguish between the different dimensions. The different cylinders are composed into a single icon that is mapped on top of selected cluster peaks within the graphical representation. Thus SOMs are suitable for providing an overview of the entire data space by revealing clusters and cluster properties.

Dynamic Hierarchies The dynamic hierarchy computation is one possible method to achieve predictable representations of given data. If an abstraction is used to organize unstructured data, it is important to remember that users may have different requirements when merging objects into groups. Thus we do not compute a fixed number of static groups. Instead, a nested sequence of groups is determined and organized into a hierarchy, whereby the requirements according to the homogeneity of those groups increase as the hierarchy is descended. In order to support the analysis of data at arbitrary levels of detail the computation of the hierarchy can be controlled interactively. An overview is provided by calculating hierarchies with only a few levels. These hierarchies can be refined for further investigations in order to reveal more subtle patterns and to identify smaller subclusters in the data. The hierarchy computation is carried out by adapted agglomerative hierarchical clustering algorithms, whereby objects are merged into groups according to their similarities in the information space. We provide several different similarity measures in order to adapt the clustering process according to exploration tasks and data characteristics. Furthermore it is our objective to generate dynamic hierarchies under different aspects from the same data set. Therefore, we need a basis which can be used effectively for the dynamic refinement of the hierarchy. This basis is provided by a binary dendrogram (cf. Figure 2).

The binary dendrogram is build up based on the calculated object similarities by using one of the hierarchical agglomerative clustering algorithms (e.g. Single Linkage; cf. Backhaus et al. (1996), Kaufman, Roussew (1990)). The values at the dendrogram nodes (cf. Figure 2) denote standardized heterogeneity values of the belonging groups. In order to control the hierarchy computation the number of desired levels, and a heterogeneity threshold for each level, can be specified interactively. In a second pass the hierarchy is derived from the binary dendrogram according to these parameters (algorithm at Kreuseler et al.(2000)).

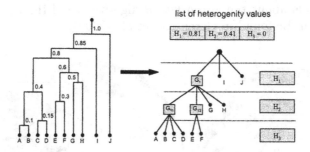

Figure 2: Construction of a Hierarchy with 3 levels based on the binary dendrogram

2.2 Visualization techniques

Magic-Eye-View
Visualizing the computed cluster hierarchies becomes complicated as the number of levels and nodes increases. Standard 2-D hierarchy browsers can typically display about 100 nodes (cf. Lamping et al.(1995)). Exceeding this number makes perceiving details difficult. Zooming and panning do not provide a satisfying solution to this drawback due to loss of context information.

In order to solve this drawback and to support navigation of large-scale information spaces, distortion oriented techniques have been developed and used, particularly in graphical applications (cf. Leung, Apperley (1994)). Typical examples of these are Focus+Context techniques such as Graphical Fisheye Views (cf. Sarkar, Brown (1994)) or the Hyperbolic Browser (cf. Lamping et al.(1995)). These techniques exploit distortion to allow the user to examine a local area in detail on a section of the screen, and at the same time, to present a global view of the space to provide an overall context to facilitate navigation (Leung, Apperley (1994)).

In order to integrate classical zooming and panning functionality and the capacity of Focus+Context approaches, we implemented the new Focus+Context technique Magic-Eye-View. Our approach maps a hierarchy graph onto the surface of a hemisphere. We then apply a projection in order to change the focus area interactively by moving the center of projection. The objective of moving the center of projection is to enlarge those parts of the graph which are in or near the focus region in order to show information details while the rest of the graph remains visible with reduced size. Rendering and navigating the projected hierarchy graph is possible in either 2D or 3D. The 2D display is realized by applying an additional projection which maps the hemisphere to a circular plane.

Further details about the graph layout algorithm and the basics of the projection can be found in (Kreuseler et al. (2000)).

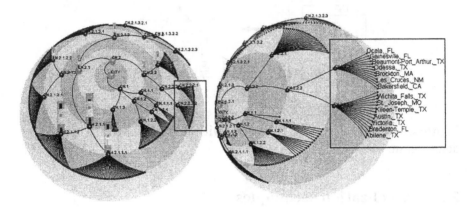

Figure 3: Complex hierarchy graph without and with focused area
(see rectangle) along with visualization of cluster properties.

Figure 3 demonstrates change of focus. The left picture shows a complex hierarchy graph mapped onto a hemisphere. The center of projection has been moved in the right picture in order to set the focus to the marked sub-graph.

Since most Focus+Context displays introduce distortion (cf. above), we have to provided mechanisms to reduce confusion and to avoid extra work for the users to interpret the visualization. In order to achieve this, colored rings are introduced. These rings minimize the amount of confusion and help to maintain users' orientation after change of focus. Furthermore it remains recognizable at which level a certain hierarchy node resides (cf. Figure 3).

Properties of the cluster hierarchy can be visualized in conjunction with the Magic-Eye-View as well. First we use different colors to distinguish between cluster nodes and object nodes within the hierarchy tree. Furthermore the size of a cluster, i.e. the number of objects is mapped to the cluster node's size and color intensity. Additional cluster properties like $t - values$[1] and $F - values$[2] can be displayed using the cylinder icons introduced in section 2.1. Figure 3 shows the t-values of selected clusters mapped onto the nodes of the hierarchy tree. Summarizing the discussion above the Magic-Eye-View provides an overview of the overall hierarchy structure in conjunction with the display of basic cluster

---

[1]The $t - value$ denotes the strength of a variable (feature) within the cluster whereby a $t - value > 0$ means a strong representation of the belonging variable.

[2]The $F - value$ denotes the variation of a single variable within a cluster compared to the variation of the variable in the overall data set.

properties.

The Magic-Eye-View has been presented at the IEEE Information Visualization Symposium 2000. Comments after presentations (cf. Kreuseler et al. (2000)) as well as user feedback[3] have shown that this technique is intuitive. Users' found that especially the colored rings help to reduce confusion and to maintain users's orientation after change of focus. Furthermore the Magic-Eye-View has been compared to the Hyperbolic Browser (cf. Lamping et al.(1995)). One of the results of this comparison has shown that the combination of classical 3D navigation such as zooming, rotating with interactively focussing arbitrary areas of the graph provides additional degrees of freedom for navigating hierarchies. However the Magic-Eye-View offers room for future work. Currently we are working on improvements in terms of increasing the number of displayable nodes and supporting change of focus depending on the underlaying data.

Visual Clustering based on an enhanced spring model (Visualization of Multi-dimensional Cluster Properties)

Computing hierarchies or using SOMs is a valid method for structuring data and identifying groups of similar data objects. However, for further analysis of those subsets e.g. revealing attribute values of the data or determining object similarities within a cluster or at certain hierarchy levels we developed ShapeVis[4] for visualizing multi-dimensional data objects. Basically ShapeVis performs visual clustering by arranging similar objects close together in 3D visualization space.

ShapeVis exploits an enhanced spring model (cf. Theisel, Kreuseler (1998)) in order to arrange objects according to their similarities. Therefore we place n-points D_i, i.e. one point for each dimension of the data set, in an equidistant way on a sphere. Small composed graphical objects (shapes) are used to depict data objects. Those shapes are attached with springs to each of the dimension points D_i. The locations of the shapes are determined by the spring model, i.e. the bigger the data value of a certain dimension the closer the shape moves towards that dimension point D_i. The shapes can be deformed in the direction of the dimension points D_i in order to depict attribute values and to solve ambiguities. The deformation is achieved by introducing small cylinders. The size of the deformation, e.g. length of a cylinder in a particular direction denotes the data value of that dimension. Thus multi-dimensional information objects are described uniquely by location, size and shape of their visual representations. More detailed information about the shape

---

[3]The technique has been applied in a project cooperation for visualizing ontologies in a WWW application named GETESS (GErman Text Exploration and Search System).

[4]We use an adapted version of our technique introduced in Theisel, Kreuseler (1998) in our work.

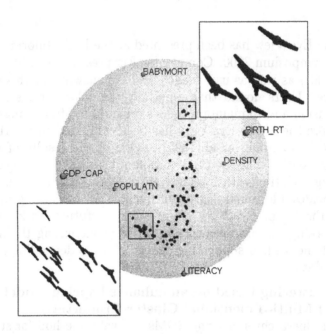

Figure 4: Visual clustering of data using an enhanced spring model

creation can by found in Theisel, Kreuseler (1998). Figure 4 illustrates this principle. Our approach is applied to a data set which measures 6 demographic parameters of 106 countries. The global clustering of the data can be obtained within the sphere. The objects in the upper right, which have big values in the dimensions *Baby mortality* and *Birthrate* move towards the corresponding dimension points D_i. Furthermore we can verify the assumption that these objects have big values in the dimensions *Baby mortality* and *Birthrate* by applying the deformation to the geometric objects. The deformations (cylinders) which point towards the *Baby mortality* and *Birthrate* dimension points are much longer than the deformation which point towards the remaining D_i (cf. Figure 4 magnification of the upper cluster). In contrast to that the cluster lower left is characterized by countries with much bigger values with respect to the dimensions *Literacy* and *Gross Domestic Product* while the values of *Baby mortality* and *Birthrate* are rather small.

Use of SOM-based clustering for data record arrangement in a visualization technique

We propose an other visualization tool for displaying multivariate data sets which we named the Data-Table-View. This method is very similar to the Table Lens introduced by (Rao, Card (1994)). The Table Lens integrates a common table with graphical representations for depicting patterns and outliers in multi-dimensional data sets. Therefore the Table Lens offers several graphical mapping schemes along with a focus+context technique for exploring large tables effectively (cf. Rao, Card (1994)).

The Data-Table-View extends the Table Lens by introducing additional features for organizing cases (data records) within the table. This principle of reordering data in order to reveal hidden patterns is similar to Bertin's reorderable matrix (cf. Bertin (1981)). We provide several mechanisms to rearrange the data. Depending on data characteristics and exploration tasks, users can choose one of the following ordering strategies:

- sort by row sum (i.e., sort table rows based on the sum of the data values of a row)

- permute rows and columns with respect to maximum (or minimum) data value (i.e., find the first maximum data value v_m in the data table, determine the corresponding row_m and $column_m$, permute the data table such that row_m and $column_m$ become the 1st. row/column, continue this process with the remaining rows and columns of the data table)

- sort table rows with respect to a particular variable (column)

- rearrange rows based on row similarity (i.e., all data values of a row are used to determine the similarity between rows)

The implementation of the reordering is designed flexible such that further ordering criterions can be added easily.

Especially considering all variables for similarity rearrangement of data records (cf. last bullet of the enumeration above) requires mapping of multi-dimensional data to 1D. This mapping can be done in a number of different ways.

One possible method for organizing unstructured multi-dimensional data provide Self-organizing maps (SOM) (cf. Kohonen (1995) and section 2.1). A key feature of SOMs is to extract groups of similar data records by projecting the n-dimensional input space onto two-dimensional visualization space. Thus the algorithm maps multi-dimensional data directly in an ordered fashion onto a 2D grid. Since it is our goal to arrange data records linearly for the data table view instead of organizing them on a two-dimensional grid, we are using the one-dimensional case of SOMs, which is proven (cf. Kohonen (1995)) to provide correct orderings as well. Thus we obtain a sequence of data records (table rows) depending on their overall similarity in information space, i.e., similar data records are placed in successive table rows.

In order to discover patterns in the data and relations between variables graphically, a bar representation is used where data values within table cells are mapped to the length of a small bar. This principle is illustrated in figure 5. In our example, the table contains a car data set with 392 cars by 6 variables. The left picture of figure 5 shows the data table without similarity arrangement of the data records. The focus is set to a particular data object in order to reveal detailed data values. The similarity arrangement is applied in the right picture. Trends and relations between variables (columns of the table) can be obtained much better than in the unordered table. This is shown in figure 5 where the first five variables are correlated.

Figure 5: Table based exploration of multi-dimensional data with similarity arrangement in order to reveal correlations.

3 Frameworks

3.1 The Framework *InfoVis3D*

The clustering and visualization techniques introduced in this paper are integrated in the scalable visualization system InfoVis3D. Moreover our framework contains other traditional techniques such as Scatter Plots, Histograms and Parallel Coordinates . In order to support flexible visualizations at arbitrary levels of detail, subsets of a hierarchy can be selected for further exploration. Any desired part from the SOMs can be selected for detailed exploration as well.

- *Selection of cluster nodes* - Each cluster node of a hierarchy tree can be selected. All data records of a selected cluster can be visualized with ShapeVis, Parallel Coordinates or one of the other techniques in a separate display area.

- *Selection of hierarchy levels* - A representative is determined for each cluster which resides at the selected level by calculating mean values of the data of all cluster members. ShapeVis or any other technique of our system can be used to visualize those representatives and all remaining objects at the selected level.

- *Selection of SOM areas* - Arbitrary regions of the SOMs can be selected. Therefore the underlaying data vectors of the selected grid area are determined and displayed with a suitable technique of our system.

In order to identify concrete data contents, i.e. real variable values, selected data records can be displayed with the Data-Table-View. Along with that each subset of the data can be visualized with different techniques at the same time (e.g. parallel display of selected clusters and their members with ShapeVis, Parallel Coordinates, Scatter-Plot-Matrices etc.) Thus we provide different views of the same data set in order to reveal deeper insights into the data. All active views are linked together via Brushing, i.e. each single data object that is highlighted in a particular display will be marked in all active views as well.

3.2 Framework for gathering meta data

The framework described above contains many analysis and visualization techniques which can be parameterized in many different ways. To support a tight integration of these techniques (cf. visual data mining problems (Ankerst (2001))), new mechanisms for selecting, combining and parameterizing appropriate techniques must be developed. One approach is using meta data to support, control and steer complex exploration tasks.

Meta data are defined as data about data, and cover special features of a data set. They are important for the visualization process (e.g., for selecting suitable visual representations depending on the dimensionality of the data) as well as for the selection and parameterization of cluster analysis techniques (e.g., for selecting suitable standardizations, measures or clustering methods).

We have designed general concepts for specifying and obtaining meta data. Based on these concepts, we have developed a framework for gathering different types of meta data, for instance:

meta data for describing the whole data set:
- e.g. complete, incomplete, ...

meta data for describing the variables of a data set:
- e.g. scale type, ranges of values, minimal and maximal data values
- special meta data for independent variables:
  - e.g. properties of space and time dimensions, so-called grid structures and regions of interest
- special meta data for dependent variables:
  - e.g. the data type[5]

---

[5] The data type comprises the kind of values of a dependent variable. Usual data

meta data for describing the relations between the variables and between the data records of a data set:
- correlations, (joint) information content
- outliers

In this paper we may not listen all meta data used in our framework, and may not prove their relevance for clustering and visualization techniques generally. Instead the relevance of selected meta data will be shown on the basis of examples. The scale type for instance is especially important for the selection of suitable measures and for the selection of suitable visual representation parameters. As another example the appliance of self-organizing maps is only useful for metric variables. Furthermore there are special visualization techniques and special numeric methods for special data types (e.g. flow visualization techniques for flow data). Correlations, (joint) information content and the detected outliers can be utilized especially for standardization (cf. sec. 2.1) as a preprocess before applying cluster analysis techniques. Furthermore correlations and (joint) information content can be used to extract sets of variables with valuable common information for a detailed visual analysis. If outliers are of special interest they can be visually emphasized.

The framework "Metadatum" has been developed for gathering and storing meta data. The process of obtaining different types of meta data is divided into several steps, such that a special kind of meta data is determined in each step. These steps are ordered in such a way that already obtained meta data can be used in following extraction steps. A flexible design of the framework allows meta data extractions with different degrees of user interaction. For gathering of meta data both automatic analysis algorithms[6] and interactive user input[7] are combined.

According to the degree of user interaction default values and standard routines can be applied.

For instance we implemented a dialog for definition and interactive adaption of meta data for describing the variables of a data set. In dependency of input format and of variable values for each variable default values for scale type and for further semantic information are specified. These meta data can be adapted using user knowledge, e.g., a variable with supposed nominal scale type can be interactively changed to ordinal scale type. Then the user can order the data suitably.

---

types in visualization context are *scalar*, *vector* and *tensor of n-th order*.

[6]For instance we use a key analysis technique for the variables of a data set with unknown types of these variables. The result is a set of minimal keys (Keys are combinations of variables which tuples allow an unequivocal mapping to each data record). By taking the shortest key(s) the classification of dependent and independent variables can be achieved.

[7]e.g. selection of appropriate key using user knowledge about the data set

To maintain an overview of current state of meta data gathering process alpha-numerical and visual presentations are integrated. For instance a dialog for displaying special meta data for independent variables has been implemented. Information about types and numbers of dimensions such as their kind (space, time or abstract dimension), information about grid structure and a display of regions of interest are provided. Furthermore the framework "Metadatum" includes a file format for storing meta data, that allows flexible loading and storing of meta data and their re-calculation at different steps.

4 Conclusions and Future Work

We developed a flexible visualization framework which provides a variety of clustering and new visualization techniques. Our framework is configurable in order to adapt the analysis process with respect to meta data. However, there are still challenges for future work. First of all the introduced frameworks have to be evaluated to determine their effectiveness and to verify their applicability in different application domains. Further work has to be done in order to enhance the functionality of our systems. In future research we would like to investigate methods how to improve users' support during the analysis process. Thus our work will be focused on algorithms how to configure and parameterize visual analysis frameworks automatically depending on influencing factors of the exploration process. Automatic and general solutions for these problems are still matters of research. Our actual research goal is the specification of concepts for an explicit attributions of both numerical and visualization techniques depending on meta data, exploration tasks and user profiles.

References

ANKERST, M. (2001): Visual Data Mining with Pixel-oriented Visualization Techniques. In Proceedings of ACM SIGKDD Workshop on Visual Data Mining; San Francisco

BACKHAUS, K. et al. (1996): Multivariate Analysemethoden. Eine anwendungsorientierte Einführung; Springer-Verlag

BERTIN, J. (1981): Graphics and Graphic Information Processing; Walter de Gruyter & Co, Berlin, 1981.

BOCK, H. H. (1974): Automatische Klassifikation; Vandenhoeck & Ruprecht; Göttingen

CARD, S. K. et al. (1999) : Readings in Information Visualization - Using Visions to Think; Morgan Kaufmann Publishers, Inc.; pp 7; San Francisco, California

KAUFMAN, L. and ROUSSEW, P. J. (1990): Finding Groups in Data An Introduction to Cluster Analysis. A WileyScience Publication John Wiley & Sons, Inc.; pp 4748

KOHONEN, T. (1995): Self Organizing Maps; Springer-Verlag; Berlin

KREUSELER, M. et al. (2000): A Scalable Framework for Information Visualization. In Proceedings of IEEE Information Visualization 2000; Salt Lake City; Utah

LAMPING, J. et al. (1995): A focus+context technique based on hyperbolic geometry for viewing large hierarchies. Proc. CHI'95, pp 401408, Denver, May; ACM.

LEUNG, Y.K. and APPERLEY M.D. (1994): A Review and Taxonomy of Distortion-Oriented Presentation Techniques. ACM transactions on computer human interaction. ACM Press ACM series on computing methodologies 1073-0516, New York

RAO, R. and CARD, S. K. (1994): The Table Lens: Merging Graphical and Symbolical Representations in an Interactive Focus+Context Visualization for Tabular Information. In Proceedings of the ACM SIGCHI Conference on Human Factors in Computing Systems

SARKAR, M. and Brown, M. H (1994): Graphical fisheye views. Communications of the ACM, 37 (12): 73—84, December

THEISEL, H. and KREUSELER, M. (1998): An Enhanced Spring Model for Information Visualization. Computer Graphics Forum, Vol 17, No 3, (Proceedings Eurographics 98)

WESTPHAL, C. and BLAXTON, T. (1998): Data Mining Solutions - Methods and Tools for Solving Real-World Problems; John Wiley & Sons, Inc New York

Bayesian Space-time Analysis of Health Insurance Data

S. Lang[1], P. Kragler[1], G. Haybach[2], L. Fahrmeir[1]

[1]Institut für Statistik,
Ludwig-Maximilians-Universität München,
D-80539 München, Germany

[2]Vereinte Versicherungen,
Fritz-Schäfferstr. 9,
D-81737 München, Germany

Abstract: Generalized linear models (GLMs) and semiparametric extensions provide a flexible framework for analyzing the claims process in non-life insurance. Currently, most applications are still based on traditional GLMs, where covariate effects are modelled in form of a linear predictor. However, these models may already be too restrictive if nonlinear effects of metrical covariates are present. Moreover, although data are often collected within longer time periods and come from different geographical regions, effects of space and time are usually totally neglected. We provide a Bayesian semiparametric approach, which allows to simultaneously incorporate effects of space, time and further covariates within a joint model. The method is applied to analyze costs of hospital treatment and accommodation for a large data set from a German health insurance company.

1 Introduction

Actuarial applications of generalized linear models (GLMs) have gained much interest in recent years, see Renshaw (1994), and Haberman and Renshaw (1998) for a survey. In non-life insurance, they are used as a modelling tool for analyzing claim frequency and claim severity in the presence of covariates. Knowledge about these two components of the claims process is the basis for determining risk premiums. A characteristic feature of many applications is that they rely on traditional GLMs or quasi-likelihood extensions, assuming that the influence of covariates can be modelled in the usual way by a parametric linear predictor. However, as in our application to health insurance, the data provide detailed individual information for types of covariates where influence on claims is difficult or almost impossible to assess with parametric models. Firstly, the effect of metrical covariates, such as age of the policy holder, is often of unknown nonlinear form. Generalized additive models (GAMs) with a semiparametric additive predictor provide a flexible framework for statistical modelling in this case. Secondly, the data also include information on the calendar time of claims, and on the district where the policy holder lives. Neglecting these effects in modelling the claims process will lead to biased fits, with corresponding consequences for risk premium calculation, see Brockman and Wright (1992) for a discussion in the context of calendar time.

Therefore, statistical modelling tools are required which make thorough space-time analyses of insurance regression data possible and allow to explore temporal and spatial effects simultaneously with the impact of other covariates. We present a semiparametric Bayesian approach for a unified treatment of such effects within a joint model, developed in the context of generalized additive mixed models in Fahrmeir and Lang (2001a, b) and Lang and Brezger (2001). Our application investigates costs caused by treatment and accommodation in hospitals. However, the basic concepts are transferable to other costs for medical treatment, to claim frequencies and to other non-life insurances.

2 Semiparametric Bayesian Inference for Space-Time Regression Data

2.1 Data

The space-time regression data from health insurance, which will be analyzed in the next section, consist of individual observations ($y_{it}, x_{it}, w_{it}, s_{it}$), $i = 1, \ldots, n$, $t = 1, \ldots, T$, where y_{it} are costs for hospital treatment or for accommodation of policy holder i in month t, x_{it} is the age at calendar time t, w_{it} is a vector of categorical covariates such as gender, occupation group, type of disease, and s_{it} is the district in West Germany where the insured lives in month t. In general, other types of response variables y, in particular claim frequency, might be of primary interest, and x could be a vector of several metrical covariates.

2.2 Observation Model

Since costs y_{it} are nonnegative, several distributional assumptions can be reasonable, see for example Mack (1998). We do not take into account zero-costs, so a Gamma or log-normal distribution is a common choice. While the former is often preferred in car insurance, a log-normal distribution gives a better fit to the health insurance data at hand. Therefore, we consider log-costs $z_{it} = \log(y_{it})$, and choose a Gaussian additive model $z_{it} = \eta_{it} + \epsilon_{it}$, with i.i.d. errors $\epsilon_{it} \sim N(0, \sigma^2)$, and predictor

$$\eta_{it} = f(x_{it}) + f_{time}(t) + f_{spat}(s_{it}) + w_{it}'\gamma, \quad i = 1, \ldots, n, \quad t \in T_i, \quad (1)$$

where $T_i \subset \{1, \ldots, T\}$ are the months with nonzero-costs $y_{it} > 0$. The unknown function $f(x)$ is the nonlinear effect of age x, $f_{time}(t)$ represents the calendar time trend, and $f_{spat}(s)$ is the effect of district $s \in \{1, \ldots, S\}$ in West Germany. We further split up this spatial effect into the sum

$$f_{spat}(s) = f_{struct}(s) + f_{unstr}(s)$$

of structured (spatially correlated) and unstructured (uncorrelated) effects. A rational for this decomposition is that a spatial effect is usually a surrogate of many underlying unobserved influential factors. Some of them may obey a strong spatial structure, others may be present only locally.

The last term in (1) is the usual linear part of the predictor, with fixed effects. To ensure identifiability, an intercept is always included into w_{it}, and the unknown functions are centered about zero.

Retransformation of the Gaussian additive model (1) for log-costs z_{it} gives a lognormal model for costs y_{it} with (conditional) expectation

$$E(y_{it}|\eta_{it}, \sigma^2) = \mu_{it} = \exp(\eta_{it} + \sigma^2/2), \tag{2}$$

i.e., we get a multiplicative model for expected costs.

Model (2) is closely related to a Gamma model for y_{it} with predictor (1) and an exponential link function. The models are special cases of generalized additive mixed models described in Fahrmeir and Lang (2001a).

2.3 Priors for Functions and Parameters

To formulate priors in compact and unified notation, we express the predictor vector $\eta = (\eta_{it})$ in matrix notation by

$$\eta = f + f_{time} + f_{struct} + f_{unstr} + W\gamma, \tag{3}$$

where f, f_{time} etc. are the vectors of corresponding function values and $W = (w_{it})$ is the design matrix for fixed effects. It turns out that each function vector can always be expressed as the product of a design matrix and a (high-dimensional) parameter vector. Using $f = X\beta$ as a generic notation for functions, (3) becomes

$$\eta = \cdots + X\beta + \cdots + W\gamma.$$

For fixed effects γ, we generally choose a diffuse prior, but a (weakly) informative normal prior is also possible. Constructions of the design matrix X and priors for β depend upon the type of the function and on the degree of smoothness. For metrical covariates, such as age and calendar time, random walk models, P-Splines and smoothing splines are suitable choices, structured spatial effects are modelled through Markov random field priors, and unstructured effects through i.i.d. normal random effects. In any case, priors for the vectors β have the same general Gaussian form

$$p(\beta|\tau^2) \propto \exp(-\frac{1}{2\tau^2}\beta'K\beta). \tag{4}$$

The penalty matrix K penalizes roughness of the function. Its structure depends on the type of covariate and on smoothness of the function, see Fahrmeir and Lang (2001a, b) and Lang and Brezger (2001) for details. The hyperparameter τ^2 acts as a smoothing parameter and controls the degree of smoothness. A highly dispersed inverse Gamma $IG(a,b)$ prior is a convenient choice as a hyperprior. The same choice is made for the variance σ^2 of the errors ϵ_{it} . As usual, observations and priors are assumed to be conditionally independent.

2.4   MCMC Inference

Estimation of functions and parameters is based on the posterior, which is defined by the observation model and the priors. Since the posterior is intractable analytically or numerically, inference is carried out via MCMC simulation. For the Gaussian additive model (1) for log-costs, full conditionals are (high-dimensional) Gaussian or inverse Gamma distributions, so that Gibbs sampling is possible. The full conditional for a typical β is Gaussian with precision matrix P and mean m of the form

$$P = \frac{1}{\sigma^2}X'X + \frac{1}{\tau^2}K, \quad m = P^{-1}\frac{1}{\sigma^2}X'(y - \tilde{\eta}),$$

where $\tilde{\eta}$ is the part of the predictor assossiated with the remaining effects. Efficient sampling can be achieved by Cholesky decompositions for band matrices (Rue, 2001) and is implemented in BayesX (Lang and Brezger, 2000). For non-Gaussian observation models, e.g., a Gamma model, additional Metropolis-Hastings steps are necessary.

3   Application to Health Insurance Data

The approach has been applied to a large space-time regression data set from a private health insurance company in Kragler (2000), with separate analyses for various types of health services. The data set contains individual observations for a sample of 13.000 males (with about 160.000 observations) and 1.200 females (with about 130.000 observations) in West Germany for the years 1991-1997. All analyses were carried out separately for males and females. Analyses for costs were based on Gamma or log-normal models, while frequencies of doctoral visits or of treatments in hospitals were modelled by logit regressions.

Supported by evidence from diagnostic model checks, we use Gaussian additive models (1) for the following space-time analyses of costs for health services in hospitals. In contrast to costs for doctoral visits, it turns out that the categorical covariates "occupation group" and "type of disease" are non-significant. Furthermore, separate analyses for the 3 types of health services (accommodation, treatment with operation, treatment without operation) are preferred to a joint model with type of service as a categorical covariate. Therefore, our analysis for the 6 subgroups, determined by the combinations of gender and type of service, uses a Gaussian additive model (1) for log-costs, where w_{it} contains only an intercept. The effect of age and the time trend are modelled by Bayesian P-splines (Lang and Brezger 2001), for the structured spatial effect a Markov random field prior with adjacency weights is used (Fahrmeir and Lang 2001b).

Figure 1 displays the estimated age effects (left panel) and time trends (right panel) for females. For males similar results are obtained and therefore omitted. The effect of age differs between groups, showing that separate analyses are necessary to avoid confounding. The effect on costs for accommodation increases monotonically over a wide range of age values. In contrast, the effects for treatment with or without operation have a different shape. Starting from a higher level in younger years, they decrease monotonically until about 45 years of age, where the effects starts to increase. A possible explanation might be the higher proportion of younger females staying in hospitals for births of their children: Mostly, they stay only for a few days with comparably low costs for accommodation, but with relatively higher costs for medical treatment. The estimated time trend is shown in the right panel of Figure 1. While the effect on accommodation costs is more or less continuously increasing over the years, a corresponding increase of the effect on treatment costs until about 1995 is followed by an enormous decline. The reason for this decline might be changes in regulations or laws for health insurance or health care. More discussion with experts is needed for a convincing explanation of this effect. Anyway, it becomes obvious that risk premium calculation based on data from this period has serious problems, when the calendar time trend is simply neglected. This is one of the main reasons, why we believe that careful space-time analyses of insurance data are needed, at least for monitoring purposes. The argument is confirmed by the results for regional effects. They are visualized for both males and females in Figure 2 by "significance maps", constructed as follows: For each region 10% and 90% posterior quantiles of its estimated effect are calculated from the posterior. If the 10% quantile is positive, the regional effect is significantly positive (black scaled in Figure 2); it is significantly negative if the 90% quantile is negative (white scaled), and it is non-significant otherwise (grey scaled). Again, the maps reveal distinct

138

patterns which motivate closer inspection by experts.

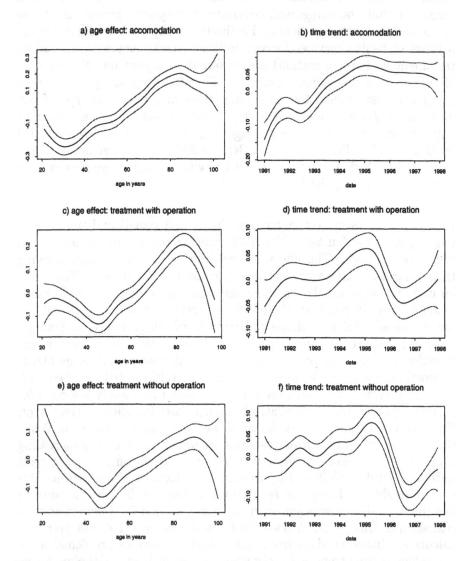

Figure 1: Estimated age effects and time trends for females. Shown is the posterior mean within 80 % credible regions.

Figure 2: Posterior probabilities of the structured spatial effect.

4 Conclusion

Our application demonstrates that a thorough space-time analysis of insurance data can reveal important features of the claims process which are not easily detected by traditional methods. Although we focussed on claim severity in health insurance, the concepts can also be used for modelling claim frequencies and for analyzing other non-life insurance data. First experience with modelling claim frequencies in health insurance shows that a direct transfer of models common in car insurance is at least problematic. We will investigate this in detail in future work.

References

Brockman, M. J and Wright, T. S. (1992): Statistical Motor Rating: Making Effective Use of Your Data. *Journal of the Institute of Actuaries 119, 457–526*

Fahrmeir, L. and Lang, S. (2001a): Bayesian Inference for Generalized Additive Mixed Models Based on Markov Random Field Priors. *Appl. Stat. (JRSS C) 50, 201–220*

Fahrmeir, L. and Lang, S. (2001b): Bayesian Semiparametric Regression Analysis of Multicategorical Time-Space Data. *Ann. Inst. Statist. Math. 53, 11–30*

Haberman, S. and Renshaw, A. E (1998): Actuarial Applications of Generalized Linear Models. In: Statistics in Finance, Ch. 3 (eds: D. Hand and S.Jacka). Arnold, London

Kragler, P. (2000): Statistische Analyse von Schadensfällen privater Krankenversicherungen. *Master Thesis, Department of Statistics, University of Munich*

Lang, S. and Brezger, A. (2000): BayesX - Software for Bayesian Inference based on Markov Chain Monte Carlo simulation techniques. *Discussion Paper 187, SFB 386,University of Munich*

Lang, S. and Brezger, A. (2001): Bayesian P-Splines. *Discussion Paper 236, SFB 386, University of Munich*

Mack, T. (1998): Schadensversicherungsmathematik. Schriftenreihe Angewandte Versicherungsmathematik, Vol. 28. Verlag Versicherungswirtschaft, Karlsruhe

Renshaw, A. E. (1994): Modelling the claims process in the presence of covariates. *Astin Bulletin, Vol. 24, 265–285*

Rue, H. (2001): Fast Sampling of Gaussian Markov Random Fields. *J. R. Statist. Soc. B 63, 325–338*

On Clustering by Mixture Models

G.J. McLachlan, S.K. Ng, D. Peel

Department of Mathematics,
University of Queensland, St. Lucia, Brisbane 4072,
Australia

Abstract: Finite mixture models are being increasingly used to model the distributions of a wide variety of random phenomena and to cluster data sets; see, for example, McLachlan and Peel (2000a). We consider the use of normal mixture models to cluster data sets of continuous multivariate data, concentrating on some of the associated computational issues. A robust version of this approach to clustering is obtained by modelling the data by a mixture of t distributions (Peel and McLachlan, 2000). The normal and t mixture models can be fitted by maximum likelihood via the EM algorithm, as implemented in the EMMIX software of the authors. We report some recent results of McLachlan and Ng (2000) on speeding up the fitting process by an an incremental version of the EM algorithm. The problem of clustering high-dimensional data by use of the mixture of factor analyzers model (McLachlan and Peel, 2000b) is also considered. This approach enables a normal mixture model to be fitted to data which have high dimension relative to the number of data points to be clustered.

1 Introduction

Finite mixtures of distributions have provided a mathematical-based approach to the statistical modelling of a wide variety of random phenomena; see, for example, McLachlan and Peel (2000a). Because of their usefulness as an extremely flexible method of modelling, finite mixture models have continued to receive increasing attention over the years, both from a practical and theoretical point of view. For multivariate data of a continuous nature, attention has focussed on the use of multivariate normal components because of their wide applicability and computational convenience. They can be easily fitted iteratively by maximum likelihood (ML) via the expectation-maximization (EM) algorithm of Dempster et al. (1977); see also McLachlan and Krishnan (1997).

With a normal mixture model-based approach to clustering, it is assumed that the data to be clustered are from a mixture of an initially specified number g of multivariate normal densities in some unknown proportions $\pi_1, \ldots, \pi_g$. That is, each data point is taken to be a realization of the mixture probability density function (p.d.f.),

$$f(\boldsymbol{y}; \boldsymbol{\Psi}) = \sum_{i=1}^{g} \pi_i \phi(\boldsymbol{y}; \boldsymbol{\mu}_i, \boldsymbol{\Sigma}_i), \tag{1}$$

where $\phi(y; \mu_i, \Sigma_i)$ denotes the p-variate normal density probability function with mean μ_i and covariance matrix Σ_i. Here the vector Ψ of unknown parameters consists of the mixing proportions π_i, the elements of the component means μ_i, and the distinct elements of the component-covariance matrices Σ_i.

Once the mixture model has been fitted, a probabilistic clustering of the data into g clusters can be obtained in terms of the fitted posterior probabilities of component membership for the data. An outright assignment of the data into g clusters is achieved by assigning each data point to the component to which it has the highest estimated posterior probability of belonging; see, for example, Bock (1996) for an account of some of the issues in cluster analysis. In this paper, we consider the use of normal and t mixture models to cluster data sets of continuous multivariate data, focussing on some of the associated computational issues.

2   Maximum Likelihood Estimation

2.1   Application of EM algorithm

We let Ψ be the vector of unknown parameters in the mixture model. It thus consists of the mixing proportions and the unspecified parameters in the component densities. The maximum likelihood estimate of Ψ is obtained as an appropriate root of the likelihood equation

$$\partial \log L(\Psi)/\partial \Psi = 0, \tag{2}$$

where $L(\Psi)$ denotes the likelihood function for Ψ formed from the observed random sample $y_1, \ldots, y_n$. Solutions of (2) corresponding to local maxima can be found by application of the EM algorithm. The latter is applied in the framework where an observation y_j is conceptualized to have arisen from one of the components and the indicator variable denoting its component of origin is taken to be missing. The E-step of the EM algorithm thus involves replacing these unobservable indicator variables by their conditional expectations given the observed data (the posterior probabilities of component membership), since the complete-data log likelihood is linear in them. For normal component densities, the estimates of their means μ_i and covariance matrices Σ_i can be updated in closed form on the M-step; see, for example, McLachlan and Peel (2000a, Chapter 3).

As the likelihood equation (2) tends to have multiple roots corresponding to local maxima, the EM algorithm needs to be started from a variety of initial values for the parameter vector Ψ or for a variety of initial partitions of the data into g groups. The latter can be obtained by randomly dividing the data into g groups corresponding to the g components of the mixture model. With random starts, the effect of the central limit theorem tends to have the component parameters initially being similar at least in large samples. One way to reduce this effect is to first select a small random subsample from the data, which is then randomly assigned to the g components. The first M-step is then performed on the basis of the subsample. The subsample has to be sufficiently large to ensure that the first M-step is able to produce a nondegenerate estimate of the parameter vector $\mathbf{\Psi}$.

Coleman et al. (1999) have considered using a combinatorial search for a good starting point from which to apply the EM algorithm. They compared two local searches with a hierarchical agglomerative approach where the objective function to be minimized was taken to be the determinant of the pooled within-cluster covariance matrix.

2.2 Mixtures of t Distributions

For many applied problems, the tails of the normal distribution are often shorter than appropriate. Also, the estimates of the component means and covariance matrices can be affected by observations that are atypical of the components in the normal mixture model being fitted. In this paper, we consider the fitting of mixtures of (multivariate) t distributions. The t distribution provides a longer tailed alternative to the normal distribution. Hence it provides a more robust approach to the fitting of normal mixture models, as observations that are atypical of a normal component are given reduced weight in the calculation of its parameters; see McLachlan and Peel (1998), Peel and McLachlan (2000), and McLachlan and Peel (2000a, Chapter 7).

The t density with location parameter μ, positive definite matrix Σ, and ν degrees of freedom is given by

$$f(\mathbf{y}; \mu, \Sigma, \nu) = \frac{\Gamma(\frac{\nu+p}{2})|\Sigma|^{-1/2}}{(\pi\nu)^{\frac{1}{2}p}\Gamma(\frac{\nu}{2})\{1 + \delta(\mathbf{y}, \mu; \Sigma)/\nu\}^{\frac{1}{2}(\nu+p)}}, \tag{3}$$

where

$$\delta(\mathbf{y}, \mu; \Sigma) = (\mathbf{y} - \mu)^T \Sigma^{-1} (\mathbf{y} - \mu) \tag{4}$$

denotes the Mahalanobis squared distance between $\mathbf{y}$ and μ (with Σ as the covariance matrix). If $\nu > 1$, μ is the mean of $\mathbf{Y}$, and if $\nu > 2$, $\nu(\nu - 2)^{-1}\Sigma$ is its covariance matrix. As ν tends to infinity, $\mathbf{Y}$ becomes marginally multivariate normal with mean μ and covariance matrix Σ. The family of t distributions provides a heavy-tailed alternative to the normal family. McLachlan and Peel (2000a, Chapter 7) have provided a detailed account how the EM algorithm and a multicycle ECM variant can be used to undertake ML estimation of a mixture of t distributions.

3 Speeding up the EM algorithm

3.1 Incremental EM (IEM) Algorithm

As the EM algorithm updates the posterior probabilities of component membership for each observation before the next M-step is performed, it can take some time to implement for large data sets. Hence variants of the EM algorithm have been considered for reducing the computation time. With the incremental EM (IEM) algorithm as proposed by Neal and Hinton (1998), the n available observations are divided into B ($B \leq n$) blocks and the E-step is implemented for only a block of observations at a time before the next M-step is performed. In this way, each data point y_j is visited after B partial E-steps and B (full) M-steps have been performed; that is, after one "pass" or scan of the IEM algorithm. The argument for improved rate of convergence is that the IEM algorithm exploits new information more quickly rather than waiting for a complete scan of the data before parameters are updated by an M-step. The theoretical justification for the IEM algorithm has been provided by Neal and Hinton (1998).

In implementing the M-step for normal components (or for other component densities belonging to the exponential family), it is computationally advantageous to work in terms of the current conditional expectations of the sufficient statistics. This is because the latter can be expressed partly in terms of their values on the previous iteration of the current scan and on the previous scan, so that their updating is confined effectively to a block of observations on a given E-step; see McLachlan and Peel (2000a, Chapter 12).

The choice of the number of blocks B so as to optimize the convergence time of the IEM algorithm is an interesting problem. McLachlan and Ng (2000) have investigated the tradeoff between the additional computation on one scan of the IEM algorithm and the fewer number of scans. Their results suggest using $B \approx n^{2/5}$ as a simple guide. The optimal choice will depend on the number of unknown parameters. McLachlan and Ng (2000) suggest modifying this guide to $B \approx n^{1/3}$ in the case of component-covariance matrices specified to be diagonal and to $B \approx n^{3/8}$ for component-covariance matrices restricted to be equal.

Thiesson et al. (1999) suggest a search method to choose the number of blocks. For a given number of blocks B, they propose to run the IEM algorithm for two scans (the first scan involves a full E-step) and calculate the ratio

$$r = (L_2 - L_1)/t,$$

where L_1 and L_2 are the log likelihood values after the first and the second complete scan of the data respectively, and t is the time required for the second scan which involves the partial E-step. The procedure is repeated for various numbers of blocks B and the choice of B is the value that maximizes r.

3.2 Sparse Versions of the EM and IEM Algorithms

In fitting a mixture model to a data set by maximum likelihood via the EM algorithm, the current estimates of the posterior probabilities for some components of the mixture for a given data point y_j are often close to zero. Neal and Hinton (1998) proposed a sparse version of the EM algorithm for which only those component-posterior probabilities that are above a specified threshold (say, C) are updated. After running this sparse EM (SPEM) algorithm a number of iterations (say, k_1), a full EM step is then performed on which the posterior probabilities of component membership of all the observations are updated. A sparse version of the IEM algorithm (SPIEM) can be formulated by combining the sparse E-step of the SPEM algorithm and the partial E-step of the IEM algorithm.

McLachlan and Ng (2000) considered the choice of values for the threshold C and the number of iterations k_1 for the SPIEM algorithm. Their simulations suggest taking $C = 0.005$ and $k_1 = 5$. In their simulations, it was found that the use of the SPIEM algorithm reduced the time to convergence of the standard EM algorithm by a factor ranging from 18% to 71%, depending on the situation. Of course in situations where the EM algorithm is quick to converge, there may be only a little reduction in the time to convergence; indeed, in some instances, the time to convergence may actually be increased.

4 Mixtures of Factor Analyzers

One approach for reducing the number of unknown parameters in the forms for the component-covariance matrices Σ_i is to adopt the mixtures of factor analyzers model, as considered in McLachlan and Peel (2000a, 2000b). This model was originally proposed by Ghahramani and Hinton (1997) for the purposes of visualizing high dimensional data in a lower dimensional space to explore for group structure; see also Tipping and Bishop (1997) who considered the related model of mixtures of principal component analyzers for the same purpose. With the mixture of factor analyzers model, the ith component-covariance matrix Σ_i has the form

$$\Sigma_i = B_i B_i^T + D_i \quad (i = 1, \ldots, g), \tag{5}$$

where B_i is a $p \times q$ matrix of factor loadings and D_i is a diagonal matrix. It assumes that the component-correlations between the observations can be explained by the conditional linear dependence of the latter on q latent or unobservable variables specific to the given component. If q is chosen sufficiently smaller than p, the representation (5) imposes some constraints on the component-covariance matrix Σ_i and thus reduces the number of free parameters to be estimated.

Note that in the case of $q > 1$, there is an infinity of choices for B_i, since (5) is still satisfied if B_i is replaced by $B_i C_i$, where C_i is any orthogonal matrix of order q. One (arbitrary) way of uniquely specifying B_i is to choose the orthogonal matrix C_i so that $B_i^T D_i^{-1} B_i$ is diagonal (with its diagonal elements arranged in decreasing order); see Lawley and Maxwell (1971, Chapter 1). Assuming that the eigenvalues of $B_i B_i^T$ are positive and distinct, the condition that $B_i^T D_i^{-1} B_i$ is diagonal as above imposes $\frac{1}{2}q(q-1)$ constraints on the parameters. Hence then the number of free parameters for each component-covariance matrix is

$$pq + p - \tfrac{1}{2}q(q-1).$$

5 Example: Clustering of Microarray Data

We now apply the t-mixture model to cluster some $n = 1290$ observations of $p = 26$ dimensions. These data were produced by DNA microarray experiments. $cDNA$ microarrays consist of thousands of different $cDNA$ clones spotted onto known locations on glass microscope slides. These slides/microarrays then are hybridized with differentially labelled DNA populations made from the $mRNAs$ of two samples. The primary data obtained are ratios of fluorescence intensity (red/green), representing the ratios of concentrations of $mRNA$ molecules that hybridized to each of the $cDNAs$ represented on the array. The data set analyzed here is a subset of the data considered in Perou et al. (2000). It contains the log ratios on 1290 gene expressions for 26 tissues on human mammary epithelial cells growing in culture and in primary human breast tumours.

In Figure 1, we display the two-cluster solution obtained by fitting a mixture of $g = 2$ t-components. The clusters are displayed in the two-dimensional space constructed using the first two principal components of the data, where we have imposed the 95% asymptotic confidence region for an observation about its component mean. The corresponding solution from fitting a two-component normal mixture model is given in Figure 2. It can be seen on comparing the two figures that the use of the t-mixture model results in a clustering that appears to be less affected by atypical observations in the data. The latter cause the estimates of the component-variances, and hence the elliptical confidence regions, to be inflated when normal components are adopted.

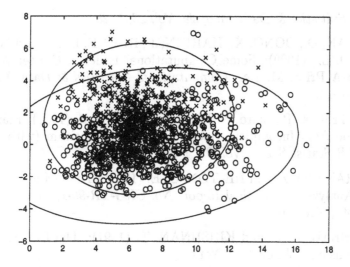

Figure 1: Plot of two-component normal mixture-based solution; (implied) cluster memberships denoted by o and x.

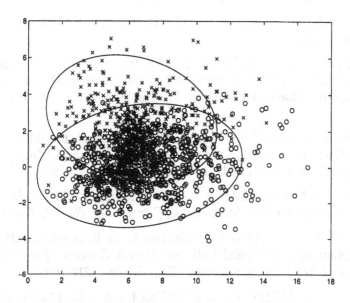

Figure 2: Plot of two-component t mixture-based solution; (implied) cluster memberships denoted by o and x.

References

BISHOP, C. M. (1998): Latent Variable Models, in Jordan (Ed.): Learning in Graphical Models, Kluwer, Dordrecht, 371-403.

BOCK, H. H. (1996): Probabilistic Models in Cluster Analysis. *Compu-*

tational Statistics & Data Analysis, Vol. 23, 5–28.

COLEMAN, D., DONG, X., HARDIN, J., ROCKE, D. M., and WOOD-RUFF, D. L. (1999): Some Computational Issues in Cluster Analysis with No A Priori Metric. *Computational Statistics & Data Analysis, Vol. 31, 1–11.*

DEMPSTER, A. P., Laird, N. M., and RUBIN, D. B. (1977): Maximum Likelihood from Incomplete Data via the EM Algorithm (with discussion). *J R Statist Soc B, Vol. 39, 1–38.*

GHAHRAMANI, Z. and HINTON, G. E. (1997): The EM algorithm for Factor Analyzers. *Technical Report No. CRG-TR-96-1*, The University of Toronto, Toronto

McLACHLAN, G. J. and KRISHNAN, T. (1997): The EM Algorithm and Extensions. Wiley, New York

McLACHLAN, G. J. and NG, S. K. (2000): A Sparse Version of the Incremental EM Algorithm for Large Databases. *Technical Report*, Centre for Statistics, University of Queensland, Brisbane

McLACHLAN, G. J. and PEEL, D. (2000a): Finite Mixture Models. Wiley, New York

McLACHLAN, G. J. and PEEL, D. (2000b): Mixtures of Factor Analyzers, in Langley (Ed.): Proceedings of the Seventeenth International Conference on Machine Learning, Morgan Kaufmann, San Francisco, 599-606.

NEAL, R. M. and HINTON, G. E. (1998): A View of the EM Algorithm That Justifies Incremental, Sparse, and Other Variants, in Jordan (Ed.): Learning in Graphical Models, Kluwer, Dordrecht, 355-368.

PEEL, D. and McLACHLAN, G. J. (2000): Robust Mixture Modelling Using the *t* Distribution. *Statistics & Computing, Vol. 10, 335–344.*

PEROU, C. M. et al. (2000): Distinctive Gene Expression Patterns in Human Mammary Epithelial Cells and Breast Cancers. *Proceedings of the National Academy of Sciences USA, Vol. 96, 9212–9217.*

THIESSON, B., MEEK, C., and HECKERMAN, D. (2000): Accelerating EM for Large Databases. *Technical Report No. MSR-TR-99-31* (revised version), Microsoft Research, Seattle

TIPPING, M. E. and BISHOP, C. M. (1997): Mixtures of Probabilistic Principal Component Analysers. *Technical Report No. NCRG/97/003*, Neural Computing Research Group, Aston University, Birmingham

TIPPING, M. E. and BISHOP, C. M. (1999): Mixtures of Probabilistic Principal Component Analysers. *Neural Computation, Vol. 11, 443–482.*

Using Additive Conjoint Measurement in Analysis of Social Network Data

A. Okada

Department of Industrial Relations, School of Social Relations,
Rikkyo (St. Paul's) University, 3-34-1 Nishi Ikebukuro, Toshima-ku Tokyo,
JAPAN 171-8501

Abstract: A procedure for analyzing social network data is introduced. The procedure can analyze social network data which are (a) not binary but discrete or continuously valued, and (b) asymmetric. The procedure utilizes the additive conjoint measurement by regarding rows and columns of a social network data matrix as two different attributes, and the (j, k) element of the matrix as the degree of preference or worth made by combining level j of the attribute corresponding to rows and level k of the attribute corresponding to columns. An application to interpersonal attraction among university students is presented.

1 Introduction

Social networks represent relationships among a set of actors which are called one-mode two-way (Carroll, Arabie (1980)) social network. Researchers on social networks have focused their attention mainly on one-mode two-way social network data which are represented by a square symmetric matrix. Usually a social network data matrix consists of 0-1 or binary elements (Wasserman, Faust (1994, p.169)). When social network data are binary, either the presence or the absence of a relationship is shown. But relationships among actors are not necessarily binary (Wasserman, Iacobucci (1986)). When social network data are discrete (not binary) or continuously valued, relationships among actors can more fully be described than when they are binary.

Although some social networks are inevitably asymmetric, relationships among actors would have been basically regarded to be symmetric (Bonacich (1972)). But several researchers have paid attention to asymmetric social networks (Carley, Krackhart (1996)). In the case of the asymmetric social network, the relationship from actors j to k and that from actors k to j can be differentiated. The two relationships are not necessarily equal to each other, nor they are not necessarily symmetric.

2 The Procedure

Let n be the number of actors. The social network data matrix among n actors has n rows and n columns. Actor j is represented by row j as well as by column j. The (j, k) element of the matrix represents the relationship from actors j to k. As mentioned earlier, the (j, k) element of the matrix is not necessarily equal to the (k, j) element of the matrix, which represents the relationship from actors k to j.

The procedure is based on additive conjoint measurement. A set of n rows of the matrix is regarded as one attribute, and a set of n columns of the matrix is regarded as the other attribute. Each row or actor represented by a row is regarded as a level of the attribute corresponding to the rows. And each column or actor represented by a column is regarded as a level of the other attribute corresponding to the columns. The (j, k) element of the matrix, which represents the relationship from actors j to k, is regarded to be the degree of the preference toward or the worth of the 'product' made by combining level j of the attribute corresponding to rows and level k of the attribute corresponding to columns.

An analysis of a social network data matrix among actors by the procedure using additive conjoint measurement results in two sets of part-worths; one consists of part-worths assigned to levels of an attribute corresponding to rows of the matrix, and the other consists of part-worths assigned to levels of the other attribute corresponding to columns. Thus two terms are assigned to each of n actors. One is the part-worth assigned to an actor represented by a row. The other is the part-worth assigned to an actor represented by a column. The part-worth assigned to an actor represented by a row is called the row factor or chooser factor of the actor, and that assigned to an actor represented by a column is called the column factor or chosen factor of the actor (Bonacich (1972); Wright, Evitts (1961)). Let r_j be the row factor of actor $j(j = 1, 2, \cdots, n)$, and c_k be the column factor of actor $k(k = 1, 2, \cdots, n)$. The analysis using the additive conjoint measurement derives r_j and c_k which minimize the discrepancy of the monotone increasing relationships of u_{jk} defined by

$$u_{jk} = r_j + c_k \tag{1}$$

with the (j, k) element of the social network data matrix.

An algorithm of the nonmetric multidimensional scaling (Kruskal (1964)) is utilized to derive r_j and c_k $(j, k = 1, \cdots, n)$, which minimize the badness-of-fit measure of u_{jk} with the (j, k) element of the social network data matrix. The badness-of-fit measure called stress is defined as

$$S = \sqrt{\left[\sum_{\substack{j=1 \\ j \neq k}}^{n} \sum_{k=1}^{n} (u_{jk} - \hat{u}_{jk})^2 \Big/ \sum_{\substack{j=1 \\ j \neq k}}^{n} \sum_{k=1}^{n} (u_{jk} - \bar{u})^2 \right]}, \tag{2}$$

where $\hat{u}_{jk}$ is the monotone transformed u_{jk} which is monotone increasingly related with the (j, k) element of the data matrix, and $\bar{u}$ is the mean of u_{jk}.

Thus the (j, k) element of the matrix, which represents the relationship from actors j to k, is summarized or approximated by u_{jk}, i.e., the sum of the row or chooser factor of actor j and the column or chosen factor of actor k. The row or chooser factor of an actor shows the strength of the outward characteristic from that actor to the other actors, and the column or chosen factor of an actor shows the strength of the inward characteristic from the other actors to that actor.

3 An Application

The present procedure was applied to the interpersonal attraction data among university students (Newcomb (1961); Nordlie (1958)) in order to show the effectiveness of the procedure. The data consist of two sets of data, i.e., Groups 1 and 2 data, each consist of the interpersonal attraction ranks among 17 university students collected once a week from weeks 0 through 15. The Group 2 data, which also had been analyzed by other researchers (e. g., Nakao, Romney (1993)), were analyzed in the present study. Attraction ranks among 17 students at each week are represented as a 17×17 matrix where the (j, k) element of the matrix represents the attraction rank from students j to k at that week. Because the attraction rank from students j to k is not necessarily equal to that from students k to j, the resulting data matrices are generally asymmetric. The data at week 6 were analyzed. The reason to choose the data at week 6 is that the attraction pattern seems to stabilize at week 6 (Nakao, Romney (1993); Nordlie (1958); Okada, Imaizumi (2000)).

Before the analysis, the attraction rank was transformed by subtracting itself from 17 so that the larger value represents the larger attraction. The transformed rank is called the attraction hereafter. Diagonal elements of each matrix are not defined in the data.

The transformed data at week 6 was analyzed by the additive conjoint measurement using rational initial values with the secondary approach to ties. The analysis resulted in the solution having the minimized stress value of 0.685. Analyses using 20 different random initial values were done. The minimum stress value obtained among these 20 results was 0.685 attained by the same result obtained by the analysis using the rational initial values. This validated the present preferred solution. The analysis yielded two sets of factors; one set for the 17 students represented by the rows of the matrix and the other set for the 17 students represented by the columns of the matrix. The obtained factors were normalized so that the sum of factors was zero for each set. Table 1 shows the resulting row and column factors given to each of the 17 students.

The row factor shown in Table 1 is small for all the 17 students. This is consistent with that each student or each row of the transformed matrix

Student	Row factor	Column factor
1	0.038	0.600
2	0.033	0.518
3	-0.095	-1.522
4	0.068	1.090
5	0.033	0.519
6	0.037	0.590
7	0.051	0.822
8	0.021	0.327
9	0.066	1.060
10	-0.248	-3.977
11	0.033	0.535
12	0.074	1.192
13	-0.006	-0.091
14	0.025	0.404
15	-0.067	-1.070
16	-0.155	-2.483
17	0.092	1.485

Table 1: Row and column factors of the solution.

of the attraction has values from 1 through 16 at off diagonal elements. Thus, all the 17 students have the same strength of the outward characteristic. On the other hand, the column factor varies from one student to another, because some students were given the higher attraction, and some students were given the lower attraction.

The row factor shows the expansiveness and the column factor shows the popularity (Wasserman, Faust (1994, p. 612)) of the corresponding student. The expansiveness is almost the same for all of the 17 students, but the popularity differs among students. Nordlie (1958) defined the popularity rank measure of a student at each week as the sum of attraction ranks given to the student (the smaller value means the larger popularity). The product moment correlation coefficient between the derived column factor in Table 1 and the popularity rank measure at week 6 was, as shown in Figure 1, -0.95. The larger column factor corresponds to the larger popularity. This validates that the derived column factor of a student shows the popularity of that student. Actor 17 has the largest column factor, and actor 10 has the smallest column factor. Thus, actor 17 is the most popular student, suggesting that the student receives the highest attraction from the other students, and actor 10 is the least popular student, suggesting that the student receives the lowest attraction from the other students.

An analysis of Group 2 data by using two-mode three-way asymmetric multidimensional scaling has been done (Okada, Imaizumi (2000)). In the analysis each student was represented as a point and a circle centered at that point in a multidimensional Euclidean space which is called the common object configuration. The radius of a circle representing a student showed the popularity of that student (the smaller radius means the larger popularity of the corresponding student). The product moment correlation coefficient between the column factor and the radius resulted in -0.96. Although the radius of a student shows the popularity of that student for week 0 through 15 (it does not show the popularity only at week 6), this figure seems to validate that the derived column factor of a student shows the popularity of that student.

4 Discussion

A procedure for analyzing the social network data matrix by using the additive conjoint measurement was introduced. The procedure was successfully applied to attraction relationships data among university students which are discrete (not binary) and asymmetric. From the analysis the row and column factors were obtained. The column factor shows the popularity of the actor which is based on the additive model of Equation (1) not just a rank order of popularity (Nakao, Romney (1993)). Because the row and column factors are based on the additive model, it is not difficult to interpret them. The result is compatible with the data, and the obtained row factor can show all the students have almost the same outward characteristic. Thus the result seems to show that the present procedure can represent the attraction relationships by the sum of row and column factors, suggesting the effectiveness of the procedure.

Figure 1 shows the scatter diagram between the derived column factor of Table 1 and the popularity rank measure of Nordlie (1958). The scatter diagram in Figure 1 also represents three categories of students suggested by Nakao and Romney. They are Groups 1 and 2, and Outlier. Each of them consists of seven, five, and five students. The cluster analysis of the obtained row factor by Ward's method resulted in clusters compatible with Groups 1 and 2, and Outlier; two clusters almost corresponding to Groups 1 and 2, and (singleton) clusters of students belonging to Outlier. Figure 1 shows that students belong to Groups 1 and 2 seems to have the larger column factor or the larger popularity than those classified into Outliers have, and that those belonging to Group 1 have the slightly larger column factor than those that belonging to Group 2. This is compatible with the result shown in Nakao and Romney.

The additive conjoint measurement used in the present procedure is based on the nonmetric multidimensional scaling algorithm, and inherits several advantages from the algorithm. Firstly, the algorithm is nonmetric which means the algorithm relies only on the rank order of the elements of the social network data matrix, suggesting the greater ver-

154

satility of the present procedure compared with other procedures for analyzing social network data which rely on numerical values of the elements. Secondly, the present procedure can analyze the data having missing entries simply by defining the stress of Equation (2) ignoring missing entries in the data. Therefore, diagonal elements of a social network data matrix, which shows the self-choice of an actor (cf. Knoke, Kuklinski (1982)), can be included in an analysis if they are necessary to be analyzed.

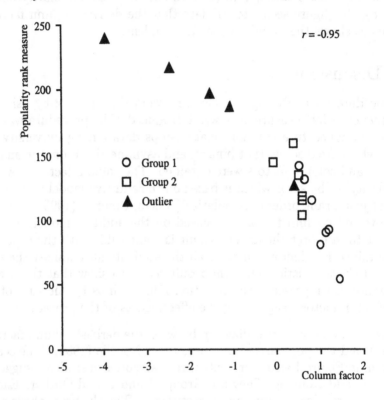

Figure 1: Relationship between the column factor and the popularity rank measure.

Bonacich (1972) introduced a notion to summarize or approximate relationships among actors or a social network data matrix by using factors assigned to actors. An element of the matrix corresponding to the relationship between actors j and k is represented by multiplying the factor assigned to actor j and that assigned to actor k (Bonacich (1972)). In Wright and Evitts (1961), the (j, k) element of the matrix corresponding to the relationship from actors j to k is represented by the product of the row or chooser factor assigned to actor j and the column or chosen factor assigned to actor k. In the present procedure, the element of the social network data matrix corresponding to the relationship from actors j to k is summarized or approximated by the sum of the row or chooser factor assigned to actor j and the column or chosen factor assigned to actor k. In other words, the relationship from actors j to k is represented by the sum of the expansiveness of actor j and the popularity of actor k. It seems that the factors which summarize or approximate the relationships among actors by the sum of the row and column factors are more easily be interpreted than those which summarize or approximate the relationships by the product of them.

In the analyses, the data were treated not as row conditional but as matrix conditional, because the row factor cannot be uniquely determined under the row conditional algorithm. The obtained row factor of a student in Table 1 is equal to (the column factor of the student)/16 to compensate the missing diagonal element of the data with the row factor. Thus the row factor is the artifact of the algorithm, and they do not represent differences in expansiveness, because as stated earlier, all 17 students have the same strength of the outward characteristic.

Social network data are sometimes represented as two-mode two-way data or as a rectangular matrix where rows correspond to a set of actors and columns correspond to a set of events, and each element of the matrix represents an actor-by-event relationship (Bonacich (1991); Bonacich, Oliver, Snijders (1998); Borgatti, Everett (1997)). The present procedure can analyze the actor-by-event relationships simply by assigning a factor to each actor and the other factor to each event. The (j, k) element of the matrix, corresponding to the relationship between actor j and event k, is represented by the sum of the row or chooser factor assigned to actor j and the column or chosen factor assigned to event k. The present procedure can also deal with three-way social networks or longitudinal social networks. The three-way social networks can be analyzed simply by deriving three sets of factors; assigning each set to each of the three ways.

Acknowledgment

The author would like to express his gratitude to two anonymous referees for their constructive and helpful reviews. He also wishes to thank Hervert A. Donovan for his help concerning English.

References

BONACICH, P. (1972): Factoring and Weighting Approaches to Status Scores and Clique Identification. *Journal of Mathematical Sociology*, Vol. 2, 113–120.

BONACICH, P. (1991): Simultaneous Group and Individual Centralities. *Social Networks, Vol. 13, 155–168.*

BONACICH, P., OLIVER, A and SNIJDERS, T. A. B. (1998): Controlling for Size in Centrality Scores. *Social Networks, Vol. 20, 135–141.*

BORGATTI, S. P. and EVERETT, M. G. (1997): Network Analysis of 2-Mode Data. *Social Networks, Vol. 19, 243–269.*

CARLEY, K. M. and KRACKHART, D. (1996): Cognitive Inconsistencies and Non-Symmetric Friendship. *Social Networks, Vol. 18, 1–27.*

CARROLL, J. D. and ARABIE, P. (1980): Multidimensional Scaling. *Annual Review of Psychology, Vol. 31, 607–649.*

KNOKE, D. and KUKLINSKI, J. H. (1982): Network Analysis. Sage, Beverly Hills

KRUSKAL, J. B. (1964): Nonmetric Multidimensional Scaling: A Numerical Method. *Psychomerika, Vol. 29, 115–129.*

NAKAO, K. and ROMNEY, A. K. (1993): Longitudinal Approach to Subgroup Formation: Reanalysis of Newcomb's Fraternity Data. *Social Networks, Vol. 15, 109–131.*

NEWCOMB, T. M. (1961): The Acquaintance Process. Holt Rinehart and Winston, New York

NORDLIE, P. G. (1958): A Longitudinal Study of Interpersonal Attraction in a Natural Group Setting. Doctoral dissertation, University of Michigan

OKADA A. and IMAIZUMI, T. (2000): Two-Mode Three-Way Asymmetric Multidimensional Scaling with Constraints on Asymmetry, in Decker, Gaul (Eds.): Classification and Information Processing at the Turn of the Millennium, Springer-Verlag, Berlin, 52–59.

WASSERMAN, S. and FAUST, K. (1994): Social Network Analysis: Methods and Applications. Cambridge University Press, Cambridge

WASSERMAN, S. and IACOBUCCI, D. (1986): Statistical Analysis of Discrete Relational Data. *British Journal of Mathematical and Statistical Psychology, Vol. 39, 41–64.*

WRIGHT, B. and EVITTS, M. S. (1961): Direct Factor Analysis in Sociometry. *Sociometry, Vol. 24, 82–98.*

Bayesian Latent Class Metric Conjoint Analysis - A Case Study from the Austrian Mineral Water Market

Th. Otter, R. Tüchler, S. Frühwirth-Schnatter

Vienna University of Economics and Business Administration (WU-Wien),
Augasse 2-6, A-1090 Vienna, Austria, e-mail: thomas.otter@wu-wien.ac.at,
regina.tuechler@wu-wien.ac.at, sfruehwi@isis.wu-wien.ac.at

Abstract: This paper presents the fully Bayesian analysis of the latent class model using a new approach towards MCMC estimation in the context of mixture models. The approach starts with estimating unidentified models for various numbers of classes. Exact Bayes' factors are computed by the bridge sampling estimator to compare different models and select the number of classes. Estimation of the unidentified model is carried out using the random permutation sampler. From the unidentified model estimates for model parameters that are not class specific are derived. Then, the exploration of the MCMC output from the unconstrained model yields suitable identifiability constraints. Finally, the constrained version of the permutation sampler is used to estimate group specific parameters. Conjoint data from the Austrian mineral water market serve to illustrate the method.

1 Introduction

The latent class model is one of several possibilities to deal with preference heterogeneity in conjoint analysis (cf. Wedel and Kamakura, 1998). We define the latent class model in a way that is well known from linear mixed modelling:

$$y_i = Z_i \alpha + W_i \beta_i + \varepsilon_i, \quad \varepsilon_i \sim N(0, R_i) \tag{1}$$

where y_i is a vector of T_i preference measurements for consumer i and $R_i = \sigma_\varepsilon^2 \cdot I$ with I being the identity matrix. Whereas parameter α contains homogeneous preferences which are fixed for all consumers, parameter β_i contains random preferences which due to heterogeneity are different for each consumer. Z_i and W_i are the design matrices for the fixed and the random preferences, respectively. The model characterizes the unobserved distribution of the hetergeneous consumer preferences by means of a finite number K of support points $\beta_1^G, \ldots, \beta_K^G$ in a multidimensional space and their respective masses $\eta = (\eta_1, \ldots, \eta_K)$:

$$\beta_i = \begin{cases} \beta_1^G, & \text{if } S_i = 1, \\ \vdots \\ \beta_K^G, & \text{if } S_i = K. \end{cases} \tag{2}$$

Here, we introduced a discrete latent group indicator S_i taking values in $\{1, \ldots, K\}$ which indicates which group consumer i belongs to. S_i has probability distribution $\Pr(S_i = k) = \eta_k$, $k = 1, \ldots, K$.

In the present paper we head for a fully Bayesian analysis of the latent class model with a priori unknown number of classes. Joint Bayesian estimation of all latent variables, model parameters, and parameters determining the probability law of the latent process is carried out by means of the Markov Chain Monte Carlo (MCMC) methods discussed in Frühwirth-Schnatter (2001). The unknown number of classes is determined by formal Bayesian model comparison through model likelihoods computed by the bridge sampling technique (Meng and Wong, 1996; Frühwirth-Schnatter, 2000).

A fully Bayesian analysis of the latent class model has – at least in principle – been tried before, as it could be viewed as a special case of the finite mixture of generalized linear models with random effects discussed in Lenk and DeSarbo (2000) and Allenby et al. (1998). Whereas Lenk and DeSarbo as well as Allenby et al. identify the model by an *a priori* order constraint on the masses $\eta = (\eta_1, \ldots, \eta_K)$ of the discrete support points $\beta_1^G, \ldots, \beta_K^G$, we handle the issue of unidentifiability of the latent class model in a completely different manner. In a first run we use the random permutation sampler discussed in Frühwirth-Schnatter (2001) to sample from the unconstrained posterior. We will demonstrate that a lot of important information, such as e.g. estimates of the subject-specific regression coefficients, is available from such an unidentified model (Frühwirth-Schnatter et al., 2000). The MCMC output of the random permutation sampler is explored in order to find suitable identifiability constraints. In a second run we use the permutation sampler to sample from the constrained posterior by imposing identifiability constraints. We illustrate the various steps toward a fully Bayesian analysis of the latent class model by a case study from the Austrian Mineral water market.

2 The Data

The data come from a brand-price trade-off study in the mineral-water category conducted as part of an ongoing research project on brand equity (Strebinger et al., 1998). Each of 213 Austrian consumers stated their likelihood of purchasing 15 different product-profiles offering five brands of mineral water (Römerquelle, Vöslauer, Juvina, Waldquelle, and one brand not available in Austria, Kronsteiner) at 3 different prices (2.80, 4.80, and 6.80 [all prices in ATS]) on 20 point rating scales (higher values indicate greater likelihood of purchasing). In an attempt to make the full brand by price factorial design less obvious to consumers, the price levels varied in the range of $\pm\,0.1$ ATS around the respective design levels such that mean prices of brands in the design were not affected. Additionally, every consumer provided evaluations for Römerquelle and Vöslauer at the price of 5.90, Juvina and Waldquelle at the price of 3.90 and finally Kronsteiner at the price of 3.20 in a separate task where the same 20 point rating scale was used. These evaluations were retained as holdout data. Note that none of the brand-price combinations in the holdout set appears in the calibration set.

We used a fully parameterized matrix W_i with 15 columns corresponding to the constant, four brand contrasts, a linear and a quadratic price effect, four brand by linear price and four brand by quadratic price interaction effects, respectively. Note that it is not economic theory that makes a linear and a quadratic effect necessary. The polynomial representation is just a convenient way to locally approximate nonlinear relations in this case, where three price levels in the design provide two degrees of freedom. The brand by linear price and brand by quadratic price interactions capture brand specific price effects. Increasing price by one unit is likely to affect the stated purchase likelihoods of the brands differently. However, the brand specific effect of a price increase is also likely to depend on the absolute price level which results in brand by quadratic price interactions. We used dummy-coding for the brands. The unknown brand Kronsteiner was chosen as the baseline. We subtracted the smallest price from the linear price column in matrix W_i, and computed the quadratic price contrast from the centred linear contrast. Therefore, the constant corresponds to the purchase likelihood of Kronsteiner at the lowest price level, if quadratic price effects are not present. Theory neither suggested that any effect should be excluded for all consumers nor that any effect is fixed in the population a priori. Therefore, we start with a model that treats all effects as random and then refine the model.

At the level of an individual consumer the model would be saturated since only 15 data points are available to estimate 15 parameters leaving zero degrees of freedom. In Frühwirth-Schnatter and Otter (1999) a random coefficient model was fitted to the data. Here, we discuss modelling of heterogeneity by a latent class model with a priori unknown numbers K of groups.

3 Empirical Results

3.1 Selecting the Number of Classes

The unknown number of classes is determined by formal Bayesian model comparison through exact model likelihoods. Model selection is based on the Bayesian model discrimination procedure where various models $\mathcal{M}_1, \ldots, \mathcal{M}_K$ are compared through the posterior probability of each model (Bernardo and Smith, 1994):

$$P(\mathcal{M}_l|y^N) \propto f(y_1, \ldots, y_N|\mathcal{M}_l)P(\mathcal{M}_l). \tag{3}$$

The factor $L(y^N|\mathcal{M}_l) := f(y_1, \ldots, y_N|\mathcal{M}_l)$ is called model likelihood and quantifies evidence in favour of a model given the data, with evidence being higher the higher the model likelihood. Alternatively, we could use the Bayes factor defined by the difference of the logarithm of the model likelihoods, $\log L(y^N|\mathcal{M}_l) - \log L(y^N|\mathcal{M}_m)$, to compare two models.

Let $\phi^K = (\beta_1^G, \ldots, \beta_K^G, \alpha, \eta_1, \ldots, \eta_K, \sigma_\varepsilon^2)$. For a latent class model the model likelihood is given by the following integral of the marginal likelihood $L(y_1, \ldots, y_N|\phi^K)$ with respect to the prior $\pi(\phi^K)$:

$$L(y^N) = \int L(y_1, \ldots, y_N|\phi^K)\pi(\phi^K)d\phi^K, \tag{4}$$

where an explicit formula for the marginal likelihood $L(y_1, \ldots, y_N|\phi^K)$ is available:

$$L(y_1, \ldots, y_N|\phi^K) = \prod_{i=1}^N \left(\sum_{k=1}^K f_N(y_i|\beta_k^G, \alpha, \sigma_\varepsilon^2) \cdot \eta_k \right). \tag{5}$$

$f_N(y_i|\cdot)$ is the density of the normal distribution with mean $Z_i\alpha + W_i\beta_k^G$ and variance $\sigma_\varepsilon^2 \cdot I$. Estimation of model likelihoods has proven to be a challenging problem for models including latent structures. Here, we apply the method of bridge sampling to compute model likelihoods for the latent class model (Meng and Wong, 1996; Frühwirth-Schnatter, 2000).

We have to use a proper prior $\pi(\phi^K)$ to compute the model likelihoods. Furthermore an improper prior for group specific parameters is likely to cause instability in MCMC sampling. We use weakly informative normal priors on β_k^G, $N(b_0, B_0)$, as well as on the fixed effects α, $N(a_0, A_0)$. The prior means a_0 and b_0 are chosen to be equal to the posterior mean of the corresponding effects estimated within the random-effects model in Frühwirth-Schnatter and Otter (1999). If such results were not available, we could have centered the priors on the estimates obtained from a model with one class which is a simple regression model. The precision matrices of the normal priors are equal to $A_0^{-1} = 0.04 \cdot I$ and $B_0^{-1} = 0.04 \cdot I$, respectively. The prior on η is the commonly used Dirichlet distribution $D(1, \ldots, 1)$. The prior on σ_ε^2 is chosen to be the inverted Gamma distribution $IG(0, 0)$.

The estimated log model likelihoods together with standard errors are reported in Table 1 for models from two to ten classes. The model likelihood increases till $K = 9$ with the Bayes factor $\log L(y^N|K = 9) - \log L(y^N|K = 8) = 4.71$ clearly favouring a model with nine classes

K	$\log \hat{L}(y^N|K)$	K	$\log \hat{L}(y^N|K)$	K	$\log \hat{L}(y^N|K)$
2	-9892.88 (0.022)	5	-9636.50 (0.294)	8	-9586.99 (1.118)
3	-9756.02 (0.048)	6	-9620.82 (0.393)	9	-9582.28 (0.720)
4	-9701.54 (0.144)	7	-9610.11 (0.396)	10	-9582.89 (1.105)

Table 1: Estimates of the logarithm of the model likelihood $\log L(y^N|K)$ for various values of K (standard errors are given in parenthesis)

compared to a model with eight classes. Increasing the number of classes to $K = 10$ reduces the model likelihood, the difference $\log L(y^N|K = 10) - \log L(y^N|K = 9) = -0.61$, however, being not significant given the standard errors. We choose the more parsimonious model with $K = 9$. This choice is supported further by a plot of the posterior densities of σ_ε^2. The observation error variance σ_ε^2 significantly decreases each time a new class is added up to nine classes. The addition of the tenth class does not help to reduce the unexplained variance σ_ε^2 indicating that the number of classes is too big.

It turns out that the prior B_0^{-1} influences the model choice. Substantially less precision as well as substantially more precision tends to favour a model with eight classes. Our prior $B_0^{-1} = 0.04 \cdot I$ is informative enough to avoid choosing too simple a model due to Lindley's paradoxon. On the other hand the prior is weakly informative enough to be dominated by the likelihood.

3.2 Estimation within Unidentified Models

It is well known that the latent class model, like any model including discrete latent variables, is only identified up to permutations of the labelling of the groups. The full unconstrained posterior of the latent class model with K classes is multimodal with at most $K!$ modes. When applying MCMC methods to such a posterior, we have to be aware of the problem of label switching which might render estimation of group specific quantities meaningless. Nevertheless, a lot of useful information may be retrieved from an unidentified model. Subject specific estimates of the random effects β_i, for instance, may be written as:

$$\beta_i = \sum_{k=1}^K \beta_k^G I_k(S_i),$$

where $I_k(S_i) = 1$ if $S_i = k$. This functional is obviously invariant to relabelling the groups and can therefore be derived from an unidentified model.

During MCMC sampling we carried out $M = 30000$ iterations with a burn-in phase of 20000 iterations. Based on these estimates that correctly reflect all sources of uncertainty given the model specification, we illustrate how the latent class model captures consumer heterogeneity for different numbers of classes. We take advantage of our Bayesian approach and investigate the posterior densities of implied choice probabilities for different offers. Throughout this section it is assumed that individual consumer i's choice probabilities may be derived from preferences directly using a multinomial logit model:

$$ P(alternative_h)_i = \left(\exp(U_h) / \sum_{l=1}^{P} (\exp(U_l)) \right)_i , $$

where U_h corresponds to the utility of alternative h as a function of brand and price: $(U_h)_i = w_h \beta_i$. Here, the utility scale is directly related to the purchase likelihood scale employed to measure consumer preferences. We use simulations of each consumer's utilities for specified combinations of brand and price to derive the densities of the respective choice probabilities. Important properties of these densities of highly nonlinear combinations of model parameters are easily estimated from the simulations. The following is based on a choice set that resembles the holdout task (cf. Section 2).

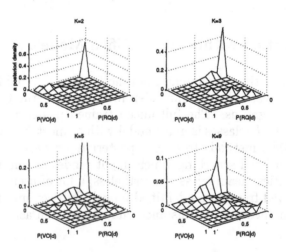

Figure 1: Choice probabilities from a logit model - bivariate representations

Figure 1 contains bivariate marginal densities for the choice probabilities of the Römerquelle and the Vöslauer offer. Notice the 45 degree frontier due to the fact that we are dealing with choice probabilities from a multinomial logit model. The scaling of the density axis is adapted to the number of classes assumed in the three dimensional plots. It is obvious that too low a number of classes will lead to very different conclusions than the optimal choice. The solution with two classes suggests that most of the mass is concentrated at the point of near zero choice probabilities for the Römerquelle and the Vöslauer offer. However, there seems to be some mass in the area of high choice probabilities for the Römerquelle offer accompanied by low choice probabilities of the Vöslauer offer. Increasing the number of classes to three changes the picture dramatically. Again most of the distributional mass is concentrated at the zero/zero point. In contrast to the solution with only two classes, the distribution indicates some support for the combination of high choice probability for the Vöslauer offer accompanied by lower choice probabilities for the Römerquelle offer. Moreover, there also is some support for the combination of small choice probabilities for the Römerquelle offer and a near zero probability for the Vöslauer offer. Naturally, the optimal solution with nine classes offers the most detailed picture. Notice that here some support for the combination of choice probabilities near one for the Vöslauer offer accompanied by such near zero for the Römerquelle offer can be found. This feature of the distribution is not present in the solution relying on five classes only.

3.3   Model Identification

So far the models are identified only up to permutations of the labels of the groups. In order to identify group specific parameters, we need to introduce identifiability constraints that guarantee a unique labelling. We use the following notation for the group specific parameters: const_k refers to the constant, RQ_k, VO_k, JU_k, and WA_k refer to the main effects for the various brands and p_k to the linear price effect for group k. We carefully searched for data-driven identification constraints by visually inspecting plots of the marginal parameter densities as well as two- and three-dimensional scatterplots of MCMC simulations from the unidentified model. As one might expect, simple constraints will only suffice to identify models with few classes.

For instance in the case of a model with three classes, we found from the scatter plots in Figure 2 that the linear price effect differentiates one class from the remaining two and that the constant is useful to tell these apart: $p_1 < \min(p_2, p_3)$, $\text{const}_2 < \text{const}_3$. Sensible identifiability constraints are not necessarily unique. For the model with three classes, an equivalent set of constraints turned out to be $p_1 < \min(p_2, p_3)$, $\text{VO}_2 < \text{VO}_3$.

Estimation of the model likelihoods pointed to a model with nine classes (see Table 1). Due to the relatively high number of classes we proceed in

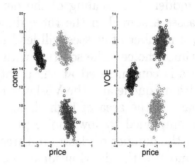

Figure 2: Scatter plots of the MCMC simulations for $K = 3$

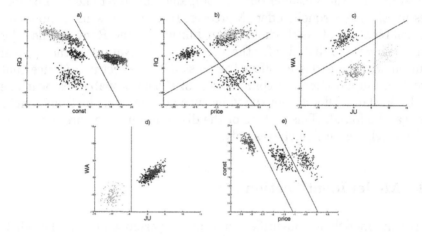

Figure 3: Scatter plots of the MCMC simulations for $K = 9$

a stepwise manner to identify this model. The constraint $\max_{k=1,\ldots,5}(4 \cdot \text{const}_k + \text{RQ}_k) < \min_{k=6,\ldots,9}(4 \cdot \text{const}_k + \text{RQ}_k)$ divides the nine classes into two subgroups of five and four classes, respectively (see Figure 3a). These two subgroups are treated separately now, the one with the five classes first. In Figure 3b we see that the constraint $\max_{k=1,\ldots,4}(10 \cdot p_k - 2 \cdot \text{RQ}_k) < 10 \cdot p_5 - 2 \cdot \text{RQ}_5$ splits off the first group and $10 \cdot p_1 + \text{RQ}_1 < \min_{k=2,3,4}(10 \cdot p_k + \text{RQ}_k)$ splits off the second group. In Figure 3c the third group is separated by $3 \cdot \text{JU}_2 - 5 \cdot \text{WA}_2 < \min_{k=3,4}(3 \cdot \text{JU}_k - 5 \cdot \text{WA}_k)$ and finally, the fourth and the fifth group are identified by $\text{JU}_4 < \text{JU}_5$. To split off one group from the subgroup with the four classes we use the constraint $\text{JU}_6 < \min_{k=7,8,9} \text{JU}_k$ (see Figure 3d). The restriction $5 \cdot \text{price}_7 + \text{const}_7 < 5 \cdot \text{price}_8 + \text{const}_8 < 5 \cdot \text{price}_9 + \text{const}_9$ separates the three classes that are still left. These constraints were imposed to obtain the group specific estimates that are not reported here due to space limitations.

3.4 Testing for Heterogeneity and Variable Selection

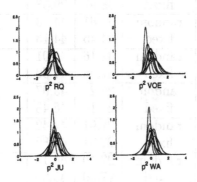

Figure 4: Marginal denities of all quadratic interaction effects

From all results reported above heterogeneity in consumer preferences is present without doubt. It is, however, not clear *a priori* that heterogeneity affects all components of the group specific parameters. As the marginal densities of the quadratic interaction effects plotted in Figure 4 overlap for all groups to a high degree, we formulate the hypothesis that the quadratic interaction effects $RQ \cdot p^2$, $VO \cdot p^2$, $JU \cdot p^2$, and $WA \cdot p^2$ are fixed rather than random. As the marginal density of the quadratic interaction effects in Figure 4 not only overlap for all groups, but most of them also cover 0, we further hypothesize that the quadratic interaction effects are not significant and should be deleted from the model: $RQ \cdot p^2 = VO \cdot p^2 = JU \cdot p^2 = WA \cdot p^2 = 0$.

All hypotheses are tested against the full model by comparing the model likelihoods. The log of the model likelihood for a model with fixed quadratic interaction effects is equal to −9521.8 with a standard error of 0.96, whereas the log of the model likelihood for a model without quadratic interaction effects is equal to -9580.76 with a standard error of 0.31. In comparison to the full model (see Table 1) assuming fixed rather than random quadratic interaction effects increases the model likelihood substantially. It turned out that the same identifiability constraints that were formulated for the full model, where all components were heterogeneous, applied to this mixed effects latent class model.

3.5 Holdout Analysis

Table 2 summarizes the performance of the various models on the holdout data. The mean squared error is obtained by averaging over every consumer's five holdout evaluations and over all consumers. The first choice hit rate corresponds to the relative frequency of consumers where

Model	$p^2\cdot$ brand	MSE		First Choice hits (in %)
Latent Class	random	38.99	29.58	21.33
($K = 1$)	fixed	38.97	29.58	21.33
Latent Class	random	34.61	44.13	32.01
($K = 2$)	fixed	34.45	44.13	32.01
Latent Class	random	31.16	46.01	32.72
($K = 3$)	fixed	31.08	44.60	31.15
Latent Class	random	29.21	45.07	32.17
($K = 4$)	fixed	29.16	46.48	33.42
Latent Class	random	27.91	47.42	33.89
($K = 5$)	fixed	27.81	47.89	34.36
Latent Class	random	27.10	44.13	31.31
($K = 6$)	fixed	27.90	46.95	34.12
Latent Class	random	27.06	46.48	33.42
($K = 7$)	fixed	26.32	47.42	33.73
Latent Class	random	26.29	46.48	33.42
($K = 8$)	fixed	26.24	46.95	33.26
Latent Class	random	25.42	51.17	38.11
($K = 9$)	fixed	25.43	50.70	37.72
Latent Class	random	25.34	50.70	37.64
($K = 10$)	fixed	24.92	49.77	36.39

Table 2: Performance on holdout data (Mean squared error, First choice hits, First choice hits corrected for tied preferences)

the model predicted the highest utility for one of the brand-price combinations actually rated highest. In the last column of Table 2 we report a first choice hit rate that corrects for tied holdout evaluations. When any consumer rates n of the five holdout combinations equally highest the correct prediction of one of these is weighted by the factor $1/n$. The holdout performance of the various models basically confirms the decisions based on the model likelihoods. We find a steady decrease in MSE when going from one to nine classes. Further increasing the number of classes to ten yields marginally smaller MSE values but decreases first choice hit rates. However, the increase in model likelihood as a consequence of assuming homogeneous $p^2\cdot$ brand interactions is not reflected in the holdout analyses. Interestingly, the relationship between the number of classes K, the associated model likelihood (see Table 1) and the first choice hit rates is not strictly monotonic in Table 2. This might be due to the implicit change in the loss function when predicting first choices. Another possible reason is that we based our point estimates of consumer utilities on 100 simulations per consumer, only.

3.6 Interpretation of the Nine Classes Model from a Marketing Point of View

An obvious starting point for the interpretation of the classes would be the class specific parameter estimates that can be obtained from the authors. However, due to the dimensionality of the designmatrix and the presence of quadratic price and interaction effects it is not an easy task to derive a coherent interpretation. Therefore, we settled for the following procedure: Given the class specific parameters we formulated three designs. The three designs offered all five brands at a low price (ATS 2.7), a medium price (ATS 4.8) and a high price (ATS 6.9), respectively. Then we computed the purchase likelihood ratings to be expected in the 9 classes for all three designs. Figure 5 illustrates the result.

Class 7 with the highest a posteriori size of all classes (approximately 33%) and class 1 are very price sensitive with only minor brand differentiation. Whereas class 7 would still accept a medium price class 1 shows a strong tendency to avoid all offers but the cheapest. Interestingly there is some differentiation between brands offered at the lowest price in class 1. The dummy brand Kronsteiner is evaluated less favourably. Classes 2, 4 and 8 are moderately price sensitive. Again, there is little brand differentiation in class 8 with a slight advantage of Römerquelle over its competitors. Class 2 clearly dislikes the Juvina brand and prefers Römerquelle, Vöslauer and Waldquelle to the dummy brand Kronsteiner. A price increase seems to affect Römerquelle to a lesser extent than Vöslauer and Waldquelle in this class. Class 4 prefers Römerquelle and Vöslauer to the other brands and clearly disapproves of the Kronsteiner brand. Class 3 again favours Römerquelle and Vöslauer over the other brands. Interestingly the advantage of Römerquelle and Vöslauer diminishes substantively at higher price levels. Classes 5, 6 and 9 reveal only little sensitivity to price. In the case of the very small class 5 the Juvina brand even is evaluated more favourably at the higher price levels. Moreover this class is the only one to clearly reject the Römerquelle brand. Finally, class 6 again favours Römerquelle and Vöslauer and clearly disapproves of Juvina. Also, the dummy brand Kronsteiner is preferred to the established brand Waldquelle in this class.

Overall, a major portion of consumers seems to be very price sensitive with only little brand differentiation. Despite intensive marketing activity especially by Römerquelle and Vöslauer these consumers behave like in a commodity market. Even an up to the interview unknown dummy brand would be readily accepted. Römerquelle and Vöslauer seem to be generally accepted (with the exception of Römerquelle in one very small class). Juvina and Waldquelle did not succeed in establishing classes that favoured their brand over the competitors.

168

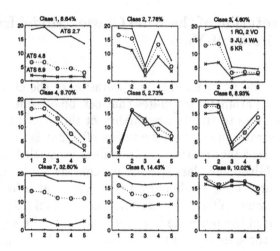

Figure 5: Class specific purchase likelihoods for three designs: all brands 2.7, 4.8 and 6.9 ATS; 1:RQ, 2:VOE, 3:JU, 4:WA, 5:KR

4   Concluding Remarks

The suggested approach was applied to data from a brand-price trade off conjoint study. We did not compare the latent class model to a parametric random effects model which is beyond the scope of this paper. The commonly used parametric random effects model assumes multivariate normally distributed random effects *a priori*. However, we would like to add that this model outperformed even the optimal latent class solution with nine classes in terms of the model likelihood. Nevertheless, the latent class approach could be used in a similar way as nonparametric maximum likelihood estimation to parsimoniously account for multimodal preference distributions. The characteristics of such a distribution can be sensibly described *a posteriori* without the need to identify a unique labelling. Finally, this paper should also be useful in the context of Bayesian estimation of general finite mixtures that in a way combine the latent class approach and parametric models for consumer heterogeneity (Allenby *et al.*, 1998; Lenk and DeSarbo, 1999).

Acknowledgments

Special thanks go to Josef Mazanec, Günter Schweiger, and Römerquelle Ges.m.b.H. This piece of research was supported by the Austrian Science Foundation (FWF) under grant SFB 010 ('Adaptive Information Systems and Modelling in Economics and Management Science') and Project No. 12025.

References

ALLENBY, G. M., ARORA, N. and GINTER, J. L. (1998): On the Heterogeneity of Demand. *Journal of Marketing Research, Vol. 35, 384-89.*

BERNARDO, J. M. and Smith, A. F. M. (1994), Bayesian Theory. Wiley, New York

FRÜHWIRTH-SCHNATTER, S. (2001): Markov Chain Monte Carlo Estimation of Classical and Dynamic Switching and Mixture Models. *Journal of the American Statistical Association, Vol. 96, 194-209.*

FRÜHWIRTH-SCHNATTER, S. (2000): Model Likelihoods and Bayes Factors for Switching and Mixture Models. Preprint, Vienna University of Economics and Business Administration.

FRÜHWIRTH-SCHNATTER, S. and OTTER, Th. (1999): Conjoint-Analysis Using Mixed Effect Models, in Friedl, H., Berghold, A. and Kauermann, G. (Eds.): Statistical Modelling, Proceedings of the Fourteenth International Workshop on Statistical Modelling. Graz, 181-191.

FRÜHWIRTH-SCHNATTER, S., OTTER, Th., and TÜCHLER, R. (2000): A Fully Bayesian Analysis of Multivariate Latent Class Models with Applications to Metric Conjoint Analysis. Technical report 2000-74, Department of Statistics, Vienna University of Economics and Business Administration.

LENK, P. J. and DE SARBO, W. S. (2000): Bayesian Inference for Finite Mixtures of Generalized Linear Models with Random Effects. *Psychometrika, Vol. 65, 93-119.*

MENG, X. L. and WONG, W. H. (1996): Simulating Ratios of Normalising Constants via a Simple Identidy. *Statistica Sinica, Vol. 6, 831-60.*

STREBINGER, A., SCHWEIGER, G., and OTTER, Th. (1998), Brand Equity and Consumer Information Processing: A Proposed Model, paper presented at AMA's Marketing Exchange Colloquium, Vienna.

WEDEL, M. and KAMAKURA, W. A. (1998): Market Segmentation: Conceptual and Methodological Foundations. Kluwer, Boston

Diagnostics in Multivariate Data Analysis: Sensitivity Analysis for Principal Components and Canonical Correlations

Y. Tanaka[1], F. Zhang[2], W. Yang[3]

[1] Okayama University, Faculty of Environmental Science & Technology, Tshusima, Okayama 700-8530, Japan

[2] Bellsystem24, Inc., 2-16-8 Minami-Ikebukuro, Toshima-Ku, Tokyo 171-0022, Japan

[3] Okayama University, Graduate School of Natural Science and Technology, Tshusima, Okayama 700-8530, Japan

Abstract: Sensitivity analysis procedures are formulated for principal component and canonical correlation analyses based on Cook's local influence(Cook, 1986). The relationships are discussed between the results of the procedures based on the local influence and those based on the influence functions. A numerical example is shown to illustrate the procedure for canonical correlation analysis.

1 Introduction

Multivariate data analysis (MDA) provides powerful tools for evaluating the relationship among variables and/or observations. However, it is well known that the result of MDA is sometimes affected seriously by a small number of outlying observations and therefore it is dangerous to interpret or use the result without checking its reliability or stability. Thus it is valuable to develop methodologies of sensitivity analysis (SA) in MDA.

There are two major approaches for SA in MDA. One is the influence function or influence curve(Hampel, 1974), which is exclusively used for evaluating the influence of observations. The other is Cook(1986)'s local influence, which gives a wider framework for SA. It can deal with not only the influence of observations but also the influence of any perturbation on general perturbation parameters. At first the former approach was adopted for developing SA procedures in various MDA techniques, but later the latter approach became the mainstream of SA. It is unquestioned that the latter provides a more general framework than the former. But, if our concern is to evaluate the influence of observations, we can show that we can get essentially the same information by adopting any one of these two approaches(Tanaka, 1994; Tanaka and Zhang, 1999; Zhang and Tanaka, 2001). In the present paper we formulate SA procedures in principal component analysis (PCA) and canonical correlation analysis (CCA) based on the idea of Cook's local influence, and discuss their relations with those based on the influence function approach.

2 Influence Function Approach

Suppose that we have p-dimensional observation vectors $\underline{x}_i$, $i = 1, \cdots, n$, which are taken independently from a population with the cumulative distribution function(cdf) $F(\underline{x})$, and that we consider a statistical model with an m-dimensional parameter vector $\underline{\theta}$. We often meet parameters which are not in vector forms, like covariance matrices. But, in such cases we can rewrite them in vector forms by applying appropriate operators such as vec and $vech$(vector-half) (see, e.g., Searle, 1982). In SA we are interested in the influence of each observation on estimate $\hat{\underline{\theta}}$ of the parameter vector, and the empirical influence function(EIF) can be used to measure the influence of each observation on $\hat{\underline{\theta}}$. It is defined as

$$EIF(\underline{x}_i; \hat{\underline{\theta}}) = \lim_{\epsilon \to 0} [\underline{\theta}((1 - \epsilon)\hat{F} + \epsilon \delta_{\underline{x}_i}) - \underline{\theta}(\hat{F})]/\epsilon,$$

assuming that the parameter vector is expressed by a functional of the cdf, where $\hat{F}$ is the empirical cdf and $\delta_{\underline{x}_i}$ indicates the cdf of a unit point mass at $\underline{x}_i$. This assumption is satisfied for most parameters in MDA. The above expression with $\underline{x}_i$, $\hat{\underline{\theta}}$ and $\hat{F}$ replaced by $\underline{x}$, $\underline{\theta}$ and F gives the definition of the theoretical influence function(TIF). It is well-known that the EIFs for the sample mean vector $\hat{\underline{\mu}} = \bar{\underline{x}} \equiv n^{-1}\Sigma_i \underline{x}_i$ and the sample covariance matrix $\hat{\Sigma} = S \equiv n^{-1}\Sigma_i(\underline{x}_i - \bar{\underline{x}})(\underline{x}_i - \bar{\underline{x}})^T$ are given by

$$EIF(\underline{x}_i; \hat{\underline{\mu}}) = \underline{x}_i - \bar{\underline{x}}, \quad EIF(\underline{x}_i; \hat{\Sigma}) = (\underline{x}_i - \bar{\underline{x}})(\underline{x}_i - \bar{\underline{x}})^T - S,$$

respectively(see, e.g., Cook and Weisberg, 1982).

The influence of observations can also be evaluated by introducing the so-called case-weight perturbation. Let us consider the following two types of case-weight perturbation.

(i) Type 1

Case-weights, which are assigned to n observations, are changed from $\underline{w}_0 = (1, 1, \cdots, 1)^T$ (unperturbed) to $\underline{w} = (w_1, \cdots, w_n)^T$ (perturbed), and the log likelihood function $L(\underline{\theta}) = \Sigma_i L_i(\underline{\theta})$ is changed to $L(\underline{\theta}|\underline{w}) = \Sigma_i w_i L_i(\underline{\theta})$, where $L_i(\underline{\theta})$ is the log likelihood for observation $\underline{x}_i$. In other words, when we can assume the normality of the distribution, we may also consider that the covariance matrices are changed from $V(\underline{x}_i) = \Sigma$ to $V(\underline{x}_i) = w_i^{-1}\Sigma$, $i = 1, \cdots, n$. Then, the log likelihood function satisfies the above condition and therefore both perturbations have the same effect on the estimate. In particular, the partial derivatives of the maximum likelihood(ML) estimates $\hat{\underline{\mu}}$ and $\hat{\Sigma}$ with respect to w_i are given by

$$\partial \hat{\underline{\mu}}_w / \partial w_i \Big|_{w_0} = n^{-1}(\underline{x}_i - \bar{\underline{x}}), \quad \partial \hat{\Sigma}_w / \partial w_i \Big|_{w_0} = n^{-1}(\underline{x}_i - \bar{\underline{x}})(\underline{x}_i - \bar{\underline{x}})^T.$$

In Belsley et al.(1980, p.25) the partial derivatives with respect to w_i are called "influence functions".

(ii) Type 2

Case-weights are changed from $\underline{w}_0 = (1, 1, \cdots, 1)^T$ (unperturbed) to $(nw_1/\sum_j w_j, \cdots, nw_n/\sum_j w_j)$(perturbed), or the covariance matrices are changed from $V(\underline{x}_i) = \Sigma$ to $V(\underline{x}_i) = (nw_i/\sum_j w_j)^{-1}\Sigma$ (Kwan and Fung, 1988). This type of perturbation is characterized that the sum of case-weights is unchanged by the perturbation. In this case the partial derivatives of $\hat{\underline{\mu}}$ and $\hat{\Sigma}$ with respect to w_i are given by

$$\left.\frac{\partial \hat{\underline{\mu}}_w}{\partial w_i}\right|_{w_0} = n^{-1}(\underline{x}_i - \bar{\underline{x}}), \quad \left.\frac{\partial \hat{\Sigma}_w}{\partial w_i}\right|_{w_0} = n^{-1}\left\{(\underline{x}_i - \bar{\underline{x}})(\underline{x}_i - \bar{\underline{x}})^T - S\right\}.$$

It is noticed that the partial derivatives in the above are n^{-1} times the EIFs for $\hat{\mu}$ and $\hat{\Sigma}$. Similar relations hold between partial derivatives and EIFs for most parameters in MDA.

In the present paper we use the term "influence function" in broad sense so that it also contains the partial derivatives $\partial \underline{\theta}_w/\partial w_i$ for type 1 and type 2 case-weight perturbations, and we simply write $\hat{\underline{\theta}}_i^{(1)}$ as the influence function vector for the i-th observation. For so-called single-case diagnostics scalar influence measures in the form of $\hat{\underline{\theta}}_i^{(1)T} M \hat{\underline{\theta}}_i^{(1)}$ are often used, where M indicates an appropriate metric. Among them Cook's D defined by

$$D_i = \hat{\underline{\theta}}_i^{(1)T} V^{-1} \hat{\underline{\theta}}_i^{(1)},$$

may be the most popular measure.

In most MDA techniques it is known that an approximation formula

$$\hat{\underline{\theta}}_{(A)} \cong \hat{\underline{\theta}} - c \sum_{\underline{x}_i \in A} \hat{\underline{\theta}}_i^{(1)}$$

holds, where $\hat{\underline{\theta}}_{(A)}$ indicates the estimate based on the sample without subset A of observations and c indicates a constant. On the basis of this additivity property Tanaka and his coworkers have proposed a general procedure of sensitivity analysis for detecting jointly as well as singly influential observations in various techniques of MDA(see, e.g., Tanaka and Zhang, 1999). The basic idea is to find observations which are located far and on similar directions from the origin in the influence function space and regard them as candidates for influential subsets of observations. For this purpose they reduce the dimensionality by applying PCA with metric V^- to the influence functions, where V indicates a consistent estimate for $acov(\hat{\underline{\theta}})$, the asymptotic covariance matrix of $\hat{\underline{\theta}}$. The PCs are obtained by solving the eigenvalue problem(EVP)

$$\left\{ n^{-1} \sum_{i=1}^{n} \hat{\underline{\theta}}_i^{(1)} \hat{\underline{\theta}}_i^{(1)T} - \eta' V \right\} \underline{a} = 0. \tag{1}$$

Let $\underline{a}_j$ be the eigenvector associated with the j-th largest eigenvalue η_j', and let $z_{ji} = \underline{a}_j^T \hat{\underline{\theta}}_i^{(1)}$ be the j-th PC score of $\hat{\underline{\theta}}_i^{(1)}$. Then it can be verified that the relation $D_i = z_{1i}^2 + \cdots + z_{mi}^2$ holds between Cook's D_i and PC scores $(z_{1i}, \cdots, z_{mi})$. It suggests that PCA of the influence functions provides the information of the multidimensional structure of the influence which is summarized into a scalar measure D_i in the single-case diagnostics. If V is degenerated, we may use a g-inverse V^- or Moore-Penrose inverse V^+ instead of V^{-1}. In such cases D_i is decomposed into $D_i = z_{1i}^2 + \cdots + z_{m'i}^2$, where m' indicates the rank of V.

3   Local Influence Approach

Let $L(\underline{\theta})$ be the log likelihood function and $\hat{\underline{\theta}}$ be the ML estimate. Cook(1986) uses an $r \times 1$ vector $\underline{w}$ to represent a wider class of perturbation to the model or observations. If we restrict our concern on the influence of observations, we may fix as $r = n$. Let $L(\underline{\theta}|\underline{w})$ be the perturbed log likelihood function and $\hat{\underline{\theta}}_w$ be the perturbed ML estimate. Let $\underline{w}_0$ denote the unperturbed weight vector and $L(\underline{\theta}) = L(\underline{\theta}|\underline{w}_0)$ denote the unperturbed log likelihood function. He has proposed a general method to evaluate the local influence of the perturbation parameters $\underline{w}$ as below.

The parameters are changed from $\hat{\underline{\theta}}$ to $\hat{\underline{\theta}}_w$ due to the perturbation from $\underline{w}_0$ to $\underline{w}$, and the amount of the change is measured with the likelihood displacement $D(\underline{w}) = 2\left[L(\hat{\underline{\theta}}|\underline{w}_0) - L(\hat{\underline{\theta}}_w|\underline{w}_0)\right]$. A graph of $D(\underline{w})$ versus $\underline{w}$, $\underline{\alpha}(\underline{w}) = (\underline{w}^T, D(\underline{w}))^T$, is introduced and called "influence graph". This graph reflects the effect of any perturbation imposed on $\underline{w}$. But, when the dimension of $\underline{w}$ is high, it is difficult to study the graph directly. Instead, a perturbation $\underline{w}_0 \to \underline{w} = \underline{w}_0 + t\underline{d}$, $\|\underline{d}\| = 1$, along an arbitrary direction $\underline{d}$ is introduced, a curve $(t, D(\underline{w}_0 + t\underline{d}))$ is studied, and the direction $\underline{d}_{max}$ is searched, where the curvature C_d of the curve at $t = 0$, which is equivalent to the normal curvature of the surface at $\underline{w}_0$, takes the largest value. The normal curvature is given by

$$C_d = 2 \left| \underline{d}^T \left[\partial \hat{\underline{\theta}}_w^T / \partial \underline{w} \right] (-\ddot{L}) \left[\partial \hat{\underline{\theta}}_w / \partial \underline{w}^T \right] \underline{d} \right|,$$

where $\ddot{L} = \partial^2 L/\partial\underline{\theta}\partial\underline{\theta}^T|_{\hat{\theta}}$, and the most influential direction $\underline{d}_{max}$ is obtained as the eigenvector $\underline{d}$ associated with the largest eigenvalue η_1 of the EVP

$$\left(2\left[\partial\hat{\underline{\theta}}_w^T/\partial\underline{w}\right](-\ddot{L})\left[\partial\hat{\underline{\theta}}_w/\partial\underline{w}^T\right] - \eta I\right)\underline{d} = \underline{0}. \tag{2}$$

In case we are interested in a subset of m_1 parameters $\underline{\theta}_1$ among $\underline{\theta}^T = (\underline{\theta}_1^T, \underline{\theta}_2^T)$, similar discussions can be made by using the profile log likelihood function. The likelihood displacement is defined by $D_s(\underline{w}) = 2\left[L(\hat{\underline{\theta}}|\underline{w}_0) - L(\hat{\underline{\theta}}_{1w}, \underline{\theta}_2(\hat{\underline{\theta}}_{1w})|\underline{w}_0)\right]$, where $\hat{\underline{\theta}}_{1w}$ is defined by partition $\hat{\underline{\theta}}_w^T = (\hat{\theta}_{1w}^T, \hat{\theta}_{2w}^T)$ and $\underline{\theta}_2(\underline{\theta}_1)$ is the value of $\underline{\theta}_2$ which maximizes $L(\underline{\theta}_1, \underline{\theta}_2(\underline{\theta}_1)|\underline{w}_0)$ for fixed $\underline{\theta}_1$, and the influence graph is defined by $\underline{\alpha}_s(\underline{w}) = (\underline{w}^T, D_s(\underline{w}))^T$. In this case the normal curvature at $\underline{w}_0$ is given by

$$C_d(\underline{\theta}_1) = 2\left|\underline{d}^T\left[\partial\hat{\underline{\theta}}_{1w}^T/\partial\underline{w}\right](-L_{11\cdot2})\left[\partial\hat{\underline{\theta}}_{1w}/\partial\underline{w}^T\right]\underline{d}\right|,$$

where $L_{11\cdot2}^{-1}$ is the upper left part of $\ddot{L}^{-1}$, which corresponds to $\underline{\theta}_1$, and it is known from the theory of ML estimation that $-L_{11\cdot2}^{-1}$ gives a consistent estimate V_{11} for $acov(\hat{\underline{\theta}}_1)$.

Suppose that there exist equality constraints $h_j(\underline{\theta}) = 0$, $j = 1, \cdots, r$ among parameters and that the constraints do not depend on the perturbation parameters $\underline{w}$. Also suppose that we are interested in parameters $\underline{\theta}_1$ among $\underline{\theta} = (\underline{\theta}_1^T, \underline{\theta}_2^T)^T$. Then, as shown by Tanaka and Zhang(1999) and Zhang and Tanaka(2001), the normal curvature is given by

$$C_d(\underline{\theta}_1) = 2\left|\underline{d}^T[\partial\hat{\underline{\theta}}_{1w}^T/\partial\underline{w}]V_{11}^-[\partial\hat{\underline{\theta}}_{1w}/\partial\underline{w}^T]\underline{d}\right|,$$

under the assumption that the subspace spanned by $\partial\hat{\underline{\theta}}_{1w}/\partial w_i$ is contained in the column space of V_{11}, a consistent estimate for $acov(\hat{\underline{\theta}}_1)$.

4 Relationship Between the Influence Function Approach and the Local Influence Approach

The relationship between the two approaches have been discussed by Tanaka (1994), Lee and Wang(1996), Tanaka and Zhang(1999), Zhang and Tanaka(2001) among others. Tanaka(1994) and Lee and Wang(1996) discuss the case where the statistical model does not contain any constraint among parameters and we are interested in all parameters. Tanaka and Zhang(1999) and Zhang and Tanaka(2001) discuss the relationship in detail in each of the four cases defined by whether we are interested in all parameters or a subset of parameters and whether there exist equality constraints or not. They show that the two approaches provide essentially equivalent information about the influence of observations in all cases. As an example, in the case where there exists no equality constraint and we are interested in all parameters, the following relations hold between the eigenvalues and eigenvectors of two EVPs (1) and (2).

$$\eta_j = (2n)\eta_j', \ \sqrt{n\eta'}\underline{d}_j = (\underline{a}_j^T\hat{\underline{\theta}}_1^{(1)}, \cdots, \underline{a}_j^T\hat{\underline{\theta}}_n^{(1)})^T, \tag{3}$$

where $\underline{a}_j^T\hat{\underline{\theta}}_i^{(1)}$ indicates the j-th PC score of the influence function $\hat{\underline{\theta}}_i^{(1)} = \partial\hat{\underline{\theta}}_w/\partial w_i$ for the i-th observation. Similar relations hold in other cases where there exist equality constraints among parameters and/or where we are interested in a subset of parameters.

5 Sensitivity Analysis in PCA

SA procedures have been proposed for PCA by Radhakrishnan and Kshirsagar(1981), Critchley(1985), Tanaka(1988) among others using the influence function approach. They derived influence functions for the eigenvalues, eigenvectors and some statistics characterizing the subspace spanned by the PCs. Shi(1997) proposed a modified Cook's local influence in PCA, and Tanaka et al.(1998) gave a formulation based on the original Cook's local influence. Here we briefly discuss the formulation based on Cook's local influence and its relation with the result of the influence function approach.

Let $\underline{x}_1, \cdots \underline{x}_n$ be n independent p-dimensional random vectors such that $\underline{x}_i \sim N(\mu, w_i^{*-1}\Sigma)$, $w_i^* = nw_i/\sum_j w_j$, $i = 1, ...n$. Let the covariance matrix be decomposed as $\Sigma = \Gamma\Lambda\Gamma^T$, where $\Lambda = diag(\lambda_1, \cdots \lambda_p), \Gamma = (\underline{\gamma}_1, \cdots, \underline{\gamma}_p)$, λ_j and $\underline{\gamma}_j$ indicating the j-th eigenvalue and the associated eigenvector of Σ, respectively. Then the profile log likelihood function for $\underline{\theta} = (\Lambda, \Gamma)$ can be expressed by

$$L(\underline{\theta}|\underline{w}) = -\frac{n}{2}\sum_i (\log\lambda_i + \underline{\gamma}_i^T S_w \underline{\gamma}_i/\lambda_i) + \frac{p}{2}\sum_i \log(nw_i/\sum_j w_j) - \frac{pn}{2}\log 2\pi,$$

with constraints

$$\underline{h}(\underline{\theta}) = (\underline{\gamma}_1^T\underline{\gamma}_1 - 1, 2\underline{\gamma}_1^T\underline{\gamma}_2, \cdots, 2\underline{\gamma}_1^T\underline{\gamma}_p; \cdots; \underline{\gamma}_p^T\underline{\gamma}_p - 1)^T = \underline{0}.$$

Suppose, for example, we are interested in $\underline{\gamma}_j$. Then, after some algebraic computation the normal curvature can be derived as

$$C_d(\underline{\gamma}_j) = 2\left|\underline{d}^T\left(\frac{\partial\hat{\underline{\gamma}}_{jw}^T}{\partial\underline{w}}\right)\left[\sum_{k \neq j}^p \frac{(\hat{\lambda}_j - \hat{\lambda}_k)^2}{\hat{\lambda}_j\hat{\lambda}_k}\hat{\underline{\gamma}}_k\hat{\underline{\gamma}}_k^T - 2\hat{\underline{\gamma}}_j\hat{\underline{\gamma}}_j^T\right]\left(\frac{\partial\hat{\underline{\gamma}}_{jw}}{\partial\underline{w}^T}\right)\underline{d}\right|,$$

and it can be verified that the quantity in the bracket is a g-inverse of a consistent estimate for the asymptotic covariance matrix for $\hat{\underline{\gamma}}_j$ given by

$$acov(\hat{\underline{\gamma}}_j) = n^{-1} \sum_{k \neq j} \lambda_j \lambda_k (\lambda_j - \lambda_k)^{-2} \underline{\gamma}_k \underline{\gamma}_k^T$$

(see, Siotani et al., 1985, p.453). Therefore, the influential directions $\underline{d}_j$ are given by the PC scores of the PCA with metric V^- of $\{\partial \hat{\underline{\gamma}}_{jw}/\partial w_i,\ i = 1, \cdots, n\}$ using the plug-in estimate V for $acov(\hat{\underline{\gamma}}_j)$. Thus, we can get the information in Cook's local influence by applying the influence function approach described in section 2 in the case where we are interested in $\hat{\underline{\gamma}}_j$. Also we can verify the equivalence of both approaches in general cases using a similar technique as shown in the next section.

6 Sensitivity Analysis in CCA

There are only a small number of papers which discuss SA in CCA. Radhakrishnan and Kshirsagar(1981), Romanazzi(1992) derived influence functions of important parameters in CCA. Gu and Fung(1998) derived $\partial \hat{\underline{\theta}}_w/\partial \underline{w}^T$ and proposed as a diagnostic statistic the direction which maximizes some distance of $\partial \hat{\underline{\theta}}_w/\partial \underline{w}^T$ other than the likelihood distance. In the present paper we consider an SA procedure based on the original Cook's local influence and discuss the relation with the result of the influence function approach.

Let $(\underline{x}_1^T, \underline{y}_1^T), \cdots, (\underline{x}_n^T, \underline{y}_n^T)$ be n sets of p- and q-dimensional random vectors such that $(\underline{x}_i^T, \underline{y}_i^T)^T \sim N(\underline{\mu}, w_i^{*-1}\Sigma)$, $w_i^* = nw_i/\sum_j w_j$, $i = 1, ..., n$. Let $\underline{\theta} = vech(\Sigma)$ and denote the set of eigenvalues(squared canonical correlations) and canonical vectors by φ. It is known that φ is ∞ times differentiable with respect to $\underline{\theta}$(see, Magnus and Neudecker, 1988, p.158). Then, the maximum likelihood estimation can be formulated by the minimization problem of $G = -L(\underline{\theta}) + \underline{\nu}^T(\varphi - g(\underline{\theta}))$, where $L(\underline{\theta})$ is the profile log likelihood function for the covariances and φ is a vector which consists of the eigenvalues of interest and/or normalized eigenvectors of interest and $g(\underline{\theta})$ is defined in the neighborhood of $\hat{\underline{\theta}}$. As the ML estimates we obtain $\hat{\underline{\theta}}$(ordinary ML estimates for the variance and covariances), $\hat{\varphi} = g(\hat{\underline{\theta}})$(eigenvalues and vectors as functions of the ML estimates of the covariance matrix) and $\hat{\nu} = 0$. Using the second derivatives of G with respect to $\underline{\theta}, \varphi$ and $\underline{\nu}$ the normal curvature is given by

$$C_d(\varphi) = 2 \left| \underline{d}^T \left(\frac{\partial \hat{\varphi}^T}{\partial \underline{w}} \right) \left(\frac{\partial \hat{\varphi}}{\partial \underline{\theta}^T} (-\ddot{L}^{-1}) \frac{\partial \hat{\varphi}^T}{\partial \underline{\theta}} \right)^- \left(\frac{\partial \hat{\varphi}}{\partial \underline{w}^T} \right) \right|.$$

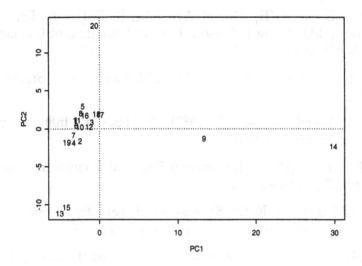

Figure 1: Scatter plot of PC2 vs. PC1

Since $\left(\partial\hat{\varphi}/\partial\underline{\theta}^T\right)\left(-\ddot{L}^{-1}\right)\left(\partial\hat{\varphi}^T/\partial\underline{\theta}\right)$ gives a consistent estimate $\widehat{acov}(\hat{\varphi})$ for the asymptotic covariance matrix of $\hat{\varphi}$, we can get the information in Cook's local influence by applying PCA with metric $[\widehat{acov}(\hat{\varphi})]^-$ to the influence function vectors $\{\partial\hat{\varphi}/\partial w_i,\ i=1,\cdots n\}$.

7 Numerical Example

Now let us apply our method of sensitivity analysis to the CCA of the fitness measurements data which are used for illustrating CCA in the SAS manual(1990). The data set consists of the measurements of three variables of X and three variables of Y for 20 individuals. The obtained canonical correlations are $0.7956 > 0.2006 > 0.0073$ in order of the magnitudes. Suppose we are interested in the first canonical vector $(\underline{\alpha}_1)$ for the X variables. At the first stage of SA Cook's D is computed for each observation. The obtained D values are $2.48209(\#14) > 0.53806(\#20) > 0.50659(\#9) > 0.42254(\#13) > 0.40884(\#15) > \cdots$. This result suggests observation #14 as a candidate for singly influential observation. Then for multiple-case diagnostics PCA with metric V^- $(V = \widehat{acov}(\hat{\underline{\alpha}}_1))$ is applied to the influence functions for $\hat{\underline{\alpha}}_1$. The eigenvalues are $60.26973 > 22.99275 > 12.80658$. The PC scores are plotted in Fig. 1. Looking at this scatter plot, we can find observations $\{14,9\}$, $\{13,15\}$ and 20 as candidates for influential subsets of observations.

References

BELSLEY, D. A., KUH, E. and WELSCH, R. E. (1980): Regression Diagnostics : Identifying Influential Data and Scources of Collinearity. John Wiley & Sons.

COOK, R. D. (1986): Assessment of Local Influence. *J. R. Statist Soc.*, *B48, 133–169.*

COOK, R. D. and Weisberg, S. (1982): Residuals and Influence in Regression. Chapman and Hall.

CRITCHLEY, F. (1985): Influence in Principal Component Analysis. *Biometrika, 72, 627–636.*

GU, H. and FUNG, W. K. (1998): Assessing Local Influence in Canonical Analysis. *Ann. Inst. Statist. Math., 69, No. 4, 755–772.*

HAMPEL, F. R. (1974): The Influence Curve and its Role in Robust Estimation. *J. Amer. Statist. Assoc., 69, 383–393.*

KWAN, C. W. and FUNG, W. K. (1998): Assessing Local Influence for Specific Restricted Likelihood: Application to Factor Analysis. *Psychometrika, 63, 35–46.*

LEE, S. Y. and WANG, S. J. (1996): Sensitivity Analysis of Structured Equation Models. *Psychometrika, 61, 93–108.*

MAGNUS, J. R. and NEUDECKER, H. (1988): Matrix Differential Calculus with Applications in Statistics and Econometrics, John Wiley & Sons.

RADHAKRISHNAN, R. and KSHIRSAGER, A. M. (1981): Influence Functions for Certain Parameters in Multivariate Analysis. *Comm. Statist. -Theory and Method, 10, 515–529.*

ROMANAZZI, M. (1992): Influence in Canonical Correlation Analysis, *Psychometrika, 57, 237–259.*

SAS/STAT User's Guide (1990): Version 6, Volume 1, SAS Institute Inc. Cary, NC, USA

SEARLE, S. R. (1982): Matrix Algebra Useful for Statistics. John Wiley & Sons.

SHI, L. (1997): Local Influence in Principal Component Analysis. *Biometrika, 84, 175–186.*

SIOTANI, M., Hayakawa, T. and FUJIKOSHI, Y (1985): Modern Multivariate Statistical Analysis: A Graduate Course and Handbook. American Science Press, Inc.

TANAKA, Y. (1988): Sensitivity Analysis in Principal Component Analysis: Influence on the Subspace Spanned by Principal Components. *Comm. Statist. -Theory and Method, 17, 3157–3175.*

TANAKA, Y. (1994): Recent Advance Sensitivity Analysis in Multivariate Methods. *J. Jpn. Soc. Comp. Statist., 7, 1–25.*

TANAKA, Y. and Zhang, F. H. (1999): R-mode and Q-mode Influence Analysis in Statistical Modeling: Relationship between Influence Function Approach and Local Influence Approach *Comm. Statist. & Data Analysis, 32, 197–218.*

TANAKA, Y., Zhang, F. H. and Mori, Y. (1998): Influence in Principal Component Analysis Revisited, In Proceedings of the Third Conference on Statistical Computing of the Asian Regional Section of IASC, 319–330.

ZHANG, F. H. and TANAKA, Y. (2001): A Note on the Assessment of Local Influence in Statistical Models with Equality Constraints. *Technical Report, No. 74, Okayama Statistical Association, Okayama, Japan*

Classification of Natural Languages by Word Ordering Rule

S. Ueda, Y. Itoh

The Instiute of Statistical Mathematics,
4-6-7 Minami-Azabu Minatoku Tokyo 106-8569, Japan

Abstract:

In Tsunoda's table for 130 languages, the word ordering rule of each language is represented by 19 values for 19 qualitative variables. Making the hierarchical clustering from the table, the 130 languages are divided into two groups, (a) prepositional languages, and (b) postpositional languages and adpositionless languages. Assuming the adposition is the most important variable, the numeral and noun, is the second most important variable for the hierarchical clustering. It seems that each language in the world fluctuates between two structures, prepositional languages and postpositional languages.

1 Tsunoda's Table for Word Ordering Rule

In Tsunoda's table (1991) for 130 languages, the word ordering rule of each language is represented by 19 values for 19 qualitative variables (parameter in linguistic typology), taking Japanese as the standard for comparison for each of 130 languages (Table 1). The 19 parameters are taken based on the Greenberg's work (1966). He used the position of adpositions in seven of the forty five universals. With overwhelmingly greater than chance frequency, languages with normal SOV order are postpositional (Greenberg (1966)), which can be seen also from Tsunoda's table of 130 languages.

We make hierarchical clustering from Tsunoda's table and find that the 130 languages are divided into two groups, (a) prepositional languages, and (b) postpositional languages and adpositionless languages (Tsunoda, Ueda and Itoh (1994)), (Fig.1).

What parameters other than the adposition are working for the hierarchical clustering ? We use Sakamoto and Akaike method (1978) which is based on the idea of variable selection (Akaike (1973), Mallows (1973)). Assuming the parameter adposition is the most important variable, the parameter, numeral and noun, is the second most important variable for the hierarchical clustering.

We discuss a stochastic model, to understand Tsunoda's data, that each language in the world fluctuates between two structures, prepositional languages and postpositional languages.

No.	parameter	Japanese	English
1.	S, O and V	SOV, etc.	SVO
2.	Noun and Adposition	+	−
3.	Genitive and Noun	+	+, −
4.	Demonstrative and Noun	+	+
5.	Numeral and Noun	+	+
6.	Adjective and Noun	+	+
7.	Relative Clause and Noun	+	−
8.	Proper Noun and Common Noun	+	−, +
9.	Comparison of superiority	+	−
10.	Main Verb and Auxiliary Verb	+	−
11.	Adverb and Verb	before V	various
12.	Adverb and Adjective	+	+, −
13.	Question Marker	sentence-final	none
14.	S,V Inversion in General Questions	none	present
15.	Interrogative Word	declarative sentence type	sentence-initial
16.	S,V Inversion in Special Questions	none	present
17.	Negation Marker	verbal suffix	immediately after V
18.	Conditional Clause and Main Clause	+	+, −
19.	Purpose Clause and Main Clause	+	−

No.	German	Thai	Mandarin Chinese
1.	V 2nd(SVO,etc.)	SVO	SVO,etc
2.	−, +	−	+
3.	−, +	−, +	+
4.	+	-	+
5.	+	-	+
6.	+	−	+
7.	−	−	+
8.	−	−	+
9.	−	−	−,+
10.	−	−,+	−
11.	various	various	various
12.	+	−	+
13.	none	sentence-final, immediately after focus of question	sentence-final
14.	weakly present	none	none
15.	sentence-initial	declarative sentence type	declarative sentence type
16.	weakly present	none	none
17.	immediately before focus of negation;sentence-final,etc.	immediately before focus of negation	between S and V
18.	+	−, +	+
19.	−	−	unknown

Table 1. Table of parameters for five languages (Tsunoda (1991)).

2 Hierarchical Clustering and Adposition

We make hierarchical clustering by group average method digitizing the table. We begin with n clusters, each containing just one object, and $n \times n$ matrix (d_{rs}) of dissimilarities, for d_{rs} a measure of distance between two clusters r and s. Fuse the two nearest objects into a single cluster.

182

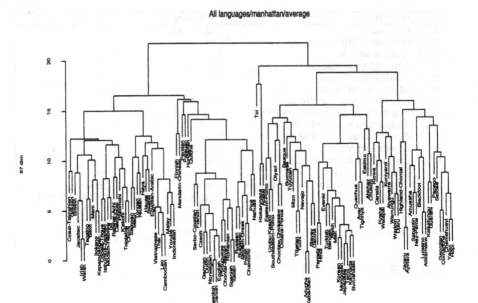

All languages/manhattan/average

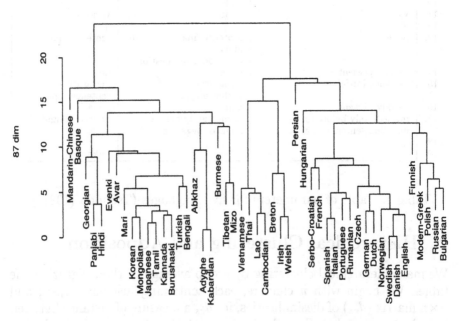

Eurasian languages(49)/manhattan/average

Figure 1: Hierarchical clustering (Tsunoda, Ueda, and Itoh (1994))

We now have $n-1$ clusters containing one object each and a single cluster of two objects. Next fuse the two nearest of the $n-1$ clusters to form $n-2$ clusters. Continue in this manner until at the $(n-1)$ th step. The distance for the group average method is defined by the arithmetic mean of the $n_1 n_2$ dissimilarities between all pairs, where C_1 and C_2 include n_1 and n_2 members, respectively.

We digitize the table as the following (Tsunoda, Ueda, and Itoh (1994)). The orders SOV, SVO, OSV, VSO, OVS, and VOS, are represented by 6-dimensional vector as $(1,0,0,0,0,0)$ for Japanese, which has 'SOV, etc.('etc.' is ignored), and $(0,1,0,0,0,0)$ for English, which has 'SVO'. For the 'parameter' 18, possible cases given in the table are $+$, $-$. For the parameter 18 in English, $(0.6, 0.4)$ is assigned. In this way we get for Japanese,

$$(1,0,0,0,0,0/1,0,0,0/1,0,0,0/1,0,0,0,0/1,0,0/1,0,0,0/$$
$$1,0,0/1,0/1,0,0,0/1,0,0/1,0,0,0,0/1,0,0/$$
$$1,0,0,0,0,0,0,0,0,0,0,0,0,0,0/1,0/1,0,0,0,0,0,0/$$
$$1,0/1,0,0,0,0,0,0,0,0,0,0,0,0,0/1,0/1,0)$$

for English,

$$(0,1,0,0,0,0/0,1,0,0/0.6,0.4,0,0/1,0,0,0,0/1,0,0/1,0,0,0/$$
$$0,1,0/0.4,0.6/0,1,0,0/0,1,0/0.2,0.2,0.2,0.2,0.2/0.6,0.4,0/$$
$$0,0,0,0,0,0,0,0,0,0,0,0,0,1,0/0,1/0,1,0,0,0,0/$$
$$0,1/0,0,1,0,0,0,0,0,0,0,0,0,0/0.6,0.4/0,1)$$

Manhattan distance between two languages i and j is defined where $x_{i,k,m}$ is the m th component for the parameter k of the language i and n_k is the number of coordinates for the parameter k, as

$$d(i,j) = \sum_{k=1}^{19} \sum_{m=1}^{n_k} \mid x_{i,k,m} - x_{j,k,m} \mid \leq 38. \tag{1}$$

The distances from Japanese are 28.2 to Thai, 27.2 to English, 25.8 to German, and 11.6 to Mandarin-Chinese. The distances from Thai are 20.0 to English, 20.2 to German, and 20.4 to Mandarin-Chinese. We use a software S language (Becker, Chambers and Wilks (1988)) to make Fig.1 where the missing values are treated in a reasonable way.

3 Classification of Parameters

What are the important parameters other than the adposition? At the distance 14.53, eliminating the clusters with the size not larger than 3, we get the 7 clusters with the size 21, 15, 22, 18, 19, 9, 15 respectively.

3) Name the 7 clusters (Fig.1) as Mi, which includes Middle American languages, as Th, which includes Thai and as En which includes European languages. as No, which includes North American languages, as Ja, which includes Japanese, as Au, which includes Australian languages, and as So, which includes South American languages.

We simplify the 87-dimensional to 66-dimensional vector. Give 2 or 3 values to each of the 19 parameters, if it is possible. The first element of a parameter of a language is 1 if it has the same word order to Japanese, the second element is 1 if it has the opposite word order to Japanese and otherwise the third element is 1. We put a numerical value between 0 and 1 based on the Tsunoda's table for the intermediate case as the parameter 18 of the English. For example the 6 cases SOV, SVO, OSV, VSO, OVS and VOS, are simplified to OV and VO. The cases +, −, adpositionless, and others for adposition are simplified to the cases +, − and others.

We analyze 71 languages without 'unknown' by the principal component analysis, making 38-dimensional vectors taking the values of the first two elements for each parameter. We calculate factor loadings to classify the parameters.

Let the group $G1$ be a set of variables of parameters 1, 2, 3, 7, 8, 9, 10, 19 with the factor loading (> 0.5), to the first principle component. Let the group $G2$ be a set of variables of parameters 4, 5, 6, 12, 14, 16, with the factor loading (> 0.5), to the second principal component Let the group $G3$ be a set of variables of parameters 3,8,14,15,16,18,19, which is the variables not included neither $G1$ nor $G2$, and the variables with the factor loading less than 0.7 and larger than 0.5 to the first principal component and the variables with the factor loading less than 0.7 and larger than 0.5 to the second principal component, where the parameters which have more than 3 elements in spite of the simplification are neglected.

4   Optimal Explanatory Parameters

The discrepancy for a model fitted to a set of observed data by the method of maximum lilelihood is evaluated by the statistics AIC defined by the following, taking the natural logarithm,

$$AIC = (-2)log(maximum \quad likelihood) + 2k$$

where k is the number of parameters (in statistics sense) in the model which are adjusted to attain the maximum of the likelihood.

Let I_0 be for a classification of languages and I_j $(1, ..., k)$ be for explanatory variables (parameters in linguistic typology). Let I_j $(j = 0, 1, ..., k)$

take values i_j ($i_j = 1, \cdots, C_j$) and let $n(i_0, i_1, ..., i_k)$ be the corresponding cell frequencies. The model that the response variable I_0 is explained by conditioning on explanatory variables I_j $(1, ..., k)$ as

$MODEL(I_0; I_1, ..., I_k)$: $p(i_0 \mid i_1, ..., i_k) = a(i_0 \mid i_1, ..., i_k)$

for $\quad i_j = 1, \cdots, C_j, \quad j = 0, 1, ..., k.$

The AIC for this model is

$AIC(I_0; I_1, ..., I_k) =$

$(-2) \sum_{i_0, i_1, ..., i_k} n(i_0, i_1, ..., i_k) \log \frac{n \cdot n(i_0, i_1, ..., i_k)}{n(i_0) n(i_1, ..., i_k)} + 2(C_0 - 1)(C_1 C_2 ... C_k - 1).$

If we assume that the information on I_j, $j = m+1, ...k$, can be ignored in the determination of the conditional probability $p(i_0 \mid i_1, ..., i_k)$ namely

$MODEL(I_0; I_1, ..., m)$: $p(i_0 \mid i_1, ..., i_k) = a(i_0 \mid i_1, ..., i_m)$

for $\quad i_j = 1, \cdots, C_j, \ j = 0, 1, ..., k$.

The AIC is

$AIC(I_0; I_1, ..., I_m) =$

$(-2) \sum_{i_0, i_1, ..., i_m} n(i_0, i_1, ..., i_m) \log \frac{n \cdot n(i_0, i_1, ..., i_m)}{n(i_0) n(i_1, ..., i_m)} + 2(C_0 - 1)(C_1 C_2 ... C_m - 1).$

Consider the response variable I_0 for a classification of languages and 19 explanatory variables. What is the optimal m? Minimize the AIC. We find a system of the response variable I_0 of values $No \cup Au$, $Ja \cup So$, Th, $Mi \cup Eu$ and explanatory variables ,I_1 for parameter 2 and, I_2 for parameter 5, gives $AIC(I_0; I_1, I_2) = -175.71$ (Table 2). Neither the model with one parameter nor the model with more than three parameters give better value of AIC, considering the four cases of taking parameters, i) one parameter from $G1$, ii) one from $G1$ and one from $G2$, iii) one from $G1$, one from $G2$ and one from $G3$, and iv) one from $G1$, one from $G2$ and two from $G3$.

parameter 2	parameter 5	4 groups(118 languages)			
		No.Au	Ja.So	Th	Mi.En
	+	4.56	30.6	0	1.6
postposition, none	−	19.04	2.4	0	0.0
	others	2.00	0.0	0	0.0
	+	0.24	1.0	5	38.4
preposition	−	0.16	0.0	10	0.8
	others	0.00	0.0	0	2.2

Table 2. Classification for the minimum AIC

5 Fluctuation Between Two Structures

We find the ternary relations from Tsunoda's table.

(1) If a language has preposition, and if the language is the VO language, then the genitive tends to follow the noun. If a language has postposition, and if it is the OV language, the genitive tends to precede the noun.

(2) If the main verb of a language follows the auxiliary verb, and if the relative clause follows the noun, the purpose clause tends to follow the noun. If the main verb of a language precedes the auxiliary verb, and if the relative clause precedes the noun, the purpose clause tends to precede the noun.

Let $y_1(j, L)$ and $y_2(j, L)$ be the first two elements of a parameter j of a language L. Put $y_1(j, L) = 1$ and $y_2(j, L) = 0$ for the same word order to Japanese. Let $y_1(j, L) = 0$ and $y_2(j, L) = 1$ for the opposite word order to Japanese. We put value betwween 0 and 1 for intermediate cases. Let $G1(L)$ be $G1$ without the parameters of NA (not available) of the language L. We define the measure of "postpositionness" of the language L as,

$$Po(L) = \frac{1}{|G1(L)|} \sum_{j \in G1(L)} (y_1(j, L) - y_2(j, L)).$$

where $| G1(L) |$ be the number of the element of the set $G1(L)$.

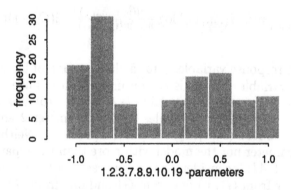

Figure 2: Postpositionness for the 130 languages (Ueda and Itoh (1995))

$Po(L) = 1$ for Japanese, $Po(L) = -0.9$ for Thai, and $Po(L) = 0.75$ for English. If we take Thai as the standard of comparison to make the table for word ordering rule, the corresponding histogram is bimodal. If we take Hungarian as the standard, the histogram is unimodal.

A pair of very similar word orders seems to appear just by chance. The distance between Tamil and Japanese is as small as the distance between Italian and Spanish. The changes in the values of the parameters of each language may depend on the values of the other parameters as in the above ternary relations (1), and (2). Consider a stochastic particle system of ternary interactions of 8 particles of two types + and −. At each step, three particles are taken at random. If two of the three particles are of the type +(−) and one is of the type −(+), the three particles become three particles of the type +(−). If the three particles taken are of the same type, no change takes place. A mutation of to the opposite sign takes place for each particle by some rate at each step. We continue the steps calculating the measure of postpositionness. The stationary probability fits the histogram in Fig.2. It seems that each language in the world fluctuates between two structures, prepositional languages and postpositional languages. We need more statistical evidence for this hypothesis.

References

AKAIKE, H. (1973) Information theory and an extension of the maximum likelihood principle, 2nd international symposium on information theory, (Petrov ,B.N. and Csaki, F. eds.) Akademiai Kiado, Budapest, 267-281.

BECKER, R.A. CHAMBERS, J.M. and WILKS, A.R.(1988) The New S Language, Pacific Grove, CA, Wadsworth.

GREENBERG, J.H.(1966) Some universals of grammar with particular reference to the order of meaningful elements, in J.H.Greenberg(ed.), Universals of Language, 73-113, MIT Press, Cambridge, Massachusetts.

ITOH, Y. and UEDA, S. (2000) The Ising model for changes in word ordering rule for natural languages, Research Memorandum 753, The Institute of Statistical Mathematics.

MALLOWS, C. L. (1973) Some comments on C_p, Technometrics, 15. 661-675.

SAKAMOTO, Y. and AKAIKE, H.(1978) Analysis of cross classified data by AIC, Ann. Inst. Statist. Math., 30, B, 185-197.

TSUNODA, T. (1991) The world languages and Japanese, Tokyo, Kuroshio (in Japanese).

TSUNODA, T., UEDA, S. and ITOH, Y.(1994) Adpositions in word order typology, LINGUISTICS, Vol.33, No.4, 741-761

UEDA, S. and ITOH, Y. (1995) The classification of languages by the two parameter model for word ordering rule, Proceedings of the Institute of Statistical Mathematics, 43. 341-365 (in Japanese with English summary).

Combining Mental Fit and Data Fit for Classification Rule Selection

C. Weihs, U. M. Sondhauss

Fachbereich Statistik and SFB 475 Universität Dortmund, D-44221
Dortmund, Germany

Abstract: Mental fit of classification rules is lately introduced to judge the adequacy of such rules for human understanding. This paper first discusses the various criteria introduced in relation to mental fit in the literature. Based on this, the paper derives a general criterion for the interpretability of partitions generated by classification rules. We introduce interpretability as a combination of mental fit and data fit, or more specifically, as a combination of comprehensibility and reliability of a partition. We introduce so-called prototypes to improve comprehensibility, and the so-called reliability of such prototypes as a measure of data fit.

1 Introduction

In the days of data mining, the number of competing classification techniques is growing steadily. Thus, it is a worthy goal to rate the classification rules from a wide range of techniques. An ideal measure for the selection of the best classification rule from candidates should combine two aspects, namely mental fit and data fit.

There are various possibilities to define data fit, which is often also called consistency or accuracy (cp. e.g. Hand (1997), p. 100). In principle we will think about data fit as predictive power measured by misclassification rate, in Section 3.2, however, we will derive another data fit measure.

In contrast, mental fit has much more recently come into play. Before we will derive a criterion for classification rule selection which combines mental fit and data fit we will first discuss the different concepts introduced in the literature to characterize mental fit. The main 'definitions' are:

- Classifiers should constitute explicit concepts meaningful to humans and evaluable directly in mind (Feng and Michie (1994)).

- The most important elements of mental fit are coverage, simplicity, and explainability (v. den Eijkel (1999)).

In the literature, comparisons of the mental fit of classification rules are often either too general or too method-specific. Surely, what is ideally needed is a unique accepted general formalization of mental fit which is always relevant and measurable. Typically, however, ratings of mental fit are inevitably method, project and customer dependent. For some

---

[1]This work has been supported by the Deutsche Forschungsgemeinschaft, Sonderforschungsbereich 475.

techniques it is nevertheless argued that they outperform all others with respect to certain aspects of mental fit as interpretability or comprehensibility in general, e.g. for production rules (Weiss and Kulikowski (1991)), decision trees (Michie et al. (1994)), and fuzzy systems (Bodenhofer and Bauer (1999)). Often however, such statements are related to specific concepts for mental fit. Examples for such 'local measures' of mental fit used with certain methods are 'low tree complexity' for decision trees (Breiman et al.(1984)), and 'small number of involved original features in discriminant coordinates' for discriminant analysis (Weihs (1992, 1993)).

In this paper we will first discuss various aspects of mental fit (section 2), and will then propose a general strategy to formalize mental fit locally w.r.t. certain representations. In that spirit, we introduce a new simple very general rule selection method for partitions, a typical way of interpreting classification rules. To this purpose we will standardize the representation of partitions and define performance criteria that can be calculated for a wide variety of techniques.

2 Formalizations of Mental Fit

2.1 Conceptual Partitioning

"The main advantage that rule induction offers for decision-making problems is what is sometimes called a mental fit to the problem" (v. den Eijkel (1999)). In contrast, it can be argued that many statistical and neural net techniques partition the feature space into one region per class by using some kind of discriminant function difficult to understand in form and content. Moreover, often the sole goal is to optimize the (predictive) accuracy of the classification.

Rule induction classification techniques like e.g. CART (Breiman et al. (1984)), however, partition the feature space into multiple regions per class, where each region is associated with one and only one class *and has a simple shape*. This way, the partitioning of feature space can be regarded as generating 'explicit knowledge' describing the data. Rule induction is thus concerned with finding explicit reasons for partitions in that the rules are derived as simple conditions of original features. This is often called conceptual partitioning or concept description of rules.

Here is an example for conceptual partitioning. In business cycle phase prediction (see e.g. Weihs et al. (1999)), we got the following partitions characterizing classification rules. By linear discriminant analysis one region per class was produced. In Figure 1, in addition, the course of projected observations into the 2D-discriminant space of a full business cycle not used for learning is indicated by connected arrows. *Solid* arrows represent correct classifications, *broken* ones misclassified examples.

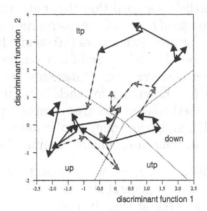

Figure 1: Linear Discriminant Analysis - one region per class

The problem with this partition is twofold. The axes are not interpretable, since they are arbitrary linear combinations of the original features, and the partition is even not easy to be described in discriminant
coordinates since the borders of the regions are not parallel to the axes.

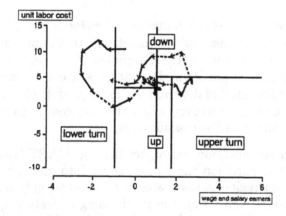

Figure 2: CART - multiple regions per class

Both problems do not exist for CART. In Figure 2 a corresponding partition is shown for two factors only. CART delivers multiple regions per
class. Here, two regions are found for the phases UPswing and DOWN-
swing of the business cycle. Since the region borders are parallel to
the axes representing original features, interpretation is easy. The illustration obviously indicates a classification problem near 0% increase of
WAGE AND SALARY EARNERS and a simultaneous 4% increase of
UNIT LABOR COSTS.

2.2 Simplicity

A second advantage of rule induction techniques comes from the partitioning process as well. The simplicity of rules, i.e. the low dimensionality of conditions, often intrinsically leads to dimension reduction by feature selection (v. den Eijkel (1999)). E.g., in decision trees with nearly the same data fit, a system of classification rules is chosen with low tree complexity. Here, pruning leads to a low number of features together with a low number of regions (Breiman et al. (1984)).

Corresponding simplicity criteria were developed for other methods too. For linear discriminant analysis, e.g., the discriminant functions may be simplified by minimizing the number of original features involved, retaining the discriminative power (Weihs (1992, 1993)). For neural networks a similar criterion is proposed. Network pruning aims at removing redundant links and nodes without increasing the error rate. A smaller number of nodes and links left in the network after pruning provides rules that are more concise and simple in describing the classification function (Lu et al. (1995)). A forefather of all these methods is the idea of finding classification rules with minimum description length (Rissanen (1978)). Here, the idea is obvious, the shorter the rule description, the easier to understand the rule, i.e. the bigger mental fit.

In Figure 3 you see an example for simplicity induction. The efficacy of a drug for impaired brain functions in geriatric patients is measured by a class score EFF at 4 levels determined by medical doctors, as well as by five efficacy measures (DBCRS, DIADLP, DSCAG, DSKTR, DZVTT). Linear discriminant analysis shows that one dimension can nearly perfectly describe the variation in the data (cp. Figure 3). This first discriminant function is of the following form: CD1 = 0.26*DBCRS - 0.10*DIADLP + 0.07*DSCAG + 0.02*DSKTR + 0.01*DZVTT.

Simplification of this function can be obtained by regressing the corresponding discriminant scores on the original predictors. Greedy forward selection first chooses DBCRS with $R^2 = 0.71$, then DSCAG leading to $R^2 = 0.85$, and at last DSKTR so that $R^2 = 0.95$. Linear discriminant analysis based on these predictors leads to a nearly equivalent simplified version of the first discriminant function, where only three predictors are involved.

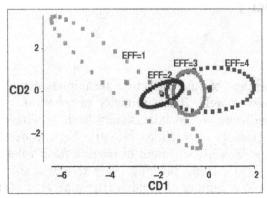

Figure 3: Linear Discriminant Analysis - confidence regions for classes

2.3 Coverage

Another term related to mental fit is coverage, often also called generality. Coverage of a partition is defined to be the higher the more powerful
the regions in a partition are, i.e. the more instances (observations, examples) the regions contain. A related concept is the height of (data)
density (or data non-sparseness) in the regions. Obviously, coverage influences simplicity.

2.4 Comprehensibility (Explainability)

This is the first really subjective measure in that comprehensibility is
(obviously) user dependent, since the rule should be understandable to
the user. Possible comprehensibility criteria for a rule formulated in
certain coordinates are:

- Original coordinates are preferred to latent ones,

- small number of coordinates to many coordinates,

- explicit rules to trees,

- trees to functions,

- conditions formulated as qualifications ('high', 'low') over thresholds '($<$, $>$)',

- conjunctions over conjunctions plus disjunctions.

- non-disjoint rules over disjoint rules, or vice versa, dependent on
  the application.

Other criteria might be important in specific applications.

2.5 Interestingness

A side step and at the same time an extension to mental fit is the term interestingness. There are two kinds of formalizations: objective and subjective ones.

Objective Interestingness Measures

For an association rule $Y \to X$, $p(X|Y)$ is called confidence of the rule. Note that we have $X = C =$ 'one of the classes' when using association rules for classification. The larger the confidence $p(C|Y)$, the more interesting the rule. Wang et al. (1998) propose a 'onesided' variant of the J-measure of Smyth and Goodman (1992) for interestingness, namely:

$$\mathbf{J_1}(C; Y) = \mathbf{p}(Y) \times \left[\mathbf{p}(C|Y)\mathbf{log_2}\left(\frac{\mathbf{p}(C|Y)}{\mathbf{p}(C)} \right) \right]$$

$$= \text{generality of rule} \times [\text{discrimination power of rule}].$$

A second interestingness measure is

$$\mathbf{J_2}(C; Y) = \mathbf{p}(Y)\left(\mathbf{p}(C|Y) \right) - \mathbf{miniconf})$$

where **miniconf** is a specified minimum confidence. Obviously, $\mathbf{J_2}(C; Y)$ is the 'surplus' of the association in the rule $Y \to C$ relative to the association level specified to be the minimum confidence.

The larger the value of $\mathbf{J_1}$, $\mathbf{J_2}$, the more interesting the rule. $\mathbf{J_1} < 0$ corresponds to negative association. $\mathbf{J_2} < 0$ corresponds to an association level below minimum confidence.

Subjective Interestingness Measures

Objective measures of interestingness, although useful in many respects, usually do not capture all the complexities of the rule discovery process. Therefore, subjective measures of interestingness are needed to define interestingness of a rule (Silberschatz and Tuzhilin (1996)). Subjective measures do not depend only on the structure of a rule and on the data used in the discovery process, but also on the beliefs of the user who examines the rule. Two major reasons why a rule is interesting from the subjective (user-oriented) point of view may be:

- unexpectedness - a rule is interesting when it is 'surprising' to the user,

- actionability - a rule is interesting if the user can act on it to his advantage.

Since actionability appears to be hard to be formalized, we will concentrate on unexpectedness and its relation to beliefs following Silberschatz and Tuzhilin (1996). They distinguish two kinds of beliefs: Hard beliefs are constraints that should not be changed with new evidence. Soft beliefs are beliefs the user is willing to change with new evidence.

One possibility of belief adjustment is using the Bayesian approach. In this case, the 'degree of a belief α' is defined as $\mathbf{P}(\alpha|\xi)$, given some previous evidence ξ, supporting that belief. With new evidence E, the update of the degree of belief in α, $\mathbf{P}(\alpha|E;\xi)$, is then obtained by using Bayes rule.

Interestingness of rules is then defined as follows. A rule is interesting relative to some belief system if it 'affects' this system, and the more it 'affects' it, the more interesting the rule is. If a rule contradicts the set of hard beliefs of the user then this rule is always interesting to the user. In case of soft beliefs, interestingness of rule $\mathbf{r}$ relative to a (soft) belief system $\mathbf{B}$ and previous evidence ξ is defined by

$$\mathbf{I}(\mathbf{r};\mathbf{B};\xi) = \sum_{\alpha_i \in \mathbf{B}} w_i \left| \mathbf{P}(\alpha_i|r;\xi) - \mathbf{P}(\alpha_i|\xi) \right|,$$

where w_i is the importance weight for the soft belief α_i in the belief system $\mathbf{B}$, and $\sum_{\alpha_i \in \mathbf{B}} w_i = 1$. This definition of interestingness measures of how much degrees of believe are changed as a result of a new rule $\mathbf{r}$.

Interestingness appears to be an extension to mental fit since mental fit is obviously a prerequisite to subjective interestingness, especially to the decision whether a hard belief is contradicted.

2.6 List of Important Aspects of Mental Fit

In summary, objective measures of mental fit are:

- Completeness, not yet mentioned, but obvious: the instance space should be described completely;

- Coverage: rules should be as powerful as possible;

- Simplicity: rules should concisely describe reality, the less variables, the better.

Subjective measures are:

- Comprehensibility: rules should be understandable to the user;

- Interestingness: the more unexpected the rule is in relation to a belief system, the better.

3   Interpretabiliy - a Meneral Approach to Rule Selection

Based on the above discussion of mental fit criteria in the literature, in what follows we would like to introduce interpretabiliy as the combination of mental fit and data fit. More specifically, we will define interpretability of classification rules as *Comprehensibility plus Reliability* since unreliable statements are not interpretable, even if they would be very much comprehensible.

We see mainly two typical ways rules are interpreted: the use of partitions to describe classes and to deduce a measure of importance of predictors to detect main influences. Here we want to formalize mental fit and data fit, i.e. comprehensibility and reliability, for partitions.

Regarding reliability, we distinguish two cases. If we interpret the rule as such, without deducing additional entities from it, it simply gains its reliability from the correctness rate. This is the case in so-called "conventional" partitions that we treat in subsection 3.1. When rules with acceptable data fit are too complex to be comprehensible, though, we have to do further transformations to get comprehensible information from them. Such transformations are realized by the **standardized** partitions that we introduce in subsection 3.2 where we use rules to derive prototypes for interpretation. Naturally, the outcome of this process also has to be reliable. For standardized partitions we introduce a special reliability measure.

3.1   Interpretability of Conventional Partitions

Our minimum requirements for interpretability of partitions resulting from classification rules are:

- Partitions should be formulated in 2 or 3 maximum 7 dimensions, more dimensions are not cognitively manageable;

- dimensions should relate to interpretable features;

- rules should have acceptable prediction accuracy.

The first two requirements relate to comprehensibility, the third to reliability. Ranking of classification rules according to the above definition of interpretability of the resulting partitions should be done as follows:

- the lower the dimension of the partition, the better;

- in case of equal dimension, choose the rule with the lowest error rate.

3.2 Standardized Partitions

What should one do, however, if these requirements cannot be fulfilled? In what follows we will standardize partitions and develop related measures for comprehensibility and reliability.

There is only one condition a classification method is subjected to so that any of its rules can be used to generate a standardized partition. The final decision $\mathbf{cl} : \mathbf{X} \to 1, ..., G$ has to be an argmax rule based on (rule-specific) transformations of observations:

$$\mathbf{cl}(x) \;:=\; \arg \max_{c=1,...,G} m(x, c).$$

For all classifiers that fulfill this condition, including e.g. all probabilistic classifiers, support vector machines, and neural networks, we can formulate partitions in a coordinate system, where axes have a common interpretation. Let

$$m(x, c) \;:=\; \text{strength of membership of } x \text{ in } c.$$

We denote the space of corresponding membership vectors $\vec{m}(x) := (m(x, 1), ..., m(x, G))$ by $\mathcal{M} \subset \mathbb{R}^G$. For probabilistic classifiers, we use

$$m(x, c) \;:=\; \hat{p}(c|x), x \in \mathbf{X}, c = 1, ..., G.$$

Thus, these membership values all lie in the interval $[0, 1]$ and sum up to one. We denote this space of membership vectors by $\mathcal{M}^s \subset [0, 1]^G$ which will be our space for standardized partitions in future. For $G = 3$ or $G = 4$ membership vectors in $\mathcal{M}^s$ can be visualized in a barycentric coordinate system called **simplex**, see e.g. Figure 4. Such a diagram is well known in experimental design to represent mixtures of components see, e.g., Cornell (1990), and is used e.g. by Anderson (1958) to display regions of risk for Bayes classification procedures. Fukunaga (1990) presents the simplex as a visualization of $\mathcal{M}^s$ what he calls the 'ideal feature space' in which the 'ideal features' $p(c|x)$ lie.

Example *For illustration, we generated small data sets (27 observations each) for the training and the testing of a quadratic discriminant classifier with Bayes rule (QDA). Observations are generated according to three $\chi^2(\nu)$-distributions with parameters $\nu = 2$, $\nu = 8$, and $\nu = 16$, respectively. The simplex on the left hand-side in Figure 4 presents the vectors of true conditional class probabilities of the observations in the test set (True Bayes classifier). On the right hand-side you see the estimated conditional probability vectors of the same observations of the QDA classifier.*

*Solid borders in Figure 4 separate regions for observations that get assigned to the same class. We call these regions **assignment areas** of the corresponding classes. Dashed borders within these regions separate observations that differ in the class with second highest (estimated) class probability that we call **preference areas**.*

The closer the marker of an object is to the class corner the higher is its (estimated) probability in that class. In general, most markers lie at the outer borders of the simplexes, where either the membership in $\chi^2(16)$ is very low or in $\chi^2(2)$. This reflects the fact that random variables from a $\chi^2(8)$ distribution lie in probability between variables from $\chi^2(2)$ and $\chi^2(16)$ distributions.

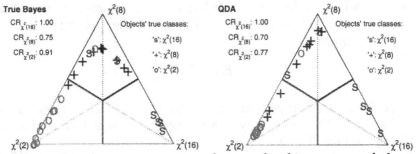

Figure 4: Simplexes representing the membership vectors of the test set observations for the True Bayes and the QDA classifier. $\mathbf{CR}_{\chi^2(2)}$, $\mathbf{CR}_{\chi^2(8)}$, and $\mathbf{CR}_{\chi^2(16)}$ denote the correctness rates for the assignments to the corresponding classes.

The QDA classifier assigns some objects to $\chi^2(8)$ with a very high membership in that class. This is visible in the simplex as the corresponding markers in the QDA simplex are closer to that corner than any marker in the simplex of the True Bayes. One can say the classifier is surer than the True Bayes that these objects really belong to $\chi^2(8)$ and in that sense it is surer that the resulting assignment is correct. A comparison of the correctness rates in the different regions, however, reveals that the QDA classifier performs worse in the assignment to $\chi^2(8)$ and also to $\chi^2(2)$. Both, True Bayes and QDA, only assign objects to $\chi^2(16)$ that really lie in that class.

In contrast to probabilistic classifiers, membership values generated by support vector machines or neural networks can be any real number. In order to compare membership vectors of different methods, we scale them into the space $\mathcal{M}^s$.

In a first step we do some monotone transformation on membership values $m(x, c)$ into the space $\mathcal{M}^s$. In the second step, we scale these transformations $m^*(x, c)$, such that - on average - the scaled membership values $m^s(x, c)$ of all observations that get assigned into a certain class can be interpreted as the probability that this assignment is correct.

The scaling is based on the comparison of average **confidence** with actual **competence** that are defined as follows:

- A classifier's **confidence** $\mathbf{CF}(x)$ in an assignment of an observation x into class $\mathbf{cl}(x)$ is reflected by its **assignment value** $\max_{c=1,...,G} m^*(x, c) = m^*(x, \mathbf{cl}(x))$. Confidence is thus defined as

$$\mathbf{CF}(x) \ := \ \max_{c=1,...,G} m^*(x, c), \ x \in \mathbf{X}.$$

- A classifier's **average confidence** in an assignment into class c, $c = 1, ..., G$, is the expected value of the confidence in such an assignment:

$$\mathbf{CF}_c \ := \ \mathbf{E}_X(\mathbf{CF}(X) \mid \mathbf{cl}(X){=}c).$$

- A classifier's **competence** to assign observations to class c is reflected by the probability of a correct assignment to this class for any random $X \in \mathbf{X}$:

$$\mathbf{P}_{C|\mathbf{cl}(X)}(C{=}c \mid \mathbf{cl}(X){=}c), \ c = 1, ..., G.$$

We estimate average confidence and competence empirically by means of observed confidences and class frequency, i.e. correctness rates $\mathbf{CR}_c$ on some test set $\mathbf{T}$.

The details of the process of scaling is published in Sondhauß and Weihs (2001) and is based on approximations of the empirical distribution of assignment values within assignment areas with the beta-distribution.

The thus scaled membership vectors $\vec{m}^s(x) := (m^s(x, 1), ..., m^s(x, G))'$ have the following properties:

- Scaled membership values are directly comparable in size.

- The average confidence of observations equals the actual competence on the test set.

- The empirical distribution of scaled membership vectors reflects the distribution of the original membership vectors corrected for the information in the test set.

- Scaled membership vectors of observations reflect as much as possible the position of the original membership vectors among each other within assignment areas.

We then support interpretation by prototyping reliable rules, where **reliability** relates to high ability to separate in the standardized partition space, and a **prototype** is the most typical instance (observation, example) in terms of scaled membership vectors.

Prototype

An observation where the rule has a justified high confidence in its decision, as well as no clear preference for the membership in any of the other classes, obviously has properties - from the perspective of the rule - that are quite specific for the assigned class. This is our motivation to define a prototype to be the correctly assigned observation the scaled membership vector of which is nearest to the class corner - denoted as $\vec{e}(c)$ - using euclidean distance:

$$x_{c,\mathbf{T}}^* \quad := \quad \arg\min_{x \in \mathbf{x}_{c,\mathbf{T}}} \left\| \vec{e}(c) - \vec{m}^s(x) \right\|, \tag{1}$$

where $\mathbf{x}_{c,\mathbf{T}}$ consists of all observations in the test set $\mathbf{T}$ in the assignment area of $c, c = 1, ..., G$. The prototype $\mathbf{x}_{c,\mathbf{T}}^*$ is the observation in the test set the unscaled membership vector of which is nearest to the corner.

Reliability

As stated above, for the reliability of deduced entities from classifiers, the process of their derivation has to be checked. The selection of a prototype is based on the measure of typicalness (1) in terms of euclidean distance of scaled membership vectors to class corners. This means here, we need a definition of the reliability of the interpretation of the prototypes in terms of the reliability of the rule's induced measure of typicalness.

If scaled membership vectors reflect the important features of the observations that help to discriminate classes, we say the selected prototype is **reliable for interpretation**. We check reliability by the typicalness of the centers of scaled membership vectors within assignment areas. According to our scaling process these centers reflect the actual competence of the classifier to assign correctly. If these centers are not typical for the assignment areas then obviously (scaled) membership vectors do not reflect the important features for the discrimination of observations in the assignment areas.

For the combination of the typicalness of the centers in all areas, we use their mean, weighted by the size of the groups. We additionally standardize the measure, such that:

- A value of one corresponds to the highest possible typicalness, where all centers lie in the assigned corners. This happens, iff there is no misclassification on the test set.

- A value of zero occurs, when all vectors are equal and lie in the barycenter of the simplex.

- Negative values indicate that the rule could be improved by an interchange of the assignment of two assignment areas.

The definition is as follows:

$$\mathbf{AS_T} := \frac{\sqrt{\frac{G-1}{G}} - \frac{1}{N_T} \sum_{x \in \mathbf{T}} \|\vec{e}(\mathbf{cl}(x)) - \tilde{\mathbf{p}}_{\mathbf{T}}(\vec{n} \mid \mathbf{cl}(x)=c)\|}{\sqrt{\frac{G-1}{G}}}, \quad (2)$$

where $\vec{n}$ denotes the vector $(1, 2, ..., G)$ of all classes and $\tilde{\mathbf{p}}_{\mathbf{T}}(\vec{n} \mid \mathbf{cl}(x)=c)$ is the vector of relative class frequencies for all test set observations assigned to class c. As the center of the assignment values reflects the actual competence of the rule on the test set, we can conclude that the rule does a good job in discriminating the classes on basis of its membership vectors, if $\mathbf{AS_T}$ is high and the centers lie near to the corresponding correct corner of the simplex. In other words, $\mathbf{AS_T}$ is also a measure for another known goodness aspect of classifiers, we call it **ability to separate**, which is the antonym of what Hand (1997) terms **resemblance**. The ability to separate is a characteristic of the classifier which should not be mistaken for **separability**, which is a characteristic of the problem that tells us how different the "true" probabilities of belonging to each class are.

Continuation of the Example *Using the same data as before, we demonstrate the scaling process for the QDA classifier. In the simplex on the left hand-side of Figure 5 you see the original, and on the right hand-side the scaled membership vectors. For each area in each simplex we present mean confidences.*

At first, observing that $\mathbf{CF}_{\chi^2(2)}=0.85$ and $\mathbf{CF}_{\chi^2(8)}=0.79$ are bigger than the actual competence of the QDA classifier ($\mathbf{CR}_{\chi^2(2)}=0.77$, $\mathbf{CR}_{\chi^2(8)}=0.70$ in Figure 4) reveals that the classifier is over-confident in its assignment to these classes, whereas it is under-confident ($\mathbf{CR}_{\chi^2(16)}=1$, $\mathbf{CF}_{\chi^2(16)}=0.72$) in its assignment to class $\chi^2(16)$. Consequently, on the one side, scaling moves membership vectors away from the class corners $\chi^2(2)$ and $\chi^2(8)$. And on the other side, scaled membership vectors of all objects in the assignment area of $\chi^2(16)$ lie exactly (except for jittering) in the corner, due to the 100% correctness rate there. By construction, resulting confidences $\mathbf{CF}^s_{\chi^2(\nu)}$, $\nu = 2, 8, 16$ are almost equal to the correctness rates $\mathbf{CR}_{\chi^2(\nu)}$, $\nu = 2, 8, 16$. Here they are all equal at least up to the second decimal place.

Two objects get moved out of their region: one from $\chi^2(8)$ to $\chi^2(2)$ and another vice versa. In both cases, they are moved into the region of their true class, but it can also happen in the other direction.

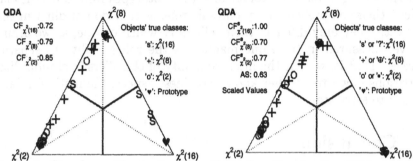

Figure 5: Simplexes illustrating the effect of scaling for the QDA classifier on the test set. Mean confidences $\mathbf{CF}_{\chi^2(\nu)}$ and $\mathbf{CF}^s_{\chi^2(\nu)}$ for $\nu = 2, 8, 16$, are given, based on the original and the scaled assignment values respectively. Scaling moves some objects out of their assignment area. These are marked using an alternative symbol for their true class.

The prototypes $x^*_{\nu,\mathbf{T}}$ for $\chi^2(\nu)$ according to the QDA classifier are $x^*_{2,\mathbf{T}} = 0.22$, $x^*_{8,\mathbf{T}} = 8.64$, and $x^*_{16,\mathbf{T}} = 18.12$. These are not very different from the prototypes of the True Bayes classifier $x^*_{2,\mathbf{T}} = 0.05$, $x^*_{8,\mathbf{T}} = 6.45$, and $x^*_{16,\mathbf{T}} = 18.12$.

Whether or not the size of 0.63 in the actual ability to separate $\mathbf{AS_T}$ of the QDA classifier makes the interpretation reliable, is, as usual, dependent on the problem and thus, can only be evaluated in comparison with other classifiers. Clearly, in comparison with the actual ability to separate of 0.76 of the best classifier - the True Bayes - we would prefer to rely on the prototypes of the True Bayes.

Interpretability of Standardized Partitions

The use of prototypes for the description of a collection of observations is quite common, both in statistics and in machine learning. In statistics arithmetic mean, median, and mode are the most basic features that are reported when describing subpopulations of any kind. In machine learning, case-based reasoners claim that what people store for future problem solving are examples rather than rules, from which a high mental fit of examples may be induced (c.p. Aamodt and Plaza (1994)).

This justifies our approach in general, since we deduced from a rule with a conventional partition with acceptable data fit but without acceptable mental fit an entity that has a high mental fit, namely prototypes

$x_{c,\mathbf{T}}^*, c = 1, ..., G$. Thus, we improved mental fit by deriving prototypes, and we defined a corresponding new measure of reliability - $\mathbf{AS_T}$ - that goes beyond the correctness rate in conventional partitions.

For the rating of interpretability of these prototypes from different classifiers, their comprehensibility plays no role, as they do not differ in that respect. They only differ in their reliability which can be, as we argued above, measured in the same manner as the rule's ability to separate classes. That is, we put the interpretability of classifiers in standardized partitions down to another known goodness aspect, the method-related ability to separate.

4 Conclusion

This paper developed a general criterion for the interpretability of partitions generated by classification rules. Based on a discussion of mental fit criteria from the literature, we introduced interpretability as a combination of mental fit and data fit, or more specifically, as a combination of comprehensibility and reliability of a partition. For cases where the partition as such is not comprehensible we developed a standardized partition which is used to derive so-called prototypes to improve comprehensibility. The reliability of such prototypes is then used as a measure of data fit, and for the rating of different classifiers.

References

AAMODT, A., and PLAZA, E (1994): Case-Based Reasoning: Foundational Issues, Methodological Variations, and System Approaches *AI Communications, 7(1):39-59.*

BODENHOFER, U., and BAUER, P. (1999): Towards an Axiomatic Treatment of "Interpretability", in Proceedings of the 6th International Conference on Soft Computing (IIZUKA2000), Iizuka, Japan, 334-339.

BREIMAN, L., FRIEDMAN, J. H., OLSHEN, R. A., and STONE, C. J. (1984): Classification and Regression Trees, Wadsworth, Belmont.

CORNELL, J. A. (1990): Experiments with mixtures. 2nd edition, Wiley.

EIJKEL, VAN DEN G. (1999): Rule Induction, in Berthold, Hand (Eds.): Intelligent Data Analysis: An Introduction, Springer, Berlin, 195-216.

FENG, C. and MICHIE, D. (1994): Machine Learning of Rules and Trees, in Michie, Spiegelhalter, and Taylor (Eds.): Machine Learning, Neural and Statistical Classification, Ellis Horwood, New York, 50-83.

FUKUNAGA, K. (1990): Introduction to Statistical Pattern Recognition, Academic Press, New York.

HAND, D. J. (1997): Construction and Assessment of Classification Rules, Wiley, Chichester

LU, H., SETIONO, R., and LIU, H. (1995): NeuroRule: A Connectionist Approach to Data Mining, in Proceedings of the 1st VLDB Conference Zürich, Switzerland, 1995

MICHIE, D., SPIEGELHALTER, D. J., and TAYLOR, C. C. (1994): Conclusions in Michie, Spiegelhalter, and Taylor (Eds.): Machine Learning, Neural and Statistical Classification, Ellis Horwood, New York, 213-227.

RISSANEN, J. (1978): Modeling by Shortest Data Description. *Automatica 14, 465-471.*

SILBERSCHATZ, A., and TUZHILIN, A. (1996): What makes patterns interesting in knowledge discovery systems. *IEEE Transactions on Knowledge and Data Engineering, 8,970-974.*

SMYTH, P. and GOODMAN, R. (1992): An information theoretic approach to rule induction from databases. *IEEE Transactions on Knowledge and Data Engineering, 4, 301-316*

SONDHAUß, U. and WEIHS, C. (2001): Standardizing the Comparison of Partitions, submitted to Journal of Computational Statistics.

WANG, K., TAY, W., and LIU, B. (1998): An interestingness-based interval merger for numeric association rules, in Proceedings of the International Conference on Knowledge Discovery and Data Mining, August 1998, New York City, AAAI, 121-128.

WEIHS, C. (1992): Vorhersagefähigkeit multivariater linearer Methoden: Simulation und Grafik, in Enke, Gölles, Haux and Wernecke (Eds.): Methoden und Werkzeuge für die exploratorische Datenanalyse in den Biowissenschaften, Fischer, Stuttgart, 111-127.

WEIHS, C. (1993): Multivariate Exploratory Data Analysis and Graphics: A Tutorial. *Journal of Chemometrics 7, 305-340.*

WEIHS, C., ROEHL, M. C., and THEIS, W. (1999): Multivariate classification of business phases, Technical Report 26/1999, SFB 475, Universität Dortmund.

WEISS, S. M. and KULIKOWSKI, C. A. (1991): Computer Systems that Learn; Morgan Kaufmann, San Francisco, 114, 168/9.

Bayesian Analysis of Econometric Models for Count Data: A Survey

R. Winkelmann

Socioeconomic Institute, University of Zurich, Switzerland

Abstract:

The paper reviews recent advances in analyzing complex count data models using computer intensive Bayesian methods, such as Gibbs sampling or Markov Chain Monte Carlo. These methods provide a powerful estimation tool for models characterized by an underlying latent structure. Examples are count data with random effects, correlated count data, counterfactuals in treatment models, latent class models, endogenous switching models and models with underreporting, to name but a few. In all of these cases, the Bayesian approach can be implemented by a simple extension of the parameter space to include the latent effects, a case of data augmentation. As a by-product of the MCMC simulation, the posterior distribution of the latent effects is obtained, which may be very useful in some problems.

1 Introduction

The econometric analysis of count data to date has been dominated by the sampling theory approach to estimation and inference. Parameters are estimated by maximum likelihood or method of moments, and inference is based on the hypothetical behavior of the estimator in repeated samples of sufficient size. Much less has been written about the Bayesian analysis of count data models. In this short survey, I present and illustrate some of the conceptual and practical advantage it has to offer.

I first outline Bayesian and frequentist estimation of the Poisson regression model. Then, I present a class of interesting count data models that cannot be easily dealt with in the sampling theory approach, and I summarize the key elements of posterior analysis using Markov Chain Monte Carlo.

2 The Poisson Model

The standard probability distribution for count data is the Poisson distribution:

$$P(y_i|\lambda_i) = \frac{\exp(-\lambda_i)\lambda_i^{y_i}}{y_i!} \tag{1}$$

where

$$\mathrm{E}(y_i|\lambda_i) = \mathrm{Var}(y_i|\lambda_i) = \lambda_i \tag{2}$$

In a regression model, the population is assumed heterogeneous with covariates x_i, specifying $\lambda_i = \exp(x_i'\beta)$ where $i = 1, \ldots, N$ indexes observations in the sample. Under random sampling

$$P(y|x) = \exp\left[-\sum_{i=1}^{N} \exp(x_i'\beta)\right] \prod_{i=1}^{N} \frac{[\exp(x_i'\beta)]^{y_i}}{y_i!} \qquad (3)$$

The distributional assumption for the observables, here the Poisson regression model, may or may not be justified. Indeed, several alternative models will be presented below. It is important to realize, however, that a specification of the likelihood function is needed in, and central to, both Bayesian and non-Bayesian methods. The differences arise in the way inference is undertaken conditional on the data model (3).

In the sampling theory approach, the maximum likelihood estimator (MLE) of β is obtained by maximizing the log-likelihood function

$$\ln L(\beta|y) = \sum_{i=1}^{N} (-\exp(x_i'\beta) + y_i \ln x_i'\beta - \ln y_i!) \qquad (4)$$

Maximization of (4), although requiring a numerical algorithm, does not pose great difficulties, as the log-likelihood is globally concave. The approximate distribution of the MLE for β is given by

$$\hat{\beta}_{ML} \overset{\text{app}}{\sim} \mathcal{N}(\beta, [-\text{EH}(\beta)]^{-1}) \qquad (5)$$

The Bayesian approach, in contrast, adds to the model a prior distribution $P(\beta)$. From the joint distribution of observables and unobservables, $P(y, \beta) = P(y|\beta)P(\beta)$, one obtains the exact distribution of the unobservables (i.e., parameters) conditional on observables, $P(\beta|y) = P(y|\beta)P(\beta)/P(y)$, where $P(y) = \int P(y|\beta)P(\beta)d\beta$ is the normalizing constant. For example, with prior distribution $P(\beta) \sim \mathcal{N}(\beta_0, B_0)$

$$P(\beta|y) \propto \prod_{i=1}^{N} \exp[-\exp(x_i'\beta)] \, [\exp(x_i'\beta)]^{y_i} \exp[-1/2(\beta-\beta_0)'B_0^{-1}(\beta-\beta_0)]$$

$$(6)$$

This posterior distribution is of unknown form, and the normalizing constant cannot be evaluated analytically or numerically (the same would be true with a flat instead of normal prior for β). Hence, the direct evaluation of quantities of interest, such as $\text{E}[\beta|y]$, is not possible.

Two solutions exist. One is through the use of asymptotic approximations (Lindley, 1961). In particular, let $\hat{\beta}$ be the mode of the posterior density. Expansion of the log density around $\hat{\beta}$ yields

$$\ln P(\beta|y) \approx \ln P(\hat{\beta}|y) - \frac{1}{2}(\beta - \hat{\beta})'V(\beta - \hat{\beta}) \tag{7}$$

where V is minus the Hessian matrix of the log posterior evaluated at $\hat{\beta}$:

$$V = B_0^{-1} + \sum_{i=1}^{N} \exp(x_i'\hat{\beta})x_i x_i' \tag{8}$$

$\ln P(\hat{\beta})$ can be obsorbed into the normalizing constant and we see that the posterior of β is approximately multivariate normal with mean $\hat{\beta}$ and covariance matrix V^{-1}. In large samples, the second term in (8) dominates the prior variance, and the posterior distribution and the classical asymptotic approximation converge. Arguably, one approximation is as good as the other, and in large samples, the choice between the classical view and the Bayesian view is a matter of taste rather than substance. The MLE and the sample mean of the posterior density are asymptotically equivalent estimators of β.

2.1 Markov Chain Monte Carlo

An alternative approach is the simulation of the posterior density $P(\beta|y)$ – i.e., the representation of the densities by related finite samples – using MCMC techniques, and the Metropolis-Hastings (MH) algorithm in particular (See Chib and Greenberg, 1996).

For a given target density $f(\psi)$ such as (6), the MH algorithm is defined by

(1) a proposal density $q(\psi^{\dagger}|\psi)$ that is used to supply a proposal value $\psi^{\dagger}$ given the current value ψ, and

(2) a probability of move that is defined as

$$\alpha(\psi, \psi^{\dagger}) = \min\left\{\frac{f(\psi^{\dagger})q(\psi|\psi^{\dagger})}{f(\psi)q(\psi^{\dagger}|\psi)}, 1\right\}. \tag{9}$$

Hence, if $f(\psi^{\dagger})q(\psi|\psi^{\dagger}) > f(\psi)q(\psi^{\dagger}|\psi)$ the chain moves to $\psi^{\dagger}$. Otherwise, it moves with probability $0 < \alpha(\psi, \psi^{\dagger}) < 1$. If rejected, the next sampled value is taken to be ψ.

For the MH algorithm to work efficiently, the choice of proposal density q is critical. Following Chib, Greenberg, and Winkelmann (1998), the proposal distribution for the Poisson regression model can be based on the mode $\hat{\beta}$ and curvature $V_\beta = [-H_\beta]^{-1}$ of $\ln P(\beta|y)$ where these quantities are found using a few Newton-Raphson steps with gradient vector

$$g_\beta = B_0^{-1}(\beta - \beta_0) + \sum_{i=1}^N [y_i - \exp(x_i'\beta)]x_i \qquad (10)$$

and Hessian matrix

$$H_\beta = -B_0^{-1} - \sum_{i=1}^N \exp(x_i'\beta)x_i x_i' \qquad (11)$$

In practice, three or four steps of the Newton-Raphson algorithm are sufficient to locate the mode of the target density.

From here, one can obtain proposals in a number of ways. One possibility is a Gaussian proposal with reflection:

$$q(\beta, \beta^\dagger|y) = \phi(\beta^\dagger|\hat{\beta} - (\beta - \hat{\beta}), \tau V_\beta) \qquad (12)$$

where τ is a scalar that is adjusted in trial runs in order to obtain acceptance rates between 40 and 60 percent. To draw from this proposal density, one simply computes

$$\beta^\dagger = \hat{\beta} - (\beta - \hat{\beta}) + \tau P' \mathrm{rndn}(k, 1) \qquad (13)$$

where $P = \mathrm{chol}(V_\beta)$ gives the Cholesky decomposition of V_β, and $\mathrm{rndn}(k, 1)$ is a vector of standard normal pseudo-random numbers. In this case, the probability of move simplifies to the ratio of density ordinates

$$\alpha(\beta, \beta^\dagger|y) = \min\left\{ \frac{P(\beta^\dagger|y)}{P(\beta|y)}, 1 \right\}, \qquad (14)$$

since the proposal density is symmetric in $(\beta, \beta^\dagger)$ and hence cancels.

As an alternative, one can draw proposal values from a multivariate -t density with ν degrees of freedom, $f_T(\beta|\hat{\beta}, V_{\hat{\beta}}, \nu)$, where ν is a tuning parameter, and use (9). This proposal density tends to be easier to adjust.

In either case, assume that R draws from $P(\beta|y)$ are available. The quantity of interest is then calculated as

$$E[h(\beta)|y] = \frac{1}{R} \sum_{r=1}^R h(\beta_r) \qquad (15)$$

2.2 Marginal Likelihood

It is a feature, and indeed a strength, of the posterior simulation approach, that it does not require the computation of the normalizing constant in (6). This constant, the *marginal likelihood*, is the integral of the likelihood with respect to the prior density of the parameters:

$$m(y) = \int P(y|\beta)P(\beta)\mathrm{d}\beta$$

Several methods to evaluate the marginal likelihood have been proposed, among them those based on Laplace approximations and Importance Sampling. Chib (1995) proposed a method that makes use of reduced MCMC runs in order to evaluate the components on the right-hand side of the identity

$$m(y) = \frac{P(y|\theta^*)P(\theta^*)}{P(\theta^*|y)}$$

where θ^* is an arbitrary point in the parameter space, typically chosen to be a high posterior density point. The difficulty lies in the evaluation of the denominator, the posterior density function at θ^* that is intractable to start with. In many cases, the problem can be simplified by factoring the joint posterior into conditional densities of known form.

3   Count Data Models with Random Effects

Analysis of count data via hierarchical Poisson regression models is conceptually straightforward and has been described in a number of textbooks (e.g., McCullagh and Nelder, 1989). The model consists of a first stage Poisson regression model, conditioned on the parameters and random effects – $P(y|\varepsilon, \beta)$ – supplemented by a second stage distribution for the random effects – $P(\varepsilon|\gamma)$. Let β and γ have joint prior $P(\beta, \gamma)$. The full posterior density of the stage 1 parameters β and the stage 2 parameters γ, augmented by the random effects, is then given by

$$P(\varepsilon, \beta, \gamma|y) = P(y|\varepsilon, \beta)P(\varepsilon|\gamma)P(\beta, \gamma) \tag{16}$$

Practical implementation of this model requires some computational ingenuity and expertise if there are a large number of random effects such as may be the case for panel or multivariate count data. It would appear that MCMC methods are eminently suitable (at least in principle) to remove the computational constraint.

3.1 Unobserved Heterogeneity

The simplest example of a hierarchical model is provided by the Poisson regression model with i.i.d. random effects that account for unobserved heterogeneity

$$y_i | \beta, \varepsilon_i \sim \text{Poisson}(\lambda_i),$$

$$\lambda_i = \exp(x_i' \beta + \varepsilon_i)$$

It is frequently assumed that $\exp(\varepsilon_i)$ is gamma distributed, or, equivalently, that ε_i has a log-gamma distribution. While a convincing statistical justification for this assumption is hard to find, it offers mathematical convenience for likelihood based inference, as the marginalized likelihood, i.e., the integral

$$\int P(y|\varepsilon, \beta) P(\varepsilon|\gamma) d\varepsilon$$

can be expressed in closed form, a negative binomial distribution.

From a Bayesian viewpoint, the marginalized likelihood (with respect to the random effects) is of no particular interest, and it is more plausible to assume normality

$$\varepsilon_i \sim \mathcal{N}(0, \sigma^2)$$

The normal distribution arises, for instance, if ε_i captures the influence of numerous independent omitted regressors. In contrast to the gamma distribution, the normal distribution directly generalizes to multivariate settings with a general covariance structure amongst multiple random effects, a feature that is exploited in the next section.

In the Bayesian approach, the model is completed by specifying the prior distributions. A standard assumption is the independence prior

$$\beta \sim \mathcal{N}_k(\beta_0, B_0), \quad \sigma^{-2} \sim \Gamma(\nu_0, r_0).$$

For simulation, one can use a MH-within-Gibbs approach and sample iteratively from the full conditionals of

$$[\varepsilon|\beta, \sigma^2] \; [\beta|y, \varepsilon]; \; [\sigma^{-2}|\varepsilon],$$

where each conditional distribution is sampled using a MH step. Alternatively, one can bring the model into the form of a hierarchical generalized linear model with $\lambda_i = \exp(\theta_i)$ and $\theta_i \sim \mathcal{N}(x_i' \beta, \sigma^2)$. In this case, the full conditional of $\beta|\theta, \sigma^2$ has a closed form (see below).

3.2 Correlated Random Effects

The basic approach can be easily extended to panel or multivariate data with correlated random effects. One possible model for multivariate data is (see Chib and Winkelmann, 2001)

$$y_{ij}|\eta_{ij} \sim \text{Poisson}(\eta_{ij}) \tag{17}$$

$$\eta_{ij} = \exp(x'_{ij}\beta + b_{ij}) \tag{18}$$

$$b_i = (b_{i1}, \ldots, b_{iJ})' \sim \mathcal{N}_J(0, D) \tag{19}$$

with prior

$$\beta \sim \mathcal{N}_k(\beta_0, B_0), \quad D^{-1} \sim \text{Wishart}(\nu_0, R_0), \tag{20}$$

Note that J can be large. Chib and Winkelmann (2001) discuss two high-dimensional applications where J is six and sixteen, respectively. Using the law of the iterated expectation, one can show that the covariance amongst the counts implied by this model is given by

$$\text{Cov}(y_{ij}, y_{ik}) = \exp(x'_{ij}\beta + 0.5d_{jj})(\exp(d_{jk}) - 1)\exp(x'_{ik}\beta + 0.5d_{kk}), \ j \neq k$$

which can be positive or negative depending on the sign of d_{jk}, the (j, k) element of D. Moreover, the model allows for overdispersion for $d_{jj} > 0$. If j indexes time, this is a panel model. It is more general than the standard panel model with individual specific random effect where

$$\lambda_{ij} = \exp(x'_{ij}\beta + b_i) \qquad b_i \sim \mathcal{N}(0, \sigma^2),$$

a special case of the above model for $\text{E}(bb') = \rho \mathbf{1}\mathbf{1}'$.

The full posterior distribution of the model with correlated random effects is proportional to

$$\pi(\beta)\pi(D^{-1}) \times \prod_{i=1}^{N} |D|^{-\frac{1}{2}} \times$$

$$\exp\left[-\frac{1}{2}b'_i D^{-1} b_i\right] \prod_{j=1}^{J} \exp(\exp(x'_{ij}\beta + b_{ij}))(\exp(x'_{ij}\beta + b_{ij}))^{y_{ij}}$$

To estimate the model, a Markov chain can be constructed using the blocks of parameters $\{b_i\}$, β, and D with full conditional distributions

$$[b|y, \beta, D]; \ [\beta|y, b]; \ [D^{-1}|b], \tag{21}$$

where $b = (b_1, ..., b_n)$. The simulation output is obtained by recursively simulating these distributions, using the most recent values of the conditioning variables at each step. In this formulation, both b and β are sampled from MH steps, whereas

$$D^{-1}|b \sim \text{Wishart}\left(N + \nu_0, [R_0^{-1} + \sum_{i=1}^{N}(b_i b_i')]^{-1}\right).$$

This was the approach employed by Chib and Winkelmann in the aforementioned paper. While feasible, it turns out that the algorithm can be improved by adopting a reparameterization. If the algorithm is applied to the reparameterized model, one MH step becomes redundant. As a consequence, the algorithm is faster. Moreover, it has better mixing properties. To see this, note that the model can also be written as follows

$$y_{ij}|\eta_{ij} \sim \text{Poisson}(\eta_{ij})$$

$$\eta_{ij} = g^{-1}(\theta_{ij}) = \exp(\theta_{ij})$$

$$\theta_i = (\theta_{i1}, \ldots, \theta_{iJ})' \sim \mathcal{N}(X_i'\beta, D)$$

$$\beta \sim \mathcal{N}_k(\beta_0, B_0), \quad D^{-1} \sim \text{Wishart}(\nu_0, R_0),$$

The sole change is that the Poisson model conditions of θ_{ij} rather than b_{ij}. The two formulations are equivalent since $\theta_{ij} = x_{ij}'\beta + b_{ij}$. However, the full posterior distribution is now proportional to

$$\pi(\beta)\pi(D^{-1}) \times \prod_{i=1}^{N} |D|^{-\frac{1}{2}} \times$$

$$\exp\left[-\frac{1}{2}(\theta_i - X_i'\beta)'D^{-1}(\theta_i - X_i'\beta)\right] \prod_{j=1}^{J} \exp(\exp(\theta_{ij}))(\exp(\theta_{ij}))^{y_{ij}}$$

and the MCMC samples recursively from

$$[\theta|y, \beta, D]; \quad [\beta|\theta, D]; \quad [D^{-1}|\theta, \beta].$$

An interesting aspect of this formulation is that, assuming again a normal prior, the full conditional posterior distribution $\beta|\theta, D$ is now proportional to

$$|B_0|^{-\frac{1}{2}} \exp\left[-\frac{1}{2}(\beta - \beta_0)'B_0^{-1}(\beta - \beta_0)\right]$$

$$\times \exp\left[-\frac{1}{2}\sum_{i=1}^{N}(\theta_i - X_i'\beta)'D^{-1}(\theta_i - X_i'\beta)\right]$$

Therefore, β can be sampled directly from the full conditional distribution without MH step:

$$\beta|\theta, D \sim \mathcal{N}(\hat{\beta}, B_N)$$

where $\hat{\beta} = B_N(B_0^{-1}\beta_0 + \sum_i X_i'D^{-1}\theta_i)$ and $B_N = (B_0^{-1} + \sum_i X_i'D^{-1}X_i)^{-1}$. In this formulation, only $\theta|y, \beta, D$ is sampled by MH. The other full conditional distributions are from standard distributions. A re-estimation of the application to health care utilization in Chib and Winkelmann (2001) showed that computation time could be reduced by almost 50 percent. Mixing was also superior. For example, the average autocorrelation at lag ten for β was reduced from 0.3 to 0.17. The sensitivity of the algorithm with respect to parameterization deserves further scrutiny. While the described approach works well, further possibilities to enhance the performance may exist.

Extensions

Similar considerations apply in extensions of the basic random effects model. For example, consider the following model for longitudinal multivariate data. Let y_{itk} denote the observation for individual i at time t for count k, and let

$$\lambda_{itk} = \exp(x_{it}'\beta_k + \varepsilon_{ik} + \nu_{it})$$

where $\varepsilon_{ik} = (\varepsilon_{i1k}, \ldots \varepsilon_{iTk})' \sim \mathcal{N}_T(0, D_\varepsilon)$ and $\nu_{it} = (\nu_{it1}, \ldots \nu_{itK})' \sim \mathcal{N}_K(0, D_\nu)$. Such a model was considered by Million (1998). Since he uses numerical integration techniques, only low dimensional settings are feasible.

Correlation across time between observations of a given person is captured by the outcome specific covariance matrix D_ε. Contemporaneous correlation between outcomes for a given individual is captured by the covariance matrix D_ν. By assumption, individuals are sampled independently and there is no correlation between individual observations.

Yet another model with a random effects structure is the random coefficients model by Chib, Greenberg and Winkelmann (1998). In this case, (18) and (19) are replaced by

$$\lambda_{it} = \exp(x_{it}'\beta + w_{it}'b_i)$$

$$b_i|\eta, D \sim \mathcal{N}_q(\eta, D) \text{ where } b_i = (b_{i1}, \ldots, b_{iq})'$$

Note that in each of the models presented in this section, maximum likelihood estimation would require the numerical evaluation of highly dimensional integrals, which is prohibitive if the dimension becomes large. For instance, Chib and Winkelmann (2001) implement a model with 16 correlated counts.

4 Selection Models

There is a large microeconometric literature on estimating causal effects from non-experimental data. The fundamental feature of such "sample selectionmodels is the necessity to account for the partial observability of the outcome (the counterfactual is unobserved) and the non-random selection into the sample (or treatment). Such models come in various flavours. In their simplest form, they consist of one or two outcome equations and a selection equation. Consider the following structure for cross-sectional data

$$y_i | \beta, \varepsilon \sim \text{Poisson}(\lambda_i), \tag{22}$$

$$\lambda_i = \exp(x_i'\beta + \varepsilon) \tag{23}$$

$$d_i^* = z_i'\gamma + u \tag{24}$$

$$[\varepsilon, u] \sim \mathcal{N}_2(0, D) \tag{25}$$

(24) is the selection equation. Estimating the outcome model (22) needs to account for sample selection unless ε and u are uncorrelated (the case of random selection). This holds regardless of whether or not z_i and x_i overlap.

Some models of interest are as follows (Winkelmann, 1998):

Truncation

$$y_i \text{ is } \begin{cases} \sim \text{Poisson}(\lambda_i) & \text{if } d_i^* \geq 0 \\ \text{unobserved} & \text{else} \end{cases}$$

An example would be non-random item non-response in surveys, where y_i is the sought information and d_i indicates whether a response was given.

Correlated Hurdle

$$y_i = \begin{cases} \sim \text{truncated-at-zero Poisson}(\lambda_i) & \text{if } d_i^* \geq 0 \\ 0 & \text{else} \end{cases}$$

Such models are used for instance in the analysis of health care utilization where y_i measures the number of doctor visits in a fixed period and the decision of being a user or not is governed by a different process than the extent of use (given that one is a user).

Switching regression

$$y_i \sim \begin{cases} \text{Poisson}(\exp(x_i'\beta_1 + \varepsilon)) & \text{if } d_i^* \geq 0 \\ \text{Poisson}(\exp(x_i'\beta_2 + \varepsilon)) & \text{else} \end{cases}$$

The treatment model with endogenous dummy variable

$$f(y_i|x_i, d_i, \varepsilon_i) = \exp(x_i'\beta + \alpha d_i + \varepsilon_i)$$

is a special case of switching regression. An example here is the number of trips taken over the previous 24 hours time period and its dependence on car ownership which may be endogenous. An MCMC algorithm for such a model is discussed by Kozumi (1999).

In none of these cases can the likelihood function be evaluated in closed form. While numerical integration has been used in some instances, the Bayesian approach can be implemented easily. One particularly attractive feature is the availability of the posterior distribution of the counterfactual outcomes that can be used for further analyses.

The methodological considerations of this section largely remain unchanged if the selection variable d_i^* itself is observed. In this case, one has a mixed discrete-continuous model that is linked through the correlated random effects. Also, none of the models depend critically on the assumption of joint normality. For instance, related models with multivariate t-distributed random effects have been successfully applied. In both cases, the appropriate posterior simulator routines require relatively few modifications relative to the basic set-up.

References

CHIB, S. (1995): Marginal Likelihood from the Gibbs Output, *Journal of the American Statistical Association*, Vol. 90, 1313 - 1321.

CHIB, S. and GREENBERG, E. (1996): Markov Chain Monte Carlo Simulation Methods in Econometrics, *Econometric Theory*, Vol. 12, 409–431.

CHIB, S., GREENBERG, E. and WINKELMANN, R. (1998): Posterior simulation and Bayes factors in panel count data models, *Journal of Econometrics*, Vol. 86, 33–54.

CHIB, S., and WINKELMANN, R. (2001): Modeling and Simulation-Based Analysis of Correlated Count Data, *Journal of Business and Economic Statistics*, Vol. 19, 428-435.

KOZUMI, H. (1999): A Bayesian Analysis of Endogenous Switching Models for Count Data. mimeo. Hokkaido University.

LINDLEY, D.V. (1961): Introduction to Probability and Statistics, Part 2: Inference. Cambridge University Press: Cambridge.

MCCULLAGH, P., NELDER, J.A. (1989): Generalized Linear Models, 2nd ed. Chapman and Hall, London.

MILLION, A. (1998): Zähldatenmodelle für korreliert abhängige Daten, unpublished Ph.D. Thesis, University of Munich.

WINKELMANN, R. (1998): Count data models with selectivity, *Econometric Reviews, Vol. 17, 339–359.*

k-Means Clustering with Outlier Detection, Mixed Variables and Missing Values

D. Wishart

Department of Management,
University of St. Andrews, St. Katharine's West,
The Scores, St. Andrews, Fife KY16 9AL, Scotland
Email: d.wishart@st-andrews.ac.uk
Website: www.clustan.com
Tel: +44 131 337 1448

Abstract: This paper addresses practical issues in k-means cluster analysis or segmentation with mixed types of variables and missing values. A more general k-means clustering procedure is developed that is suitable for use with very large datasets, such as arise in data mining and survey analysis. An exact assignment test guarantees that the algorithm will converge, and the detection of outliers allows the densest regions of the sample space to be mapped by tessellations of tightly-specified spherical clusters. A summary tree is obtained for the resulting k-cluster partition.

1 Introduction

Clustering problems routinely occur in survey analysis and data mining where incomplete data and different types of variables are present. A popular method of classification is k-means analysis, which partitions a set of cases into k clusters so as to minimise the "error" or sum of squared distances of the cases about the cluster means. However, k-means analysis is usually only implemented with quantitative variables and complete data. This paper extends Gower's General Similarity Coefficient to k-means analysis so that mixed data types, missing values and differential weighting of cases or variables can be handled, with an exact assignment test that guarantees the algorithm will converge.

2 k-Means Cluster Analysis

The conventional k-means clustering procedure is as follows:

1. Choose an initial partition of the cases into k clusters. This can be a tree section, k selected cases, or a random assignment to k clusters.

2. Compute the distance from every case to the mean of each cluster, and assign the cases to their nearest clusters.

3. Re-compute the cluster means following any change of cluster membership at step 2.

4. Repeat steps 2 and 3 until no further changes of cluster membership occur in a complete iteration. The procedure has now converged to a stable k-cluster partition.

The k-means procedure was first proposed by Thorndike (1953) in terms of minimizing the average of all the within-cluster distances. Forgey (1965), Ball (1965), Jancey (1966), Ball and Hall (1967), MacQueen (1967) and Beale (1969) implemented it computationally to minimise the distances from the cluster means. It is provided with Clustan (Wishart (1970, 1984, 1999)).

Some programs re-compute the cluster means only at the end of each completed iteration, after all changes in cluster membership have been made, and not following each change of cluster membership as in step 3. This has the advantage that the resulting partition is independent of the case order; but it is generally slower to converge than the progressive re-calculation of means at step 3, sometimes referred to as "drifting means".

The object of k-means analysis is to arrive at a stable k-cluster partition in which each case is closest to the mean of the cluster to which it is assigned. In essence, the "model" is the final set of k cluster means, and the procedure seeks to minimise the "error" in the model, as measured by the sum of the squared distances from the cases to the cluster means. Some authors advocate imputing missing values prior to clustering, because their programs only work with complete data. However, the k-means algorithm developed below allows for missing values by estimating cluster means, distances to cluster means, and criterion function values from the complete data, ignoring any missing values.

3 General Similarity Coefficient

The starting point is a General Similarity Coefficient s_{ij} for the comparison of two cases i and j, proposed by Gower (1971) that has been widely implemented and used:

$$s_{ij} = \sum_k w_{ijk} s_{ijk} \Big/ \sum_k w_{ijk} \qquad (1)$$

where s_{ijk} is a similarity component for the kth variable (defined below), and the weight w_{ijk} is 1 if the comparison is valid for the kth variable, or 0 otherwise. Thus $w_{ijk} = 0$ where one or both of the observations on the kth variable is missing.

For quantitative variables, Gower standardises s_{ijk} by the range r_k of the kth variable:

$$s_{ijk} = 1 - \mid x_{ik} - x_{jk} \mid / r_k \tag{2}$$

where x_{ik} and x_{jk} are the values for the kth variable in cases i and j. Thus s_{ijk} varies between 1, for identical values $x_{ik} = x_{jk}$, and 0, for any two extreme values $x_{max} - x_{min} = r_k$.

For binary variables, the component of similarity s_{ijk} and the weight w_{ijk} are defined according to the following table, where + denotes that attribute k is "present" and - denotes that attribute k is "absent". Thus $s_{ijk} = 1$ if cases i and j both have attribute k "present" or $s_{ijk} = 0$ otherwise, and the weight w_{ijk} causes negative matches to be ignored.

	value of attribute k			
Case i	+	+	-	-
case j	+	-	+	-
s_{ijk}	1	0	0	0
w_{ijk}	1	1	1	1

For nominal or categorical variables, $s_{ijk} = 1$ if $x_{ik} = x_{jk}$, or $s_{ijk} = 0$ if $x_{ik} \neq x_{jk}$. Thus $s_{ijk} = 1$ where two cases i and j have the same value for attribute k, or zero if they differ, and $w_{ijk} = 1$ if both cases have observed values for attribute k, or 0 if either value is missing.

Gower's General Similarity Coefficient for mixed data types and missing values is provided in Clustan for the computation of a similarity matrix (Wishart (2002)).

4 General Distance Coefficient

Because k-means analysis is defined in terms of minimizing the distances of cases from cluster means, a general distance measure for the comparison of two cases i and j, following Gower (1971), can be derived from (1) as follows:

$$d_{ij}^2 = \sum_k w_{ijk} d_{ijk}^2 \bigg/ \sum_k w_{ijk} \tag{3}$$

where d_{ijk}^2 is a squared distance component for the kth variable (defined below), and w_{ijk} is 1 or 0 depending upon whether or not the comparison is valid for the kth variable. If differential weights w_k are assigned to the variables then $w_{ijk} = w_k$ if the comparison is valid, or zero otherwise.

Following (2), ordinal and continuous variables can be standardised by the range r_k:

$$d_{ijk}^2 = (x_{ik} - x_{jk})^2 / r_k^2 \tag{4}$$

so that d_{ijk}^2 varies between 0 for identical values $x_{ik} = x_{jk}$, and 1 for $x_{max} - x_{min} = r_k$, as for Gower's coefficient. Because the transformed values are wholly determined by the two extremes x_{max} and x_{min} it is difficult to justify range standardisation for continuous variables, and the same criticism applies to Gower's range standardisation (2). However, range standardisation might be appropriate where the data consist wholly of ordinal variables.

A better approach is to standardise ordinal and continuous variables by the variance σ_k^2, i.e. $x^* = (x_{ik} - \mu_k)/\sigma_k$. The transformed values x^* then have a mean μ_k of zero and variance σ_k^2 of one, and thus:

$$d_{ijk}^2 = (x_{ik} - x_{jk})^2 / \sigma_k^2 \tag{5}$$

If an ordinal or continuous variable is not transformed, the squared distance component is $d_{ijk}^2 = (x_{ik} - x_{jk})^2$. This may be appropriate where all the data consist of ordinal variables on the same range.

For binary variables, $d_{ijk}^2 = 0$ if cases i and j both have attribute k "present" or both "absent", or $d_{ijk}^2 = 1$ if attribute k is "present" in one case and "absent" in the other case.

For nominal variables, $d_{ijk}^2 = 0$ if cases i and j have the same value for attribute k, i.e. if $x_{ik} = x_{jk}$, or $d_{ijk}^2 = 1$ if they have different values, i.e. if $x_{ik} \neq x_{jk}$.

As in Gower's coefficient, $w_{ijk} = w_k$ if both x_{ik} or x_{jk} are valid observations on the kth variable, or $w_{ijk} = 0$ if x_{ik} or x_{jk} are missing. Unlike Gower's coefficient, however, (3) does not ignore negative matches on binary variables. For k-means analysis, the mean of a binary variable within a cluster is expressed as a probability ϕ_{pk} that the kth attribute is "present" in a cluster. The computation of ϕ_{pk} must therefore take into account "absent" states within a cluster, and hence matches on "absent" characters should not be ignored.

Euclidean distance d_{ij} is obtained simply by taking the square root $\sqrt{d_{ij}^2}$ in (3) after summation across all variables. City block distance, or the "Manhatten" metric, is the sum of the distances on each variable $d_{ij} = \sum_k w_{ijk} | x_{ik} - x_{jk} | / \sum_k w_{ijk}$.

Maximum distance is the maximum distance score for any one variable $d_{ij} = max_k | x_{ik} - x_{jk} |$ where x_{ik}, x_{jk} may be standardised by the range, as in (4), or variance (5). This measure may be appropriate if all the cases in a cluster are to be located within a hyperbox of side 2δ, where δ could be an outlier deletion threshold in k-means analysis (section 8). This criterion is comparable to CHAID analysis where hyperbox segments are formed for quantitative variables (Chi-square Automatic Interaction Detector, Kass (1980)).

The above general squared distance and dissimilarity coefficients for mixed data types and missing values are provided in Clustan for the computation of a proximity matrix (Wishart (2002)). Other dissimilarity coefficients between cases are possible following (3).

5   k-Means Cluster Analysis

As stated in section 2, the object of k-means analysis is to arrive at a stable k-cluster partition by which each case is closest to the mean of the cluster to which it is assigned. It seeks to minimise the "error", or the sum of the squared distances between the cases and the clusters, sometimes referred to as minimizing trace($\mathbf{W}$), where $\mathbf{W}$ is the within-groups dispersion matrix. The squared distance d_{ip}^2 between any case i and the mean μ_p of a cluster p is:

$$d_{ip}^2 = \sum_k w_{ipk}(x_{ik} - \mu_{pk})^2 \bigg/ \sum_k w_{ipk} \qquad (6)$$

where x_{ik} is the value of the kth variable for case i; μ_{pk} is the mean of the kth variable for cluster p; and w_{ipk} is a weight of 1 or 0 depending on whether or not the comparison between case i and cluster p is valid for the kth variable. If differential weights w_k are assigned to the variables, then $w_{ipk} = w_k$ if the comparison is valid, or zero otherwise.

For quantitative variables the mean μ_{pk} is estimated from the observed values in cluster p, i.e. ignoring any missing values. It is therefore assumed that the distribution of missing values for a cluster has the same mean as the observed values for that cluster. The mean μ_{pk} is treated as missing if all the values for the kth variable in cluster p are missing.

For binary variables $\mu_{pk} = \phi_{pk}$, the probability that attribute k is present in cluster p. Thus $d_{ipk}^2 = (1 - \phi_{pk})^2$ if attribute k is present in case i, or ϕ_{pk}^2 if attribute k is absent in case i.

For nominal variables, μ_{pk} is a vector $\{\phi_{pks}\}$ of probabilities for each state s of the kth variable within cluster p. In k-means analysis, the squared distance d_{ipk}^2 between a case and a cluster is computed between these vectors. In the following example of a nominal variable having four states and a case taking the value 2, the squared distance is 0.11139, being the sum of the 4 squared distance components.

Nominal state:	1	2	3	4
Cluster mean $\{\varphi_{pks}\}$:	0.034	0.721	0.088	0.157
Case i:	0	1	0	0
Difference:	0,034	0,279	0,088	0,157
$Distance^2$:	0.001156	0.077841	0.007744	0.024649

6 Euclidean Sum of Squares

The object of k-means analysis is to minimise the Euclidean Sum of Squares (ESS) for a k-cluster partition. The Euclidean Sum of Squares E_p for a cluster p can be generalised as follows:

$$E_p = \sum_{i \epsilon p} n_i \sum_k w_k (x_{ik} - \mu_{pk})^2 \bigg/ \sum_k w_k \qquad (7)$$

where x_{ik} is the value of the kth variable for case i in cluster p, μ_{pk} is the weighted mean of the kth variable for cluster p, n_i is a differential weight for case i (normally 1), and w_j is an differential weight for the kth variable (normally 1). Missing values are allowed for by setting $w_k = 0$ if x_{ik} or μ_{pk} are missing for the kth variable, following Gower (1971).

The total ESS for a k-cluster partition is thus $E = \sum_p E_p$. To minimise E, a case i assigned to a cluster p should be relocated to another cluster q *only* if the move reduces E, that is if:

$$E_p + E_q > E_{p-i} + E_{q+i} \qquad (8)$$

We call (8) the "exact assignment test" for minimum E. It is not the same as assigning a case i to the nearest cluster mean, as described in section 2, because any relocation from cluster p to cluster q also changes the means of p and q; and in certain circumstances, these changes may actually increase E (discussed below). Rearranging (8) is equivalent to moving a case i from p to q if:

$$I_{p-i,i} > I_{q,i} \qquad (9)$$

where I_{pq} is the increase in E on the union of two clusters p and q. Wishart (1978) gives an unbiased estimate of I_{pq} in the presence of missing values:

$$I_{pq} = \frac{n_p n_q}{(n_p + n_q)} \sum_k w_k (\mu_{pk} - \mu_{qk})^2 \bigg/ \sum_k w_k \qquad (10)$$

where $n_p = \sum_{i \epsilon p} n_i$ is the sum of the case weights n_i for cluster p; and $\mu_{pk} = \sum_{i \epsilon p} n_i x_{ik} \Big/ \sum_{i \epsilon p} n_i$ is the mean of the kth variable for cluster p, estimated from the complete observations. Applying (10) to (9), after some arithmetic, a case i should move from cluster p to cluster q iff:

$$\frac{n_p n_i}{(n_p - n_i)} \frac{\sum_k w_{ipk}(x_{ik} - \mu_{pk})^2}{\sum_k w_{ipk}} > \frac{n_q n_i}{(n_q + n_i)} \frac{\sum_k w_{iqk}(x_{ik} - \mu_{qk})^2}{\sum_k w_{iqk}}$$

(11)

where μ_{pk} is the mean of the kth variable in cluster p *including* case i; n_i is the weight of case i; and $w_{ipk} = w_k$, the weight of the kth variable, if the comparison between x_{ik} and μ_{pk} is valid, or zero if x_{ik} or μ_{pk} is missing. We call (11) the "exact assignment test" for k-means analysis, as implemented in Clustan (Wishart (1970, 1984, 1999)), because it satisfies (8) and (9) and therefore reduces E. Since E is a sum of squares and each reassignment that satisfies (8) *actually reduces E*, the algorithm is guaranteed to converge in finite time. This is because E, being bounded by zero, cannot be indefinitely reduced. It should be noted that the procedure may not converge for incomplete data, because the cluster means and criterion values are estimated. However, it appears to work reliably with incomplete data in practise.

When the procedure converges the resulting k-cluster partition satisfies the conventional k-means goal (section 2) because every case is closest to the mean of the cluster to which it is assigned. However, (11) is *not* equivalent to moving a case i from cluster p to cluster q if $d_{ip} > d_{iq}$ as implemented in most software packages. Put another way, moving a case i from p to q pulls the mean $\mu_{q+i,k}$ of q towards it and pushes the mean $\mu_{p-i,k}$ of p away from it. This can cause the distances from the means to other cases in p and q to increase such that E increases. With very large datasets, boundary cases can oscillate between two or more clusters in successive iterations causing the conventional k-means procedure to fail to converge. This is presumably why some software packages require additional *ad hoc* convergence criteria to be specified, such as stopping the process when a complete iteration fails to move a cluster mean by more than a specified percentage of the smallest distance between any of the means. MacQueen (1967) recognises this defect in his k-means algorithm but failed to identify the exact assignment test (11) that corrects it.

7 Differential Weighting

It is interesting to compare the contribution to E (7) of the different types of variables. Range standardisation (4) is difficult to justify for continuous variables because the transformed values are wholly determined by two extremes x_{max} and x_{min}, and the same criticism applies to Gower's range standardisation (2). Standardisation by the variance (5) contributes $\sum_i n_i$ to E, or n if the cases have equal weight. A binary variable contributes $(1 - \phi_k)\phi_k \sum_i n_i$ to E, where ϕ_k is the probability that the kth attribute is present. It can therefore be argued that a binary variable should be given a weight of $1/((1 - \phi_k)\phi_k)$ to make its contribution to E comparable to a continuous variable standardised by its variance. Similarly, a nominal variable contributes $\sum_s(1 - \phi_{ks})\phi_{ks} \sum_i n_i$ to E where $\{\phi_{ks}\}$ are probabilities for each state s occurring in the kth variable. It should therefore be given a weight of $1/\sum_s(1 - \phi_{ks})\phi_{ks}$ to make it comparable to a standardised quantitative variable.

The use of differential weights n_i for cases in (7) and (11) provides extra flexibility. For example, in a customer segmentation survey the cases might be weighted by customer transaction volumes, thereby giving greater emphasis to those who contribute most to business turnover. Such case weighting also provides a surrogate cluster analysis of individual business transactions.

8 Outlier Detection and Deletion

k-means analysis is almost always expressed in terms of finding a partition of all the cases into k clusters. Such a classification is illustrated for the Hertzsprung-Russell diagram of visual stars shown in Fig. 1 (left). In this example of a 60-cluster partition, every point is forced to join a cluster including all the outliers outside the dwarf and giant star sequences. However, we often wish to exclude outliers from a classification to focus on describing the densest part of the sample space. This can be done by specifying an outlier deletion threshold δ to exclude any case that is remote from a cluster by a distance greater than δ to the mean. The resulting clusters then have a maximum radius from the mean of δ. They are not distorted by the presence of outliers and are thus compact groups of points and hence much more descriptive. Fig. 1 (right) illustrates a 40-cluster partition of the H-R diagram where the outliers have been deleted. The result is a tessellation of tight spherical clusters that maps the two main sequences of giant and dwarf stars, while the outliers are identified separately as remote points lying outside the two main tessellations.

Outlier detection and deletion is important when extreme or inconsistent cases would otherwise unduly influence the criterion function and hence distort the resulting cluster model. Recalibration of variables following outlier deletion may also be necessary, for example to remove the influence of extreme outliers on the range r_k or variance σ_k^2 where they are used to standardise quantitative variables (4) and (5).

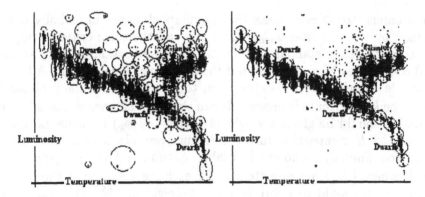

Figure 1: Hertzsprung-Russell plot of visual stars, with the giant and dwarf star sequences described by tessellations of 60 clusters (left) without outlier deletion, and 40 clusters (right) with outliers deleted.

9   Summary Tree

A k-cluster partition can now be summarised by classifying the cluster means hierarchically. However, cluster means may not always be the best way of describing a classification, particularly where a nominal variable is expressed as a vector of probabilities corresponding to the attribute states (section 5). We look for the largest probability as illustrating the response that typifies the cluster, and this is equivalent to selecting the modal or most typical case. For a k-cluster partition this involves selecting the cases that are nearest to the cluster means, as discussed by Diday and Simon (1976). It is similar to the method of medoids proposed by Kaufman and Rousseeuw (1960) but simpler to compute in a k-means context.

With Clustan (Wishart (2002)) it is also possible to obtain a summary tree for a k-cluster partition that displays both the relationships between the k clusters and also the composition of each cluster. It is obtained by classifying the membership of each cluster separately to show how the cases combine hierarchically in a sub-tree to constitute each cluster. A discussion with examples can be found at www.clustan.com/ kmeanstree.html

10   Modified k-Means Clustering Algorithm

The k-means clustering procedure is now extended for mixed data, missing values and outlier deletion as follows:

1. Choose an initial partition of the cases into k clusters. This can be a tree section, k selected cases, or a random assignment to k clusters.

2. Compute the smallest distance d_i from each case i to the means of the clusters. If $d_i > \delta$, where δ is the outlier deletion threshold, assign case i to a residue set of outliers.

3. If $d_i \leq \delta$, move case i from its current cluster p to another cluster q iff $I_{p-i,i} > I_{q,i}$ i.e. if test (11) is satisfied.

4. Re-compute the cluster means following any change of cluster membership at steps 2 and 3.

5. Repeat steps 2, 3 and 4 until no further changes of cluster membership occur in a complete iteration. The procedure has now converged to a stable k-cluster partition.

This revised k-means algorithm requires the maximum cluster radius δ to be specified. If it is not specified, then no outliers are identified and removed, as in Fig 1 (left). Alternatively, with Clustan (Wishart (2002)) a percentage p of outliers can be specified. For example, $p = 5\%$ would mean that 5% of the cases are to be treated as outliers. The program then computes δ such that the 5% of cases having the largest d_i at step 2 are treated as outliers and excluded. The algorithm has been tested on a million cases and shown to converge in finite time on a standard PC - see www.clustan.com for further details.

References

BALL, G. H. (1965): Data analysis in the social sciences: What about the details?. *Proc. Fall Joint Computer Conf., Spartan Books, Washington D.C., Vol. 27 (1), 533–539.*

BALL, G. H. and HALL, D. J. (1967): A clustering technique for summarizing multivariate data. *Behavioral Science, Vol. 12, 153–155.*

BEALE, E. M. L. (1969): Euclidean cluster analysis. *Bull. I. S. I., Vol. 43 (2), 92–94.*

DIDAY, E., and SIMON, J. C. (1976): Cluster analysis, in Fu, K. S. (Ed): Digital pattern recognition. Springer, Berlin, 47–94.

FORGEY, E. W. (1965): Cluster analysis of multivariate data: efficiency versus interpretability of classifications. *Biometrics, Vol. 21, 768–769.*

GOWER, J. C. (1971): A general coefficient of similarity and some of its properties. *Biometrics, Vol. 27, 857–874.*

JANCEY, R. C. (1966): Multidimensional group analysis. *Austral. J. Botany, Vol. 14 (1), 127–130.*

KASS, G. V. (1980): An exploratory technique for investigating large quantities of categorical data. *Applied Statistics, Vol. 29, 119–127.*

KAUFMAN, L. and ROUSSEEUW, P. J. (1960): Finding groups in data. Wiley, New York.

MacQUEEN, J. (1967): Some methods for classification and analysis of multivariate observations. *Proc. 5th Berkeley Symp., Vol. I, 281–297.*

THORNDIKE, R. L. (1953): Who belongs in the family. *Psychometrika, Vol. 18, 267–276.*

WISHART, D. (1970): Some problems in the theory and application of the methods of numerical taxonomy. Ph.D. dissertation, University of St. Andrews.

WISHART, D. (1978): Treatment of missing values in cluster analysis. *Proc. Compstat 1978, Physica-Verlag, Wien, 281–287.*

WISHART, D. (1984): Clustan Benutzerhandbuch. Gustav Fischer Verlag, Stuttgart, 46–54.

WISHART, D. (1986): Hierarchical cluster analysis with messy data, in: Gaul, Schader, (Eds.): Classification as a Tool of Research. North-Holland, Amsterdam, 453–460.

WISHART, D. (1999): ClustanGraphics Primer. Clustan, Edinburgh, 37–38.

WISHART, D. (2002): Clustan Professional User Guide. Clustan, Edinburgh (in preparation).

Web Mining, Data Mining and Computer Science

XX

Web Mining, Data Mining and Computer Science

Repeat-buying Theory and its Application for Recommender Services

W. Böhm[1], A. Geyer-Schulz[2], M. Hahsler[3], and M. Jahn[3]

[1]Mathematische Methoden der Statistik, WU-Wien, A-1090 Wien, Austria

[2]Informationsdienste und Elektronische Märkte, Universität Karlsruhe (TH), D-76128 Karlsruhe, Germany

[3]Informationswirtschaft, WU-Wien, A-1090 Wien, Austria

Abstract: In the context of a virtual university's information broker we study the consumption patterns for information goods and we investigate if Ehrenberg's repeat-buying theory which successfully models regularities in a large number of consumer product markets can be applied in electronic markets for information goods, too. First results indicate that Ehrenberg's repeat-buying theory succeeds in describing the consumption patterns of bundles of complementary information goods reasonably well and that this can be exploited for automatically generating anonymous recommendation services based on such information bundles. An experimental anonymous recommender service has been implemented and is currently evaluated in the Virtual University of the Vienna University of Economics and Business Administration at http://vu.wu-wien.ac.at.

1 Introduction

In this article we study anonymous recommender services based on consumption patterns for information goods as presented in figure 1 showing a list of recommended web-sites of courses which are recommended because students usually use them together with M. Hahsler´s Introduction to C++. For a discussion of the design space for recommender services we refer the reader to Resnick et al. (1997).

In this setting we consider an information broker with a clearly defined system boundary. Clicking on an external link is equated as "purchasing an information product". The rationale for this stems from an analysis of the transaction costs of a user of an information broker. Even "free" information products burden the consumer with search, selection, and evaluation costs. Therefore, in this article we derive recommendations from products which have been repeatedly used (= purchased) together in the same sessions (= purchase occasions). The following advantages make such recommendations attractive for information brokers:

- Observed consumer purchase behavior is the most important information for predicting consumer behavior online and offline as claimed in Bellmann et al. (1999).

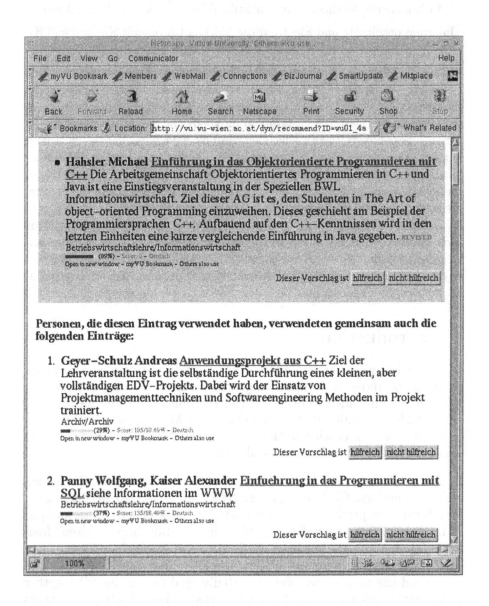

Figure 1: Example: An Anonymous Recommender Based on "Observed Purchase Behavior"

- Market basket analysis shows up to 70 percent cross-selling potential. See, e.g. Blischok (1995). Such recommendations facilitate
  "repeat-buying", as suggested in Bellmann et al. (1999).
- Such recommendations are not subject to several incentive problems found in systems based on explicit recommendations as free-
  riding,
  bias, ... See Avery and Zeckhauser (1997). Faking such recommendations leads to high transaction costs because only a single
  co-occurrence of products is counted per user-session. Free-riding
  is impossible because each user automatically contributes usage
  data for the recommendations. The user's privacy is preserved.
- The transaction costs for the broker are low because such recommendations can be generated without editor, author, or web-scout.

Anonymous recommendations based on consumption patterns have been
made famous by Amazon.com and e.g. the first phase of the a-priori algorithm for association rules of Agrawal and Srikant (1994) can compute
anonymous recommendations – even without a model and its underlying
behavioral assumptions – quite well. The reason for this is that the first
few entries of such frequency sorted lists of recommendations usually are
good candidates for recommendations. Provided usage is counted in the
same way and the thresholds for the support and confidence parameters
of the a-priori algorithm are set to 0 the raw frequency-sorted recommendation lists produced by phase 1 of the a-priori algorithm will be
the same as in this article. However, all such approaches still have the
following two problems which we address in this article with Ehrenberg's
repeat-buying theory (see Ehrenberg (1988)): Which co-occurrences of
products are regarded as non-random? And how many products should
we recommend? Showing random co-occurrences of products runs a high
risk of giving bad recommendations (type I error), whereas suppressing
non-random co-occurrences of products implies that possible useful recommendations are not given (type II error).

Ehrenberg's repeat-buying theory is a descriptive theory based on consumer panel data well suited for this task because of the strong stationarity and independence assumptions in the theory discussed in section
2 and because it has been supported by strong empirical evidence in
consumer product markets since the late 1950's. Although quite sophisticated and general models of the theory (e.g. the Dirichlet model
(Goodhardt et al. (1984)) exist, for our purposes the simplest model –
the logarithmic series distribution (LSD) model – will be sufficient. For
a survey see e.g. Wagner and Taudes (1987).

The careful and experienced reader certainly will ask at this point, how
we can apply a theory for analyzing purchase histories from consumer
panels to mere market baskets. A market basket contains all information

products which a user has visited (= purchased) in a session (= purchase occasion), but not the identity of the consumer. In a consumer panel, in addition, the identity of each user is known. The purchase history of a consumer is just the sequence of the purchases in his market baskets. Well, the answer is simple: We consider anonymous market baskets as consumer panels with **unobserved consumer identity** – and as long as we work only at the aggregate level everything works out fine. And keep in mind that for repeat-buying analysis we count all occurrences of an information product in a market basket just once.

2 Ehrenberg's Repeat-Buying Theory for Bundles of Information Goods

Of the thousand and one variables which might affect buyer behavior, it is found that nine hundred and ninety-nine usually do not matter. Many aspects of buyer behavior can be predicted simply from the penetration and the average purchase frequency of an item, and even these two variables are interrelated. A.S.C. Ehrenberg (1988).

The key result of Ehrenberg's repeat-buying theory which we exploit in this paper for anonymous recommender services is that the logarithmic series distribution (LSD) describes the following frequency distribution of purchases (Ehrenberg (1988)), namely how many buyers buy a specific product $1, 2, 3, \ldots$ times (without taking into account the number of non-buyers)?

$$P(r \quad \text{purchases}) = \frac{-q^r}{r \ln(1 - q)}, \quad r \geq 1, |q| < 1 \tag{1}$$

$$\text{Mean purchase frequency} \quad w = \frac{-q}{(1 - q) \ln(1 - q)} \tag{2}$$

In purchasing a product a consumer basically makes two decisions: when does he buy a product of a certain product class (purchase incidence) and which brand does he buy (brand choice)? Ehrenberg claims that almost all aspects of repeat-buying behavior can be adequately described by formalizing the purchase incidence process for a single brand and to integrate these results later. The logarithmic series distribution results from the following assumptions about the consumers' purchase incidence distributions:

1. The share of never-buyers in the population is not specified. In our setting this definitely holds.
2. The purchases of a consumer in successive periods follow a Poisson distribution with a certain long-run average μ.

The purchases of a consumer follow a Poisson distribution in subsequent periods, if a purchase tends to be independent of previous purchases (as is often observed) and a purchase occurs in such an irregular manner that it can be regarded as random (see Wagner and Taudes (1987)).

3. The distribution of μ in the population follows a truncated Γ-distribution, so that the frequency of any particular value of μ is $(ce^{-\mu/a}/\mu)d\mu$, for $\delta \leq \mu \leq \infty$, where δ is a very small number, a a parameter of the distribution, and c a constant, so that $\int_\delta^\infty (ce^{-\mu/a}/\mu)d\mu = 1$.

A Γ-distribution of the μ_i in the population may result from the following independence conditions (see Ehrenberg (1988)): For different products $P, Q, R, S, \ldots$ the average purchase rate of P is independent of the purchase rates of the other products, and $\frac{P}{(P+Q+R+S+\ldots)}$ is independent of a consumer's total purchase rate of buying all the products. These independence conditions are likely to hold approximately in practice.

4. The market is in equilibrium (stationary).

For the sake of completeness (and to correct a persistent typesetting error in the original proof), we include the following short proof which is due to Chatfield (see Chatfield et al. (1966) and Ehrenberg (1988)):

1. The probability p_r of r purchases is Poisson distributed: $p_r = \frac{e^{-\mu}\mu^r}{r!}$
2. We integrate over all buyers in the truncated Γ-distribution:

$$p_r = c\int_\delta^\infty \left(\frac{e^{-\mu}\mu^r}{r!}\right)\left(\frac{e^{-\mu/a}}{\mu}\right)d\mu = \frac{c}{r!(1+\frac{1}{a})^r}\int_\delta^\infty e^{-(1+\frac{1}{a})\mu}\left((1+\frac{1}{a})\mu\right)^{r-1}$$

$d\left((1+\frac{1}{a})\mu\right)$. Since δ is very small, for $r \geq 1$ this is approximately $\left(\frac{c}{r!(1+\frac{1}{a})^r}\right)\Gamma(r) = \frac{c}{(1+\frac{1}{a})^r r} = c\frac{q^r}{r} = qp_{r-1}(r-1)/r$, with $q = \frac{a}{1+a}$.

3. If $\sum p_r = 1$ for $r \geq 1$, we get $p_1 = \frac{-q}{\ln(1-q)}$ and $p_r = \frac{-q^r}{r\ln(1-q)}$.

Next, consider for some fixed information product x in the set X of information products in the broker the purchase frequency of pairs of (x, i) with $i \in X - \{x\}$. The reason for considering pairs of information products is that we expect complementarities between information products because Internet users usually tend to use several information products for a task. Because of the independence assumptions outlined above the frequency distribution that such pairs occur $1, 2, 3, \ldots -$ times follows a logarithmic series distribution by the same line of reasoning as above. And we expect that non-random occurrences of such pairs occur more often than predicted by the logarithmic series distribution. A *recommendation* in this setting simply implies that co-occurrences occur more often than expected from independent random choice acts and that a recommendation reveals a complementarity between information products. We use the stochastic model as a benchmark for discovering regularities.

Finally, we present a short overview of the algorithm for computing recommendations:

1. *Compute for all information products x in the market baskets the frequency distributions for repeat-purchases of the co-occurrences of x with other information products in a session, that is of the pair (x, i) with $i \in X - \{x\}$.*
2. *Discard all frequency distributions with less than l observations. (We set $l < 10$ in order to prune frequency distributions which are unlikely to lead to a significant LSD model. More than 80% of these frequency distributions contain no repeat-buys in our data. For the rest, a χ^2-goodness-of-fit test should not be used.)*
3. *For each frequency distribution:*
   (a) *Compute the **robust** mean purchase frequency w by trimming e.g. the 2,5 percentil of the high repeat-buy pairs.*
   (b) *Estimate the parameter q for the LSD-model from*
      $$w = \frac{-q}{(1-q)(\ln(1-q))} \text{ with either a bisection or Newton method.}$$
   (c) *Apply a χ^2-goodness-of-fit test with a suitable α (e.g. 0.01 or 0.05) between the observed and the expected LSD distribution with a suitable partitioning.*
   (d) *Determine the outliers in the tail. (We suggest to be quite conservative here: Outliers at r are above $\sum_r^\infty p_r$.)*
   (e) *Finally, we prepare the list of recommendations for information product x if we have a significant LSD-model with outliers.*

3   Data Set and Results

The data set used for the example shown in figures 2 and 3 is from the anonymous recommender services of the Virtual University of the Vienna University of Economics and Business Administration (http://vu.wu-wien.ac.at) for the observation period from 1999-09-01 to 2001-03-05. The agent architecture as well as the data collection techniques have been described in Geyer-Schulz et al. (2001a). Personalized recommendations based on self-selection combined with interactive evolutionary algorithms can be found in Geyer-Schulz et al. (2000). The potential of recommender systems in education and scientific research are discussed in Geyer-Schulz et al. (2001b). A revised and expanded version of this contribution was presented at the WEBKDD2001 conference and is under review (Geyer-Schulz et al. (2001c)). This (later) version presents the model in more detail. Special emphasis in Geyer-Schulz et al. (2001c) is on the identification of non-random outliers, on the discussion of detailed results, and on the validation of recommendations.

Intelligent Software Agents (CMU)

Persons using the above entry
used the following entries too:
  1. Intelligent Software Agents (Sverker Janson)
  2. agent (Definition and Links from webopedia) META
  3. Intelligent Software Agents on the Internet
     (Björn Hermans) META
  4. Mobasher - List of Publications Personalization
     and Adaptive Web Sites from Web-Usage Patterns.
  5. Intelligent Software Agents and New Media
=== Cut ==
  6. Geyer-Schulz Intelligente Internet Agenten
  7. Agent Technology Projects
     in the Stanford Digital Library
  8. vista's virtual friends
  9. Books on Software Agents
  10. AVALANCHE - Agent-Based Value Chain Experiment
  11. The Zeus Agent Building Toolkit
=== Cut ==
  12. Let's Browse: A Collaborative Web Browsing Agent
  13. German Agent Pages
  14. Mobile Service Agents
=== Cut ==
  15. Foundation for Intelligent Physical Agents
  16. Auctions and Bargaining in Electronic Commerce
  . . .

Figure 2: List of entries with cuts

In figure 2 we show the first 17 recommendations generated for the research site **Intelligent Software Agents (CMU)** by the method described in the previous section. 101 other information products have been found in market baskets together with this research site. The (robust) mean purchase frequency is 1.556, the parameter q of the LSD-model is 0.562. A χ^2 goodness-of-fit test is highly significant ($\chi^2 = 10.763$ which is considerably below 30.144, the critical value at $\alpha = 0.05$).

We regard outliers whose observed repeat-purchase frequency is above the theoretically expected frequency as recommendations. In figure 3 we explore three options of determining the cut-off point for such outliers:

  1. Without doubt, as long as the observed repeat-purchase frequency is above the cumulated theoretically expected frequency, we have detected outliers. In our example, this holds for all observations of more than 5 co-purchases which correspond to the top 5 sites shown

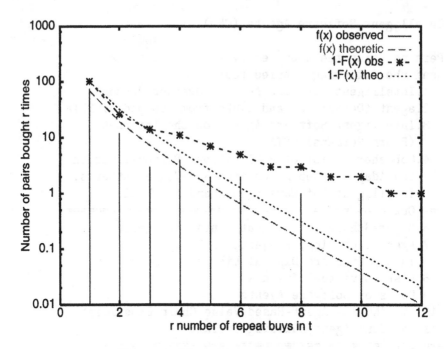

Figure 3: Plot of log distributions

as recommendations in figure 2. (This is the most conservative choice.)

2. Discounting any model errors, as long as the observed repeat-purchase frequency is above the theoretically expected frequency is a less conservative option. For the example, we select all co-purchases with more than 3 occurrences as recommendations. See the top 11 sites in figure 2.

3. If we take the cut, where both cumulative purchase frequency distributions cross, we get 14 recommendations regarding all co-purchases occurring more than twice as nonrandom – the top 14 sites in figure 2.

However, we recommend that the most conservative approach should be implemented. This recommendation is based on a check of the face validity of the recommendations for a small sample of information products (25 products).

As summarized in table 1 we fitted a LSD-model for the frequency distributions of co-occurrences for 1300 information products. For 675 information products, that is more than 50 percent, the estimated LSD-models pass a χ^2 goodness-of-fit test at $\alpha = 0.01$. A small scale face validation experiment of inspecting every entry in 100 randomly selected recommendation lists for plausibility led to a quite satisfactory result: 87.71 % good recommendations (LSD significant, 31 lists), and 89, 45

	Number	%
Number of information products	9498	100.00
Number of products bought together with other products	7150	75.28
Not a uniform distribution and $n > 9$	4582	48.24
Enough repeat buys to compute LSD parameter and χ^2 test	1300	13.69
LSD with $\alpha = 0.01$ (robust)	675	7.11
LSD not significant	625	6.58
LSD fitted, no χ^2 test	703	7.40

Table 1: Summary of results. (Observation period: 1999-09-01 – 2001-05-07)

% good recommendations (LSD not significant, 42 lists). Only 75.74 % good recommendations were found for those LSD-models where no χ^2 test could be computed (25 lists), which is significantly lower.

Surprisingly, the class of models where the LSD model was not significant contains a slightly higher number of recommendations evaluated as good. A close inspection of frequency distributions for these lists revealed the quite unexpected fact that several of these frequency distributions were for information products which belong to the oldest in the data set and which account for many observations. The reasons for this may be explained by a shift in user behavior (non-stationarity) or too regular behavior (e.g., for cigarettes in consumer markets).

Also, the fact that the data set contains information products with different age may explain some of these difficulties. However, to settle this issue further investigations are required.

Finally, table 1 indicates that identification of non-random outliers is important for the perceived quality of recommender systems because of the high risk of recommending random co-occurrences of products. The fact that recommendations for LSD-models for which no χ^2 test could be computed were considered as containing significantly more bad recommendations supports this conclusion.

4 Further Research

The main contribution of this paper is that Ehrenberg's classical repeat-buying models describe – despite their strong independence and stationarity assumptions – the consumption patterns of information products surprisingly well. For anonymous recommender services they do a remarkable job of identifying non-random repeated-choice acts of consumers of information products which serve as the base of automatically generated recommendations of high-quality.

However, the current version of the anonymous recommender services (and the analysis in this article) still suffers from two deficiencies. The first is that new information products are daily added to the information broker's data base so that the stationarity assumptions for the market are violated and the information products in the data set are of non-homogenous age. The second drawback is that testing the behavioral assumptions of the model as well as validation either by studying user acceptance or by controlled experiments still has to be done.

In addition, we expect Ehrenberg's repeat-buying models to be of considerable help to create anonymous recommender services for recognizing emerging shifts in consumer behavior patterns (fashion, emerging trends, moods, new subcultures, ...) and imbedded marketing research services which provide forecasts and classical consumer panel analysis in a cost-efficient way.

5 Acknowledgement

The financial support of the Jubiläumsfonds of the Austrian National Bank under Grant No. 7925 is gratefully acknowledged.

References

AGRAWAL, R. and SRIKANT, R. (1994): Fast Algorithms for Mining Association Rules, Proceedings of the 20th VLDB Conference, Santiago, 487–499.

AVERY, C. and ZECKHAUSER, R. (1997): Recommender Systems for Evaluating Computer Messages. *Communications of the ACM, 40(3)*, 88–89.

BELLMANN, S., LOHSE, G.L., and JOHNSON, E.J. (1999): Predictors of Online Buying Behavior. *Communications of the ACM, 42(12)*, 32–38.

BLISCHOK, T.J. (1995): Every Transaction Tells a Story. *Chain Store Age Executive, 71(3)*, 50–62.

CHATFIELD, C., EHRENBERG, A.S.C., and GOODHARDT, G.J. (1966): Progress on a Simplified Model of Stationary Purchasing Behavior. *Journal of the Royal Statistical Society A*, Vol. 129, 317-367.

EHRENBERG, A. S. C. (1988): *Repeat-Buying: Facts, Theory and Applications*. Charles Griffin & Company Limited, London.

GEYER-SCHULZ, A., HAHSLER, M., and JAHN, M. (2000): myVU: A Next Generation Recommender System Based on Observed Consumer Behavior and Interactive Evolutionary Algorithms. In: W. Gaul, O. Opitz, M. Schader (Eds.): *Data Analysis – Scientific Modeling and Practical Applications*, Studies in Classification, Data Analysis, and Knowledge Organization, Vol. 18, Springer, Heidelberg, 447-457.

GEYER-SCHULZ, A., HAHSLER, M., JAHN, M. (2001a): Recommendations for Virtual Universities from Observed User Behavior. In: W. Gaul, Ritter, M. Schader (Eds.): Studies in Classification, Data Analysis, and Knowledge Organization, Springer, Heidelberg, to appear.

GEYER-SCHULZ, A., HAHSLER, M., JAHN, M. (2001b): Educational and Scientific Recommender Systems: Designing the Information Channels of the Virtual University. *International Journal of Engineering Education*. Vol. 17, N. 2, 153-163.

GEYER-SCHULZ, A., HAHSLER, M., JAHN, M. (2001c): A Customer Purchase Incidence Model Applied to Recommender Services. Submitted Procs. WEBKDD2001, LNCS, Springer, 20 pages.

GOODHARDT, G.J., EHRENBERG, A.S.C., COLLINS, M.A. (1984): The Dirichlet: A Comprehensive Model of Buying Behaviour. *Journal of the Royal Statistical Society*, A, 147, 621-655.

RESNICK, P. and VARIAN, H.R. (1997): Recommender Systems. *Communications of the ACM, 40(3)*, 56-58.

WAGNER, U. and TAUDES, A. (1987): Stochastic Models of Consumer Behaviour. *European Journal of Operations Research, 29(1)*, 1-23.

Joker – Visualization of an Object Model for a Cost Accounting Educational Software

K. Friesen[1,2], H. Schmitz[2]

[1]Lehrstuhl für Wirtschaftsinformatik III
University of Mannheim
D-68131 Mannheim, Germany

[2]Lehrstuhl für Allgemeine Betriebswirtschaftslehre
und Industrie, insb. Produktionswirtschaft und Controlling
University of Mannheim,
D-68131 Mannheim, Germany

Abstract: The development of educational software faces different problems. One of the major problems is how to find an appropriate way of visualization. The problem discussed in this paper is the visualization of a complex data structure for learners. For this problem, two requirements can be considered to be most important from a semantic point of view: a learner should be able to easily navigate the data structure of a learning software, and the visualization should improve the understanding of the underlying model.

A suitable technical solution has to provide the necessary flexibility. For applicability in different problem domains, the separate implementation of semantic and visualization aspects is required. In addition, a visualization technique should be able to support database interfaces, as databases are an efficient way to store complex data.

The solution presented in this paper is developed for a common technical environment. It is implemented by a framework of Java classes and tested in practice. The solution can be re-used for other problem domains. The first part of the paper deals with semantic modeling issues in cost accounting and presents some aspects of the underlying object model of the educational software Joker. The second section describes the software environment of the solution. The following two parts explain the concept and implementation.

1 A Short Introduction to the Cost Accounting Object Model

Joker ("Javabasierte objektorientierte Lernumgebung zur Kosten- und Erlösrechnung"[1], Friesen, Schmitz (2001)) is an educational software that can be used for cost accounting courses. It is designed for both demonstration purposes in class and for computer-based learning at home. The educational software includes a fully functional cost accounting software. Its functionality can be used in all kinds of learning modules. Joker trains the understanding of cost accounting processes, not arithmetic skills and has been used in practice for two years in undergraduate and graduate courses at different universities. Continuous improvements are part of

---

[1]"Java-based object-oriented learning environment for cost accounting"

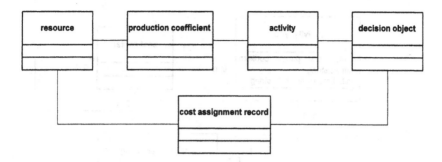

Figure 1: Basic cost accounting pattern

the development process, based on the feedback of the several hundred users.

The software is written in the Java programming language and is based on an object model including 40 different classes in the analysis model. The model is the result of a construction approach used for analysis purposes.

Construction approaches are starting with a general analysis of the problem domain on an abstract level. Thus their starting point is not a single modeling task, e.g. the cost accounting system of a specific enterprise. This approach leads to highly reusable models. The REA Accounting model is an example for the use of construction methods in Financial Accounting. It shows that a double entry bookkeeping system can be modeled by sets of economic resources, economic events and economic agents plus relationships among those sets (McCarthy (1982)).

The core of the object model is based on a relatively simple pattern. Cost accounting itself models the production and sales of goods and services by the valuation of quantities of resources, goods and services. Cost accounting introduces a specific view on existing data structures, which adds new aspects to this data. Thus a basic pattern that includes resources, objects that induce resource consumption and linking coefficients can be used as a starting point. The pattern contains a class **Resource** as the input element, the **Decision Object** class and **Activity** class as output elements and a linking class incorporating the input-output-ratio. To derive a fundamental cost accounting pattern from this first simple model, a class has to be added, that records the specific cost accounting data, the value of a transaction.

Figure 1 shows the basic cost accounting pattern in the unified modeling language notation (OMG (2000)).

In order to implement dynamic aspects of cost planning and more detailed requirements, the basic pattern shown in figure 1 has to be ex-.

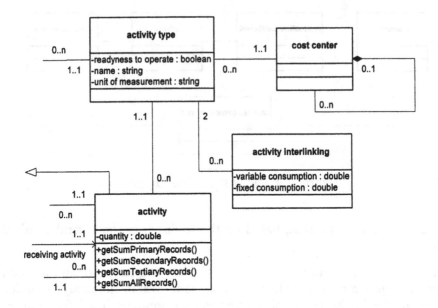

Figure 2: Detail of the Joker object model

tended. This leads to the 40 classes mentioned earlier.

For visualization purposes the most important features of the completely elaborated object model are:

- The model contains highly networked structures, especially directed acyclic graphs.

- The visualization of the model requires several views representing only details of the whole system.

- An object often appears in different contexts.

A detail of the Joker object model is shown in figure 2. It is the extension of the Activity class shown in figure 1. The differentiation of Activity Type and Activity is sensible in order to separate period-dependent data from data that depends on certain periods. Relationships to cost centers and among activity types have to be added to fulfill certain information needs.

Cost centers and activities form a tree structure. This structure has to be combined with a resource hierarchy that forms another part of the complete model.

The detail in figure 2 is a typical structure to be visualized. The learner has to deal with multiple interlinked hierarchies.

2   Covered Implementation Aspects

The object model introduced in the last section is implemented as the core part of Joker. Nevertheless, because of its extensive functionality, it may also serve as a base for a cost accounting software in small- and medium-sized enterprises.

For this software, a Java-based graphical user interface (GUI), operating independently on an ODMG (Cattell, Barry (2000)) compliant database management system was developed. It contains standard components such as trees, lists, and tables, but is enhanced by database sensitivity.

Before going deeper into the details, let us define the goals associated with a database driven GUI toolkit.

- The toolkit should allow the user and the developer to have transparent access to the underlying database. The user, in the best case, should not even notice he is using a database.

- It should be easy for non-programmers to customize the data sensitive components, e.g., by using an integrated developing environment or by editing text files.

- The new components have to be stable and independent of each other in a multi-threaded environment. Thus they can be used simultaneously (thread safety). If necessary, they have to update themselves in order to avoid the presentation of outdated information.

- The GUI should flexibly support multiple display representations for database objects. Ergonomic software often associates a given object type with an icon. Therefore the GUI should allow an object to be represented by a string or an icon.

- All the functionality listed above should be achieved with minimal performance penalties.

3   The Extension of the Model-View-Controller Pattern

Known as the "Model-View-Controller" pattern, the separation of model and view is one of the most important patterns in software engineering. According to Gamma et al. (1995, pp. 4–5), this pattern "consists of three kinds of objects. The model is the application object, the view is its screen, and the controller defines the way the user interface reacts to user input." A vast number of sub-patterns is derived from this fundamental technique. Normally the view and the controller parts of a component require a tight coupling, so these two entities are often "collapsed into a single UI (user-interface) object" (Fowler (1998)).

This applies to the Java Swing toolkit, which is used by Joker. It supports this pattern by defining a model interface for almost every visual component. The model supplies the data whereas the visual component displays them.

To keep the examples and diagrams simple, we will only refer to a data sensitive *tree* component in the following. It can be used to display the detail of the Joker object model shown in figure 2 and is a sufficient example for showing how we achieve our goals using a "double concatenation" of the Model-View-Controller pattern. Nevertheless all components for Java have already been implemented and are working. They include trees, tables, tree tables, lists, combo boxes, text fields, text labels, and a number of minor components.

For example, an object of the `javax.swing.JTree` class shown in figure 3 obtains its data from an object which implements the `javax.swing.-tree.TreeModel` interface. Often the standard implementations provided by the Java Swing toolkit are sufficient and widely used. The advantage is that one model may have multiple view representations, so the same data can be displayed in different ways.

In the database context, the usage of this pattern is less obvious. We suppose that all the relevant objects of the data model are persistent and can be stored in a database. For simplification reasons, any persistent object is subsumed under `java.lang.Object` in figure 3. Most database vendors provide specialized operation listeners that can be registered to any object to be notified on update, insert, or delete actions. Basically, an operation listener does not do more than to observe a database for potential changes. So one may consider the persistent, database-stored objects as "model" objects and any non-persistent object that has registered itself to possible database operations using an operation listener as "view". The coupling is done by the database management system.

For example, to provide a data-sensitive tree component that obtains its data from an ODMG compliant database, one basically has to concatenate two model-view patterns:

1. Connect a `javax.swing.JTree` with a specialized `javax.swing.-tree.TreeModel` (DBTreeModel here) that acts as a data supplier.

2. Connect `DBTreeModel` to the database notification mechanism. The class can, for example, implement the `OperationListener` interface.

So, according to figure 3, it is clear that the class responsible for coupling the database objects with the user interface is `DBTreeModel`. If a change

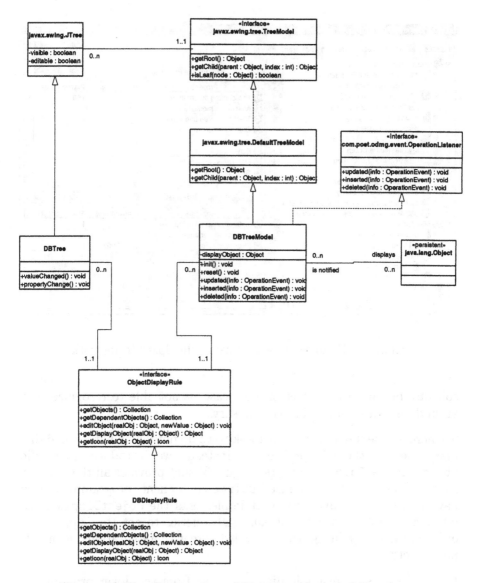

Figure 3: The Connection of User Interface and Database

in a persistent object occurs, the DBTreeModel object is notified, the DBTreeModel checks if their views have to change themselves, and if so, it notifies them. In the opposite direction, the JTree component forwards the user input to the DBTreeModel component, so that the DBTreeModel can alter all the corresponding persistent objects directly. As a consequence, other OperationListeners (if they exist) will then be notified, so they can update themselves accordingly. The results are visual components that provide an up-to-date partial view of an ODMG compliant database. This is a very important feature for educational

246

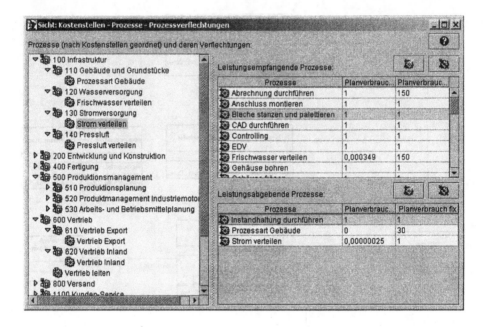

Figure 4: Example of a tree using the Joker framework

software because the learner in most cases is not able to recognize that an update of a visualization is necessary.

To increase the flexibility of our toolkit, we do not define hard-coded display rules in the **DBTreeModel**. Instead, we introduce a specific **ObjectDisplayRule** interface (see figure 3) that provides all the relevant rules. To use a fully customized **DBTree** component, programmers only have to write one Java class that implements the **ObjectDisplayRule** interface. This class may contain OQL queries to get the appropriate objects, rules for editing, rules to get different representations, icon definitions etc.

Figure 4 shows an example dialog window from the Joker program visualizing the detail of the object model presented in figure 2. The tree component on the left displays the cost center and activity structure whereas the two tables list the activity interlinkings of an activity already selected in the tree. The behavior of the dialog window can be configured with three Java classes implementing the **ObjectDisplayRule** interface.

Future **ObjectDisplayRule**s may even include parsers for simple text files. In this case the programmer/user does not have to program his rules in a Java class, but has to write only a few entries in a plain text or XML file. This is a very useful feature for teachers who want to customize Joker for their own courses. They can easily hide aspects of a visualized object that are unimportant for the current topic, or add details that are not provided from the start.

4 Summary

Joker is an example for an educational software with a complex, highly interconnected object model. In order to solve the problem of presenting complex data structures in an appropriate way for learners, an extension of the Model-View-Controller pattern is presented. It combines database sensitivity and flexibility for different kinds of visualizations. The solution enables an educational software to simultaneously provide different partial views of one model to the learner. Because of the integration of widely used standards (ODMG-Database, Java) the suggested solution is applicable to a wide range of similar problems, not only to visualization tasks in educational software.

References

CATTELL, R. G. G. and BARRY, D. K.(Eds.) (2000): The Object Data Standard ODMG 3.0. Morgan Kaufmann Publishers, San Francisco, USA

FOWLER, A. (1998): A Swing Architecture Overview, *The Swing Connection Website*. Available from http://java.sun.com/products/jfc/tsc/articles/architecture/index.html

FRIESEN, K. and SCHMITZ, H. (2001): Joker Summer Term Release 2001, University of Mannheim. http://joker.uni-mannheim.de/downld/joker/

GAMMA, E., HELM, R., JOHNSON, R., and VLISSIDES, J. (1995): Design Patterns: Elements of Reusable Object-Oriented Software. Addison Wesley, Reading/Mass., USA

McCARTHY, W. E. (1982): The REA accounting model: A generalized framework for accounting systems in a shared data environment. *The Accounting Review, Vol. 57, 554–578.*

OBJECT MANAGEMENT GROUP (OMG) (2000): OMG Unified Object Modeling Language Specification (Version 1.3). Available from ftp://ftp.omg.org/pub/docs/formal/00-03-01.pdf

Data Preparation in Large Real-World Data Mining Projects: Methods for Imputing Missing Values

Th. Liehr

EX-A-MINE Center, Infratest Burke,
Landsberger Str. 338, 80 607 Munich, Germany.
The EX-A-MINE Center is the competence center for data mining and data
matching within NFO Europe (in Germany: Infratest Burke)

Abstract: One of the most important aspects in data preprocessing for data
mining concerns the handling and imputation of missing values. While differences in the performance of varying state-of-the-art algorithms on the same
dataset remain usually rather small, the quality of missing value handling can
have dramatic consequences and is often crucial for the success of the following
model building.

This paper explores the consequences of two major missing value replacement
strategies (replace-with-mean and multivariate regression) for the performance
of classification models: By using a complete real-world dataset for a binary
classification problem (churn in financial services), the hit rates of different
data mining algorithms are benchmarked for the case of no missing values
being present. Then, different missing value patterns (MCAR, MAR and IM)
are simulated by deleting predictor values from the training samples following
those patterns. After this, the two imputation strategies (replace with mean
and regression) are used to recreate complete training datasets, in order to
build classification models on them. Finally, the hit rates of the models are
determined on (the original complete, not imputed) hold-out test sets and the
performances of the models are compared.

It is clearly shown, that the regression strategy outperforms by far the simpler replace-with-mean imputation by introducing much less artificial bias in
the data and thus enabling better models to be built. The results underline
the performance advantages of more complex and time-consuming multivariate imputation schemes over the straightforward replace-with-mean techniques
unfortunately implemented in many commercial data mining packages.

1 Introduction

The preparation of the analysis data plays a key role for any complex
real-world data mining project. While differences in the performance
of varying state-of-the-art algorithms on the same dataset remain usually rather small, the quality of data preparation can have dramatic
consequences and is often crucial for the success of the following model
building.

It is agreed upon, that in every large-scale data mining project, different
algorithms should be tested, as even small improvements in performance
can have a large monetary consequence when deploying the model (e.g.,

when used for selection for a mailing). But nevertheless, data preparation has clearly the bigger impact and thus far more project time and effort should be dedicated to it.

The basic idea in pre-processing consists in preparing the data in a way, that the information is most clearly exposed, and thus algorithms have an easier task in finding accurate an robust models. The major components of pre-processing relevant for this goal are:

- Handling missing values: replacement strategies [see Little and Rubin (1987)]

- Handling categorical variables: remapping and "numeration" [see Pyle (1999)]

- Transforming metric variables: distribution and outliers [see Chaterjee (1991) and Tukey (1977)]

- "The curse of dimensionality" in data mining: compressing the variable space [see Pyle(1999)]

The current paper will be limited in dealing with the effect of handling missing values on model performance. The basic reason for realizing imputation of missing values consists in getting a complete dataset for further multivariate analysis, as some of the most widely used algorithms in data mining like (logistic) regression and neural networks need complete datasets. Some (modern) algorithms such as different decision trees and MARS [Friedman(1991)] are able to deal with missing values directly, either by treating them as separate categories (decision trees) or by imputing the missing entries with the constant "0" and capturing the missing value patterns with indicator variables, having value "1" when the entry of the original variable is missing and value "0" elsewhere (MARS).

As imputation of missing values adds information to the dataset, that was not present before, care has to be taken not to bias the previously available information in any direction. This bias - besides from being captured directly by comparing the statistical properties of the original dataset with the imputed dataset - can show its effects in the further modeling phase by affecting model accuracy.

This paper shows the effects of using two different popular techniques for imputing missing values - "replace with mean" and regression imputation - on model accuracy (classification) by using a simulation dataset. In the first step, a complete dataset serves as benchmark for classifier accuracy in presence of complete information. Then, different patterns of missing values are simulated and the missing values are replaced by using the two above mentioned methods (mean and regression). Finally a classifier model using logistic regression is build on the imputed datasets and their accuracy on a hold-out validation sample is measured and compared.

2 Types of Missing Values

Rubin (1976) founded the theory on missing values and identified three types (patterns) of missing values: MCAR, MAR and IM.

MCAR - missing completely at random - describes the situation, where the fact that the entry of a variable is missing does not depend on any other variable in the dataset.

MAR - missing at random - describes the case, where the fact that the entry of a variable is missing depends on the observed values of other variables present in the dataset.

IM - informatively missing - stands for the pattern, where the fact that the entry of a variable is missing depends on the values of other variables present and not present in the dataset.

When the missing value pattern is either MCAR or MAR, the information which entry is missing does not introduce additional information in the dataset. For the IM case, the information which entry is missing depends on other values, not yet present in the dataset and does therefore introduce new information in the dataset.

The simulation of the three patterns of missing values can be realized as follows:

For MCAR, the missing value generating mechanism is a simple random function not being conditional on any other variables. For MAR, the missing value generating mechanism is applied to only a subset of the cases, the selection of this subset being conditional on the observed values of other variables in the dataset. Finally for IM, the missing value generating mechanism is also applied to only a subset of the cases, but this time the selection of the subset is dependent on observed and unobserved values of the dataset. This can be simulated by introducing an additional dummy random variable, that triggers the missing value mechanism, but is ignored for the following imputation process and the model building.

3 Imputation Techniques

Different techniques for the imputation of missing values were proposed in the literature, see e.g., Little and Rubin (1987).

The most simple and thus most widely used approach consists in replacing the missing values with the mean of the remaining observed values of the respective variable. This is often the default handling method for missing values implemented into statistical software packages. Other, more complex imputation techniques take into account the multivariate structure of the dataset by using imputation techniques such as regression, decision trees, neural networks or the EM-algorithm. This results in making the imputed value conditional on observed values of other variables and tends to conserve the multivariate structure of the dataset. The bias introduced in the dataset by the multivariate imputation mechanism tends to be lower than for the unconditional "replace with mean mechanism", that ignores the available information of the inter-variable relationships in the data.

The question is, if this bias is a problem in real-world data mining model building. The major aim of most real-world data mining projects consists in building models with maximum accuracy on validation hold-out data. Our experiences in many database marketing applications tell us, that using "replace with mean" or multivariate imputation techniques does matter in terms of model accuracy. To confirm those practical experiences, a simulation will try to systematically explore these effects. For this purpose, in order to keep the simulation simple, we concentrate on only one multivariate imputation algorithm which is stepwise regression. In our practical projects we have used regression as well as decision tree based imputation methods.

The reason for not choosing any of the more complex imputation techniques like EM-algorithm or neural networks lies in the "real-world" project background of our applications: in real-world data mining (database marketing) projects we often face situations with hundreds of variables where a high proportion of the values are missing. Because in multivariate imputation, at least one model per variable having missing values must be computed, the computational complexity of the algorithm used is an important consideration for project duration. Considering training times for neural networks multiplied by the number of variables with missing values in the dataset (or the time of the convergence process of the EM-algorithm in large data sets) makes the decision for choosing a computationally simpler algorithm straightforward.

Another reason for not considering the more complex algorithms lies in the fact, that in many data mining projects, the model code has to be delivered to the client and is implemented and run on-site in client's analytical systems. Obviously, the whole pre-processing has to be included in the code in order to apply the model on data having undergone the same transformations as the data used for model building. Under this perspective, the volume of code becomes an important criteria in terms of debugging effort and maintenance. Thus algorithms resulting in more compact code have clear advantages in those real-world applications where the deployment of the code is needed.

4 Simulation Results

A reduced dataset from a database marketing application is chosen as the basic analysis dataset. It has been "reduced" in two dimensions: 1) the number of variables has been reduced to the 9 predictors having explanatory power concerning the response variable, and 2) a stratified sample of 10.000 cases has been drawn from the universe, having no missing values in any predictor nor the response variable. The stratified sample has been drawn in a way, that each of the binary response variable's outcomes account for roughly 5.000 cases. The selection of a balanced sample for binary classification problems is a widely used heuristic and is appropriate for this simulation study, but one should keep in mind that recent research [Weiss and Provost (2001)] has shown, that this is not the optimal choice for all datasets and under all circumstances.

This basic dataset is called the "original" dataset in the following text. As a first step, different popular data mining algorithms are tested upon the original dataset, in order to get an idea of how the choice of different algorithms would affect accuracy, always measured on a validation hold-out sample (Table 1). The measure used here for the comparison of model accuracy is: percentage of the validation cases assigned to the right class. I am aware, that especially for database marketing applications, other accuracy measures are better suited like ROC [Weiss and Provost (2001)], LIFT [Berry and Linoff (1999)] or L-Quality [Piatetsky-Shapiro and Steingold (2000)], but their specific strengths are not relevant for the effects I want to show here.

Neural Network (Cascade Correlation)	83,1 %
Neural Network (MLP)	82,9 %
Logistic Regression	82,1%
Linear Discriminant Analysis	82,0 %
CART	80,1 %
CHAID	79,3 %
QUEST	78,8 %

Table 1: Hit rates of different algorithms on the validation hold-out sample

The difference in accuracy between the "best" and the "worst" algorithm for this specific dataset lies in 4.3 percentage points accuracy. For reasons of training time and robustness, I will choose logistic regression for the following comparison of imputation methods, and not one of the slightly better performing types of neural networks. For measuring the bias introduced into datasets by imputation, I look at the mean, standard deviation and correlation matrix of the 9 predictors of the original dataset as a benchmark (Tables 2 and 3). Later on, I will measure the difference from those original dataset benchmarks for the "imputed" datasets. This will give an idea of the biasing effect on the (inter-)variable information of the two imputation techniques under investigation here. For the original datasets, the benchmark measures are:

	P1	P2	P3	P4	P5	P6	P7	P8	P9
Mean	1,96	2,34	2,04	2,45	2,17	2,33	2,33	2,51	1,96
Std Deviation	0,86	1,12	0,89	1,04	0,99	0,98	1,06	1,16	0,84

Table 2: Mean and standard deviations of the 9 predictors of the original dataset

Correlations	P1	P2	P3	P4	P5	P6	P7	P8	P9
P1	1,00								
P2	0,30	1,00							
P3	0,34	0,38	1,00						
P4	0,24	0,27	0,36	1,00					
P5	0,22	0,22	0,32	0,45	1,00				
P6	0,23	0,29	0,37	0,32	0,43	1,00			
P7	0,22	0,26	0,27	0,43	0,55	0,44	1,00		
P8	0,25	0,24	0,26	0,39	0,35	0,32	0,38	1,00	
P9	0,29	0,33	0,38	0,34	0,43	0,39	0,41	0,38	1,00

Table 3: Correlation matrix of the original dataset

As a next step, I construct a MCAR dataset, which is derived from the original dataset by assigning 10% of the entries of each variable missing values, following a MCAR pattern. Then, the missing values are replaced with the means of the remaining valid entries of each variable. The means, standard deviations and correlations of the imputed variables are measured, and the mean of the absolute differences against the respective benchmark measures of the original dataset is computed. These differences serve as indicators of the amount of bias introduced into the dataset. In this case (MCAR, replace with mean), the values are shown in Table 4.

	BIAS
Mean	0,01
Std. Dev.	0,05
Correlation	0,03

Table 4: Bias of the "replace with mean" imputation on the MCAR dataset

A logistic regression model being built on this imputed dataset achieves a validation set accuracy of 80.42%.

In the next step, the MCAR missing values are imputed with the regression method, the respective "BIAS values" are shown in Table 5.

	BIAS
Mean	0,001
Std. Dev.	0,03
Correlation	0,02

Table 5: Bias of the regression imputation on the MCAR dataset

The logistic regression model accuracy on the validation sample is this time 81.15%.

This procedure is repeated for the two remaining missing value patterns, MAR and IM. At each time, both the "replace with mean" and the regression imputation methods are used to reconstruct a complete dataset, and then, the "imputation bias" and model accuracy is measured. Table 6 gives an overview of the results.

	"Replace with mean"				"Regression"			
	BIAS			ACCURACY	BIAS			ACCURACY
	Mean	Std.	Cor.	Log.Regr.	Mean	Std.	Cor.	Log.Regr
MCAR 10%	0,01	0,05	0,03	80,42	0,00	0,03	0,02	81,15
MAR 10%	0,06	0,06	0,07	77,56	0,02	0,04	0,02	80,31
IM 10%	0,12	0,12	0,06	74,61	0,00	0,04	0,03	79,85

Table 6: Overview of bias measures and model accuracy

Compared with the benchmarks of the original dataset, the results show that the more complex (conditional upon other observed or not observed values) the missing value generating mechanism is, the more bias tends to be introduced by the imputation process. This increasing bias is reflected by the decrease in model accuracy on the validation sample of the respective logistic regression models having been built on the imputed datasets. On the other hand, the multivariate imputation method (regression) tends to introduce much less bias and thus enables a systematically better model accuracy than the "replace with mean" method. The difference between the two imputation methods is the more pronounced, the more complex (in the above mentioned sense) the missing value generating mechanism was.

5 Discussion and Conclusion

The simulation results confirm our practical experiences, that multivariate imputation methods like regression (or decision trees) tend to better preserve the existing (inter-)variable information by introducing only minimal bias, and thus enable algorithms to find better models in terms of accuracy.

It should be noted, that for this specific IM case, the difference in terms of model accuracy for the two imputation methods lies in 5.2 percentage points, whereas the difference between the two "best and worst" algorithms (measured on the original dataset) was 4.3 percentage points. This reflects only the impact of missing value handling as just one (important) part of data pre-processing. Our practical project experiences tell us, that the remaining parts of data preprocessing (handling categorical variables, transforming metric variables and dealing with high-dimensional datasets) also have a clear impact on the resulting model accuracy. Thus, the data preprocessing phase in a real-world data mining project plays a crucial role for the final model performance.

References

BERRY, M. J. A. and LINOFF, G. (1999): Mastering Data Mining: The Art and Science of Customer Relationship Management. John Wiley and Sons, New York.

CHATERJEE, S. and PRICE B. (1991): Regression analysis by example. John Wiley and Sons, New York.

FRIEDMAN,J.H. (1991): Multivariate adaptive regression splines. *Annals of Statistics 19, 1.*

KENNEDY, R. L. et al. (1997): Solving Data Mining Problems through Pattern Recognition. Prentice Hall, Upper Saddle River.

LITTLE , R. J. A. and RUBIN, D.B. (1987): Statistical analysis with missing data. John Wiley and Sons, New York.

PAR RUD, O. (2001) : Data mining cookbook . John Wiley and Sons, New York.

PIATETSKY-SHAPIRO, G. and STEINGOLD, S. (2000): Measuring lift quality in database marketing. *SIGKDD Explorations, Dec. 2000, Vol. 2, Issue 2, 76 - 80.*

PYLE, D. (1999): Data preparation for data mining. Morgan Kaufmann Publishers, San Fransisco

RUBIN, D.B. (1976): Inference and missing data. *Biometrika, 63, 581 - 592.*

RUBIN, D.B. (1987): Multiple imputation for nonresponse in surveys. John Wiley and Sons, New York.

TUKEY, J.W. (1977): Exploratory data analysis. Addison Wesley, New York.

256

WEISS, G.M. and PROVOST, F. (2001): The effect of class distribution on classifier learning. Technical report ML-TR-43, Department of computer science, Rutgers University, January 11, 2001.

Clustering of Document Collections to Support Interactive Text Exploration

A. Nürnberger, A. Klose, R. Kruse[1], G. Hartmann, M. Richards[2]

[1]Institute for Knowledge and Language Processing,
University of Magdeburg, Germany

[2]Max-Planck-Institut für Aeronomie, Katlenburg-Lindau, Germany

Abstract: In this contribution an approach for document retrieval is presented which groups (pre-processed) documents using a similarity measure. The methods were developed based on self-organising maps to realise interactive associative search and visual exploration of document databases. This helps a user to navigate through similar documents. The navigation, especially the search for the first appropriate document, is supported by conventional keyword search methods.

1 Introduction

The results of the work presented in this paper have been achieved as part of a pilot project for the validation of atmospheric data (DUST), carried out at the Max-Planck-Institut für Aeronomie, Germany. The presented part of the project was motivated by the limitations of standard text retrieval methods which usually just make use of specific keywords provided by the user, but neglect the potential of state-of-the-art document preprocessing techniques or document similarities. The prototypical implementation of this approach was not intended to replace professional search engines, but to complement them. For evaluation purposes it is available on a CD-ROM (Hartmann et al. (2000)), that contains – among additional material on other parts of the DUST project – two sample datasets which allow the user to validate the usability of the proposed methods.

In the following, the concepts of self-organising systems will be briefly reviewed. The implemented document preprocessing and the methods used for grouping the text documents and the prototypical implementation of the tool will be discussed.

1.1 Self-Organising Maps

Self-organising maps (Kohonen (1982)) are a special architecture of neural networks that cluster high-dimensional data vectors according to a similarity measure. The clusters are arranged in a low-dimensional topology that preserves the neighbourhood relations in the high dimensional data. Thus, not only objects that are assigned to one cluster are similar to each other (as in every cluster analysis), but also objects of nearby clusters are expected to be more similar than objects in more distant clusters. Usually, two-dimensional grids of squares or hexagons are used (e.g. Fig. 1). Although other topologies are possible, two-dimensional maps have the advantage of an intuitive visualisation and thus good exploration possibilities.

Self-organising maps are trained in an unsupervised manner (i.e. no class information is provided) from a set of high-dimensional sample vectors. The network structure has two layers. The neurons in the input layer correspond to the input dimensions. The output layer (map) contains as many neurons as clusters needed. All neurons in the input layer are connected with all neurons in the output layer. The weights of the connection between input and output layer of the neural network encode positions in the high-dimensional data space. Thus, every unit in the output layer represents a prototype. After learning, arbitrary vectors (i.e. vectors from the sample set or prior 'unknown' vectors) can be propagated through the network and are mapped to the output units. For further details on self-organising maps see, e.g, Kohonen (1984).

2 Building a Map of a Document Collection

The main idea of the presented approach is to use a self-organising map as a tool to arrange similar documents. After training of the map, documents with similar contents are assigned to nearby neurons. So, when a user has discovered a document of interest on the map, the surrounding area can also be inspected.

For the creation of a self-organising map, the documents must be encoded in form of vectors. To be suited for the learning process of the map, similar vectors must be assigned to similar documents, i.e. the vectors have to represent the document content. As common in document retrieval, our approach is based on statistical evaluations of word occurrences. We do not use any information on the *meaning* of the words.

In domains like scientific research we are confronted with a wide and (often rapidly) changing vocabulary, which is hard to catch in fixed structures like manually defined thesauri or keyword lists. However, it is important to calculate significant statistics. Therefore, the number of considered words must be kept reasonably small, and the occurrences of words sufficiently high. This can be done by either removing words or by grouping words with equal or similar meaning. The number of words can be reduced by filtering so-called *stop words* (i.e. words that bear no content information, like articles, conjunctions, prepositions) and to build the stems of the words (see, e.g. Frakes and Baeza-Yates (1992)).

For further reduction of relevant words we use two alternatives. The first approach reduces the vocabulary to a set of index words. These words are automatically chosen by an information theoretic measure. The second approach is based on the work of Honkela et al. (1996). It uses a self-organising map to build clusters of similar words, where *similarity* is defined with a statistical measure over the word's context.

2.1 Selection of Index Words Based on Their Entropy

For each word a in the vocabulary we calculate the entropy as defined by Lochbaum and Streeter (1989):

$$W(a) = 1 + \frac{1}{\ln(m)} \sum_{i=1}^{m} p_i(a) \cdot \ln(p_i(a)) \text{ with } p_i(a) = \frac{n_i(a)}{\sum_{j=1}^{m} n_j(a)},$$

where $n_i(a)$ is the frequency of word w in document i, and m is the number of documents.

The entropy can be seen as a measure of importance of words in the given domain context. We choose a number of words that have a high entropy relative to their overall frequency (i.e. from words occurring equally often we prefer those with the higher entropy). This procedure has empirically been found to yield a set of relevant words that are suited to serve as index terms.

2.2 The Word Category Map

This approach does not reduce the number of words by removing irrelevant words from the vocabulary, but by building groups of words which are frequently used in similar (three-word-)contexts. A self-organising map is used to find appropriate clusters of words. To be able to use words for training of a self-organising map, the words have to be encoded. Therefore, to every word a high dimensional random vector $\vec{w}$ is assigned (90 dimensions were used as proposed by Honkela (1997)). This encoding does not imply any word ordering, as random vectors of dimensionalities that high can be shown to be "quasi-orthogonal": the scalar product for nearly every pair of words is approximately zero. Then, the three-word-context of a word a is encoded by calculating the element-wise mean vectors of the words before $\vec{w}_{\text{before}}$ and after $\vec{w}_{\text{after}}$ the considered word over all documents and all occurrences of a. These mean (or expectation value) vectors $\langle \vec{w}_{\text{before}} \rangle$ and $\langle \vec{w}_{\text{after}} \rangle$ over the random vectors of enclosing words are used to define the context vector $\vec{w}_c$ of the considered word: $\vec{w}_c = (\langle \vec{w}_{\text{before}} \rangle, \langle \vec{w}_{\text{after}} \rangle)$.

The obtained context vectors have $180(= 2 \times 90)$ dimensions. Words a, b that often occur in similar contexts have similar expectation values and therefore similar context vectors $\vec{w}_c^{(a)}$, $\vec{w}_c^{(b)}$. The vectors $\vec{w}_c$ are finally clustered on a two-dimensional hexagonal grid using a self-organising map. Words that are used in similar contexts are expected to be mapped to the same or to nearby neurons on this so-called word category map. Thus, the words in the vocabulary are reduced to the number of clusters given by the size of the word category map. The word categories (*buckets*) are then used for the document indexing.

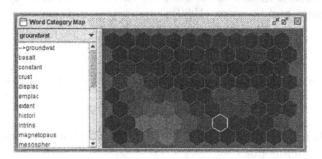

Figure 1: A word category map and the fingerprint of a document

2.3 Generating Characteristic Document vectors

Fig. 2 shows the realized document encoding. First, the original documents are pre-processed, i.e. they are split into words, then stop words are filtered and word stems are generated. Afterwards the considered vocabulary is reduced to a number of groups or *buckets*. These buckets are the index words from Sect. 2.1 or the word categories from Sect. 2.2. The words of every document are then sorted into the buckets, i.e. the occurrences of the word stems associated with the buckets are counted. Each of the n buckets builds a component in a n-dimensional vector that characterises the document. These vectors can be seen as the *fingerprints* of each document.

For every document in the collection such fingerprints are generated. Using self-organising maps, these document vectors are then clustered and arranged into a hexagonal grid, the so-called *document* map. Thus we obtain one map based on the index term approach and one using the word categories. Furthermore, each grid cell is labeled by a specific keyword which describes the content of the assigned documents as described in Lagus and Kaski (1999).

The most apparent advantage of the word category map approach over the index term approach is that no words are removed from the vocab-

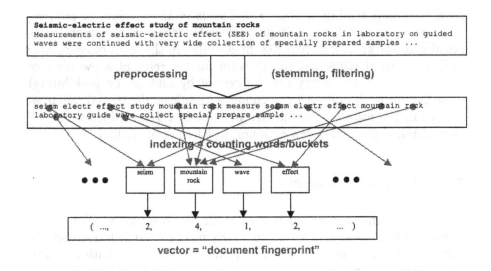

Figure 2: Document preprocessing and coding

ulary. Thus, all words are considered in the document clustering step. Furthermore, the word category map can be used as an expedient for the visual exploration of the document collection, because one often finds related words clustered together in the same or adjacent neurons of the map. However, due to the statistical peculiarities of the approach and the rather weak semantic clues the context vectors give, there are often additional words in the clusters that stand in no understandable relation to the others. The main drawback of this approach is that the words in one cluster become indistinguishable for the document indexing. Thus, the document map trained by use of the index terms frequently seems to represent the document collection more appropriate than the map trained by use of the word categories.

2.4 Using Maps to Explore Document Collections

To assess the usability of this approach a software prototype has been developed (Nürnberger et al. (2000)). In Fig. 3 an overview of the system structure is given. In contrast to the approach presented in Honkela et al. (1996), which only provides a labelled map for navigation and search, we implemented a number of functionalities to support iterative search processes. Thus, queries can be performed on the full text database using *keyword search* to find initial interesting subsets of the document collection. Then the *document map* can be visually inspected, e.g. to see how wide the documents are spread over the map. Our tool allows to associate the terms of the keyword search with colors. The nodes of the document map are highlighted with blends of these colors to indicate how well they match the individual search terms. Furthermore, it is possible to browse the documents' fingerprints in the *word category map* to discover similarities and get ideas for further relevant keywords (see Fig. 1). Thus, the query can be *refined* by adding (or prohibiting) these keywords. If interesting documents have been discovered, *adjacent nodes* can be inspected for further potentially relevant documents. A more detailed discussion can be found in Klose et al. (2000).

2.5 Dynamical Aspects

Generally, scientific research is a dynamic, evolving area. New documents will become available and obviously be of relevance to researchers. Furthermore, new topics will come up in scientific communities. Traditional document retrieval, which is based on manually defined thesauri or fixed sets of index words, always suffered from its implicit inflexibility. Our implemented tool also represents a 'snap-shot' of a document collection. However, the approach has the potential to meet the challenges of changing document material.

First of all, keyword queries on full texts do not depend on predefined index terms or thesauri. Furthermore, as the system is built with very little user intervention and directly from the documents themselves, it can always be completely reconstructed for a changed document database. However, more sophisticated methods are possible:

- For small changes we can keep the learned maps and just add the new documents and word stems to the nearest map nodes.

- If we expect extensive changes, we could re-train the document map by adding the new documents to the training data but by keeping the old map as an initialisation. Thus, we just slightly rearrange the nodes to better fit the new collection. Thus, we avoid that a user who is familiar with the data distribution in the old map has to deal with a new one. Alternatively, we can also relearn a new, perhaps bigger document map from scratch. However, we can still use the old document encoding, i.e. the old word category map or automatically chosen set of index words.

- We should additionally recompute the vector encodings, when the changes in the collection are more severe. In the case of the word category map we have again the possibility of incremental adaptation. The index terms have to be chosen independently of the old terms. In both cases – word category map or index terms – an analysis of the changes might yield interesting hints on current developments of research topics, e.g. upcoming 'buzz words'.

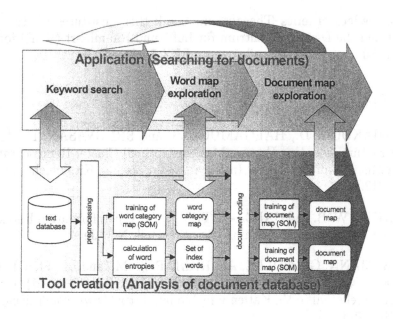

Figure 3: System overview

3 Conclusions

The methods proposed in this article combine (iterative) keyword search with grouping of documents based on a similarity measure in an interactive environment. This enables a user to search for specific documents, but also to enlarge obtained result sets (without the need to redefine search terms) by navigating through groups of documents with similar contents surrounding the search hits. Furthermore, the user is supported in finding appropriate search keywords to reduce or increase the documents under consideration by using a word category map, which groups together words used in similar contexts.

Nevertheless, the proposed method has still some insufficiencies. The main problem is that the size of the used self-organising maps has to be defined manually. Therefore, the training process has to be done several times with modified map sizes, until an 'optimal' size has been discovered by manual inspection of the maps. Methods that adapt the size of the map with repect to a similarity measure, e.g. growing self-organizing maps (for example Alahakoon (1998)), are currently analysed in the continuation of this project.

Furthermore, to improve the clustering of words in the word category map the use of thesauri to define e.g. synonyms should be considered and dictionaries may be used to extend the method for multi-lingual databases.

Acknowledgements The authors thank their institutes for the support and the Deutsche Zentrum für Luft- und Raumfahrt (DLR) for the financial support by the pilot study FKZ 50 EE 98038.

References

ALAHAKOON, D., HALGAMUGE, S. and SRINIVASAN, B. (1998): A Structure Adapting Feature Map for Optimal Cluster Representation. Proc. Int. Conf. On Neural Information Processing, Kitakyushu, Japan, 809–812.

FRAKES, W. B. and BAEZA-YATES, R. (1992): Information Retrieval: Data Structures & Algorithms. Prentice Hall, New Jersey.

HARTMANN, G. , NÖLLE, A. , RICHARDS, M. and LEITINGER, R. (2000): Data Utilization Software Tools 2 (DUST-2 CD-ROM). Copernicus Gesellschaft e.V., Katlenburg-Lindau, http://www.copernicus.org/EGS/EGS.html.

HONKELA, T. (1997): Self-Organizing Maps in Natural Language Processing. Ph.D. thesis, Helsinki University of Technology, Neural Networks Research Center, Espoo, Finland.

HONKELA, T. , KASKI, S. , LAGUS, K. and KOHONEN, T. (1996): Newsgroup Exploration with WEBSOM Method and Browsing Interface. Technical Report, Helsinki University of Technology, Neural Networks Research Center, Espoo, Finland.

KLOSE, A., NÜRNBERGER, A., KRUSE, R., HARTMANN, G., and RICHARDS, M. (2000): Interactive Text Retrieval Based on Document Similarities. Physics and Chemistry of the Earth, 25:8, 649-654.

KOHONEN, T. (1982): Self-Organized Formation of Topologically Correct Feature Maps. Biological Cybernetics, 43, 59–69.

KOHONEN, T. (1984): Self-Organization and Assoziative Memory. Springer Verlag, Berlin.

LAGUS, K. and KASKI, S. (1999): Keyword Selection Method for Characterizing Text Document. Proc. International Conference on Artificial Neural Networks (ICANN'99), Edinburgh, UK.

LOCHBAUM, K. E. and STREETER, L. A. (1989): Comparing and Combining the Effectiveness of Latent Semantic Indexing and the Ordinary Vector Space Model for Information Retrieval. Information Processing and Management, 25(6), 665–676.

NÜRNBERGER, A., KLOSE, A., KRUSE, R., HARTMANN, G., and RICHARDS, A. (2000): Interactive Text Retrieval Based on Document Similarities. Data Utilization Software Tools 2 (DUST-2 CD-ROM). Copernicus Gesellschaft e.V., Katlenburg-Lindau.

Web Usage Mining - Languages and Algorithms

J. R. Punin, M. S. Krishnamoorthy, M. J. Zaki

Computer Science Department
Rensselaer Polytechnic Institute, Troy NY 12180, USA

Abstract: Web Usage Mining deals with the discovery of interesting information from user navigational patterns from web logs. While extracting simple information from web logs is easy, mining complex structural information is very challenging. Data cleaning and preparation constitute a very significant effort before mining can even be applied. We propose two new XML applications, XGMML and LOGML to help us in this task. XGMML is a graph description language and LOGML is a web-log report description language. We generate a web graph in XGMML format for a web site and generate web-log reports in LOGML format for a web site from web log files and the web graph. We show the simplicity with which mining algorithms can be specified and implemented efficiently using our two XML applications.

1 Introduction

Recently XML has gained wider acceptance in both commercial and research establishments. In this paper, we present two XML languages and a web data mining application which utilizes them to extract complex structural information. Extensible Graph Markup and Modeling Language (XGMML) is an XML application based on Graph Modeling Language (GML) which is used for graph description. XGMML uses tags to describe nodes and edges of a graph. The purpose of XGMML is to make possible the exchange of graphs between different authoring and browsing tools. The conversion of graphs written in GML to XGMML is straight forward. Using Extensible Stylesheet Language (XSL) with XGMML allows the translation of graphs to different formats.

Log Markup Language (LOGML) (Punin & Krishnamoorthy 2000) is an XML application designed to describe log reports of web servers. Web data mining is one of the current hot topics in computer science. Mining data that has been collected from web server logfiles, is not only useful for studying customer choices, but also helps to better organize web pages. This is accomplished by knowing which web pages are most frequently accessed by the web surfers. When mining the data from the log statistics, we use the web graph for annotating the log information. Further we produce summary reports, comprising of information such as client sites, types of browsers and the usage time statistics. We also gather the client activity in a web site as a subgraph of the web site graph.

This subgraph can be used to get better understanding of general user activity in the web site. In LOGML, we create a new XML vocabulary to structurally express the contents of the logfile information.

Recently web data mining has been gaining a lot of attention because of its potential commercial benefits. For example, consider a web log database at a popular site, where an object is a web user and an attribute is a web page. The mined patterns could be the sets or sequences of most frequently accessed pages at that site. This kind of information can be used to restructure the web-site, or to dynamically insert relevant links in web pages based on user access patterns. Furthermore, click-stream mining can help E-commerce vendors to target potential online customers in a more effective way, at the same time enabling personalized service to the customers. Web mining is an umbrella term that refers to mainly two distinct tasks. One is web content mining (Cooley, Mobasher, & Srivastava 1997), which deals with problems of automatic information filtering and categorization, intelligent search agents, and personalize web agents. Web usage mining (Cooley, Mobasher, & Srivastava 1997) on the other hand relies on the structure of the site, and concerns itself with discovering interesting information from user navigational behavior as stored in web access logs. The focus of this paper is on web usage mining. While extracting simple information from web logs is easy, mining complex structural information is very challenging. Data cleaning and preparation constitute a very significant effort before mining can even be applied. The relevant data challenges include: elimination of irrelevant information such as image files and cgi scripts, user identification, user session formation, and incorporating temporal windows in the user modeling. After all this pre-processing, one is ready to mine the resulting database.

The proposed LOGML and XGMML languages have been designed to facilitate this web mining process in addition to storing additional summary information extracted from web logs. Using the LOGML generated documents the pre-processing steps of mining are considerably simplified. We also propose a new mining paradigm, called Frequent Pattern Mining, to extract increasingly informative patterns from the LOGML database. Our approach and its application to real log databases are discussed further in Section 5. We provide an example to demonstrate the ease with which information about a web site can be generated using LOGML with style sheets (XSLT). Additional information about web characterization can also be extracted from the mined data.

The overall architecture of our system is shown in Figure 1. The two inputs to our web mining system are 1) web site to be analyzed, and 2) raw log files spanning many days, months, or extended periods of time.

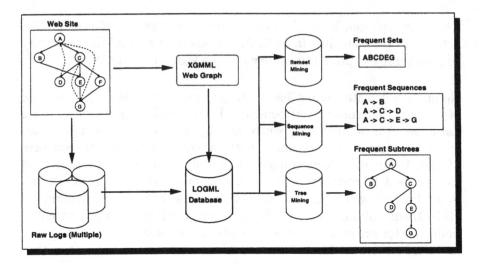

Figure 1: Web Usage Mining Architecture

The web site is used to populate a XGMML web graph with the help of a web crawler. The raw logs are processed by the LOGML generator and turned into a LOGML database. This processed database contains log information that can be used to mine various kinds of frequent pattern information such as itemsets, sequences and subtrees. The LOGML database and web graph information can also be used for web characterization, providing detailed statistics on top k pages, addresses, browsers, and so on.

It should be noted that association and sequence mining have also been applied to web usage mining in the past. Chen et al. (Chen, Park, & Yu 1996) introduced the notion of a maximal forward chain of web pages and gave an algorithm to mine them. The WUM system (Spiliopoulou & Faulstich 1998) applies sequence mining to analyze the navigational behavior of users in a web site. WUM also supports an integrated environment for log preparation, querying and visualization. Cooley et al. (Cooley, Mobasher, & Srivastava 1999) describe various data preparation schemes for facilitating web mining. Recent advances and more detailed survey on various aspects of web mining spanning content, structure and usage discovery can be found in (Masand & Spiliopoulou 2000; Kosala & Blockeel 2000). Our work differs in that our system uses new XML based languages to streamline the whole web mining process and allows multiple kinds of mining and characterization tasks to be performed with relative ease.

2 XGMML

A graph, G= (V,E), is a set of nodes V and a set of edges E. Each edge is either an ordered (directed graph) or unordered (undirected) pair of nodes. Graphs can be described as data objects whose elements are nodes and edges (which are themselves data objects). XML is an ideal way to represent graphs. Structure of the World Wide Web is a typical example of a graph where the web pages are "nodes," and the hyperlinks are "edges." One of the best ways to describe a web site structure is using a graph structure and hence XGMML documents are a good choice for containing the structural information of a web site. XGMML was created for use within the WWWPal System (Punin & Krishnamoorthy 1998) to visualize web sites as a graph. The web robot of W3C (webbot), a component of the WWWPal System, navigates through web sites and saves the graph information as an XGMML file. XGMML, as any other XML application, can be mixed with other markup languages to describe additional graph, node and/or edge information.

Structure of XGMML Documents: An XGMML document describes a graph structure. The root element is the **graph** element and it can contain **node**, **edge** and **att** elements. The **node** element describes a node of a graph and the **edge** element describes an edge of a graph. Additional information for graphs, nodes and edges can be attached using the **att** element. A **graph** element can be contained in an **att** element and this graph will be considered as subgraph of the main graph. The **graphics** element can be included in a **node** or **edge** element, and it describes the graphic representation either of a node or an edge.

Resource Description Framework (RDF) is one way to describe metadata about resources. XGMML includes metadata information for a graph, node and/or edge using the **att** tag. The example below is part of a graph describing a web site. The nodes represent web pages and the edges represent hyperlinks. The metadata of the web pages is included as attributes of a node. RDF and Dublin Core (DC) vocabularies have been used to describe the metadata of the nodes.

```
<?xml version="1.0"?>
<graph xmlns = "http://www.cs.rpi.edu/XGMML"
       xmlns:xsi="http://www.w3.org/2000/10/XMLSchema-instance"
       xsi:schemaLocation="http://www.cs.rpi.edu/XGMML
       http://www.cs.rpi.edu/~puninj/XGMML/xgmml.xsd"
       directed="1" >
<node id="3" label="http://www.cs.rpi.edu/courses/" weight="5427">
<att>
<rdf:RDF
  xmlns:rdf="http://www.w3.org/1999/02/22-rdf-syntax-ns#"
  xmlns:dc="http://purl.org/dc/elements/1.0/">
  <rdf:Description about="http://www.cs.rpi.edu/courses/"
    dc:title="Courses at Rensselaer Computer Science Department"
    dc:subject="M.S. requirements; Courses; People;
```

```
      Graduate Program; Computer  Algorithms; Programming in Java;
      Research; Course Selection
      Guide; Programming in Java; Models  of Computation"
      dc:date="2000-01-31"
      dc:type="Text"
      >
   </rdf:Description>
 </rdf:RDF>
 </att>
 </node>
 ....
 <edge source="1" target="3" weight="0" label="SRC IMG gfx/courses2.jpg" />
 <edge source="7" target="3" weight="0" label="SRC IMG ../gfx/courses2.jpg" />
 </graph>
```

3   LOGML (Log Markup Language)

Log reports are the compressed version of logfiles. Web masters in general save web server logs in several files. Usually each logfile contains a single day of information. Due to disk space limitation, old log data gets deleted to make room for new log information. Generally, web masters generate HTML reports of the logfiles and do not have problems keeping them for a long period of time as the HTML reports are an insignificant size. If a web master likes to generate reports for a large period of time, he has to combine several HTML reports to produce a final report. LOGML is conceived to make this task easier. Web masters can generate LOGML reports of logfiles and combine them on a regular basis without much effort. LOGML files can be combined with XSLT to produce HTML reports. LOGML offers the flexibility to combine them with other XML applications, to produce graphics of the statistics of the reports. LOGML can also be combined with RDF to provide some metadata information about the web server that is being analyzed. LOGML is based on XGMML. LOGML document can be seen as a snapshot of the web site as the user visits web pages and traverses hyperlinks. It also provides a succinct way to save the user sessions. In the W3C Working Draft "Web Characterization Terminology & Definitions Sheet", the user session is defined as "a delimited set of user clicks across one or more Web servers".

Structure of LOGML Documents: A typical LOGML document has three sections under the root element `logml` element. The first section is a graph that describes the log graph of the visits of the users to web pages and hyperlinks. This section uses XGMML to describe the graph and its root element is the `graph` element. The second section is the additional information of log reports such as top visiting hosts, top user agents, and top keywords. The third section is the report of the user sessions. Each user session is a subgraph of the log graph. The subgraphs are reported as a list of edges that refer to the nodes of the log graph. Each edge of the user sessions also has a timestamp for when the edge was traversed. This timestamp helps to compute the total time of the user session. LOGML files are large files.

LOGML Elements and Attributes: The root element of a LOGML document is the `logml` element. The rest of the elements are classified with respect to the three sections of the LOGML document. The first section is the report of the log graph and we use the XGMML elements to describe this graph. The second section report the general statistics of the web server such as top pages, top referer URLs, top visiting user agents, etc. And, the last section reports the user sessions.

The following global attributes are used by most of the LOGML elements: `id` - unique number to identify the elements of LOGML document. `name` - string to identify the elements of LOGML document. `label` - text representation of the LOGML element `access_count` - number of times the web server has been accessed. For example, the number of times of a specific user agent accessed the web server. `total_count` - total number of times that an element is found in a logfile. For example, the total count of a keyword. `bytes` - number of bytes downloaded. `html_pages` - number of HTML pages requested from the web server. For example, the number of html pages requested by a specific site.

The XGMML elements that we use to describe the log graph are `graph`, `node`, `edge` and `att`. We add the `hits` attribute to the `node` and `edge` elements to report the number of visits to the node (web page) and the number of traversals of the edge (hyperlink). The `att` element is used to report metadata information of the web page such as mime type and size of the file. The elements of the second section include hostname, IP, domains, directories, user agents, referers, host referers, keywords, http code, method and version, summary information like the the total number of requests, user sessions or bytes transferred, and statistics by date of requests.

The third section of the LOGML document reports the user sessions and the LOGML elements are:

- `userSessions, userSession` - The `userSessions` element is the container element for the set of the user sessions. Each user session is described using the `userSession`, `path` and `uedge` elements where a `path` is the collection of hyperlinks that the user has traversed during the session.
- `path` - The `path` element contains all hyperlinks that the user has traversed during the user session.
- `uedge` - The `uedge` element reports a hyperlink that has been traversed during the user session. The `source` and the `target` attributes are reference to nodes of the Log Graph in the first section and the `utime`

attribute is the timestamp where the user traversed this hyperlink. Example below is the report of one user session in a LOGML document:

```
<userSession name="proxy.artech.com.uy" ureferer="No referer"
entry_page="www.cs.rpi.edu/~puninj/XGMML/" start_time="12/Oct/2000:12:50:11"
access_count="4">
<path count="3">
<uedge source="3" target="10" utime="12/Oct/2000:12:50:12"/>
<uedge source="3" target="21" utime="12/Oct/2000:12:51:41"/>
<uedge source="21" target="22" utime="12/Oct/2000:12:52:02"/>
</path>
</userSession>
```

LOGML Generator: We have written a simple LOGML Generator as part of our WWWPal System. The LOGML Generator reads a common or extended log file and generates a LOGML file. The LOGML Generator also can read the webgraph (XGMML file) of the web site being analyzed and combine the information of the web pages and hyperlinks with the log information.

The information that we extract from the common log files include host name or IP, date of the request, relative URI of the requested page, HTTP version, HTTP status code, HTTP method and the number of bytes transferred to the web client. The extended log files additionally contain the absolute URI of the referer web page and a string that describes the User Agent (web browser or web crawler) that has made the request. This information is saved in a data structure to generate the corresponding LOGML document. The LOGML Generator also can output HTML reports making this module a powerful tool for web administrators.

Several algorithms have been developed to find the user sessions in the log files (Cooley, Mobasher, & Srivastava 1999; Wu, Yu, & Ballman 1997). A simple algorithm uses the IP or host name of the web client to identify a user. SpeedTracer System (Wu, Yu, & Ballman 1997) also checks the User Agent and date of the request to find the user session. Straight ways to find user session requires "cookies" or remote user identification (Cooley, Mobasher, & Srivastava 1999). The LOGML Generator algorithm, to find user sessions, is very similar to the algorithm used by SpeedTracer System.

4 Using LOGML for Web Data Mining

In this section, we propose solving a wide class of mining problems that arise in web data mining, using a novel, generic framework, which we term Frequent Pattern Mining (FPM). FPM not only encompasses important data mining techniques like discovering associations and frequent sequences, but at the same time generalizes the problem to include more complex patterns like tree mining and graph mining. These patterns arise in complex domains like the web. Association mining, and frequent subsequence mining are some of the specific instances of FPM that have been studied in the past (Agrawal *et al.* 1996; Zaki 2000; Srikant & Agrawal 1996; Zaki 2001b). In general, however, we can discover increasingly complex structures from the same database. Such complex patterns include frequent subtrees, frequent DAGs and frequent directed or undirected subgraphs. Mining such general patterns was also discussed in (Schmidt-Thieme & Gaul 2001). As one increases the complexity of the structures to be discovered, one extracts more informative patterns.

The same underlying LOGML document that stores the web graph, as well as the user sessions, which are subgraphs of the web graph, can be used to extract increasingly complex and more informative patterns. Given a LOGML document extracted from the database of web access logs at a popular site, one can perform several mining tasks. The simplest is to ignore all link information from the user sessions, and to mine only the frequent sets of pages accessed by users. The next step can be to form for each user the sequence of links they followed, and to mine the most frequent user access paths. It is also possible to look at only the forward accesses of a user, and to mine the most frequently accessed subtrees at that site. Generalizing even further, a web site can be modeled as a directed graph, since in addition to the forward hyperlinks, it can have back references, creating cycles. Given a database of user accesses one can discover the frequently occurring subgraphs.

In the rest of this section, we first formulate the FPM problem. We show how LOGML facilitates the creation of a database suitable for web mining. We illustrate this with actual examples from RPI logs (from one day). Using the same example we also describe several increasingly complex mining tasks that can be performed.

4.1 Frequent Pattern Mining: Problem Formulation

FPM is a novel, generic framework for mining various kinds of frequent patterns. Consider a database $\mathcal{D}$ of a collection of structures, built out of a set of primitive *items* $\mathcal{I}$. A structure represents some relationship among items or sets of items. For a given structure G, let $S \preceq G$ denote the fact that S is a substructure of G. If $S \preceq G$ we also say that G *contains* S. The collection of all possible structures composed of the set of items $\mathcal{I}$ forms a partially ordered set under the substructure relation

$\preceq$. A structure formed from k items is called a *k-structure*. A structure is called *maximal* if it is not a substructure of any other in a collection of structures. We define the *support* of a structure G in a database $\mathcal{D}$ to be the number of structures in $\mathcal{D}$ that contain G. Alternately, if there is only one very large structure in the database, the support is the number of times G occurs as a substructure within it. We say that a structure is *frequent* if its support is more than a user-specified *minimum support (min_ sup)* value. The set of frequent k-structures is denoted as $\mathcal{F}_k$.

A *structural rule* is an expression $X \Rightarrow Y$, where X and Y are structures. The *support* of the rule in the database of structures is the joint probability of X and Y, and the *confidence* is the conditional probability that a structure contains Y, given that it contains X. A rule is *strong* if its confidence is more than a user-specified *minimum confidence (min_ conf)*.

The frequent pattern mining task is to generate all structural rules in the database, which have a support greater than *min_ sup* and have confidence greater than *min_ conf*. This task can be broken into two main steps: 1) *Find all frequent structures having minimum support and other constraints.* This step is the most computational and I/O intensive step, since the search space for enumeration of all frequent substructures is exponential in the worst case. The minimum support criterion is very successful in reducing the search space. In addition other constraints can be induced, such as finding maximal, closed or correlated substructures. 2) *Generate all strong structural rules having minimum confidence.* Rule generation is also exponential in the size of the longest substructure. However, this time we do not have to access the database; we only need the set of frequent structures.

4.2   Database Creation: LOGML to Web Mining

We designed the LOGML language to facilitate web mining. The LOGML document created from web logs has all the information we need to perform various FPM tasks. For structure mining from web logs, we mainly make use of two sections of the LOGML document. As described above, the first section contains the web graph; i.e., the actual structure of the web site in consideration. We use the web graph to obtain the page URLs and their node identifiers. For example, the example below shows a snippet of the (node id, URL) pairs (out of a total of 56623 nodes) we extracted from the web graph of the RPI computer science department:

```
1 http://www.cs.rpi.edu/
6 http://www.cs.rpi.edu/courses/
```

```
8 http://www.cs.rpi.edu/current-events/
12 http://www.cs.rpi.edu/People/
14 http://www.cs.rpi.edu/research/
...
```

For enabling web mining we make use of the third section of the LOGML document that stores the user sessions organized as subgraphs of the web graph. We have complete history of the user clicks including the time at which a page is requested. Each user session has a session id (the IP or host name), a path count (the number of source and destination node pairs) and the time when a link is traversed. We simply extract the relevant information depending on the mining task at hand. For example if our goal is to discover frequent sets of pages accessed, we ignore all link information and note down the unique source or destination nodes in a user session. For example, let a user session have the following information as part of a LOGML document:

```
<userSession name="ppp0-69.ank2.isbank.net.tr" ...>
<path count="6">
<uedge source="5938" target="16470"
utime="24/Oct/2000:07:53:46"/>
<uedge source="16470" target="24754"
utime="24/Oct/2000:07:56:13"/>
<uedge source="16470" target="24755"
utime="24/Oct/2000:07:56:36"/>
<uedge source="24755" target="47387"
utime="24/Oct/2000:07:57:14"/>
<uedge source="24755" target="47397"
utime="24/Oct/2000:07:57:28"/>
<uedge source="16470" target="24756"
utime="24/Oct/2000:07:58:30"/>
```

We can then extract the set of nodes accessed by this user:

```
#format: user name, number of nodes accessed, node list
ppp0-69.ank2.isbank.net.tr 7 5938 16470 24754 24755 47387 47397 24756
```

After extracting this information from all the user sessions we obtain a database that is ready to be used for frequent set mining, as we shall see below. On the other hand if our task is to perform sequence mining, we look for the longest forward links, and generate a new sequence each time a back edge is traversed. Using a simple stack-based implementation all maximal forward node sequences can be found. For the example user session above this would yield:

```
#format: user name, sequence id, node position, node accessed
ppp0-69.ank2.isbank.net.tr 1 1 5938
ppp0-69.ank2.isbank.net.tr 1 2 16470
ppp0-69.ank2.isbank.net.tr 1 3 24754
ppp0-69.ank2.isbank.net.tr 2 1 5938
ppp0-69.ank2.isbank.net.tr 2 2 16470
ppp0-69.ank2.isbank.net.tr 2 3 24755
ppp0-69.ank2.isbank.net.tr 2 4 47387
ppp0-69.ank2.isbank.net.tr 3 1 5938
```

276

```
ppp0-69.ank2.isbank.net.tr 3 2 16470
ppp0-69.ank2.isbank.net.tr 3 3 24755
ppp0-69.ank2.isbank.net.tr 3 4 47397
ppp0-69.ank2.isbank.net.tr 4 1 5938
ppp0-69.ank2.isbank.net.tr 4 2 16470
ppp0-69.ank2.isbank.net.tr 4 3 24756
```

For more complex mining task like tree or graph mining, once again the appropriate information can be directly produced from the LOGML user sessions.

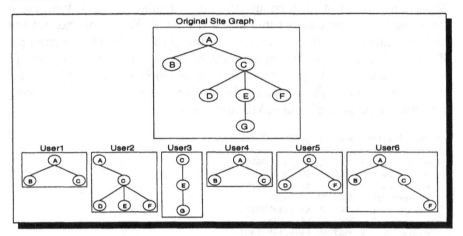

Figure 2: LOGML Document: Web Site Graph and User Sessions

We will illustrate various instances of the FPM paradigm in web mining using the example in Figure 2, which pictorially depicts the original web graph of a particular web site. There are 7 pages, forming the set of primitive items $\mathcal{I} = \{A, B, C, D, E, F, G\}$ connected with hyperlinks. Now the LOGML document already stores in a systematic manner the user sessions, each of them being a subgraph of the web graph. The figure shows the pages visited by 6 users. We will see below how this user browsing information can be used for mining different kinds of increasingly complex substructures, starting with the frequently accessed pages and frequently traversed paths, to the frequent subtrees, etc.

4.3 Web Data Mining

Frequent Sets: This is the well known association rule mining problem(Agrawal *et al.* 1996; Zaki 2000). Here the database $\mathcal{D}$ is a collection of *transactions*, which are simply subsets of primitive items $\mathcal{I}$. Each structure in the database is a transaction, and $\preceq$ denotes the subset relation. The mining task, then, is to discover all frequent subsets in $\mathcal{D}$. These subsets are called *itemsets* in association mining literature.

Consider the example web logs database shown in Figure 3. For each user (in Figure 2) we only record the pages accessed by them, ignoring the path information. The mining task is to find all frequently accessed sets of pages. Figure 3 shows all the frequent k-itemsets $\mathcal{F}_k$ that are contained in at least three user transactions; i.e., $min\_sup = 3$. ABC, AF and CF, are the maximal frequent itemsets.

We applied the Charm association mining algorithm (Zaki & Hsiao 2002) to a real LOGML document from the RPI web site (one day's logs). There were 200 user sessions with an average of 56 distinct nodes in each session. It took us 0.03s to do the mining with 10% minimum support. An example frequent set found is shown below:

```
FREQUENCY = 22 , NODE IDS =  25854 5938 25649 25650 25310 16511
        http://www.cs.rpi.edu/ sibel/poetry/poems/nazim_hikmet/turkce.html
        http://www.cs.rpi.edu/ sibel/poetry/sair_listesi.html
        http://www.cs.rpi.edu/ sibel/poetry/frames/nazim_hikmet_1.html
        http://www.cs.rpi.edu/ sibel/poetry/frames/nazim_hikmet_2.html
        http://www.cs.rpi.edu/ sibel/poetry/links.html
        http://www.cs.rpi.edu/ sibel/poetry/nazim_hikmet.html
```

Figure 3: Set Mining   Figure 4: Sequence Mining

Frequent Sequences: The problem of mining sequences (Srikant & Agrawal 1996; Zaki 2001b) can be stated as follows: An *event* is simply an itemset made up of the items $\mathcal{I}$. A *sequence* is an ordered list of events. A sequence α is denoted as $(\alpha_1 \to \alpha_2 \to \cdots \to \alpha_q)$, where α_i is an event; the symbol $\to$ denotes a "happens-after" relationship. We say α is a *subsequence* of another sequence β, denoted as $\alpha \preceq \beta$, if there exists a one-to-one order-preserving function f that maps events in α to events in β, such that, 1) $\alpha_i \subseteq f(\alpha_i)$, and 2) if $\alpha_i < \alpha_j$ then $f(\alpha_i) < f(\alpha_j)$.

The structure database $\mathcal{D}$ consists of a collection of sequences, and $\preceq$ denotes the subsequence relation. The mining goal is to discover all frequent subsequences. For example, consider the sequence database shown in Figure 4, by storing all paths from the starting page to a leaf (note that there are other ways of constructing user access paths; this is just one example). With minimum support of 3 we find that $A \to B$, $A \to C$, $C \to F$ are the maximal frequent sequences.

We applied the SPADE sequence mining algorithm (Zaki 2001b) to an actual LOGML document from the RPI web site. From the 200 user sessions, we obtain 8208 maximal forward sequences. It took us 0.12s to do the mining with minimum support set to 0.1%. An example frequent sequence found is shown below:

```
Let Path=http://www.cs.rpi.edu/~sibel/poetry
FREQUENCY = 21, NODE IDS =  37668 -> 5944 -> 25649 -> 31409
  Path/ -> Path/translation.html ->
  Path/frames/nazim_hikmet_1.html -> Path/poems/nazim_hikmet/english.html
```

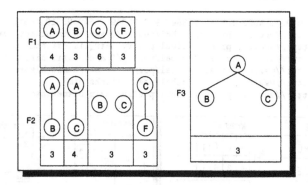

Figure 5: Frequent Tree Mining

Frequent Trees: We denote an ordered, labeled tree as $T = (V_t, E_t)$, where V_t is the vertex set, and E_t are the edges or branches. We say that a tree $S = (V_s, E_s)$ is a subtree of T, denoted as $S \preceq T$, if and only if $V_s \subseteq V_t$, and for all edges $e = (v_1, v_2) \in E_s$, v_1 is an ancestor or descendant of v_2 in T. Note that this definition is different from the usual definition of a subtree. In our case, we require that for any branch that appears in S, the two vertices must be on the same path from a root to some leaf. For example, in Figure 2 the tree S, with $V = \{C, G\}$ and $E = \{CG\}$ is a subtree of the site graph.

Given a database $\mathcal{D}$ of trees (i.e., a forest) on the vertex set $\mathcal{I}$, the frequent tree mining problem (Zaki 2001a) is to find all subtrees that appear in at least min_sup trees. For example, for the user access subtrees shown in Figure 2, we mine the frequent subtrees shown in Figure 5. There are two maximal frequent subtrees, $(V = \{C, F\}, E = \{CF\})$ and $(V = \{A, B, C\}, E = \{AB, AC\})$ for $min\_sup = 3$.

We applied the TreeMinerV algorithm (Zaki 2001a) to the same RPI LOGML file used above. From the 200 user sessions, we obtain 1009 subtrees. It took us 0.37s to do the mining with minimum support set to

5% (or a frequency of at least 50). An example frequent subtree found is shown below (-1 denotes a back edge):

```
Let Path=http://www.cs.rpi.edu/~sibel/poetry
Let Poet = Path/poems/orhan_veli
FREQUENCY = 65, NODE IDS = 16499 31397 37807 -1 37836 -1 -1 25309
            Path/orhan_veli.html
           /                    \
          /                      \
  Poet/turkce.html       Path/frames/orhan_veli_2.html
   /        \
  /          \
Poet/golgem.html   Poet/gunes.html
```

Source	Raw Logs		LOGML		LOGML Breakdown		
	Regular	Gzip	Regular	Gzip	Webgraph	Sessions	Other
RPI1(14Jun)	52.4MB	5.5MB	19.9MB	2.1MB	88.3%	8.3%	3.4%
RPI2(15Jun)	52.4MB	5.5MB	19.4MB	2.1MB	88.2%	8.4%	3.4%
CS1 (28Jun)	10.5MB	1.1MB	4.6MB	0.5MB	74.6%	16.7%	8.7%
CS2 (29Jun)	10.3MB	1.1MB	5.3MB	0.6MB	75.8%	16.6%	7.6%

Table 1: Size of Raw Log Files versus LOGML Files (Size is in Bytes)

Size of LOGML Documents: Since raw log files can be large, there is a concern that the LOGML files will be large as well. Table 1 shows the observed size of raw log files compared to the LOGML documents (with and without compression), the number of requests and user sessions, and the breakdown of LOGML files for the CS department (*www.cs.rpi.edu*) and RPI web site (*www.rpi.edu*). For example, for RPI1 (logs from 14th June, 2001) there were about 275,000 request for different nodes comprising 6,000 user sessions. The LOGML file is more than 2.5 times *smaller* than the raw log file. Same trends are observed for other sources.

The benefits of LOGML become prominent when we consider the breakdown of the LOGML files. For the RPI site we find that about 88% of the LOGML file is used to store the webgraph, while the user sessions occupy only 8% (the other elements to store statistics, etc. use up 3.4% space). For the CS department site, we find that the webgraph takes about 75% space, while the user sessions occupy 17%. In general, the webgraph is not likely to change much from one day to the next, and even if it does, one can always store a master webgraph spanning several days or months separately. Then on a per day basis we need only store the user sessions (and the other LOGML sections if desired). For example for the RPI site this would require us to store 174,573 bytes per day, while for the CS site is comes to 86,888 bytes per day for storing only the user sessions (with compression). Thus, not only does LOGML facilitate web usage mining, it also can drastically reduce the amount of daily information that needs to be stored at each site.

Conclusion: In this paper, we defined two new XML languages, XG-MML and LOGML, and a web usage mining application. XGMML is a graph file description format, and an ideal candidate to describe the structure of web sites. Furthermore XGMML is a container for meta-data information. LOGML, on the other hand, is an extension of XGMML to collect web usage. Future work includes mining user graphs (structural information of web usages), as well as visualization of mined data using WWWPal system (Punin & Krishnamoorthy 1998). To perform web content mining, we need keyword information and content for each of the nodes. Obtaining this information will involve analyzing each of the web pages and collecting relevant keywords. Work is under way to accomplish this task.

References

AGRAWAL, R.; MANNILA, H.; SRIKANT, R.; TOIVONEN, H.; and VERKAMO, A. I. (1996): Fast discovery of association rules. In Fayyad, U., and et al. (Eds.), *Advances in Knowledge Discovery and Data Mining*, 307–328. AAAI Press, Menlo Park, CA.

CHEN, M.; PARK, J.; and YU, P. (1996): Data mining for path traversal patterns in a web environment. In *International Conference on Distributed Computing Systems*.

COOLEY, R.; MOBASHER, B.; and SRIVASTAVA, J. (1997): Web mining: Information and pattern discovery on the world wide web. In *8th IEEE Intl. Conf. on Tools with AI*.

COOLEY, R.; MOBASHER, B.; and SRIVASTAVA, J. (1999): Data preparation for mining world wide web browsing pattern. *Knowledge and Information Systems* 1(1).

KOSALA, R., and BLOCKEEL, H. (2000): Web mining research: A survey. *SIGKDD Explorations* 2(1).

MASAND, B., and SPILIOPOULOU, M. (Eds.). (2000): *Advances in Web Usage Mining and User Profiling: Proceedings of the WEBKDD'99 Workshop*. Number 1836 in LNAI. Springer Verlag.

PUNIN, J., and KRISHNAMOORTHY, M. (1998): WWWPal System - A System for Analysis and Synthesis of Web Pages. In *Proceedings of the WebNet 98 Conference*.

PUNIN, J., and KRISHNAMOORTHY, M. (2000): Log Markup Language Specification. http:// www.cs.rpi.edu/ ~puninj/ LOGML/ draft- logml.html.

SCHMIDT-THIEME, L., and GAUL, W. (2001): Frequent substructures in web usage data - a unified approach. In *Web Mining Workhsop (with 1st SIAM Int'l Conf. on Data Mining.*

SPILIOPOULOU, M., and FAULSTICH, L. (1998): WUM: A Tool for Web Utilization Analysis. In *EDBT Workshop WebDB'98, LNCS 1590.* Springer Verlag.

SRIKANT, R., and AGRAWAL, R. (1996): Mining sequential patterns: Generalizations and performance improvements. In *5th Intl. Conf. Extending Database Technology.*

WU, K.; YU, P.; and BALLMAN, A. (1997): Speed Tracer: A Web usage mining and analysis tool. *Internet Computing* 37(1):89.

ZAKI, M. J., and HSIAO, C.-J. (2002): CHARM: An efficient algorithm for closed itemset mining. In *2nd SIAM International Conference on Data Mining.*

ZAKI, M. J. (2000): Scalable algorithms for association mining. *IEEE Transactions on Knowledge and Data Engineering* 12(3):372-390.

ZAKI, M. J. (2001a): Efficiently mining trees in a forest. Technical Report 01-7, Computer Science Dept., Rensselaer Polytechnic Institute.

ZAKI, M. J. (2001b): SPADE: An efficient algorithm for mining frequent sequences. *Machine Learning Journal* 42(1/2):31-60.

Thesaurus Migration in Practice

U. Rist

Head of Metadata Systems and Linguistics,
KirchMedia GmbH und Co. KGaA, Ismaning, Germany
ulfert.rist@kirchgruppe.de

Abstract: The development of operatively used documentation languages is normally carried out on an evolutionary basis, i.e. in steps in relation to changing environmental data or practical requirements. As part of the merger of departments or companies or, in general, following a desire to achieve an exchange of existing heterogeneous documentation stocks with the minimum losses, however, an approach is necessary which basically allows as many documentation languages or systems to be linked with each other as required. The process of descriptor mapping developed in media documentation practice consists of treating given documentation systems as part-representations of a dynamic metadocumentation system or metadata system. By referencing, mapping tables can then be created which will allow content description elements of concrete documentation systems to be mapped on to content description elements of other documentation systems (e.g. documentation standards such as IPTC/4). Old stocks in which the content is opened up in this way can thus be updated for retrodocumentation. In addition, the automation of coarse documentation with suitable computer-language methods is also planned. Using descriptor mapping and auto-classification, it will be possible in the future to automatically prepare undocumented data stocks in documentary form with a view to achieving a concrete documentation system.

1 Introduction

The domain in which the measures described here for the migration of thesauri or documentation languages in general are to be carried out on an ongoing basis is to be assigned to the private commercial media world, in particular the media segments television (e.g. documentation centres of the German-speaking television channels ProSieben, Sat.1, N24 and local television channels), brand extension or Internet pages relating to e-commerce and B2B communication[1], news (e.g. the German news agency ddp) and so on. These sectors are covered by background conditions that are pragmatic/commercial, related to corporate policy and, in some cases, determined by mergers.

From the point of a media group, it seems a good idea strategically, when designing a *metadata management system* (MDMS), to apply an

---

[1]e.g. http:\\www.prosieben.com, http:\\www.sat1.de, http:\\www.n24.de.

integrative solution which i) meets the above-mentioned framework conditions, ii) is basically designed according to standards to avoid any interface problems from the start and iii) is scalable, so that even highly differentiated material can be made available through updating of the descriptors.

The MDMS developed under these conditions was designed to be output-neutral as regards content and media types, so that there is neither a contents-related domain restriction nor is it bound to certain formal data, such as data relating to the medium of television. The system can be updated in that the strategy proposed in the German standard (DIN 1987a) of incorporating specialists on particular topics into the work is being applied, so that use can be made of their explicit implicit knowledge[2].

2 Representation of the System

The specialist documentation terms used here are based on the corresponding DIN standards relating to systems of concepts and to the study of terminology (see DIN (1979), DIN (1987a), DIN (1987b), DIN (1987c), DIN (1988), DIN (1992), DIN (1993)). The MDMS is designed to manage documentation languages. According to DIN (1987a), the term *documentation language* means "a quantity of linguistic expressions (designations) which, applied according to particular rules, serve to describe documents for the purposes of storage and carefully targeted retrieval. Documentation languages may be represented by keyword systems, thesauri and classifications". The term *thesaurus* is understood to be a system of concepts, i.e. an "ordered composition of terms and their (predominantly natural language) designations that are used in an area of documentation for indexing, storage and retrieval" (DIN 1987a, translated), which can be linked with other systems of concepts such as a *classification system*, i.e. an "aid to the ordering of objects or knowledge about objects" (DIN 1987b, translated). As an internal representation, the mono-hierarchical list (DIN 1987b) was selected, which uses, at every level, the class notation, by default the German-language class names, plus additional information such as explanations, alias designations (spelling variants, synonyms), status indications. Because of the clarity of the terminological relations in relation to hierarchical structuring, a thesaurus representation can be created by computer if necessary, since the necessary information for creating a thesaurus in accordance with DIN (1987a) exists.

---

[2]From the company viewpoint, the systematic collation of the knowledge that is "only kept in the heads of the specialist staff", and which is then lost when they change jobs or are not there, represents a valuable asset.

2.1 Requirements and Objectives

We will now look at the environment of the MDMS and the ensuing problems. The situation, objectives, solutions and development methods can be broken down roughly as follows:

1. **Existing situation**

   (a) Documentation languages (DS_1 ... DS_n) with varying characteristics: i. conceptual-lexical, relating to the different use of *descriptor*[3] and *non-descriptor*[4], covering ambiguous designations reaching into millions (e.g. proper names of people, companies, geographical areas, products, etc.), ii. conceptual-relational, relating to the types and quantity of conceptual relationships, iii. formal-notational, relating to breakdown logic, iv. temporal, i.e. relating to changes over time;

   (b) Distributed data pools with document collections reaching into multiple millions with their documentation language specific content access;

   (c) Varying documentation conventions, caused by i. system necessity through the working interfaces in the documentation departments, ii. changes in staff, iii. management requirements in relation to content marketing which have an effect on documentation and iv. a continuously growing number of multi-media document types such as sound documents, AV documents, animation documents, composite documents (cf. DIN 1987c).

2. **System requirements**

   (a) Use of the system's own system of concepts in order to exclude copyright problems[5] as regards the marketing and sale of the MDMS or its descriptors;

---

[3]"Descriptor is a designation (preferential designation) which is permitted for describing the content" (DIN 1987a, translated).

[4]"Non-descriptor is a designation that is not approved for indexing but which is listed in the thesaurus and marked accordingly in order to make it easier for the user to access the descriptors of the thesaurus." (DIN 1987a, translated)

[5]In our case, a precise check was carried out to see how far external systems of concepts could be incorporated, showing that this method is not viable. This is why the development and establishment of our own systems of concepts was basically proposed.

(b) Focus on intuitive logic tending towards the semantics of natural language, and so i. must be communicable and comprehensible e.g. in presentations; not much familiarization time for documentary assistants and temporary staff, ii. can be used by both documentary laypersons such as editors and by documentation specialists such as picture documentary experts as part of editorially specified workflow processes, and iii. tendency-neutral subject access which can be applied both for stock documentation and for subject-specific contents access;

(c) Support of subscriptions to descriptors in order to be able to provide small or specialist documentation departments with only the fragments of the concept system that they need for their work;

(d) Tools like parsers to check the well-formedness of output documents – an important work sequence to ensure that the concept system can be converted into exchange formats such as XML – and workflow incorporation as regards distributed databases;

(e) Scalability i.e. i. incorporation of additional information technologies from document management, computer linguistics and information linguistics, ii. unchanging descriptors which can be modulated according to status information and iii. capacity for continuous development.

3. Implementation

(a) Explicit project management;

(b) Continuous workflow definition, including incorporation of the selected migration path into the operating system (workflow modification) and coordination of several workflows in a workflow metaconcept;

(c) Advising and task-oriented involvement of the related companies and company divisions in work groups, information meetings and individual discussions relating to the management or decision-making level, documentation, system users and technical staff.

4. Methods of development

(a) Updating, continuous maintenance;

    (b) Development of output formats, including supporting various layout representations (e.g. manual reference catalogues according to the customer's corporate design, plus multilingual output formats, i.e. translation of natural-language designation names or verbalization in other languages (cf. DIN 1993));

    (c) Documentation language clipping, i.e. provision of several parallel views of data stocks;

    (d) Development of documentary conventions that are accepted by all for minimal contents access;

    (e) Incorporation of computer-linguistic and information-linguistic technologies such as autoclassification;

    (f) Monitoring on data flows through use of abstraction functions.

2.2 System Data

The MDMS is divided into the following sections: 1. Contents descriptions, 2. Unambiguous entities such as GIS descriptors e.g. to support standardized national catalogues, 3. Formal descriptions such as document types, journalistic forms of representations, descriptors for picture, AV and sound documentation, time descriptors, 4. Customer-specific areas e.g. for the 1:1 incorporation of own systems such as keyword systems and for indexing by keywords of non-public material, 5. Internal coding descriptors for the coding of particular matters. Further sections will be kept open for future codings such as the incorporation of lexicons and domain-specific descriptions. At the moment, around 12,000 descriptors are being used operatively. With the incorporation of various comprehensive data pools for the proper names of people, companies and products, however, this figure will move into the millions in the medium term.

3 Documentation Language Migration

Only in exceptional cases will a situation occur in which the leading documentation expert has the possibility of creating a new documentation language. As a rule, mature systems exist which have to be linked with the MDMS in order to guarantee a data-related merger of data pools through the translation of metadata from the existing individual systems to the metadata of the MDMS, so-called *descriptor mapping* or general *metadata mapping*. In this, data stocks or *data islands* accessed by proprietary means can be processed and made accessible on a standardized basis. The existing system can then continue to be used. In some cases, a gradual introduction of fragments of the MDMS documentation language is required ("subscriptions" to the documentation language). In both cases, however, there is the problem of the retrodocumentation of old data stocks from the viewpoint of the MDMS documentation language.

3.1 Metadata Mapping

Metadata mapping has emerged as a key technology in the field of documentation system migration. One of the crucial elements is the quality with which descriptors of a documentation language are mapped onto descriptors of another documentation language on a table-oriented basis[6], since automatic processes depend on the results of this basic instruction.

The following example shows the mapping of german slugs of a documentation language (field 1) onto MDMS descriptors, i.e. codes (field 2) and german designations (field 3; the symbol '|' is used as field seperator).

HEILKUNDE/GESUNDHEITSWESEN|100110140|Gesundheit, Medizin
Behandlung|100110140030|Heilungsmaßnahmen, Vorbeugemaßnahmen
Geburt/Schwangerschaft/§218|100110140015020|Schwangerschaft
Geburt/Schwangerschaft/§218|100110140015030|Geburt
Geburt/Schwangerschaft/§218|100110140015020020|Abtreibung
Geburt/Schwangerschaft/§218|500500900020|Nachkodierung erforderlich
Krankenhaus|100110140060040|Krankenhäuser, Kliniken
Krankheit|100110140020|Krankheiten
Medizintechnik|100110140032|Medizintechnik
PatientIn|100110140070020030|Patienten

Since in general data stocks are functionally incorporated into operative workflows, it is of particular importance to know and support the special requirements of the relevant processes. For this reason, the proposals for mapping tables are being discussed in working groups in order to properly support the special characteristics of the various documentation departments. Since an explicit notation for all types of mapping

---

[6]Descriptor mapping is carried out on a rule basis, and complex rules in a formal logical style are perfectly possible. Table-oriented display can be seen as a simple special instance of rule-based mapping in which a projection of source descriptors onto destination descriptors takes place.

possibilities with the relevant quality assessments through which parameters for quality can be calculated is only planned for the next release of MDMS, no statement on quality key figures can be made at this point. However, so far the table-oriented method has proved to be stable, since the descriptor mappings are generated partly automatically and all critical cases are carefully checked.

3.2 Retrodocumentation

As already mentioned above, retrodocumentation cannot be carried out generally for time, costs and staff reasons. The documentation language also changes over time within a single data stock, so that old stocks may sometimes be problematic in retrieval, since some descriptors, for example, are no longer in use or are no longer part of the vocabulary of the current documentation language. The desire for a computerized indexing facility has often been expressed in documentation circles. The corresponding theoretical principles from information retrieval are sufficiently well-known. However, one must bear in mind the textual reality, which affects the morphology of words and spelling variants just as much as misspellings or interpretations of composites. When observing the index lists for large data volumes, it becomes clear that lexicon-based algorithms for the pre-processing of these lists must be used and solutions can only be used operatively if the results are fault-free. Otherwise, situations would occur in the area of post-stored automatic content distribution which would be difficult to control.

4 Summary and Outlook

The development and introduction of a metadata management system seems to be a suitable instrument for the combining of disparate data stocks. Using metadata mapping, documentation languages can be mapped to each other. In this way, data stocks can also be mapped from the viewpoint of standard documentation languages such as IPTC/4, from the International Press Telecommunication Council, or geographical information according to the countries in accordance with ISO standards. The MDMS acts here as a translation tool for documentation languages. In the area of content syndication in the B2B sector in particular, the provision of standard-oriented mapping tables for contents brokers is proving to be a desirable instrument for the channelling of information. It is clear that against the background of the above-mentioned commercial framework condition, a system at a permanently scientific level is neither desirable nor viable. Nonetheless, the creation of development routes in the system that are possible in principle has created the possibility of being able to use coding options into areas such as the linguistic coding of expressions of natural language (cf. DIN 1986) - a part of text mining - to carry out corresponding work in the lexicon area or other special codings at any time. The possibility of library access according to the rules for alphabetical cataloguing in scientific libraries (RAK-WB) as in Haller/Pobst (1991), such as that used in the Munich State Library, can be integrated in MDMS if necessary, e.g. in the form of an XML grammar. Through the incorporation of computer-linguistic and information-linguistic processes, an intensive networking of otherwise distributed data is planned, and, in the near future, computerized indexation of text messages in an editorial system.

References

DEUTSCHES INSTITUT FÜR NORMUNG E.V. (DIN) (Ed.) (1979): Begriffe und Benennungen. DIN 2330. Berlin

DEUTSCHES INSTITUT FÜR NORMUNG E. V. (DIN) (Ed.) (1986): Format für den Austausch terminologischer/lexikographischer Daten – MATER. Kategorienkatalog. DIN 2341 Part 1. Berlin

DEUTSCHES INSTITUT FÜR NORMUNG E. V. (DIN) (Ed.) (1987a): Erstellung und Weiterentwicklung von Thesauri. Einsprachige Thesauri. DIN 1463 Part 1. Berlin

DEUTSCHES INSTITUT FÜR NORMUNG E. V. (DIN)(Ed.) (1987b): Klassifikationssysteme. Erstellung und Weiterentwicklung von Klassifikationssystemen. DIN 32705. Berlin

DEUTSCHES INSTITUT FÜR NORMUNG E. V. (DIN) (Ed.) (1987c): Kategorienkatalog für Dokumente. Codes für Einträge zu Datenkategorien. DIN 31631 Part 4. Berlin

DEUTSCHES INSTITUT FÜR NORMUNG E. V. (DIN) (Ed.) (1992): Begriffe der Terminologielehre. Grundbegriffe. DIN 2342 Part 2. Berlin

DEUTSCHES INSTITUT FÜR NORMUNG E. V. (DIN) (Ed.) (1993): Erstellung und Weiterentwicklung von Thesauri. Mehrsprachige Thesauri. DIN 1463 Part 2. Berlin

HALLER, K. and POBST, H. (1991): Katalogisierung nach den RAK-WB. Eine Einführung in die Regeln für die alphabetische Katalogisierung in wissenschaftlichen Bibliotheken (RAK-WB). Saur, München, New York et al.

Relational Clustering for the Analysis of Internet Newsgroups

T. A. Runkler[1], J. C. Bezdek[2]

[1]Siemens AG, Corporate Technology,
Information and Communications, D–81730 München, Germany

[2]University of West Florida, Computer Science Department,
11000 University Parkway, Pensacola, FL 32514, U.S.A.

Abstract: Clustering is used to determine partitions and prototypes from pattern sets. Sets of *numerical* patterns can be clustered by alternating optimization (AO) of clustering objective functions or by alternating cluster estimation (ACE). Sets of *non–numerical* patterns can often be represented numerically by (pairwise) relations. For text data, relational data can be automatically computed using the Levenshtein (or edit) distance. These *relational* data sets can be clustered by relational ACE (RACE). For text data, the RACE cluster centers can be used as keywords. In particular, the cluster centers extracted by RACE from internet newsgroup articles serve as keywords for those articles. These keywords can be used for automatic document classification.

1   Introduction

Clustering is an unsupervised learning method that partitions a set of patterns into groups (clusters) (Jain et al. (1999)). The partitions can be generated by objective function, cluster estimation, mixture resolving (e.g. expectation maximization), and other types of algorithms. Here we focus on objective function and cluster estimation algorithms. *Objective function* algorithms minimize square error criteria based on distances between patterns and clusters. This minimization is often done by alternatingly updating the partitions and the cluster prototypes. Alternating optimization is a clustering algorithm specified by an objective function. *Alternating cluster estimation* is a generalization of the alternating optimization scheme. It is specified by user–defined partition and prototype functions.

In this paper we focus on the application of clustering methods to text documents from the internet. An important focus in this area is clustering text documents (Steinbach et al. (2000)), i.e. to find groups of documents that are related to similar topics and to extract the most important keywords that are considered in this classification (Han and Karypis (2000)). In many articles, large word corpuses with hundreds of thousands of documents are used (Larsen and Aone (1999)) that allow

---

[2]Research supported by ONR grant # N00014-96-1-0642.

the use of probabilistic features like (relative) term frequencies (Yang and Liu (1999)) and allow to represent clusters as large groups of similar text documents. In this paper we focus on the analysis of much smaller sets of text documents (up to hundreds) that are collected from postings to internet newsgroups or from individual e–mail messages. Due to the relatively small size of the data sets, term frequencies are not applicable here, since most non–trivial terms occur only once. However, it can be observed that the significant keywords for individual texts often repeatedly appear in *modified* versions. For example, an article about electronics contains the words *conduct, conductors, semiconductor*, which will remain separate even after preprocessing with stemmer algorithms. Each of these words appears only once, but the complete cluster of similar words is apparently significant for the document. In this paper we try to find these clusters and use them for the analysis of the document contents. Thus, we break down the document clustering problem to a smaller scale: Instead of searching for clusters of similar *documents* in the *whole data set*, we are searching for clusters of similar *words* within *individual documents*.

Many clustering algorithms assume that the patterns are real vectors. However, the data in text documents are text strings that are non–numerical patterns but that can be represented numerically using (pairwise) relation matrices. Clusters in this kind of relational data sets can be found with *relational alternating cluster estimation* (Runkler et al. (1998, 2000)). For many text data sets it is difficult if at all possible to define semantically meaningful relations. In these cases relations can be computed based on comparing individual characters of pairs of strings, e.g. using the Levenshtein (or edit) distance (Levenshtein (1966)). In Runkler et al. (2000) it was found that cluster centers obtained from Levenshtein relations can be used as text keywords. In this article, we apply this method to automatically extract keywords for internet newsgroups.

2 Objective Function Clustering

Numerical clustering algorithms partition a data set $X = \{x_1, \ldots, x_n\} \in \mathcal{R}^p$ into $c \in \{2, \ldots, n-1\}$ clusters. The clusters are described by a $c \times n$ partition matrix U and a set V containing c cluster prototypes. Objective function clustering algorithms compute U and V from X by optimization of an objective function $J(U, V; X)$. The *c–means model* (CM, Ball et al. (1965)), for example, is specified by

$$J_{CM}(U, V; X) = \sum_{i=1}^{c} \sum_{k=1}^{n} u_{ik} \|x_k - v_i\|^2, \tag{1}$$

where $u_{ik} \in \{0,1\}$ for all $i = 1,\ldots,c$, $k = 1,\ldots,c$,

$$\sum_{i=1}^{c} u_{ik} = 1, \quad \text{for each } k = 1,\ldots,n, \text{ and} \tag{2}$$

$$\sum_{k=1}^{n} u_{ik} > 0, \quad \text{for each } i = 1,\ldots,c. \tag{3}$$

The necessary conditions for optima of $J(U,V;X)$ are $\partial J(U,V;X)/\partial U = 0$ and $\partial J(U,V;X)/\partial V = 0$. For the c–means model (1) this leads to

$$u_{ik} = \begin{cases} 1 & \text{if } \|x_k - v_i\| = \min_{j=1,\ldots,c}\{\|x_k - v_j\|\}, \\ 0 & \text{otherwise, and} \end{cases} \tag{4}$$

$$v_i = \frac{\sum_{k=1}^{n} u_{ik} x_k}{\sum_{k=1}^{n} u_{ik}}. \tag{5}$$

Notice that in case of multiple minima in (4) condition (2) can not be satisfied, since each point either belongs completely to a cluster or does not at all belong to it. To avoid this, we allow $u_{ik} \in [0,1]$ for all $i = 1,\ldots,c$, $k = 1,\ldots,c$, and define the *fuzzy c–means model* (FCM, Bezdek (1981)) using

$$J_{FCM}(U,V;X) = \sum_{i=1}^{c} \sum_{k=1}^{n} u_{ik}^m \|x_k - v_i\|^2, \tag{6}$$

where $m \in (1,\infty)$ is a fuzziness parameter. From the necessary conditions for optima of $J_{FCM}(U,V;X)$ we obtain (5) and

$$u_{ik} = 1 \Big/ \sum_{j=1}^{c} \left(\frac{\|x_k - v_i\|}{\|x_k - v_j\|} \right)^{\frac{2}{m-1}}. \tag{7}$$

To reduce the effect of outliers in the data set, condition (2) was dropped and the *possibilistic c–means model* (PCM, Krishnapuram et al. (1993)) was defined using

$$J_{PCM}(U,V;X) = \sum_{k=1}^{n} \sum_{i=1}^{c} \left(u_{ik}^m \|x_k - v_i\|^2 + \eta_i (1 - u_{ik})^m \right), \tag{8}$$

input $X = \{x_1, \ldots, x_n\} \in \mathcal{R}^p$	input $D \in \mathcal{R}^{n \times n}$
initialize $V = \{v_1, \ldots, v_c\} \in \mathcal{R}^p$	initialize $V = \{v_1, \ldots, v_c\}$ $\subset \{1, \ldots, n\}$
repeat	for $j = 1, \ldots, t$
store $V' = V$	store $V' = V$
compute $U(V, X)$	compute $U(V, D)$
compute $V(U, X)$	compute $V(U, D)$
until $\|V - V'\| < \varepsilon$	end for
output U, V	output U, V

Figure 1: Alternating cluster estimation (left) and relational alternating cluster estimation (right)

with $\eta_i \in \mathcal{R}^+$, $i = 1, \ldots, c$. Here, the necessary conditions for optima lead to (5) and

$$u_{ik} = \frac{1}{1 + \left(\frac{\|x_k - v_i\|_A}{\sqrt{\eta_i}} \right)^{\frac{2}{m-1}}} \tag{9}$$

The optimization of CM, FCM, PCM, and other clustering methods can be done by alternatingly computing $U(V, X)$ and $V(U, X)$ using the equations derived from the necessary conditions for optima of $J(U, V; X)$. This algorithm is called *alternating optimization* (AO).

3 Alternating Cluster Estimation

Alternating optimization computes membership functions that have the shape given in $U(V, X)$. For example, for CM we obtain rectangular functions (4), and for PCM, we obtain Cauchy functions (9). If the user wants to obtain a partition with a specific shape, e.g. with triangular membership functions, finding the corresponding objective function might be difficult if possible at all. Instead, the user might abandon the objective function $J(U, V; X)$ and directly specify the clustering algorithm by desired membership functions $U(V, X)$ and prototype functions $V(U, X)$. This leads to the *alternating cluster estimation* algorithm (ACE, Runkler et al. (1997)) that is shown in the left view of Fig. 1.

Notice that AO uses the same algorithmic scheme as ACE, whereas AO is specified by $J(U, V; X)$, and ACE can be specified by $U(V, X)$ and

$V(U, X)$, for example (with $\alpha, \gamma \in \mathcal{R}^+$) by

$$v_i = \frac{\sum\limits_{k=1}^{n} u_{ik}^{\gamma} x_k}{\sum\limits_{k=1}^{n} u_{ik}^{\gamma}}, \quad u_{ik} = \begin{cases} 1 - \left(\frac{\|x_k - v_i\|}{r_i}\right)^{\alpha} & \text{for } \|x_k - v_i\| \leq r_i, \\ 0 & otherwise. \end{cases} \quad (10)$$

4 Relational ACE

The AO and ACE algorithms introduced in the previous sections require numerical input data X. However, the text strings considered in document analysis are non–numerical, but can be numerically specified by (pairwise) relation matrices. Clusters in relational data like these can be found by *relational alternating cluster estimation* (RACE, Runkler et al. (1998, 2000)).

The right view of Fig. 1 shows the RACE algorithm which is very similar to the ACE algorithm from the left view of Fig. 1. Given a set O of n objects represented by an $n \times n$ distance matrix D, RACE randomly picks (the indices of) c initial cluster centers V from O and then alternatingly computes a membership matrix U and one of the cluster centers V. This alternating computation is done t times (in this article we choose $t = c \times n$). In the general case, the RACE update equations $U(V, D)$ and $V(U, D)$ can be arbitrarily picked by the user. Here we update $U(V, D)$ by

$$u_{ik} = \frac{1}{1 + d_{ik}}, \quad i = 1, \ldots, c, \quad k = 1, \ldots, n, \quad (11)$$

To update $V(U, D)$ we choose the maximum membership operator. This operator randomly picks (say) the ith cluster at step (j), finds the object o_k in O that has the maximum membership in cluster i, and sets $v_i^{(j)} = k$. Notice that we store the object index in V instead of the object itself. Hence, the RACE algorithm does not require O itself as an input.

Successful applications of the RACE algorithm include language concordance data, the wedding table problem (Runkler et al. (1998)), and keyword extraction from plain texts (Runkler et al. (2000)). In this article we extend the keyword extraction from plain texts to the keyword extraction from internet newsgroup articles.

5 Relations on Text

Relations on text can be defined based on character distances. For two characters x and y the *character distance* is defined as

$$z(x,y) = \begin{cases} 0 & \text{if } x = y, \\ 1 & \text{otherwise.} \end{cases} \tag{12}$$

We define a text as a sequence of words that are separated by blank spaces or other special characters. Each word is a vector of characters. For two words $x = (x^{(1)}, \ldots, x^{(p)})$ and $y = (y^{(1)}, \ldots, y^{(p)})$, we define the *Hamming distance* (Hamming (1950)) as

$$H((x^{(1)}, \ldots, x^{(p)}), (y^{(1)}, \ldots, y^{(p)})) = \sum_{i=1}^{p} z(x^{(i)}, y^{(i)}). \tag{13}$$

The Hamming distance is only defined for words with the same length. For words with different lengths, the shorter word might be filled with blank spaces. For example, the Hamming distance between the words class␣␣ and cluster is 4. However, if we add blank spaces at the *beginning* of the word, we obtain ␣␣class and a Hamming distance of 7. This ambiguity is avoided by the *Levenshtein distance* (Levenshtein (1966)) that is also called *edit distance*, since it determines the number of necessary edit (delete, insert, change) steps to convert one string into the other. The Levenshtein distance L is recursively defined as

$$L((x^{(1)}, \ldots, x^{(p)}), (y^{(1)}, \ldots, y^{(q)})) =$$
$$\begin{cases} p & q = 0 \\ q & p = 0 \\ \min\{L((x^{(1)}, \ldots, x^{(p-1)}), (y^{(1)}, \ldots, y^{(q)})) \ +1, \\ \qquad L((x^{(1)}, \ldots, x^{(p)}), \ (y^{(1)}, \ldots, y^{(q-1)}))+1, & \text{otherwise.} \\ \qquad L((x^{(1)}, \ldots, x^{(p-1)}), (y^{(1)}, \ldots, y^{(q-1)}))+z(x^{(p)}, y^{(q)})\} \end{cases}$$
$$\tag{14}$$

6 Keyword Extraction for Newsgroups

Information from the internet is often available as text. Here we consider texts from an internet based distributed bulletin board system called *usenet news* (Pfaffenberger (1995)). Usenet news is hierarchically organized into several thousands of groups like comp.ai.fuzzy, comp.ai.neural-networks, alt.drinks.beer, or alt.music.zz-top. Each group contains messages discussing the topics indicated by its title. Here we attempt to automatically extract keywords for a newsgroup from the articles it contains.

We used the 20 newsgroups data set from Carnegie Mellon University (http://www.cs.cmu.edu/~TextLearning/datasets.html). This data set
consists of 20,000 articles, composed of 1000 articles of 20 different newsgroups. Since news articles contain many special characters and header information, preprocessing of the data is necessary. We chose a simple preprocessing method that considers the first 10,000 non–empty strings that satisfy the extended Backus–Naur form

$$[(\mid `] \underbrace{[*\{A\ldots Z\}]\ [*\{a\ldots z\}]}_{o_k^*} \ [)\mid `\mid .\mid ,\mid :\mid ;\mid !\mid ?]. \qquad (15)$$

The preprocessed text objects o_k, $k = 1,\ldots,10,000$, are obtained by converting the alphabetic parts o_k^* from (15), $k = 1,\ldots,10,000$, into their lower case equivalents. For each of the 20 newsgroups we performed this preprocessing, computed a Levenshtein distance matrix on O, and determined $c = 20$ cluster centers with the RACE algorithm described in Section 4. For the newsgroup sci.electronics, for example, we obtained the following 20 cluster centers:

gaithersburg f *cyanoacrylate simulatenously compatability semiconductor electrically potentiometers* massachusetts *multivibrators intelligence revolutionized asynchronously* joseph *subdirectories* birlinghoven *transportation microcontroller administration reprogramming*

We printed those keywords in italics that we consider meaningful for the contents of this newsgroup. 15 out of the 20 keywords seem meaningful to us. This is a very good result, since our keyword extraction method is only based on Levenshtein distances and does not use any additional semantic information (notice that it even ignores the misspelling in the cluster centers *simulatenously* and *compatability*). Similar results were obtained for the other 19 newsgroups.

The keywords extracted with relational clustering can be used for the classification of individual documents. We extracted 20 keywords for each of the 20 newsgroups from the first 100 articles of each group. Then we classified the next 100 articles of each group using a nearest neighbor classifier (minimum Levenshtein distance). The best classification rates were achieved for articles from the newsgroups talk.politics.misc (65.47%) and comp.graphics (60.88%). This classification rate could be increased to up to 81.58%, when only articles with more than 5000 bytes were considered. For more details about these experiments we refer to (Runkler and Bezdek (2001)).

7 Conclusions

We have shown how keywords for individual internet newsgroup articles can be automatically extracted. After some preprocessing we computed Levenshtein distances that are based on character distances only and do not use any grammatical or semantical information about the underlying words. To the resulting relational data set we applied relational alternating cluster estimation with Cauchy membership functions and the maximum membership operator. 15 out of 20 cluster centers extracted for the newsgroup sci.electronics are considered meaningful for this newsgroup, and similar results were obtained for other newsgroups. Classification using these keywords achieves a classification rate of more than 60%, and more than 80% for large text documents. This indicates that our method is well suited for keyword extraction from internet texts. We are currently working on the application of these techniques to the classification of individual e-mail documents.

References

BALL, G. H. and HALL, D. J. (1965): Isodata, an iterative method of multivariate analysis and pattern classification. In: Proceedings of the IFIPS Congress

BEZDEK, J. C. (1981): Pattern recognition with fuzzy objective function algorithms. Plenum Press, New York

DUNN, J. C. (1974): A fuzzy relative of the ISODATA process and its use in detecting compact, well separated clusters. *Journal of Cybernetics, Vol. 3, 32–57.*

HAMMING, R. W. (1950): Error detecting and error correcting codes. *The Bell System Technical Journal, Vol. 26, Number 2, 147–160.*

HAN, E.-H. and KARYPIS, G. (2000): Centroid–based document classification: Analysis & experimental results. *Technical Report 00–017, University of Minnesota, Department of Computer Science.*

JAIN, A. K., MURTY, M. N., and FLYNN, P. J. (1999): Data clustering: A review. *ACM Computing Surveys, Vol. 31, Number 3, 264–323.*

KRISHNAPURAM, R. and KELLER, J. M. (1993): A possibilistic approach to clustering. *IEEE Transactions on Fuzzy Systems, Vol. 1, Number 2, 98–110.*

LARSEN, B. and AONE, C. (1999): Fast and effective text mining using linear time document clustering. *ACM SIGKDD International Conference on Knowledge Discovery and Data Mining*, 16–22.

LEVENSHTEIN, V. I. (1966): Binary codes capable of correcting deletions, insertions and reversals. *Sov. Phys. Dokl., Vol. 6, 705–710.*

PFAFFENBERGER, B. (1995): The Usenet Book: finding, using, and surviving newsgroups on the internet. Addison–Wesley

RUNKLER, T. A. (2000): Information Mining — Methoden, Algorithmen und Anwendungen intelligenter Datenanalyse. Vieweg, Wiesbaden

RUNKLER, T. A. and BEZDEK, J. C. (1998): RACE: Relational alternating cluster estimation and the wedding table problem, in Brauer (Ed.): Proceedings of the Workshop Fuzzy–Neuro–Systems, München, Infix, Sankt Augustin, 330–337.

RUNKLER, T. A. and BEZDEK, J. C. (1999): Alternating cluster estimation: A new tool for clustering and function approximation. *IEEE Transactions on Fuzzy Systems, Vol. 7, Number 4, 377–393.*

RUNKLER, T. A. and BEZDEK, J. C. (2000): Automatic keyword extraction with relational clustering and Levenshtein distances, in Langari (Ed.): Proceedings of the IEEE International Conference on Fuzzy Systems, San Antonio, USA, IEEE Press, Piscatway, 636–640.

RUNKLER, T. A. and BEZDEK, J. C. (2001): Classification of internet newsgroup articles using RACE, in Hall (Ed.): Proceedings of the Joint IFSA World Congress and NAFIPS International Conference, Vancouver, Canada.

STEINBACH, M., KARYPIS, G., and KUMAR, V. (2000): A comparison of document clustering techniques. *KDD Workshop on Text Mining.*

YANG, Y. and Liu, X. (1999): A re–examination of text categorization methods. *International SIGIR Conference.*

The Theory of On-line Learning - A Statistical Physics Approach

D. Saad

The Neural Computing Research Group
University of Aston, Birmingham, B4 7ET, UK

Abstract: In this paper we review recent theoretical approaches for analysing the dynamics of on-line learning in multilayer neural networks using methods adopted from statistical physics. The analysis is based on monitoring a set of macroscopic variables from which the generalisation error can be calculated. A closed set of dynamical equations for the macroscopic variables is derived analytically and solved numerically. The theoretical framework is then employed for defining optimal learning parameters and for analysing the incorporation of second order information into the learning process using natural gradient descent and matrix-momentum based methods. We will also briefly explain an extension of the original framework for analysing the case where training examples are sampled with repetition.

1   Introduction

Layered neural networks are powerful nonlinear information processing systems, capable of implementing arbitrary continuous and discrete input-output maps to any desired accuracy, given a sufficient number of hidden nodes and a sufficiently large example set. They have been employed successfully in a variety of regression and classification tasks, and have been studied using a wide range of methods (for a review see Bishop (1995)). On-line learning refers to the iterative modification of the network parameters according to a predetermined training rule, following successive presentations of single training examples, each representing a specific input vector and the corresponding output. On-line learning is one of the leading techniques in training large neural networks, especially via gradient descent on a differentiable error measure.

In this review we focus on the use of methods from non-equilibrium statistical mechanics, for analysing on-line learning in multilayer neural network. We concentrate on our contribution to this area and show how these methods can be employed to monitor the learning dynamics, particularly the evolution of the generalisation error, to define optimal learning parameters and to devise and examine improved learning methods. For a general review see Saad (1998) and Mace and Coolen (1998).

The paper is organised as follows: In section 2 we will derive a compact description of the training dynamics using a set of macroscopic variables, setting up the main theoretical framework. This will then be employed to derive optimal training parameters (section 3), to examine analytically the efficacy of natural gradient descent (section 4), and to suggest and examine practical alternatives using matrix-momentum based methods. In section 5 we will explain how the method can be extended to handle scenarios where training examples are sampled with repetition. In section 6 we will point to the main remaining open questions.

2 Learning in Multilayer Neural Networks

For setting up the basic framework, as in Saad and Solla (1995a, 1995b), we consider a learning scenario whereby a feed-forward neural network model, the 'student', emulates an unknown mapping, the 'teacher', given examples of the teacher mapping (in this case another feed-forward neural network); here we restrict the derivation and the examples to the noiseless case although more general scenarios where training examples are corrupted by noise may also be considered. This provides a rather general learning scenario since both student and teacher can represent a very broad class of functions. Student performance is typically measured by the generalization error, which is the student's expected error on an unseen example. The object of training is to minimize the generalization error by adapting the student network's parameters appropriately.

We consider a student mapping from an N-dimensional input space $\boldsymbol{\xi} \in \mathbb{R}^N$ onto a scalar function $\sigma(\mathbf{J}, \boldsymbol{\xi}) = \sum_{i=1}^{K} g(\mathbf{J}_i \cdot \boldsymbol{\xi})$, which represents a soft Committee machine (SCM - Biehl and Schwarze (1995)), where $g(x) \equiv \mathrm{erf}(x/\sqrt{2})$ is the activation function of the hidden units; $\mathbf{J} \equiv \{\mathbf{J}_i\}_{1 \le i \le K}$ is the set of input-to-hidden adaptive weights for the K hidden nodes and the hidden-to-output weights are set to one. The activation of hidden node i in the student under presentation of the input pattern $\boldsymbol{\xi}^\mu$ is denoted $x_i^\mu = \mathbf{J}_i \cdot \boldsymbol{\xi}^\mu$. This configuration preserves most properties of a general multi-layer network and can be extended to accommodate adaptive hidden-to-output weights as shown by Riegler and Biehl (1995).

Training examples are of the form $(\boldsymbol{\xi}^\mu, \zeta^\mu)$ where $\mu = 1, 2, ..$ labels each independently drawn example in a sequence. Components of the independently drawn input vectors $\boldsymbol{\xi}^\mu$ are uncorrelated random variables with zero mean and unit variance. The corresponding output ζ^μ is given by a teacher of a similar configuration to the student except for a possible difference in the number M of hidden units: $\zeta^\mu = \sum_{n=1}^{M} g(\mathbf{B}_n \cdot \boldsymbol{\xi}^\mu)$, where $\mathbf{B} \equiv \{\mathbf{B}_n\}_{1 \le n \le M}$ is the set of input-to-hidden adaptive weights for teacher hidden nodes. The activation of hidden node n in the teacher under presentation of the input pattern $\boldsymbol{\xi}^\mu$ is denoted $y_n^\mu = \mathbf{B}_n \cdot \boldsymbol{\xi}^\mu$. Indices i, j, k and n, m refer to student and teacher units respectively.

The error made by the student is given by the quadratic deviation,

$$\epsilon(\mathbf{J}^\mu, \boldsymbol{\xi}^\mu) \equiv \frac{1}{2} \big[\, \sigma(\mathbf{J}^\mu, \boldsymbol{\xi}^\mu) - \zeta^\mu \,\big]^2 = \frac{1}{2} \bigg[\, \sum_{i=1}^{K} g(x_i^\mu) - \sum_{n=1}^{M} g(y_n^\mu) \,\bigg]^2 . \quad (1)$$

This training error is then used to define the learning dynamics via a gradient descent rule for the update of student weights $\mathbf{J}_i^{\mu+1} = \mathbf{J}_i^\mu + \frac{\eta}{N}\delta_i^\mu \boldsymbol{\xi}^\mu$, where $\delta_i^\mu \equiv g'(x_i^\mu)[\sum_{n=1}^{M} g(y_n^\mu) - \sum_{j=1}^{K} g(x_j^\mu)]$ and the learning rate η has been scaled with the input size N. Performance on a typical input defines the generalization error $\epsilon_g(\mathbf{J}) \equiv \langle \epsilon(\mathbf{J}, \boldsymbol{\xi}) \rangle_{\{\xi\}}$ through an average over all possible input vectors $\boldsymbol{\xi}$.

Expressions for the generalization error and learning dynamics have been obtained in the thermodynamic limit ($N \to \infty$), and can be represented by a set of macroscopic variables (order parameters) of the form: $\mathbf{J}_i \cdot \mathbf{J}_k \equiv Q_{ik}$, $\mathbf{J}_i \cdot \mathbf{B}_n \equiv R_{in}$, and $\mathbf{B}_n \cdot \mathbf{B}_m \equiv T_{nm}$, measuring overlaps between student and teacher vectors. The overlaps R and Q become the dynamical variables of the system while T is defined by the task. The learning dynamics is then defined in terms of differential equations for the macroscopic variables with respect to the normalized number of examples $\alpha = \mu/N$ playing the role of a continuous time variable:

$$\frac{dR_{in}}{d\alpha} = \eta\, \phi_{in}\,, \qquad \frac{dQ_{ik}}{d\alpha} = \eta\, \psi_{ik} + \eta^2\, v_{ik}\,, \quad (2)$$

where $\phi_{in} \equiv \langle \delta_i y_n \rangle_{\{\xi\}}$, $\psi_{ik} \equiv \langle \delta_i x_k + \delta_k x_i \rangle_{\{\xi\}}$ and $v_{ik} \equiv \langle \delta_i \delta_k \rangle_{\{\xi\}}$. The explicit expressions for ϕ_{in}, ψ_{ik}, v_{ik} and ϵ_g depend exclusively on the overlaps Q, R and T (Saad and Solla (1995a,1995b)). Equations (2), depend on a closed set of parameters and can be integrated and iteratively solved, providing a full description of the order parameters evolution from which the evolution of the generalization error can be derived.

Typical plots of the learning dynamics are presented in Fig.1. In this example the learning process prunes unnecessary nodes when the student network has excessive resources. A teacher with $M = 2$ hidden units characterized by $T_{nm} = n\, \delta_{nm}$ is to be learned by a student with $K = 3$ hidden units. The initial values of the order parameters are $R_{in} = 0$ for all i, n, $Q_{ik} = 0$ for all $i \neq k$, while the norms Q_{ii} of the student vectors are initialized independently from a uniform distribution in the $[0, 0.5]$ interval. The time evolution of the various order parameters is shown in Fig. 1a-c for $\eta = 1$. The picture that emerges is one of specialization with increasing α; asymptotically the first student node imitates the first teacher node ($Q_{11} = R_{11} = T_{11}$) while ignoring the second one ($R_{12} = 0$), the second student node imitates the second teacher node while ignoring the first one, and the third student node gets eliminated ($Q_{33} = 0$).

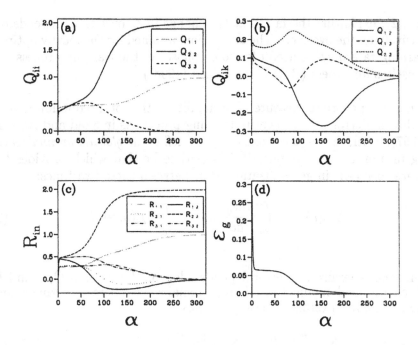

Figure 1: Dependence of the overlaps and ϵ_g on the normalized number of examples α, for $K = 3$ and $M = 2$: (a) the lengths of student vectors, (b) the correlation between student vectors, (c) the overlap between various student and teacher vectors, and (d) the generalization error.

The off-diagonal components Q_{ik} shown in Fig.1b indicate that the two surviving student vectors become increasingly uncorrelated. The overlap between student and teacher vectors (Fig.1c) displays a small α behavior dominated by an undifferentiated symmetric solution, followed by a transition onto the specialization required to obtain perfect generalization. The evolution of the generalization error is shown in Fig.1d.

3   Optimal Learning Parameters

On-line methods are often sensitive to the choice of learning parameters and in particular the choice of learning rate; if chosen too large the learning process may diverge, but if η is too small then convergence can take an extremely long time. The optimal learning rate will also vary substantially over time and may require annealing asymptotically. Most existing analytical results for defining optimal learning rates concentrate on the asymptotic regime where the system may be linearized.

The naive approach to learning rate optimization is to consider the fastest rate of decrease in generalization error as a measure of optimality. To find the locally optimal learning rate one minimizes $d\epsilon_g/d\alpha$, using

Eqs.(2), exploiting the fact that the change in ϵ_g over time depends exclusively on the overlaps. The expression obtained for the locally optimal learning rate may be useful for some phases of the learning process but is useless for others (Rattray and Saad (1998)).

A more appropriate measure of optimality is the total reduction in generalization error over the entire learning process as in Saad and Rattray (1997). With this measure one can then define the *globally optimal* learning rate in a given time-window $[\alpha_0, \alpha_1]$ to be that which provides the largest decrease in generalization error between these two times:

$$\Delta\epsilon_g(\eta) \;=\; \int_{\alpha_0}^{\alpha_1} \frac{d\epsilon_g}{d\alpha} \, d\alpha \;=\; \int_{\alpha_0}^{\alpha_1} \mathcal{L}(\eta, \alpha) \, d\alpha \;. \tag{3}$$

Since the generalization error depends solely on the overlaps Q, R and T, which are the dynamical variables (T remains fixed here), we can expand the integrand in terms of these variables,

$$\mathcal{L}(\eta, \alpha) \;=\; \sum_{in} \frac{\partial \epsilon_g}{\partial R_{in}} \frac{dR_{in}}{d\alpha} + \sum_{ik} \frac{\partial \epsilon_g}{\partial Q_{ik}} \frac{dQ_{ik}}{d\alpha} \tag{4}$$

$$-\; \sum_{in} \mu_{in} \left(\frac{dR_{in}}{d\alpha} - \eta \, \phi_{in} \right) - \sum_{ik} \nu_{ik} \left(\frac{dQ_{ik}}{d\alpha} - \eta \, \psi_{ik} - \eta^2 \, \upsilon_{ik} \right) \;.$$

The last two terms in equation (4) force the correct dynamics using sets of Lagrange multipliers μ_{in} and ν_{ik} corresponding to the equations of motion for R_{in} and Q_{ik} respectively. Variational minimization of the integral in equation (3) with respect to the dynamical variables leads to a set of coupled differential equations for the Lagrange multipliers along with a set of boundary conditions. Solving these equations over the interval $[\alpha_0, \alpha_1]$ determines necessary conditions for η to maximize $\Delta\epsilon_g(\eta)$. The theory is completely general and may be employed for different learning parameters (e.g., regularization parameters as in Saad and Rattray (1998), site dependent learning rates), various learning scenarios (structurally unrealisable or where examples are corrupted by noise) and for obtaining optimal learning rules (Rattray and Saad (1997).

4 Natural Gradient Descent

The same theoretical framework may be used for examining novel training methods. Natural gradient descent (NGD) was recently proposed by Amari (1998) as a principled alternative to standard on-line gradient descent (GD). When learning to emulate a stochastic rule with some probabilistic model, e.g. a feed-forward neural network, NGD has the desirable properties of asymptotic optimality, given a sufficiently rich model which is differentiable with respect to its parameters, and invariance to re-parameterization of our model distribution. These properties are achieved by viewing the parameter space of the model as a Riemannian space in which local distance is defined by the Kullback-Leibler divergence. The Fisher information matrix provides the appropriate metric in this space. If the training error is defined as the negative log-likelihood of the data under our probabilistic model, then the direction of steepest descent in this Riemannian space is found by premultiplying the error gradient with the inverse of the Fisher information matrix; this defines the NGD learning direction.

Studying the learning performance of NGD in the case of isotropic tasks and structurally matched student and teacher ($K = M$ and $T = T\delta_{nm}$) we determined generic behaviour in terms of task complexity K and non-linearity T (Rattray et al (1998)). An analysis of the transient, using globally optimal learning parameters reveals that trapping time in the symmetric phase for the NGD optimized system scales as K^2, compared to a scaling of $K^{8/3}$ for optimal GD. Asymptotically, NGD saturates the universal bounds on generalization performance and provides a significant improvement over optimized GD, especially for small T.

However, in practical applications there will be an increased cost required in estimating and inverting the Fisher information matrix as it requires an average over the input distribution and a matrix inversion. An on-line matrix momentum algorithm (Orr and Leen (1994)) was introduced in order to invert an estimate of the Hessian efficiently on-line. We propose to use this method to compute the inverse of the Fisher information matrix as required for NGD. This method is particularly efficient since the inversion is replaced by a matrix-vector multiplication which can be carried out by a back-propagation step. Since the true Fisher information matrix will not be known in general we use a single step approximation of it, which can be determined on-line. We compared the efficiency of the proposed matrix momentum NGD with that of standard GD and true NGD in training two-layer networks. It turns out to provide a significant improvement over gradient descent learning but with some sensitivity to parameter choice, due to noise in the Fisher information estimate (Scarpetta et al (1999)).

5   Restricted Training Sets

In a realistic scenario the number of training examples scales with the number of free parameters, and examples are therefore sampled with repetition. This gives rise to correlations between the network parameters and the training examples, which clearly affect the learning process. One of the most significant aspects of having a fixed example set is the distinction between the two key performance measures: the *training error*, measuring the network performance with respect to the restricted training set, and the *test (generalisation) error*, calculated for all possible inputs sampled from the true distribution. The former may be monitored in practical training scenarios, while the latter can only be assessed. Another important aspect of learning from restricted training sets which have been corrupted by noise is the emergence of *overfitting* and the need to employ regularization techniques (e.g., weight decay, early stopping - see Bishop (1995)).

The fundamental difference between the infinite and restricted training set scenarios is that the joint probability distribution $P(\mathbf{x}, \mathbf{y})$ for the student and teacher node activations, which is Gaussian in the former case, takes here a more general form, which depends on the training patterns and changes dynamically during the learning process. In fact, we define $P(\mathbf{x}, \mathbf{y})$ as one of the macroscopic variables to be monitored continuously, together with the overlaps R and Q (Coolen and Saad (2000)). To follow the dynamics, one derives a set of coupled differential equations describing the evolution of the macroscopic variables in the limit $N \to \infty$. This set of equations cannot be closed in general; closure is obtained by invoking the dynamical replica theory. The resulting equations can be solved numerically with some simplifications.

The solutions describe the dynamics of both training and generalization errors (and the various overlaps, Coolen et al (2000), Xiong and Saad (2000)), provide insight to the link between the number of examples and the breaking of internal symmetries as well as some asymptotic scaling laws. Our ability to provide analytical solutions is limited due to the complexity of the equations; however, such solutions are highly desirable for deriving analytically generic scaling laws in both the symmetric phase and asymptotically, and to make a quantitative link between the noise level and the optimal regularization to be used.

6 Conclusion

We showed how the methods of statistical physics can provide insight into the dynamics of on-line learning as well as play an important role in defining optimal learning parameters and in examining the properties of new learning algorithms. Several open questions remain, for instance, finding principled methods for optimising the generalisation ability in the case of restricted training sets and the dependence of the length of the symmetric phase on the number of training examples.

References

AMARI, S. (1998): Natural Gradient Works Efficiently in Learning. *Neural Computation, Vol. 10, 251–276.*

BIEHL, M. and SCHWARZE, H. (1995): Learning by Online Gradient Descent. *Jour. Phys. A, Vol. 28, 643–656.*

BISHOP, C. M. (1995): Neural Networks for Pattern Recognition. Oxford University Press, Oxford.

COOLEN, A. C. C. and SAAD, D. (2000): Dynamics of Learning with Restricted Training Sets. *Phys. Rev. E., Vol. 62, 5444–5487.*

COOLEN, A. C. C., SAAD, D. and XIONG, Y. (2000): On-line Learning from Restricted Training Sets in Multilayer Neural Networks. *Europhys. Lett., Vol. 51, 691–697.*

MACE, C. W. H. and COOLEN, A. C. C. (1998): Statistical Mechanical Analysis of the Dynamics of Learning in Perceptrons. *Statistics and Computing, Vol. 8 55–88.*

ORR, G. B. and LEEN, T. K. (1994):Using Curvature Information for Fast Stochastic Search. in Cowan, Tesauro and Alspector (Eds.): Advances in Neural Information Processing Systems, NIPS Vol. 6, Morgan Kaufmann, San Mateo CA, 477–484.

RATTRAY, M. and SAAD, D. (1997): Globally Optimal Rules for On-line Learning in Multilayer Networks. *Jour. Phys. A, Vol. 30, L771–776.*

RATTRAY, M. and SAAD, D. (1998): An analysis of on-line training with optimal learning rates. *Phys. Rev. E., Vol. 58, 6379–6391.*

RATTRAY, M., SAAD, D. and AMARI, S. (1998): Natural Gradient Descent for On-line Learning. *Phys. Rev. Lett., Vol. 81, 5461–5464.*

RIEGLER, P. and BIEHL, M. (1995): Online Backpropagation in Two Layered Neural Networks. *Jour. Phys. A, Vol. 28, L507–513.*

SAAD, D. (Editor) (1998): On-Line Learning in Neural Networks. Publications of the Newton Institute, Cambridge University Press, Cambridge.

SAAD, D. and RATTRAY, M. (1997): Globally Optimal Parameters for On-line Learning in Multilayer Networks. *Phys. Rev. Lett., Vol. 79, 2578–2581.*

SAAD, D. and RATTRAY, M. (1998): Learning with Regularizers in Multilayer Neural Networks. *Phys. Rev. E., Vol. 57, 2170–2176.*

SAAD, D. and SOLLA, S. A. (1995): Exact Solution for On-Line Learning in Multilayer Neural Networks. *Phys. Rev. Lett., Vol. 74, 4337–4340.*

SAAD, D. and SOLLA, S. A. (1995): On-Line Learning in Soft Committee Machines. *Phys. Rev. E, Vol. 52, 4225–4243.*

SCARPETTA, S., RATTRAY, M. and SAAD, D. (1999): Matrix Momentum for Practical Natural Gradient Learning. *Jour. Phys. A, Vol. 32, 4047–4059.*

XIONG, Y. and SAAD, D. (2001): Noise, Regularizers and Unrealizable Scenarios in On-line Learning From Restricted Training Sets. *submitted.*

A Framework for Web Usage Mining on Anonymous Logfile Data

F. Säuberlich, K.-P. Huber

SAS Institute GmbH,
In der Neckarhelle 162,
D-69118 Heidelberg, Germany

Abstract: In this paper we point out how important generalized views of consumer behaviour can be in real life applications. These views can be extracted by performing web usage mining on log file data from anonymous visitors. After discussing the raw data we describe how additional variables can be generated from common log files. By a combination of different data mining techniques like sequence analysis algorithms, cluster analysis, and neural networks it is possible to find out structural problems in web design, typical navigation paths of visitors who will register, or web pages visited together. We show how this information can be used to improve the important customer touch point web for anonymous visitors. We clarify this methodology by showing an example from a real life application.

1 Introduction

With the increasing competition in e-business it has become a necessity for enterprises to be able to present personalized content for different user segments. For example the proportion of visitors of online shops, who transact a purchase, is very small (according to Mowrey (2000) the conversion rate is below 5%). In order to be able to increase this rate, it is inevitable to take care of a well performing web design and to present anonymous visitors (not registered) individual content they are interested in. Web usage mining, the analysis of visitor behaviour in the WWW (see, e.g., Srivastava et al. (2000)), provides analytical techniques to perform this task.

Existing approaches for recommendations on anonymous user behaviour mainly use navigation paths (so called transactions) as data input. For example Mobasher (2000) uses clustering techniques on transactions to propose a recommendation set of objects (links) that closely match the current user transaction (an overview of recommendation systems based on the navigation history of users is given in Gaul, Schmidt-Thieme (2001)).

In this paper we use additional information from raw logfiles about the current user session of an anonymous user to make recommendations. First we describe the general framework of our methodology, consisting of preprocessing, web mining and deployment steps. We show the

310

applicability of this framework by analysing real life logfile data from an information portal provider and clarify that good predictions of user interest can be made on basis of the first clicks on a website.

2 General Framework

Figure 1 gives an overview of our general framework. Starting point of our analyses are common logfile datasets from a web server. In a first step these data have to be cleaned from entries not relevant for analysis (like entries of graphic files; accesses from spiders and so on). Then these data are aggregated on user sessions which presupposes the presence of session-ids to detect a specific visitor. Additional variables can be generated as will be shown in the following section.

In the web mining step we use sequence analysis algorithms to detect frequent naviagational paths, cluster analysis to identify user segments and predictive modelling techniques to be able to predict cluster-ids or visitor interest. The results of these analyses then are used in the deployment step to make link recommendations based on sequential patterns (similar to the Mobasher (2000) framework) or to predict user interest based on cluster profiles or interest variables respectively.

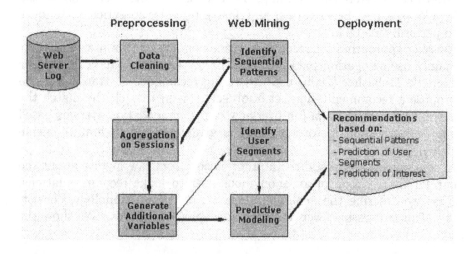

Figure 1: General framework for web usage mining on anonymous logfile data

2.1 Preprocessing (Data Model for Anonymous Web Log Data)

In the first stage of our framework we have to prepare the raw log data for the analysis. In reality a normal logfile contains a lot of entries like accesses of graphic files, which do not represent active requests from clients. In a first step we have to filter the logfile from all those entries. Apart from those graphical requests it is often necessary to remove unsuccessful status codes, request methods other than get, internal user traffic (like traffic from the own webmasters) or robot/spider entries from search engines which do not represent normal user behaviour.

Pages with frame elements have to be surveyed seperately. For pages consisting of several frame elements the web server registers seperate requests for each element. For navigational path analysis those frame requests must be identified and removed.

A very important topic in context of analysing logfiles is session identification. Heuristics like "the same IP-address and the same user agent must be the same visitor" are used, which often are false when we think of proxy servers or dynamic IP-addresses. In this case we need information like cookie- or session-ids.

We also must define a timeout interval after which we assume that a new session was started. If sessions can be identified it is possible to generate additional variables from common logfile entries (see Theusinger, Huber (2000) for a similar methodology) as shown in Table 1.

The only variables we use from an extended logfile directly are the cookie- or session-id and the information of the referrer site from the first click in a user session. From the user agent entries we can derive information about the platform and the browser a visitor is using (as far as these entries are available). Variables that come from the date and time information are hour of the day of the first click, the day of the week and whether the user session has taken place on a weekend or not.

Further variables can be computed after session identification: the number of clicks in a session, the total session length and the average length per click.

After performing a sequence analysis we can generate additional variables describing navigational behaviour of users, which means binary variables for selected paths, with the value of 1 if the user has traversed the path and the value 0 if not.

		Variable	Description
From log file directly		Cookie-ID / Session-ID	To identify hits from one user session
		Referrer	Referrer information from first page view in user session
Additional variables	Derived form agent log	Browser (+Version)	Browser (+Version) information
		Platform	Information of platform (Windows, Mac, Linux etc.)
	Derived from date and time	Hour	Hour of first page view
		Week day	Day of the week of first page view
		Weekend	Binary variable; is first page view on a weekend (yes/no)
	Derived form HTTP requests	Clicks	Number of clicks (page views) in user session
	Derived from HTTP requests and date time	Length	Length of user session
		Clicklength	Average length per click (length/clicks)
	Derived from sequence analysis	Navigational variables	Binary variables; has the user traversed a specific path in his session (yes/no)?

Table 1: Descriptive variables for user sessions

2.2 Web Mining (Applying Data Mining Techniques to Anonymous Web Log Data)

After the additional variables are generated from the logfile we propose in the second step of our framework to use different data mining algorithms (see Säuberlich (2000) for an overview of data mining techniques).

Sequence Analysis

With a sequence analysis algorithm we get for support and confidence values given a set of frequent paths. The sequence analysis algorithms implemented in the SAS Enterprise Miner^{TM} Software were used (see SAS (2001) for a description of these algorithms; for further approaches of navigation path analysis see, e.g., Büchner et al. (1999) or Spiliopoulou et al. (1999)). These frequent paths can be the basis of link recommendations in the deployment step.

We examine further for every user session and every frequent path whether the user has traversed this path or not and generate the binary variables

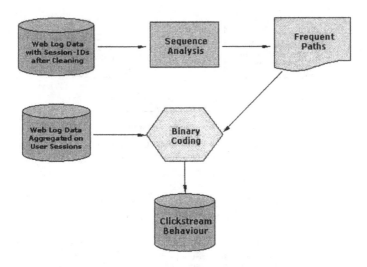

Figure 2: Process flow for sequence analysis

described in the previous section. This methodology makes it possible to incorporate navigational behaviour into the set of input variables used for the following analyses.

User Segmentation

All these variables can be used as input for cluster analysis algorithms to identify user segments with different navigational behaviour. These segments then can be examined with respect to clicks in specific website categories, banners, interest in products and so on to derive an interest profile for each cluster. The cluster-id then can be used as target variable in the predictive modelling step.

Predictive Modelling

With the descriptive session variables described earlier as input variables and a target variable which can be the cluster-id from the previous step or an interest variable (clicks in specific site categories, banners and so on) it is possible to apply predictive modelling techniques. These can be for example decision trees, neural networks or regression techniques. To apply the models generated to new user data in an online personalization/recommendation scenario we propose to transform the model results in C- or java-code to make the score code usable on any hardware platform desired.

314

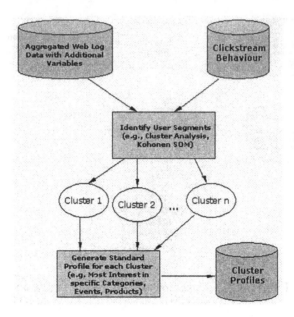

Figure 3: Process flow for user segmentation

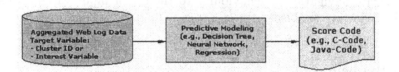

Figure 4: Process flow for predictive modelling

Deployment

In the last stage of our framework we apply the results from three scenarios for recommendations. After a specific number of clicks of a new anonymous visitor (say m clicks) we generate the variables mentioned and apply the results from a former analysis.

Three types of models can be used: the most similar path from all frequent paths computed by sequence analysis to recommend links the user is most likely to take; score code to find the cluster the user is most likely to fall in and therefore recommendations based on the corresponding cluster profile; and finally score code to predict a predefined interest variable directly to actively present the user categories, products or events he is most likely interested in. The latter form of recommendations is used in the following application example.

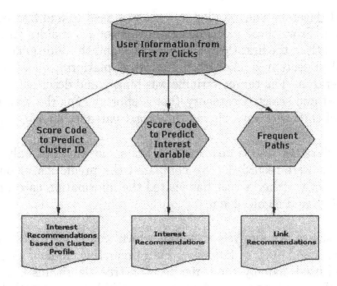

Figure 5: Process flow for the deployment stage

3 Application on a Real Life Datasets

In the following section we demonstrate the scenario of direct interest recommendations by analysing a real life data set. In this example we use the framework given by the bold arrows in Figure 1 and do not use clustering techniques. The logfiles come from an information portal which provides search engine functionality as well as an information catalogue with different categories like sports, health, science and so on.

Task of our analysis was to predict the interest of anonymous visitors in one of the information categories in an early stage of their visit. Benefit for the portal provider: Banners can be presented according to interests of visitors and links or news from the predicted category can be given very early to satisfy visitor needs.

The raw logfile dataset consisted of over 5 million hits; this resulted in 222,044 page views after filtering of graphic or spider entries and frame requests. Finally all page views from sessions with less than 4 clicks were eliminated, which resulted in 84,701 page views. After the aggregation on user sessions 11,393 user sessions existed for the analysis. The average number of page views per session was 6.87 (with a standard deviation of 5.94).

The web mining task was to learn from three clicks which category the visitor is interested in. Therefore the data was divided into a training

316

and a test dataset. The training dataset was used to find frequent paths of length 3. From these paths binary navigational variables (traversed a frequent path in the first three clicks yes/no) and the additional variables described in section 2 (clicklength, browser, platform, hour and so on) were generated. The target variable was binary and described if the user has visited one selected category (here: sports). For the training data set 26% of the users have visited the target category sports.

The test data set was transformed in the same way, but only the first three clicks were recorded. We compared the prediction of our model with the fact whether a user has visited the information category in the clicks 4, 5, 6 and so on or not.

Figure 6 shows the analysis scenario, which was implemented with the SAS Enterprise Miner^{TM} Software. Four predictive modelling algorithms have been used, namely the three decision tree algorithms C4.5, CART and ChAID as well as a neural network (one hidden layer with 10 nodes, backpropagation). These techniques have examplary been chosen to prove the applicability of our framework and were used with the standard parameter settings given by the SAS Enterprise Miner^{TM} Software.

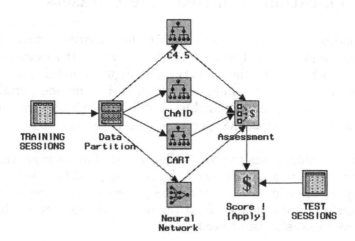

Figure 6: Analysis scenario for web mining

The resulting score code of each model (representing the predictive model) was applied on the test data to validate the predictive quality. Table 2 shows the overall accuracy, which is the percentage of visitors the target variable was predicted correctly.

For all four models an overall accuracy around 80% was reached. But with a closer look on the correct predictions of targets with value 1, which means users who are interested in the information category, the

Model	Accuracy	Accuracy Target=1	Accuracy Target=0
Neural Network	78,24	**72,54**	79,31
CART	78,03	53,97	82,56
ChAID	80,08	70,26	81,93
C4.5	80,31	67,70	82,69

Table 2: Accuracy of the models on the first three clicks of the test data

neural network has the best accuracy of 72.5 percent, whereas the CART algorithm comes out last with a accuracy of 54%. This means that the trained neural network is able to predict the visit of the sports category after the first three clicks with more than two third, which seems to be a promising result. In this case the prediction of the neural network should be used and the sports category should be recommended to the corresponding visitors.

4 Conclusion and Outlook

A general framework for performing website personalization on anonymous user data was presented. Frequent paths can be used for link recommendations, while predictive models and cluster profiles can be used for interest recommendations. These recommendations are based on the data given in Table 1 which shows variables which can be used to describe the behaviour of anonymous users apart from only using navigational path information.

By analysing common logfiles from an information portal navigational paths were combined with various variables for describing anonymous user behaviour. The predictive accuracy of the models was very good with respect to only using the first three clicks of web site visitors for prediction. Further analyses are on the way to prove the applicability of the whole framework (recommendations on frequent paths and cluster profiles) and to report experiences of a real life recommendation scenario based on this work.

References

BÜCHNER, A.G., BAUMGARTEN, M., ANAND, S.S., MULVENNA, M.D., and HUGHES, J.G. (1999): Navigation pattern discovery from internet data. Proceedings of WEBKDD'99, Springer, Berlin.

GAUL, W. and SCHMIDT-THIEME, L. (2001): Recommender Systems Based on Navigation Path Features. Proceedings of WEBKDD'2001, August 2001, San Francisco.

MOBASHER, B. (2000): Discovery of Aggregate User Profiles for Web Personalization. Proceedings of WEBKDD'2000, August 2000, Boston.

MOWREY, M. (2000): Thank You, Please Come Again. *Industry Standard*, http://www.thestandard.com/research/metrics/display/0,2799,13016,00.html

SAS (2001): SAS Enterprise Miner Procedures Guide. Cary, NC: SAS Institute Inc., 2001.

SÄUBERLICH, F. (2000): KDD und Data Mining als Hilfsmittel zur Entscheidungsunterstützung. Peter Lang Verlag, Frankfurt.

SPILIOPOULOU, M., FAULSTICH, L.C., and WINKLER, K. (1999): A Data Miner analyzing the Navigational Behaviour of Web Users. Proceedings of the Workshop on Machine Learning in User Modelling of the ACAI'99 Int. Conf., Creta, Greece, July 1999.

SRIVASTAVA, J., COOLEY, R., DESHPANDE, M., and PAN-NING, T. (2000): Web Usage Mining: Discovery and Applications of Usage Patterns from Web Data. *SIGKDD Explorations, Vol. 1/2*, January 2000.

THEUSINGER, C. and HUBER, K.-P. (2000): Analyzing the Footsteps of Your Customers–A Case Study by ASKnet and SAS Institute GmbH. Proceedings of WEBKDD'2000, August 2000, Boston.

Medicine, Biological Sciences and Health

Knee Replacement Surgery and Learning Effects - Data Evidence from a German Hospital

C. Ernst, G. Ernst, A. Szczesny

Lehrstuhl für Betriebswirtschaftslehre, insb. Controlling, Johann Wolfgang Goethe–Universität Frankfurt, D-60054 Frankfurt(Main), Germany

Abstract: In 2003, Germany will be the first country in the world to adopt a fully prospective payment system for the reimbursement of all inpatient hospital services. To face the increasing competition, hospitals can pursue either a specialization or a cost and quality leadership strategy. It stands to reason that organizational and individual learning will play an important role for both strategies. We develop a theoretical model of surgical learning and test it using detailed operating room data from the first 601 total knee replacement surgeries of a small German hospital between 1994–2000. Our results suggest that classical learning curve theory can indeed be applied to this high cost high volume procedure.

1 Introduction

In 2003, Germany will be the first country in the world to adopt a fully prospective payment system for the reimbursement of all inpatient hospital services. A very important objective of this reform is to increase competition for patients between hospitals. To face this new situation, hospitals can either specialize in services where they enjoy a comparative advantage or pursue a cost and quality leadership strategy where the latter may be more appropriate in highly competitive areas such as endoprothetic surgery. It seems likely that organizational and individual learning will play an important role for both strategies. Though, some studies show an inverse relationship between volume and time per procedure,[1] these findings have been rarely linked to traditional learning curve theory. Using data on the first 601 knee replacement surgeries of a German hospital between 1994–2000, we estimate the learning effect with respect to "skin to skin time" for this procedure. Since classical learning curve theory is based on homogenous products, whereas patients tend to be extremely heterogeneous because of differences in prior disease history and health status, we include other factors likely to have an impact on procedure time. We also address the question of a link between procedure volume and quality indicators, since reduced procedure times at the expense of quality would obviously be problematic. Section 2 briefly compares the concepts of learning used in accounting and

---

[1] See Cromwell et al. (1990)

medicine. Section 3 describes the study design, section 4 presents the results and discusses the finding's implications for hospital managerial accounting and management.

2   Economic Learning vs. Medical Learning

For more than half a century, the concept of the learning curve has played an eminent role in engineering and managerial accounting for a wide range of industries. In fact, the occurrence of learning effects has been so typical for almost any industry that their absence is often interpreted as a sign of severe inefficiencies.[2] Today, the most common type of learning curve is the so–called Crawford version[3] which states that with each doubling of production volume of a product, there is a systematic and uniform percentage reduction (the learning rate) in the time required to build one unit of the product. Let X denote the cumulative production volume for a product, where $X = 2^z$ and z is the number of doublings of volume. Let $U(X)$ be the time required to manufacture that product and $U(1)$ be the corresponding time for the first product. If α denotes the percentage reduction and κ an elasticity that states the relative reduction in time per unit for an increase in production volume, the learning curve in terms of doubling and for any quantity can be expressed as:

$$U(X) = U(1)(1 - \alpha)^z = U(1)X^\kappa \quad \text{where} \quad \kappa = \frac{\ln(1 - \alpha)}{\ln(2)}. \quad (1)$$

Some of the theory's most important uses include pricing decisions, budgeting, contract negotiations with suppliers, government procurement and government projects (NASA).[4] Knowledge about the exact relations involved in learning is still far from comprehensive. There appears to be agreement on the following issues, however. Learning effects play only a limited role in processes whose speed is mainly determined by automated machinery but exert a large influence on complex processes controlled by human skill.[5] In addition, it is claimed that learning will not "just take place" but requires a planning phase, creating incentives to realize learning potentials.[6]

---

[2]See The Boston Consulting Group (1972), p. 2
[3]See Treplitz (1991), Chapter 2, for other versions
[4]See Treplitz (1991), p. 48 f.
[5]See Hirsch (1956), critical Baloff (1966)
[6]See Coenenberg (1997), p. 209

Interest in surgical learning was almost non–existent until the advent of managed care in the late 70's, which was largely the result of the general absence of efficiency incentives in the health care industry up to that time. Another point is that the learning curve deals with homogenous products and hospital output, even for the same disease or procedure, tends to be extremely heterogeneous with respect to age, prior disease history and other risk factors of the patients.[7]

Since then, medical care in general and surgery in particular have adopted their own concept of learning. Its main concern is with the quality of medical care and seeks to document the relation between surgical volume and accepted indicators of outcome quality such as mortality, morbidity, iatrogenic injury or hospital acquired–infections.[8] These studies have consistently shown an inverse relation between the number of procedures performed and these quality indicators (volume–outcome relation).[9] More recently, the learning effect has played an important role in the evaluation and comparison of new procedures such as minimal-invasive laparoscopic surgery vs. the respective open variants.[10] Surgeons often claim that learning is likely to occur with respect to a new procedure with an accompanying impact on cost and quality of care.[11] Studies that focus explicitly on learning yield mixed results and cannot be discussed in detail here.[12] Studies that link volume to cost are extremely rare and tend to suffer from methodological problems.[13] Though many studies document that time per procedure declined and complications were reduced as volume increased, these findings have up to now rarely been linked to specific theoretical learning models.[14]

3   Study Design and Data

Our study is an attempt to address this point. We analyze the first 601 elective knee replacement surgeries of a small hospital from detailed anonymous operating room data collected by the hospital's anesthesiology department between 1988 and the present. In 1994, this procedure was newly added to the services offered, though the hospital had previously acquired considerable experience with total hip replacement

---

[7]See Breyer/Zweifel (1997), p. 331

[8]See Luft et al. (1979) and Philipps/Luft (1997)

[9]See Eisenberg (1986), p. 11 with more literature

[10]See Park et al. 1998)

[11]See Lee (1999), Piano et al. (1998), p. 616 f. and Luks et al. (1999), p. 968

[12]See Kern et al. (1998), Cromwell et al. (1990), Soot et al. (1999)

[13]These are use of charge instead of cost data, see Smith/Larsson (1998), Finkler (1994) and the use of cross–sectional data like in Guttierrez et al. (1998)

[14]See Phillips/Luft (1997), p. 345

surgery. The absence of major technical changes for this procedure during the period under consideration also rendered it an ideal candidate for the study of learning effects. Furthermore, expert surgeons remained the same for the entire study period, while several different interns also performed the operation. Based on Germany's current fixed price reimbursement for this procedure, the 601 observations constitute a considerable revenue volume of approximately 4.5 Mio. Euro.

Our estimation is based on the theoretical model for the operating time (opt) of a knee replacement procedure $opt = k \cdot R \cdot BL^{\sigma} \cdot AOP \cdot Ex \cdot TD^{\tau}$. In this model, we assume that time per procedure is influenced by a basic time requirement k, the patient's risk status R, loss of blood during the operation BL, additional operations AOP, the expert status of the surgeon Ex and the time difference between successive operations TD. Taking logarithms yields the following linear model:

$$
\begin{aligned}
\ln(opt) \;=\; & \ln(k) + \ln(R) + \sigma \ln(BL) + \ln(AOP) \\
& + \ln(Ex) + \tau \ln(TD).
\end{aligned} \tag{2}
$$

To model that the basic time requirement k is affected by a learning effect, we assume that the time for knee replacement X displays a learning effect in accordance with classical learning curve theory introduced in section 2, where operating time for the Xth procedure is determined by $opt(X) = opt(1)X^{\kappa}$. Taking the logarithm of this expression and using it in (2), yields the modified linear equation (3).

$$
\begin{aligned}
\ln(opt(x)) \;=\; & \ln(opt(1)) + \kappa \ln(X) + \ln(R) + \sigma \ln(BL) \\
& + \ln(AOP) + \ln(Ex) + \tau \ln(TD).
\end{aligned} \tag{3}
$$

Based on this theoretical model, we develop the following regression function:

$$
\begin{aligned}
\ln(opt(x)) \;=\; & \beta_0 + \beta_1 \cdot \ln(X) + \beta_2 \cdot asa_3 + \beta_3 \cdot asa_4 + \beta_4 \ln(bloss) \\
& + \beta_5 \cdot opadd + \beta_6 \cdot expert + \beta_7 \cdot \ln(mtime).
\end{aligned} \tag{4}
$$

For the continuous and unobservable variables R, AOP and Ex of the underlying theoretical model, we include the dummy variables asa_3, asa_4, $opadd$ and $expert$ to be explained below.

To incorporate the idea of patients' heterogeneity with respect to morbidity and other risk factors, we used the patient's American Association of Anesthesiologists (ASA) physical status classification[15] as an estimator for the original model's patient risk R.[16] Patients in ASA class 1 and 2 were pooled, since only one patient fell in ASA 1. Risk measured by a higher ASA score is defined by means of dummy variables (asa_3, asa_4 with asa_{12} as reference category) and serves as an estimator for $ln(R)$. For instance $asa_3 = 1$ (0) if the patient (does not) fall in class ASA 3. It is hypothesized that a higher ASA score leads to an increase in time per procedure.

The variable *bloss* denotes blood loss in ml during the operation. $ln(bloss)$ is included in our estimation and therefore corresponds with $ln(BL)$ in the original model. β_4 thereforerepresents the elasticity σ from the original model. The hypothesis is that higher blood loss will increase the time per procedure.

The dummy variable *opadd* informs about additional operations and is an estimator for $ln(AOP)$. It is hypothesized as having a positive impact on *opt*. The dummy variable *expert* signifies that the operation was performed by an expert surgeon with great experience in endoprothetic surgery and serves as an estimator for $ln(Ex)$.[17] The hypothesis is that expert will have a negative influence on *opt*. The variable *mtime* denotes time in days between procedures performed and is included as $ln(mtime)$, corresponding to $ln(TD)$ in the original model.[18] The hypothesis is that if more time elapses between procedures, *opt* will be higher.

β_0 is thus an estimator for a representative first knee replacement surgery for an ASA 1 or 2 patient, which is performed by a non–expert, without blood loss or additional operations

Table 1 presents descriptive statistics for the respective variables which give a fairly accurate idea of the estimated effects' direction. For instance, means for different ASA values differ considerably.

---

[15]See http://www.asahq.org/ProfInfo/PhysicalStatus.html (ASA homepage), ASA 1: normal healthy, ASA 2: mild systemic disease, ASA 3: severe systemic disease, ASA 4: severe systemic disease that is a constant threat to life, ASA 5: moribund patient not expected to survive without operation

[16]Numerous studies have shown ASA status to be a good predictor of postoperative outcome. See Hall, Hall (1996), Arvidsson et al. (1996), Prause et al. (1997), Wolters et al. (1996)

[17]Several studies have documented that experts have shorter procedure times. See Willeke et al. (1999); Soot et al. (1998)

[18]The reason for its addition are findings from learning theory that not only total volume but also the time between procedures performed may exert an influence on learning. See Treplitz (1991), p. 131

Table 1: Means and standard deviations of operation time

ASA	Mean	Std.Dev.	# Obs.	opadd	Mean	Std.Dev.	# Obs.
1 or 2	63.76	18.45	254	0	66.12	21.18	550
3	67.90	22.90	335	1	68.43	21.13	51
4	76.25	18.96	12	—	—	—	—
Total	66.31	21.17	601	Total	66.31	21.17	601

bloss	Mean	Std.Dev.	# Obs.	mtime	Mean	Std.Dev.	# Obs.
$\leq$100	65.36	21.96	433	1	65.09	19.31	265
100–200	62.44	17.43	39	2–5	63.90	19.08	214
200–500	68.27	17.47	101	6–10	68.11	25.74	82
>500	79.46	21.32	28	>10	83.63	25.44	40
Total	66.31	21.17	601	Total	66.31	21.17	601

expert	Mean	Std.Dev.	# Obs.
0	70.67	17.00	186
1	64.36	22.54	415
Total	66.31	21.17	601

Blood loss is measured in ml, mtime in days. If *opadd* = 1 (0) an (no) additional operation was performed. If *expert* = 1 (0) the operation was performed by an (no) expert. Observe that ASA is a categorial variable for the raw data which was used to construct the dummy variables asa_3 and asa_4 explained above.

4   Results and Discussion

After performing an OLS–regression, we obtained the following results, summarized in table 2. The learning effect based on classical learning curve theory (κ) is highly significant at the 1%–level. This is equivalent to a learning rate of $\alpha = 0.083$ or 8.3% or a 91.7 percent learning curve. All other influencing factors held constant, this implies that each doubling in procedure volume for knee replacement surgery from say 2 to 4, 4 to 8, 8 to 16 procedures is accompanied by an 8.3% reduction in operating time. Surprisingly, this factor is rather small compared to those developed for other industries even though manual skills play such an eminent role in surgery. Since $\beta_0 \equiv ln(opt(1))$, we can compute the time necessary for a representative first knee replacement surgery for an ASA 1 or 2 patient, which is performed by a non–expert, without blood loss or additional operations. For the data it is given by $exp(\beta_0) = exp(4.8942) = 133.5$ minutes.

In addition, ASA risk classification has a significant effect on procedure time. Compared to the reference category of ASA 1 or 2, there is no notable difference for a patient in ASA 3, which is not significant. ASA 4, however, is significant at the 5% level relative to the reference category.

Whether the procedure was performed by an expert has also a highly significant effect on procedure time at the 1%–level. The resulting factor is 0.84 and suggests that all other factors held constant, an expert will perform a knee replacement surgery at 84% of the time, a non–expert would need for the same procedure. This result is somewhat remarkable,

Table 2: Learning effect – estimation results

Variable	Coef.		Std.Dev.	t	P>|t|
$ln(X)$	-0.1250	***	0.0158	-7.9080	0.0000
asa_3	0.0321		0.0264	1.2160	0.2250
asa_4	0.2071	**	0.0939	2.2050	0.0280
$expert$	-0.1731	***	0.0296	-5.8500	0.0000
$opadd$	0.0032		0.0467	0.0690	0.9450
$ln(bloss)$	0.0092	*	0.0050	1.8310	0.0680
$ln(mtime)$	0.0051		0.0155	0.3280	0.7430
$const$	4.8942	***	0.1016	48.1680	0.0000
α	0.0830				
Goodness of fit					
R^2	0.1525				

Significant at the * 10 percent level / ** 5 percent level / *** 1 percent level.

since one might argue that experts will have longer procedure times as a result of treating riskier patients. Descriptive statistics showed, however, that this was not true for our data set. Experts did not treat riskier patients.[19] This may be attributable to the comparatively small number of physicians involved and data from larger institutions may well show a different result.

Additional operations ($opadd$), which were performed on 52 out of the 601 procedures analyzed, have the expected time–increasing sign but are not significant.

Loss of blood ($ln(bloss)$) is significant at the 10%–level and tends to increase procedure time. At this point it should be noted that knee replacement surgery is conducted under a blood stoppage and will routinely not lead to blood loss in the operating room. Patients will lose blood in post–operative care, however. Unrelated to complications, it may sometimes become necessary for some patients to open the blood stoppage in the operating room. What this analysis shows is that this tends increase procedure time. The impact on time can be computed as (Quantity of blood lost)$^{ln(bloss)}$. For instance, for a loss of 1000 ml we have $1000^{0.0092} = 1.065$, implying that this will increase procedure time by 6.5%, if all other factors are held constant.

Contrary to some refinements of the traditional learning curve, time between procedures denoted by $mtime$ is not significant, though it has the expected sign. This may be attributable to the fact that this variable was 0 (same day procedure) 1 or 2 (same procedure within one or two days) for more than 50% of our observations.

---

[19]We thank our session head for pointing out that this issue merited further attention, the respective ratios of non–expert vs. expert were ASA 1 or 2: 31,89%/68,11%, ASA 3: 29.55%/79.45%, ASA 4: 50%/50%, Total: 30,95%/69,05%.

Physicians might argue that operating time is a poor predictor of quality, though it has occasionally been used for this purpose.[20] It may even be possible that time reductions occur at the expense of quality. To address the question, whether reduced procedure times could possibly compromise quality, we tested an additional model for the relation between the number of procedures[21] and the probability of operating and recovery room complications. It yielded the result that a positive adverse relation exists between the number of procedures and these complications. Consequently, short–term quality measured by reduced probabilities for complications is actually improved by an increased number of procedures subject to the learning effect.[22]

This corroborates findings from the volume–outcome literature on the positive relationship between volume and the quality of care. Contrary to most volume–outcome studies, we obtained our results from a small non-teaching institution. It seems reassuring that the average patient in an average hospital is also likely to benefit from these improvements in the form of reduced operating time and less operating room complications.

Direct implications for managerial accounting in German hospitals are difficult to evaluate at present. Though some authors link reduced procedure times to better outcomes and increased efficiency and consequently lower cost, this relationship is by no means obvious.[23] Currently most German hospitals calculate their operating room costs allocating fixed costs like physician or nurse salaries to procedures using physician or nursing minutes as allocation base.[24] It is also obvious that reduced procedure times as the result of learning would suggest that "cost savings" occurred under such a system if the minute cost factor is held constant. This constitutes proportionalizing fixed costs, however, with all the associated problems, since the hospital incurs salary costs regardless of operating room activities. For a meaningful analysis of learning effects with respect to time or complications on costs, one would need to identify cost categories where a functional relationship between procedure time or complications and cost existed (variable relevant cost). Caution should therefore be exercised when proposals are made that derive a need for reimbursement cuts from learning with respect to time. A

---

[20]See Scott et al. (1998), Willeke et al. (1998) and Piano et al. (1998)

[21]Further influencing factors like patient risk etc. were also included.

[22]Some caution should be exercised in interpreting this result, since the complication variable mainly focuses on anesthesia related complications in the operating and recovery room only. Intermediate indicators of severity and quality such as time spent in intensive care or post–operative mortality are not included in the data set. Interviews with the physicians yielded the result that mortality and complications were not significantly different from reported outcomes in the literature.

[23]See Cromwell/Mitchell/Statson (1990)

[24]See DKG manual (1998), p. 75 ff.

more direct implication of the analysis would be in the field of hospital benchmarking, which enjoys considerable popularity in Germany at the moment.[25] The results stress that a comparison of similar hospitals with respect to procedure time for selected operations should take the respective hospitals' "position on the learning curve" into account. Other possible usage of the learning curve may be long–term capacity planning for the operating room or revenue management resulting from the expected time savings. The result with respect to the experts may also render heavy teaching and training obligations more transparent. These last uses would obviously require knowledge about whether our results carry over to other procedures and hospitals – a question we plan to address in future research.

References

ARVIDSSON, S., OUCHTERLONY, J. and SJOSTEDT, L. et al. (1996): Predicting postoperative adverse events. Clinical efficiency of four general classification systems. The project perioperative risk, in *Acta Anaesthesiol Scand 40, 783–791*.

BALOFF, N. (1966): The Learning Curve: Some controversial issues, in: *Journal of Industrial Economics 14, 275–282*.

BOSTON CONSULTING GROUP (1970): Perspectives on experience, 2nd ed.

BREYER, F. and ZWEIFEL, P. (1997): Gesundheitsökonomie, 2nd rev. And ext. Ed., Heidelberg, Springer.

COENENBERG, A. (1997): Kostenrechnung und Kostenanalyse, 3rd Rev. Ans ext. Ed., Landsberg/Lech, Motorbuch.

CROMWELL, J., MITCHELL, J. and STATSON, W. (1990): Learning by doing in CABG surgery, in: *Medical Care 28, 6–18*.

DEUTSCHE KRANKENHAUSGESELLSCHAFT (DKG) (1998): Handbuch zur Nach– und Neukalkulation von Fallpauschalen und Sonderentgelten, www.dkgev.de (Stand vom 20.02.2001).

EISENBERG, J.M. (1986): Doctors' Decisions and the Cost of Medical Care. The Reason for Doctors' Practice Patterns and Ways to change them, Ann Arbor MG.

---

[25]See for instance Ohm/Albers/Assmann (2001)

FINKLER, S. (1994): The Distinction between Costs and Charges, in: Finkler, S. (Ed.): Issues in Cost Accounting for Health Care Organizations, Gaithersburg MD, Aspen, 81–93.

GUTTIERREZ, B., CULLER, S. and FREUND, D. (1998): Does Hospital Procedure–Specific Volume Affect Treatment Costs? A National Study of Knee Replacement Surgery, in: *Health Services Research 33, 489–511.*

HALL, J. and HALL, J. (1996): ASA status and age predict adverse events after abdominal surgery, in: *J Qual Clin Pract 1996 Jun; 16, 103–1088.*

HIRSCH, W. (1956): Firm Progress Ratios, in: *Econometrica 24, 136–143.*

KERN, J., FRY, W. and HELMER, S. (1998): Institutional Learning Curve of Surgeon–Performed Trauma Ultrasound, in: *Archives of Surgery 133, 530–536.*

LEE, P. (1999): The Year 2000: Placing New Technology in Context Paul, in: *Archives of Ophtalmology, 117.*

LUFT, H., BUNKER, J. and ENTHOVEN, A. (1997): Should operations be regionalized? The empirical relation between surgical volume and mortality, in: *New England Journal Of Medicine 301, 1364–1369.*

LUKS, F., LOGAN, J. and BREUER, C. (1999): Cost–effectiveness of Laparoscopy in Children, in: *Pediatr Adolesc Med., 153, 965–968.*

OHM, G., ALBERS, E., ASSMANN, J. et al. (2001): Der Mut zum Vergleich lohnt, um von den Besten zu lernen, in: *führen und wirtschaften im krankenhaus f&w, 18, 38–42.*

PARK, A., MARCACCIO, M., STERNBACH, M. et al. (1999): Laparoscopic vs Open Splenectomy, im Archives of Surgery 134, 1263–1269.

PHILLIPS, K. and LUFT, H. (1997): The Policy Implications of Using Hospital and Physician Volumes as "Idicators" of Quality of Care in a Changing Health Care Environment, in: *International Journal for Quality in Health Care 9, 341–348.*

PIANO, G., SCHWARTZ, L., FOSTER, L. et al. (1998): Assessing Outcomes, Costs, and Benefits of Emerging Technology for Minimally Invasive Saphenous Vein In Situ Distal Arterial Bypasses, in: *Archives of Surgery, 133, 613–618.*

PRAUSE, G., RATZENHOFER–COMENDA, B., PIERER, G. et al. (1997): Can ASA grade or Goldman's cardiac risk index predict perioperative mortality? A study of 16,227 patients, in: *Anaesthesia; 52,* 203–206.

SMITH, D. and LARSSON, J. (1989): The Impact of Learning on Cost: The Case of Heart Transplantation, in: *Hospital & Health Services Administration 34, 85–97.*

SOOT, S., ESHRAGHI, N., FARAHMAND, M. et al. (1999): Transition From Open to Laparoscopic Fundoplication–The Learning Curve, in: *Archives of Surgery 1999; 134, 278–281.*

TREPLITZ, C. (1991): The Learning Curve Deskbook, Westport CT, Quorum.

WILLEKE, M. et al (1998): Effect of Surgeon Expertise on the Outcome in Primary Hyperparathyroidism, in: *Archives of Surgery, 133, 1066–1070.*

WOLTERS, U., WOLF, T., STUTZER, H. et. al. (1996): ASA classification and perioperative variables as predictors of postoperative outcome, in: *British Journal of Anaesthesiology 77, 217–222.*

Self-Organizing Maps and its Applications in Sleep Apnea Research and Molecular Genetics

G. Guimarães[1], W. Urfer[2]

[1]CENTRIA (Center of Artificial Intelligence), Universidade Nova de Lisboa, Portugal

[2]Department of Statistics, Universität Dortmund, Germany

Abstract: This paper presents the application of special unsupervised neural networks (Self-Organizing Maps) to different domains, such as sleep apnea discovery, protein sequence analysis and tumor classification. An enhancement of the original algorithm, as well as the introduction of several hierachical levels enables the discovery of complex structures as present in this type of applications. Finally we recommend the use of regression-type models for Kohonen's Self-Organizing Network in gene expression data analysis.

1 Introduction

The development of more and more powerful computers in recent years has lead to a recording of a great amount of data gathered from, for example, industrial processes, medical applications, meteorological phenomena, etc. Artificial Neural Networks (ANNs) and methods from statistics are particularly interesting for handling such noisy and inconsistent data. The application of ANNs and statistics often refers to problems of discrimination (supervised learning) or to clustering problems (unsupervised learning). Self-Organizing Maps (SOMs) as proposed by Kohonen (1982) are well suited for the discovery of patterns in high dimensional data, i.e. clustering problems (Kohonen, 2001; Kaski and Kohonen, 1996). In addition, SOMs have also been successful in applications, where temporal or sequential data are processed, for instance, in speech recognition, process control and time series analysis in medicine (Behme et al., 1993; Walter and Schulten, 1993; Guimarães, 2000). In this paper we give a review of SOMs to several application domains, such as sleep apnea, protein sequence analysis and tumor classification.

For the diagnosis of sleep apnea the temporal dynamics of physiological parameters such as respiration and heart rate have to be recorded and evaluated. In order to perform an automated identification of sleep apnea a simultaneous analysis of all signals is needed. Different types of sleep apnea diseases represent complex patterns in the time series that occur during one night. Those patterns may differ strongly, even for the same patient. The main aim of statistics in bioinformatics is the development and application of methods for the analysis of genomic and proteomic data in order to elucidate biological processes. Statistics and

ANNs play a major role in diagnosing diseases and developing new drugs (Brunnert et al., 2000). The most important contribution of statistics has been the development of strategies for extracting information from DNA and protein sequence databases by sequence comparison, characterization and classification. These applications have in common that complex patterns are searched for and a hierarchical segmentation of the problem is needed introducing hierarchical SOMs. In addition to the hierarchical component, both applications demonstrate a temporal or a sequential component. In section 2 SOMs and their possible extensions are introduced. Section 3 presents the application of extended SOMs to sleep apnea. The application of extended SOMs to protein sequence analysis and tumor classification is shown in section 4. Special emphasis is given to gene expression data and the possible use of regression-type models for Kohonen's Self-Organizing Networks.

2 Self-Organizing Maps for Exploratory Data Analysis

Artificial Neural Networks (ANNs) may be classified according to their learning principles mainly into two different types: ANNs with supervised learning and ANNs with unsupervised learning. ANNs with supervised learning adapt their weights to a given input-to-output relationship, for instance for the recognition of images representing handwritten characters. For this kind of pattern recognition problems, ANNs with supervised learning, such as Feed-forward Networks or Radial Basis Function Networks, have been widely studied in relation to their statistical properties. It is well-known that Feed-forward Networks can approximate, to arbitrary accuracy, any smooth function. In the context of classification problems, Feed-forward Networks with sigmoidal non-linear activation functions of the neurons can approximate arbitrarily any decision boundary. Such ANNs provide universal non-linear discriminant functions modeling posterior probabilities of class membership, permitting a probabilistic interpretation of the results (Bishop, 1995). Models in statistics strongly related to those ANNs are logistic discrimination functions, projection pursuit regression, and multivariate adaptive regression splines. In this work we will not focus on pattern recognition problems, where a classification is known a priori, but on pattern discovery problems, where the inherent patterns in the data are searched for. ANNs with unsupervised learning are suitable for such problems, since they adapt their internal structures (weights) to the structural properties (e.g. regularities, similarities, frequencies, etc.) of high-dimensional input data. This means that these models can be used for the generation of some constitutive hypothesis. Instead of trying to perform a high prediction accuracy, as would be afforded in any classification system, the main aim here is to identify relevant, meaningful, as well as "hidden" structures, i.e. patterns in the data. These approaches lie in the filed of exploratory data analysis. In combination with Machine Learning algorithms, as fostered here, they are a powerful tool for Data Mining projects. ANNs like ART (Adaptive Resonance Theory) and Self-Organizing Maps (SOMs) belong to this type of models and specially the latter are well-suited for clustering tasks (Kaski and Kohonen, 1996; Kohonen, 2001). The motivation of SOMs is strongly biology-oriented where principles concerning the generation of topographical maps in the brain through self-organization play an important role. In the following, the learning process will be described from a more algorithmic point of view. During learning SOMs adapt their weights such that a n-dimensional input space is projected onto a m-dimensional map with m<n, preserving the neighborhood of the input data on the map. Usually, a two-dimensional rectangular map is chosen. The map is formed by the properties inherent to the data itself. Consequently, no previous classification of the data is needed. The input layer has n units representing the n components of an input vector . The output layer is a two dimensional array of pxq units arranged on a map. Each unit in the input layer is connected to every unit in the output layer with a weight associated. All weights are initialized randomly. They are adjusted according to Kohonen's learning rule

$$\triangle w_i = \eta(t) \cdot h_{ir}(t) \cdot (x_k - w_i) \tag{1}$$

that uses a distance measure $\|w_r - x_k\| = \min_i \|w_i - x_k\|$ to determine the bestmatch r and a neighborhood function

$$h_{ir}(t) = e^{-\frac{(i-r)^2}{2\sigma(t)^2}} \tag{2}$$

that realizes the lateral inhibition. The learning rate $\eta(t)$ with $0 < \eta(t) < 1$ determines the strength of learning. The radius $\sigma(t)$ determines the set of neurons in a neighborhood of the bestmatch that are included into the learning process. Both functions usually decrease monotonously during learning. On the map neighboring units form regions that correspond to similar input vectors. These neighborhoods form disjoint regions, thus enabling a classification of the input vectors. However, in order to perform a classification a visualization of the network structures is needed, since the Kohonen algorithm converges to an equal distribution of the units on the output layer (Kohonen, 2001). Therefore, a visualization of the network structure may give an enhanced insight into the network sturctures. A possible and often used visualization tool is the U-Matrix (Ultsch and Siemon, 1990), a three dimensional landscape which represents structural properties of the high dimensional input space on the map. At each point of the grid the weights are analyzed with respect to their neighbors. The distance between the weights of two neighboring units then is displayed as height into the third dimension. A U-Matrix has valleys where the vectors on the map are close to each other and represent data that are in the same class. Hills or walls represent larger distances indicating dissimilarities of the input data. Alternative visualization tools for SOMs have also been proposed, such as the agglomerative clustering where the SOM neighborhood relation can be used to construct a dendrogram on the map (Vesanto and Alhoniemi, 2000) and a hierarchical clustering of the units on the map with a simple contraction model (Himberg, 2000). U-matrices have been widely used for clustering, since similar input vectors are close together on the map and fall into the same valley, i.e. cluster. In the last years SOMs together with the U-Matrix method have been successfully applied to a wide-ranging number of applications where a clustering of high-dimensional data is intended (Kaski and Kohonen, 1996; Kohonen, 2001).

The discovery of complex patterns, for instance in multivariate time series, protein sequences, and genes with SOMs is much more complex, since it demands an improvement and extension of the original SOM (Guimarães and Moura-Pires, 2001). Therefore, and for the handling of more complex and temporal patterns, SOMs with several hierarchical layers can be introduced in order to capture and discover structures at different abstraction levels. This proceeding is necessary when a segmentation of complex and structured problems is needed, for instance, in such areas as image recognition (Koh et al., 1995), temporal pattern discovery in multivariate time series (Guimarães, 2000), speech recognition (Kemke and Wichert, 1993), and protein sequence analysis (Andrade et al., 1997). In addition, for applications with temporal or sequential data a visualization of trajectories on a map enables the monitoring of such phenomena, for instance, for the recognition of misarticulations in speech (Mujunen et al., 1993) EEG signal monitoring (Joutsiniemi et al., 1995) and sleep apnea detection (Guimarães and Ultsch, 1999).

In the following section we present an application in medicine, called sleep apnoea, which is predestinated as a complex and temporal problem where SOMs can be used for temporal pattern discovery. Here, no generation of a traditional classifying system is attempted. In contrast, inherent patterns in the multivariate time series are searched for, leading to a hypothesis which can be interpreted by a domain expert. As one of the main aims of this approach in medicine not well-defined hypothesis can be proposed by the model.

3 Detection of Sleep-related Breathing Disorders

In this section, we introduce a generic method for Temporal Knowledge Conversion, named TCon (Guimarães, 1998; Guimarães, 1999) that enables the discovery of temporal patterns in multivariate time series. The main idea lies in introducing several abstraction levels such that a stepwise and successive detection of the temporal patterns becomes possible, breaking down this highly structured and complex problem into several sub-tasks. This method also performs a transition of temporal patterns in multivariate time series into a linguistic representation form in form of temporal grammatical rules, intelligible and understandable for human beings such as domain experts. Multivariate time series $Z = \vec{x}(t_1), \ldots, \vec{x}(t_n)$ with $\vec{x}(t_1) \in \mathbf{R}^m, m > 1$ with sampled at equal time steps $t_1, \ldots, t_n$ gathered from signals of complex processes are the input of the system. Results are the discovered temporal patterns as well as a linguistic description of the patterns, interpretable for human beings. An overview of the method is given by Guimarães and Ultsch (1999).

Features: First of all, a pre-processing and feature extraction for all time series is a pre-requisite for further processing (Bishop, 1995). For the feature extraction one or even more than one time series $\vec{x_S}(t_i) = x_{j1}(t_i), \ldots, x_{jS}(t_i)^T \in \mathbf{R}^S$ may be selected from the multivariate time series Z with $j_k \in S, k = 1, \ldots, s, S \subset \{1, \ldots, m\}, s = |S|$. A feature $m_S(t_i, l) = f(\vec{x_S}(t_i), \ldots, \vec{x_S}(t_{i+l}))$ then is the value of a function $f : \mathbf{R}^{s \times l} \to \mathbf{R}$ at time t_i with $l \in \{1, \ldots, n-1\}$ from selection S. In order to find a suitable representation of all time series, methods, for instance, from signal processing, statistics or fuzzy theory, can be used.

Primitive Patterns: Second, exploratory methods, in particular, SOMs together with the U-Matrix-method are used for the discovery of elementary structures in the time series, named as *primitive pattern classes* $p_j, j = 1, \ldots, k$. An element of a primitive pattern class is a primitive pattern $p_j(t_i), j = 1, \ldots, k$ that belongs to a given primitive pattern class $p_j, j = 1, \ldots, k$ and is associated to a given time point t_i. Regions on a U-Matrix that do not correspond to a specific primitive pattern class are associated to a special group, named *tacet*. We are now able to classify the whole features with primitive patterns and tacets. This will be called a primitive pattern (PP)-channel. Instead of analyzing all time series simultaneously, several selections of features are made. Consequently, several SOMs are learned . This leads to more than one PP-channels . At this level, machine learning algorithms may be used, in order to generate a rule-based description of the primitive pattern classes

Successions: In order to consider temporal relations among primitive patterns, succeeding identical primitive patterns $p_j(t_i), \ldots, p_j(t_{i+k}), i = 1, \ldots, n-k+1$, obtained from each SOM are identified as successions. Since several feature selections are possible, successions from different PP-channels may occur more or less simultaneously. Each *succession* $s_j(a, e)$ is associated to a given primitive pattern class and has a starting point $a := t_i$ an end point $e := t_{i+1}$ and, consequently, a duration $l = e - a$. Since each primitive pattern is represented through its bestmatch on a U-Matrix, trajectories of succeeding primitive patterns (best-matches) on a U-Matrix are used for the identification of successions.

Events: More or less simultaneous occurring successions $s_1(a_1, e_1), \ldots, s_q(a_q, e_q)$ that occur more than once are identified as an *event* $e(l)$. Then $A = \max(a_1, \ldots, a_q)$ is the starting point, $E = \max(e_1, \ldots, e_q)$ the end point and $l = E - A$ the duration of the event $e(l)$. Each event belongs to a given event class. In order to reduce the great amount of information, a vague simultaneity is introduced.

In addition, the significance of events (frequency of the occurrence of events) is calculated using conditional probabilities between the occurrence of simultaneous primitive patterns on different PP-channels. Histograms over the calculated probabilities enable a differentiation between significant events (very frequent events) and less significant events (less frequent events). Rare events are omitted in the sense as they are regarded as delays between events, named as event tacets. In order to join events with different significance levels, very frequent events are associated to less significant events. Therefore, similarities among significant and less significant events will be considered counting the number of equal types of successions occurring in both events. This results in an extremely reduced number of events. Consequently, each event is described by one significant event and, possibly, one or more than one less significant events. At this level, the whole multivariate time series is described by a sequence of events $F = e_{1j}, \ldots, e_{nj}, j = 1, \ldots, m$. In order to identify events with SOMs, extended hierarchical SOMs have to be used (Guimarães, 2000). For each event a temporal grammatical rule is generated .

Sequences: Subsequences of events $e_1, \ldots, e_k$ that occur more than once in F are identified as a sequence $sq(\min, \max) = e_i(\min_i, \max_i), \ldots, e_k(\min_k, \max_k)$. This means that sequences are repeated subsequences of the same type of events at different time points t_i. Since events may succeed immediately or after a time delay, i.e. an event tacet, the duration of event tacets can be used for determining the starting event or/and the end event of a sequence. This is possible, if the duration of event tacets is regarded as a transition between different sequences due to larger delays between succeeding events. In addition, probabilistic automata, can be also used for the identification of sequences. Probabilistic automata describe transition probabilities between events such that paths through such an automata describe probable subsequences of events. For each sequence a temporal grammatical rule is generated.

Temporal Patterns: Finally, small variations in the events of each sequence type lead to the identification of similar sequences. Similar sequences $sq_i(\min_i, \max_i) \vee \ldots \vee sq_v(\min_v, \max_v)$ will be joined together to a *temporal pattern* $tp(\min, \max)$, where $\min = \min(\min_i, \ldots, \min_v)$ and $\max = \max(\max_i, \ldots, \max_v)$. Temporal patterns are abstract descriptions of the main temporal structures in multivariate time series. String exchange algorithms are suitable for the identification of temporal patterns. For each temporal pattern a temporal grammatical rule is generated.

This approach was applied to an example in medicine, namely sleep-related breathing disorders (SRBDs), consisting in various types among which sleep apnea is best known (Penzel and Peter, 1992). The whole data set covered the most relevant patterns in sleep apnea, as fostered by the domain experts. As we were searching for patterns in the context of Data Mining, and not building a classifier, which would afford the knowledge about existing classes, i.e. patterns, we confined ourselves to this more restrictive sample. The main aim, here, was to show that already well-defined patterns, as well as if possible "knew" temporal patterns could be discovered by unsupervised methods, such as SOMs, taking into account several hierachical levels.

For the identification of different types of SRBD, mainly apnea and hypopnea, just the signals concerning the respiration (airflow, ribcage and abdominal movements, oxygen saturation, snoring) had to be considered (Peter et al., 1998). Severity of the disorder is calculated by counting the number of apnea events per hour of sleep. The sum of the index of apneas and hypopneas is a measure for the respiratory disturbance index (RDI). It can be seen as pathological, when the RDI exceeds 20 events per hour of sleep, while patients with more than 40 events per hour of sleep have to be referred to therapy.

Technical assistants usually make the visual classification of the different types of SRBDs based on such a recording. An automatic identification of SRBDs is a quite hard task, since a simultaneous analysis of all signals is needed. In addition, quite different patterns for the same SRBD may occur, even for the same patient during the same night, and a strong variation of the duration of each event may occur, as well.

SRBDs can be subdivided into SRBDs with and SRBDs without an obstruction of the upper respiratory tracs. The different kinds of SRBDs are identified through the signals ´airflow´, ´ribcage movements´ and ´abdominal movements´, ´snoring´ and ´oxygen saturation´, where a distinction between amplitude-related and phase-related disturbances is made. Concerning the amplitude- related disturbances, we distinguish disturbances with 50% as well as as disturbances with 10-20% of the baseline signal amplitude. Phase-related disturbances are characterized by a lag between ´ribcage movements´ and ´abdominal movements´. An interruption of ´snoring´ is present at most SRBDs as well as a drop in ´oxygen saturation´. 25 Hz sampled data from three patients having the most frequent SRBDs (altogether 27 patterns) have been used. No additional information was provided from the medical experts, since the main aim is to discover inherent structures in multivariate time series using unsupervised methods, such as SOMs.

A structured and complete evaluation of the discovered temporal knowledge at the different abstraction levels was made using a questionnaire.
All events (six) and temporal patterns (four) consisting in six different
sequences presented to the medical expert described the main properties
of SRBD as, for instance, ´hyperpnoe´, ´obstructive snoring´, ´obstruc-
tive apnoe´ or ´hypopnoe´. The generated temporal grammatical rules
described very well the domain knowledge. An evaluation of the rules
was based on an evaluation made by the domain experts and at this
level lead to an overall sensitivity of 0,762 and a specificity of 0,758.
´Event5´ was correctly identified as a special event, called ´hyperpnea´.
SRBDs always end up with a ´hyperpnea´. In some cases the duration
of ´Event5´ was too short. The duration of all other events were in a
valid range. For one of them even previously unknown knowledge was
discovered. This temporal pattern was named by the expert as ´mixed
obstructive apnoe´, distinguished into a ´mixed obstructive apnoe´ with
an interruption and snoring having a ´central´ and an ´obstructive´ part
and a ´mixed obstructive apnoe´ without an interruption and without
snoring ending in an ´hypoventilation´. A comparison to conventional
statistical methods is quite dificult, since experiments have been limited
until now to at the most two time series. This approach, in contrast,
considers a lot more time series simulataneously.

4 Classification of Protein Sequences and Tumors

A self-organizing map (SOM) can be used to classify sequences within
a protein family (ras-p 21 family) into subgroups that correspond to biological subcategories. Andrade et al. (1997) present a modified SOM-
algorithm and use the rab family of small guanosine-triphosphate-ases
to illustrate the performance of the method. In their approach each of
N protein sequences is binary coded as a sequence vector (input vector)
$f_k = (f_{k1}, \ldots, f_{kn}), n = 20 \cdot L, k = 1, \ldots, N$. Each position of the sequence is described by 20 components corresponding to all possible 20
amino acids $A, C, D, E, \ldots, S, T, V, W, Y$. The component corresponding to
the amino acid type at this position is coded by ‚one‘ and the rest of the
components are set to zero. The resulting sequence vector f_k has length
$20 \cdot L$, where L is the length of the sequence alignment.
The SOM is a two-dimensional layer of $p \times q$ units with one weight
$w_i = (w_{i1}, \ldots, w_{in}), n = 20 \cdot L, i = 1, \ldots, p \times q$ for each of the $p \times q$
units. The weights have the same number of components as the se-
quence vectors f_k and their components take real values between zero
and one. At zero time the weights $w_i, i = 1, \ldots, p \times q$ are the mean of
all sequence vectors. The distance $\delta_{i,k}$ from weight w_i to the sequence

vector f_k is given by

$$\delta_{i,k} = \sqrt{\sum_{n=1}^{20 \cdot L} |f_{k,n} - w_{j,k}|^2} \qquad (3)$$

where $n = 1, \ldots, 20 \cdot L$ is the index of the vector components. The bestmatch is identified by having the smallest distance to the sequence vector. This vector is updated with a linear combination of its previous value with the presented sequence vector as follows:

$$w_i(t + 1) = (1 - \alpha^0)w_i(t) + \alpha^0 f_k \qquad (4)$$

where α^0 is a factor that sets the weight given to the example sequence in the updating step. The update makes the weight more closer to the example presented to the network. The examples are presented to the system in random order, once for each training cycle. Then the time devoted to all cycles is $s \cdot N$, where N is the number of examples (protein sequences) and s is the number of learning epochs. These procedure adds noise to the dynamics of the weight evolution, which helps the system to avoid non-optimal classification. However, a single SOM just leads to the clustering of the family at a definite resolution level. Only several SOMs with several resolutions enable the identification of a sequence relationship not existant in a single map that means a hierarchy of sequences in the family. Therefore, a set of experiments with SOMs having different sizes are arranged in a tree-like fashion through a linkage of the clusters that contains the same sequences at successive levels. Such a tree representation can be compared with phylogenetic trees that try to accommodate the evolutionary relationsships of a group of sequences in a tree according to their sequence homology. Andrade et al. (1997) analyzed 42 proteins of the rab family and showed the power of the SOM to obtain a reliable classification that agrees with the classifications obtained by phylogenetic trees.

Another recent application is given by the molecular classification of tumors using SOMs. Specific cancer treatments (chemotherapy) try to maximize efficacy and minimize toxicity. Therefore, improvements in cancer classification are important to advances in cancer treatment therapies. Golub et al. (1999) described a statistical approach to cancer classification based on gene expression monitoring by DNA micro-arrays. A DNA-microarray or DNA-chip is a glass slide onto which single-stranded

DNA molecules are attached at fixed locations. They were called spots or probes. DNA-chips use the molecular recognition for binding complementary single-stranded nucleic acid sequences. The aim is to compare mRNA abundance in two different samples, which we call targets. RNA from both sample cells are extracted and labeled with a red dye for the RNA from one sample population and a green dye for the other sample population. Gene sequences from both extracts hybridize to their complementary sequences in the spots. Then the array is excited by a laser and the relative abundance of the hybridized RNA is measured from the fluorescence intensities and colours for each spot. The relative expression levels of the genes in the samples can be estimated in an organism under various conditions, at different developmental stages and in different tissues.

Gene expression profiles characterize the dynamic functioning of each gene in the genome. Expression data can be represented in a matrix with rows representing genes, columns representing samples (e.g. various tissues, developmental stages and treatments) and each cell in this matrix containing a number characterinzing the expression level of the particular gene in the particular sample. Such a matrix is called gene expression matrix. A database of such gene expression matrices could help to understand gene regulation, metabolic and signaling pathways, the genetic mechanism of disease and the response to drug treatments. A gene expression can be studied by comparing expression profiles of genes or by comparing expression profiles of samples. We regard these objects (rows or columns in the matrix) as points in a n-dimensional space or as n-dimensional vectors, where n is the number of samples for gene comparison, or number of genes for sample comparison.

Brazma and Vilo (2000) mention the use of a microarray database in toxicology, an area which uses special statistical methods given by Gilberg et al. (1999), Selinski et al. (2000) and Urfer (2001). Knowing the response of a gene to a certain toxin in rats allows users to predict the response of a homologous human gene. Current classification methods for patient samples rely on the cancer's tissue of origin and on the microscopic appearance and location of cancerous cells. Tumors of identical appearance can be distinguished only by observing the patient over time and discovering whether the initial treatment was successful. There is a need for new methods of classification of patient samples that might predict the course of the disease at the time of diagnosis.

Although the genome in each cell of an organism is the same, only a fraction of the proteins that can be produced are active in any particular cell. A gene's expression level indicates the approximate number of copies of that gene's RNA produced in a cell. This mRNA abundance indirectly measured by the intensity of the fluorescence of the spots on the array is thought to correlate with the amount of the corresponding protein made. DNA-chips can measure the expression of thousands of genes simultaneously. Golub et al. (1999) present a method for performing molecular classification of patient samples by gene expression analysis. Their approach is divided into class discovery and class prediction. Class discovery refers to the process of dividing samples into groups with similar behaviour or properties. In contrast, class prediction refers to the assignment of particular tumor samples to already-defined classes, which could reflect current states or future outcomes. Golub et al. (1999) developed a supervised analysis method called neighborhood analysisänd applied this approach to 38 acute leukemia samples.

5 Outlook

An interesting aspect, and also of high practical relevance, is the inclusion of environmental variables measured during the experimental phase that play an important role in the results of the gene expression data. However, therefore a modification of the original Kohonen Network is needed. Bock (2000) proposed a general statistical concept, where the data are partitioned into an explanatory part and a response part. The resulting SOMs can be applied to include the environmental variables. This approach is promising for solving molecular biological problems related to gene expression data.

6 Acknowledgements

We would like to thank Prof. Dr. J. H. Peter and Dr. T. Penzel, Medizinische Poliklinik, Philipps-University of Marburg for providing the sleep apnea time series. We would also like to thank the referees for improving this article. Especially we appreciate the help of Isabelle Grimmenstein. This research was supported by the German Research Council (DFG) through the Graduate College and the Collaborative Research Center at the University of Dortmund (SFB 475): Reduction of complexity for multivariate data structures.

References

ANDRADE, M.A., CASARI, G., SANDER, C., VALENCIA, A. (1997): Classification of protein families and detection of the determinant residues with an improved self-organizing map, *Biological Cybernetics*, 76, 441-450.

BEHME, H., BRANDT, W.D., STRUBE, H.W. (1993): Speech Recognition by Hierarchical Segment Classification, in: S. Gielen, B. Kappen

(Eds.): Proc. Intl. Conf. on Aritificial Neural Networks (ICANN 93), Amsterdam, Springer Verlag, London, 416-419.

BISHOP, C.M. (1995): Neural Networks for Pattern Recognition, Oxford, Clarendon Press.

BOCK, H.H. (2000): Regression-Type Models for Kohonen's Self-Organizing Networks, in: R. Decker, W. Gaul (Eds.): Classification and Information Processing at the Turn of the Millenium, Procs. of the 23rd Annual Conference of the Gesellschaft für Klassifikation, Bielefeld, 10-12 March, 1999, Springer, 18-31.

BRAZMA, A., VILO, J. (2000): Gene expression data analysis, *FEBS Letters, 480, 17-24.*

BRUNNERT, M., MÜLLER, O. and URFER, W. (2000): Genetical and statistical aspects of polymerase chain reactions, Technical Report 6/2000, University of Dortmund.

GILBERG, F., EDLER, L. URFER, W. (1999): Heteroscedastic Nonlinear Regression Models with Random Effects and Their Application to Enzyme Kinetic Data, *Biometrical Journal, 41, 543-557.*

GOLUB, T.R., SLONIM, D.K., TAMAYO, P., HUARD, C., GAASEN-BEEK, M., MESIROV, J.P., COLLER, H., LOH, M.L., DOWNING, J.R., CALIGIURI, M.A., BLOOMFIELD, C.D., LANDER, E.S. (1999): Molecular Classification of Cancer: Class Discovery and Class Prediction by Gene Expression Monitoring, *Science, Vol. 286, October, 531-537.*

GUIMARÃES, G. (2000): Temporal Knowledge Discovery for Multivariate Time Series with Enhanced Self-Organizing Maps, To appear in: IEEE-INNS-ENNS Intl. Joint Conf. on Neural Networks (IJCNN'2000), Como, 24-27 July, Italy.

GUIMARÃES, G. (1998): Eine Methode zur Entdeckung von komplexen Mustern in Zeitreihen mit Neuronalen Netzen und deren Überführung in eine symbolische Wissenrepräsentation, PhD Dissertation, University of Marburg, Germany.

GUIMARÃES, G., MOURA-PIRES, F. (2001): An Essay in Classifying Self-Organizing Maps for Temporal Sequence Processing. In Allison, N., Yin, H., Allison L., Slack J. (eds), Advances in self-Organizing Maps, (pp. 259-266), Springer.

GUIMARÃES, G., ULTSCH, A. (1999): A Method for Temporal Knowledge Conversion, Procs. of IDA99, The Third Symposium on Intelligent

Data Analysis, August 9-11, Amsterdam, Netherlands, *Lecture Notes in Computer Science, Springer Verlag, 369-380.*

HIMBERG, J. (2000): A SOM based cluster visualization and its application for false coloring. In: Proceedings of the IEEE-INNS-ENNS Intl. Joint Conf. on Neural Networks (IJCNN'2000), (pp. 587-592), Vol. 3, 24-27 July, Como, Italy.

JOUTSINIEMI, S.L., KASKI, S., LARSEN, T.A., (1995): Self-Organizing Map in Recognition of Topographic Patterns of EEG Spectra, *IEEE Transactions on Biomedical Engineering, Vol. 42, No. 11, 1062-1068.*

KASKI, S., KOHONEN, T., (1996): Exploratory Data Analysis by Self-Organizing Map: Structures of Welfare and Poverty in the World, in: A.P.N Refenes, Y. Abu-Mostafa, J. Moody, A. Weigend (Eds.): Neural Networks in Financial Engineering. Proc. of the Intl. Conf. on Neural Networks in the Capital Markets, London, England, 11-13 October, 1995, Singapore, 498 - 507.

KEMKE, C.,WICHERT, A., (1993): Hierarchical Self-Organizing Feature Maps for Speech Recognition, *Proc. of the World Congress on Neural Networks (WCNN 93), Hillsdale, Vol. III, 45-47.*

KOH, J., SUK, M., BHANDARKAR, S.M., (1995): A Multilayer Self-Organizing Feature Map for Range Image Segmentation, *Neural Networks, Vol.8, No. 1, Elsevier Science Publisher, 67-86.*

KOHONEN, T. (2001): Self-Organizing Maps, Springer, New York.

KOHONEN, T. (1982): Self-organized formation of topologically correct feature maps, *Biological Cybernetics 43, 141-152.*

MUJUNEN, R.,LEINONEN, L, KANGAS, J., TORKKOLA, K., (1993): Acoustic Pattern Recognition of /s/ Misarticulation by the Self-Organizing Map, *Folia Phoniatr., 45, 135-144.*

PENZEL, T., Peter, J.H.: Design of an Ambulatory Sleep Apnea Recorder, in: H.T. Nagle, W.J. Tompkins (Eds.): Case Studies in Medical Instrument Design, IEEE, New York, 1992, 171-179.

PETER, J.H., BECKER, H., BRANDENBURG, U., CASSEL, W., CONRADT, R., HOCHBAN, W., KNAACK, L., MAYER, G., PENZEL, T. (1998): Investigation and diagnosis of sleep apnoea syndrome, in: Mc-Nicholas, W.T. (ed.): Respiratory Disorders during Sleep. European Respiratory Society Journals, Sheffield, 106-143.

SELINSKI, S., GOLKA, K., BOLT, H.M. and URFER, W. (2000): Estimation of toxicokinetic parameters in population models for inhalation studies with ethylene, *Environmetrics, 11, 479-495.*

ULTSCH, A., SIEMON, H.P. (1990): Kohonen´s Self-Organizing Neural Networks for Exploratory Data Analysis, Proc. Intl. Neural Network Conf. INNC90, Paris, Kluwer Academic, 305-308.

URFER, W. (2002): Hazardous Agents. In: Encyclopedia of Environmetrics. A.H. El-Shaarawi, W.W. Piegorsch (Eds.), Wiley, Chichester, Vol. 2, 983-987.

VESANTO, J., ALHONIEMI, E. (2000): Clustering of the Self-Organizing Map. *IEEE Transactions on Neural Networks, Special Issue on Data Mining, 11 (3), 586-600.*

WALTER, J.A., SCHULTEN, K.J. (1993): Implementation of Self-Organizing Neural Networks for Visual-Motor Control of an Industrial Robot, *IEEE Transactions on Neural Networks, Vol. 4, No.1, January 86-95.*

Automated Classification of Optic Nerve Head Topography Images for Glaucoma Screening

T. Hothorn[1], I. Pal[2], O. Gefeller[1], B. Lausen[1], G. Michelson[3], D. Paulus[2]

[1]Department of Medical Informatics, Biometry and Epidemiology,

[2]Chair of Pattern Recognition, Department of Computer Science,

[3]Department of Ophthalmology and University Eye Hospital, Universität Erlangen-Nürnberg, D-91054 Erlangen, Germany

Abstract: Glaucoma prevalence is 5% and consequently it is an important public health issue to improve possibilities of an early Glaucoma treatment. We present a research strategy and first results of the development of an automated classification scheme of optic nerve head topography images for Glaucoma screening. Feature detection using the Heidelberg Retina Tomograph currently requires the manual determination of the disc margin of the optic nerve head. Procedures for the automated detection of the optic nerve head as well as vessel segmentation are discussed. Based on data accumulated in the *Erlangen Glaucoma Register* we evaluate linear and tree based discriminant analysis for the classification of glaucoma. In a cross-sectional study including 196 subjects we found that the use of stabilized methods like bagging and the incorporation of various anamnestic data decreases the estimated error rates.

1   Introduction

Glaucoma is the second leading cause of blindness worldwide and is most common in the elderly people, see e.g. (Coleman(1999)). It is an important public health issue to improve possibilities of an early glaucoma treatment, e.g. (Michelson and Groh(2001)). Several eye conditions are considered variants of glaucoma. Primary open-angle glaucoma (POAG) and primary angle-closure glaucoma (PACG) are the most common ones. POAG is characterized as an ocular disease that causes progressive changes in the optic nerve head, visual field loss or both. Traditionally, POAG has been characterized as a disease of raised eye pressure. Increased eye pressure is viewed as major risk factor for POAG and the reduction of eye pressure maintains to be the target of a treatment. Diagnosis of the ocular diseases relies on measurements and investigations of intraocular pressure (IOP), anamnestic data and on the evaluation of the morphology of the optic disc. Several medical imaging procedures are used that are surveyed in (Michelson and Groh(2001)).

Our work is based on data accumulated in the *Erlangen Glaucoma Register* which currently consists of anamnestic and diagnostic data from repeated observations of 1100 subjects starting 10 years ago. The data base contains the following entries: psycho-physiological data (e.g. visual field, contrast sensitivity), senso-physiological data (e.g. electroretinogram), intraocular pressure, ocular perfusion data (scanning laser doppler flowmetry, ultrasound doppler sonography), 24h profile of intra ocular pressure, optic nerve tomography (HRT), and reflectivity images of the optic nerve head (fundus image). More details on the variables used in our analysis are shown below.

A subset of the HRT parameters is used in another study in (Mikelberg et al.(1995)) which contains 45 normal eyes and 46 eyes of individuals with early glaucomatous visual field loss. Yet another subset of HRT parameters was used in the experiments of (Swindale et al.(2000)) on a data set of 100 subjects in both a normal and case group.

The paper is organized as follows: after a description of image acquisition in Section 2 we present image processing and image segmentation for retinal images in Section 2.2. The surface of the area around the optic nerve head is approximated by parametric curves in Section 2.3 introduced by (Swindale et al.(2000)). Features derived from the surface properties as well as those from image segmentation are subject to statistical analysis and classification in Section 3. We compare our analysis with the results of (Mikelberg et al.(1995)) and (Swindale et al.(2000)) using the same selection of HRT parameters for the classification. Finally, we give a perspective on future work in which we propose to integrate the features described in Section 2.3 into the classification procedures described in Section 3.

2   Medical Images of the Eye

2.1   The Functional Principle of the HRT

The Heidelberg Retina Tomograph (HRT) is a non-invasive confocal scanning laser system for three-dimensional imaging of the eye background. HRT provides a three-dimensional image computed from a image series of 32 images that are made from the different depth planes of the examined object. The laser beam is focused on the examined retina point and the reflected light is detected by a photo diode. The retinal tissue is scanned and digitized in two dimensions sequentially.

The important technical data of HRT are as described by (Heidelberg Engineering(1997)): the field of view can be set to $10° \times 10°$, $15° \times 15°$, or $20° \times 20°$. The total acquisition time of 32 images is 1.6 seconds. The

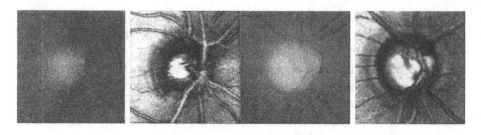

Figure 1: Topography and mean images (field of view is 10°) of a normal (left) glaucomatous eye (right).

wave length of the diode laser is 670 nm and the laser ray has 10μm diameter with 100μWatt power.

Blood vessels and the surface of the papilla of the retina are visible in the HRT images. From the image series, the so-called topography image and the mean image can be computed. The topography image is a 2.5-dimensional image in which the pixels represent the depth coordinate (z-coordinate) (Figure 1, left parts). The gray-value of pixels on the mean image corresponds to the average intensity of the 32 images (Figure 1, right parts).

On the HRT-images, the vessel system and the optic nerve head of the eye background can be examined and diagnosed, so the early detection of many diseases of the retina is possible, in particular glaucoma. Before a visual field defect can be measured, loss and damage of the nerve fiber layer can be observed in the area of optic nerve head, see (Jonas and Naumann(1993)). The important characteristics of the healthy and glaucoma eye can be observed on the Figure 1, namely the excavation of the papilla, so the growing of the two- and three-dimensional size of the optic nerve head.

2.2 Retinal Vessel Segmentation

In the first step of the vessel segmentation a Gaussian smoothing for the noise filtering is made and this is followed by the histogram equalization in order to eliminate the reflection problems. Different edge and line detection methods were examined and compared for the segmentation of the retinal vessel systems (Sobel, Previtt, Kirsch, Canny, log-linear operator, Marr, in the frequency domain, high-pass and directional filters, multi-layer perceptron approaches etc., see Figure 2) in (Pál et al.(2000)). The segmented vessel systems can be used for the determination of the diameter of the vessels and for the determination of

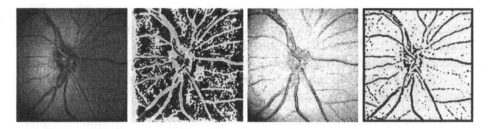

Figure 2: Vessel segmentation: original, Fourier, multilayer perceptron, Marr

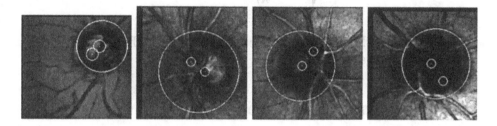

Figure 3: The automatic localization of the papilla with circle, the small circles are the gravity of white and black pixels after the segmentation of the mean image, from (Pál(2000a)), (Pál(2000b))

the artery-vein ratio, see (Pál et al.(2000)) and (Hubbard et al.(1998)). These features are important for diagnosis of stroke.

2.3 Automatic Optic Nerve Head Detection, Segmentation and Approximation

With the help of the size of the optic disk area (in 2D) after the papilla detection and with the topography images, the size of the cup volume of papilla can be determined which is an important feature of glaucoma damage. In order to derive features from the morphology of the optic disc head, a localization of this area is required in the image. Results are shown in Figure 3.

After the papilla has been found, the three dimensional surface of papilla can be determined as described in (Swindale et al.(2000)). The equation

$$z(x,y) = \frac{z_m}{1 + e^{(r-r_0)/s}} + a(x-x_0) + b(y-y_0) + c(x-x_0)^2 + d(y-y_0)^2 + z_0$$

with $r = \sqrt{(x - x_0)^2 + (y - y_0)^2}$ is used for the approximation of the papilla cup. To determine the initial values a, b, c, d, z_0 an iterative non-linear least square fitting is used, where the parameters are determined by Gauss-Jordan elimination. After this step the parameters can be refined to minimize the fitting function

$$f = \sqrt{\frac{1}{NM} \sum_{\forall ij}^{NM} [I(i,j) - z(i,j)]^2}$$

using Levenberg-Marquardt optimization technique, where $I(i,j)$ is the topography image.

3 Glaucoma Classification

(Mikelberg et al.(1995)) propose a method for the classification of glaucoma based on HRT data. The method of (Mikelberg et al.(1995)) is a linear discriminant analysis (LDA) using selected parameters of the HRT (namely area below reference, mean height in contour, mean radius, volume above reference, volume above surface and volume below reference) and requires a human input for the determination of the optic nerve head, which introduces uncontrolled variability. (Swindale et al.(2000)) propose as a solution a linear discriminant analysis on estimated parameters of a smooth two-dimensional surface model without human interaction.

As outlined in the Introduction, we use discriminant analysis for glaucoma classification and we evaluate both LDA and tree classifiers.

3.1 Case-control Design

Focusing on Glaucoma screening we use the Erlangen Glaucoma Register for a case-control study matched by age and sex. We use the data of the first observation and we define the glaucoma diagnosis of each subject by the current state (July 2001) of the diagnosis. Non glaucoma subjects, which include subjects with macro papilla and ocular hypertension, are included in the control group and subjects with high and low tension glaucoma are included in the case group. We match the control and case subjects with a maximal difference of 1 year and by sex: control group: 54.68 ± 9.29 years, $n = 98$; case group: 54.74 ± 9.27 years, $n = 98$; overall 114 female and 82 male subjects. We have complete observation for all 196 subjects with the current diagnosis and 79 diagnostic and anamnestic variables.

- Anamnestic variables: age, sex, height, body weight, body mass index, and blood pressure.

- Diagnostic (HRT) variables: mean standard deviation, height in contour, height variation contour, mean radius, mean RNFL thickness, mean depth in contour, effective mean depth and mean variability. The following variables are included as global and separated in 4 circular sectors: area global, effective area global, area below reference global, mean height contour, peak height of contour, volume below surface, volume above surface, volume below reference, volume above reference, maximum depth and third moment.

3.2 Classification Tree Methods and Bagging

(Breiman(1996b); Breiman(1998)) discusses the problem of instability in model selection and for classification. If T is a training set used to build a classifier, than a classification procedure is called instable if small changes in T might result in large changes in the classification procedure. Simulations indicate that nearest neighbors and LDA are stable in this sense, but tree classifiers and neural networks are unstable. However, trees are right 'on average' whereas LDA may lead to biased results whenever the model assumptions are not met. Bagging is proposed by (Breiman(1996a)) to stabilize trees. The procedure works as follows:

1. Draw a bootstrap sample from the training set T;

2. Form a classifier based on the bootstrap sample;

3. Classify an observation x.

K repetitions result in K classifications of x. Therefore, we do not have a single classifier based on T, but K classifiers based on independent replications of T. An overall classification of an observation x is specified by letting all K classifiers 'vote' for a class. The majority of the K classifications defines the value of the bagged classifier. We use $K = 50$ for all computations.

We report the performance of four different classifiers, namely LDA to be able to compare the results with existing proposals, classification trees

(CTREE) (e.g. (Breiman et al.(1984))) and bagged classification trees ((Breiman(1998))). We use the classification tree function of the package tree, version 1.0-2, of R, version 1.2.2, for a reference see (Ihaka and Gentlemen(1996)). We construct trees with a minimum number of observations to include in either child node of 5, a smallest node size of 10 and a within-node deviance of at least 0.01 times that of the root node for the node to be split.

Moreover, we use the P-value adjusted regression tree method (PTREE) as classifier, which allows for variables measured on different scales (see ()Lausen:1992 and ()Lausen:1994). The adjusted P-value is computed with proportions of the sample size of $\varepsilon_1 = 0.05$ and $\varepsilon_2 = 0.95$. The recursion stops for nodes with less or equal 5 observations or if the minimum P-value is greater or equal 0.01.

The performance of all methods is measured by the estimated misclassification rate. We use 10% cross-validation, which is balanced for the control and case group; i.e. 10% each.

3.3 Evaluation of Different Classifiers

We evaluate the classifiers for two sets of variables: all variables and only the 6 HRT parameters used by (Mikelberg et al.(1995)). We get for the full set of variables: the error rate for the LDA is 26.7%, pruned CTREE gives an error of 20.5%, PTREE an error of 19.8%, bagged CTREE an error of 15.9 and bagged pruned CTREE an error of 18.4%.

Following (Lausen et al.(1997)) Figure 4 is a graphical representation of the results of the PTREE approach.

(Swindale et al.(2000)) evaluate the performance of both their approach and the proposal of (Mikelberg et al.(1995)) by a leave-one-out cross-validation (jack-knife) using 100 normal and 100 glaucoma subjects and report an estimate of the misclassification error of 16.5% for a LDA of the 6 HRT variables. In contrast to our age and sex matched case-control design, one cause of different results might be that the two groups studied by (Swindale et al.(2000)) have different age distributions (normal: 53 ± 14 years, glaucoma: 61 ± 13 years). For our case-control study the results are for the reduced set of variables: the error rate for the LDA is 20.4%, for pruned CTREE 18.8% and bagged CTREE 16.8%. Using only quantitative variables we do not apply PTREE to the reduced set of variables.

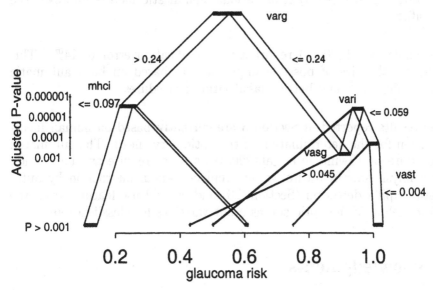

Figure 4: P-value adjusted regression tree. Stop criterion p = 0.001; y-axis adjusted p-values. Factors shown: Volume above reference global (**varg**); volume above surface global (**vasg**); volume above reference inferior (**vari**); volume above surface temporal (**vast**); mean height of contour inferior (**mhci**).

4 Conclusion and Discussion

Image processing and segmentation procedures are available for vessel segmentation and automatic optic disc detection. Our procedure allows the determination of surface-parameters of the papilla and can be used for improving the proposals of (Mikelberg et al.(1995)) and (Swindale et al.(2000)).

Using the procedure by (Mikelberg et al.(1995)) we show that the error rates of LDA computed for our case control study confirm the error rates of (Swindale et al.(2000)). We analyze a control and case group with an identical age distribution whereas (Swindale et al.(2000)) analyze groups with different age distributions and use non-selected control subjects. Consequently, we are convinced that the papilla differences for our subjects are less established compared to the subjects of (Swindale et al.(2000)).

We demonstrate that tree based methods, especially stabilized methods like bagging, decrease the misclassification error compared to LDA (based on a comparison of estimated misclassification errors). Additionally, the incorporation of various anamnestic factors improves the classifier.

(Swindale et al.(2000)) report a misclassification error of 14%. This indicates the size of possible improvements based on both automatic feature detection and better classification procedures.

The results reported in Section 3 are currently based on human incorporation for the determination of the optic nerve head. The automated procedure by (Swindale et al.(2000)) does not require any interaction by a physician. In future we will replace human interaction by automatic papilla detection (Section 2.3) and we will use features from the 3D modeling of the optic cup as input variables for classification.

Acknowledgments

We acknowledge support by Deutsche Forschungsgemeinschaft SFB 539 and in particular C.Y. Mardin, M. Korth, N. Ngugen, A. Jünemann, F. Horn and M. Dzialach. Additionally we thank N. Swindale for his assistance on 3D feature detection.

References

BREIMAN, L. (1996a): Bagging predictors, *Machine Learning*, *Vol. 26*, *123-140*.

BREIMAN, L. (1996b): Heuristics of instability and stabilization in model selection, *The Annals of Statistics*, *Vol. 24*, *2350-2383*.

BREIMAN, L. (1998): Arcing classifiers, *The Annals of Statistics*, *Vol. 26*, *801-824*.

BREIMAN, L., FRIEDMAN, J., OLSHEN, R., and STONE, C. (1984): Classification and Regression Trees, Wadsworth, California.

COLEMAN, A. (1999): Glaucoma, *The Lancet*, *Vol. 354*, *1803-1810*.

HEIDELBERG ENGINEERING (1997): Heidelberg Retina Tomograph: Bedienungsanleitung Software version 2.01, Heidelberg Engineering GmbH, Heidelberg.

HUBBARD, L., BROTHERS, R., COOPER, L., NEIDER, N., KING, W., CLEGG, L., and SHARRETT, A. (1998): Relationship of blood pressure and generalised retinal arteriolar narrowing in the aric study, in Association for Research in Vision and Ophthalmology (ARVO), Fort Lauderdale, Florida.

IHAKA, R. and GENTLEMEN, R. (1996): R: A language for data analysis and graphics, *Journal of Computational and Graphical Statistics*, *Vol. 5, 299-314*.

JONAS, J. B. and NAUMANN, G. O. H. (1993): Morphologie des gesunden und des glaukomatösen Sehnerven, in PILLUNAT, L.-E. and STODTMEISTER, R. (Eds.): Das Glaukom: Aspekte der Forschung für die Praxis, Springer-Verlag, Berlin, Vol. 15. of *Ophthalmothek*, 141-151.

LAUSEN, B., KERSTING, M., and SCHOECH, G. (1997): The regression tree method and its application in nutritional epidemiology, *Informatik, Biometrie und Epidemiologie in Medizin und Biologie*, *Vol. 28, 1-13*.

LAUSEN, B., SAUERBREI, W., and SCHUMACHER, M. (1994): Classification and regression trees (cart) used for the exploration of prognostic factors measured on different scales, in DIRSCHEDL, P. and OSTERMANN, R. (Eds.): Computational statistics, Physica Verlag, Heidelberg, 483-496.

LAUSEN, B. and SCHUMACHER, M. (1992): Maximally selceted rank statistics, *Biometrics*, *Vol. 48, 73-85*.

MICHELSON, G. and GROH, M. (2001): Screening models for glaucoma, *Current Opinion in Ophthalmology*, *Vol. 12, 105-111*.

MIKELBERG, F., PARFITT, C., SWINDALE, N., GRAHAM, S., and DRANCE, S. (1995): Ability of the heidelberg retina tomograph to detect early glaucomatous visual field loss, *J Glaucoma*, *Vol. 4, 242-247*.

PÁL, I. (2000a): Detection and Segmentation Methods for the Optic Nerve Head on Fundus Images, in J.JAN, KOZUMPLÍK, J., PROVAZNÍK, I., and SZABÓ, Z. (Eds.): Analysis of Biomedical Signals and Images, Vutium Press, Brno, Vol. 15, 207-209.

PÁL, I. (2000b): Verfahren zur Detektion der Papille auf den Fundusbildern, in HORSCH, A. and LEHMANN, T. (Eds.): Bildverarbeitung für die Medizin 2000: Algorithmen–Systeme–Anwendungen Proceedings des Münchener Workshops, Springer Verlag, Berlin, Informatik aktuell, 376-380.

PÁL, I., MICHELSON, G., and ZINSER, G. (2000): Automatische Bestimmung des Arterie-Vene Verhältnisses auf Retina-Tomograph Bildern, in HORSCH, A. and LEHMANN, T. (Eds.): Bildverarbeitung für die Medizin 2000: Algorithmen–Systeme–Anwendungen Proceedings des Münchener Workshops, Springer Verlag, Berlin, Informatik aktuell, 319–323.

SWINDALE, N. V., STJEPANOVIC, G., CHIN, A., and MIKELBERG, F. S. (2000): Automated analysis of normal and glaucomatous optic nerve head topography images, *Investigative Ophthalmology and Visual Science*, Vol. *41*, 7, *1730–1742*.

Comparing Split Criteria for Constructing Survival Trees

M. Radespiel-Tröger[1], T. Rabenstein[2], L. Höpfner[2],
H.T. Schneider[2]

[1]Department of Medical Informatics, Biometry and Epidemiology,
Friedrich-Alexander-Universität, D-91054 Erlangen, Germany

[2]Department of Medicine I, Friedrich-Alexander-Universität, D-91054
Erlangen, Germany

Abstract: Various split criteria for constructing recursively partitioned trees from censored data have been proposed. However, no uniform superior split criterion is available. We compared five different split criteria regarding the explained variation of outcome with application to a clinical data set from a study of extracorporeal shock wave lithotripsy (ESWL) for treatment of gallbladder stones in 408 patients. The main end point was time until detection of complete stone clearance. The covariates patient age, sex, body mass index, relative gallbladder volume reduction after fatty meal, stone density, stone diameter, and stone number were analysed. The split criteria relative risk, log-rank statistic, log-rank statistic adjusted for measurement scale, partial likelihood ratio, and constant hazard likelihood ratio were compared. Model likelihood and Magee's R^2 were used as measures of information content and explained variation, respectively. In the first step, trees were constructed and validated using separate learning and validation samples, respectively. In the second step, bootstrap confidence intervals of explained variation were estimated using an internal validation procedure. There was no substantial difference between splitting algorithms in terms of explained variation. Simulation studies of pruned trees are necessary to extend our results.

1 Introduction

Conventional methods of survival analysis are often complicated by the need to incorporate pre-specified interactions in the predictive model. Once a model has been specified, its application by clinicians is hampered by the fact that prediction of outcome for new patients is not intuitive and sometimes requires software programs to be carried out routinely. Risk group definitions derived from the quantiles of predicted risk scores are arbitrary. Following the seminal work of Breiman et al. (1984), tree-based models were extended to allow for analysis of censored data. Their clinical application is growing. Three major advantages of tree-based methods are their easy interpretation, the availability of methods for handling missing covariate values, and their implicit ability to model treatment-covariate interactions. Their disadvantages include the instability of split point and/or variable selection and the problems associated with the determination of optimally-sized trees. Most of the proposed splitting algorithms do not take into account any adjustment for covariates measured on different scales. According to Lausen and Schumacher (1996), this adjustment is mandatory in the evaluation of cutpoints. Based on the major computational differences between existing splitting algorithms, we find it hard to believe that all of them should be equally well-suited to discover the underlying hazard structure of a given data set. For example, one would expect logrank-based splitting algorithms to perform especially well because the logrank test has a high statistical power for detecting differences between two (or more) survival curves. Even if an overall best splitting algorithm cannot be found, there is reason to assume that not all split criteria perform equally well. Based on this assumption, we compared five different splitting algorithms, one out of which uses an adjustment for covariates measured on different scales. Using two different approaches, we applied these methods to a clinical data set from a prognostic study of extracorporeal shock wave lithotripsy (ESWL) in patients with cholecystolithiasis.

2 Patients and Methods

2.1 Patients

We analysed 408 consecutive cases of cholecystolithiasis treated with extracorporeal shock wave lithotripsy (ESWL) for the first time and followed up prospectively between 01.01.1988 and 31.01.1997 in the Department of Medicine I of the University of Erlangen-Nuremberg, Germany. The main end point for this study was the time until complete stone disappearance. The overall censoring rate was 34.8%. The following predictive factors were analyzed: patient age, body mass index (BMI), reduction in gallbladder volume [%] after diagnostic fatty meal, maximal stone diameter [mm] measured by ultrasound scan of the gallbladder, maximal radiographic gallbladder stone density [Hounsfield units] measured by CT scan, number of gallbladder stones, and gender of patient.

2.2 Methods

We built trees according to Breiman et al. (1984) without pruning. In this methodology, recursive partitioning is defined by specifying 3 elements: A way to select a split at every intermediate node, a rule for determining when a node is terminal, and a rule for assigning a value $y(\tau)$ to every terminal node τ. We used two measures for comparison

Table 1: Patient characteristics

covariate	n	mean (sd)	range
AGE[a]	408	46.2 (13.5)	15 - 82
BMI[b]	390	26.7 (4.4)	13.5 - 40.4
CONTRACT[c]	401	62.9 (19.3)	-66 - 96
DENSITY[d]	356	80.4 (76.1)	-22 - 420
SIZE[e]	406	17.8 (6.3)	6 - 40
NUMBER[f]	408	/	1 - 5/mult.
GENDER[g]	f=310; m=98	/	f/m

[a] AGE = patient age [years]

[b] BMI = body mass index

[c] CONTRACT = gallbladder ejection fraction after fatty meal [%]

[d] DENSITY = maximal gallbladder stone CT-density [HU]

[e] SIZE = maximal gallbladder stone diameter [mm]

[f] NUMBER = number of gallbladder stones

[g] GENDER = gender of patient (f=female; m=male)

NOTE: CONTRACT and DENSITY can take negative values

of split criteria: The likelihood of a fitted Weibull survival model and Magee's R_M^2 (1990), respectively. Model likelihood (usually given as log likelihood (LL)) can be seen as a measure of information content of a tree (Edwards (1992)). Magee's R_M^2 is given by:

$$R_M^2 = 1 - exp(-LR/n) \qquad (1)$$

with LR being the global log likelihood ratio statistic for testing the importance of all p predictors in the model and n being the case number. Under moderate censoring, R_M^2 can be seen as an approximation to the R^2 measure in linear regression. R_M^2 was estimated using a Weibull survival model with all terminal nodes included as dummy covariates. In the first step, we performed a split sample analysis using 275 patients who were treated between 01.01.1988 and 18.05.1992 as the training sample and the remaining 133 patients treated between 19.05.1992 and 31.01.1997 as test sample. In this approach, tree growth was stopped either if the number of terminal nodes reached 11 or if further splitting would have resulted in a terminal node containing less than $\sqrt{n}/2$ cases (with n being the number of events in the training or test sample, respectively). In the second step, we employed an internal validation procedure according to Harrell et al. (1996) by performing the following steps:

1. Build the best tree model from the complete original data set.

2. Compute explained variation for this model and denote it R_{app}^2.

3. Generate a bootstrap sample of size n with replacement from the original sample of size n.

4. Build a tree model from the bootstrap sample using the same strategy as in step 1.

5. Compute R^2_{boot} for this model from the bootstrap sample.

6. Compute R^2_{orig} for this model from the original sample.

7. The optimism in the model fit obtained from the bootstrap sample is $R^2_{boot_i} - R^2_{orig_i} = O_i$.

8. Repeat steps 3 to 7 1000 times.

9. Average the estimated optimism O_i to arrive at O.

10. The bootstrap corrected R^2_{bc} of the original tree model is $R^2_{app} - O$.

11. The upper and lower 95%-confidence limits of R^2_{bc} are given by R^2_{app} minus the 2.5^{th} and 97.5^{th} BC_a-adjusted percentiles of O_i, respectively (Efron and Tibshirani (1993)).

In this approach, tree growth was stopped either if the number of terminal nodes reached 35 or if further splitting would have resulted in a terminal node containing less than $\sqrt{n}/2$ cases (with n being the number of events in the original or bootstrap sample, respectively). Missing values were treated according to Clark and Pregibon (1992). The following split criteria were studied:

1. Log-rank statistic:
Similar to Segal (1988), the value of the log-rank statistic associated with the considered split is computed.

2. Log-rank statistic with adjustment for measurement scale:
Similar to 1., but with adjustment for measurement scale according to Lausen and Schumacher (1996).

Using criteria 1 and 2, the split point associated with the smallest p value is selected.

3. Relative risk under proportional hazards:
The relative risk between offspring nodes is given by $exp(\beta)$, where β is the maximum partial likelihood estimate of the Cox proportional hazards regression model coefficient (Cox 1972)).

Table 2: Results of data splitting approach

split criterion	training sample		test sample	
	$-2LL^a$	$R_M^{2\ b}$	$-2LL$	R_M^2
RR^c	2468	.2346	1076	.1067
$logrank_{adj}{}^d$	2424	.3481	1073	.1252
$logrank_{unadj}{}^e$	2401	.4020	1073	.1293
$LR_{PH}{}^f$	2403	.3971	1072	.1336
$LR_{exp}{}^g$	2410	.3814	1068	.1620

^{a} $-2LL$ = - 2 log likelihood under Weibull model
^{b} R_M^2 = Magee R squared
^{c} RR = relative risk under PH model
^{d} $logrank_{adj}$ = adjusted log-rank statistic
^{e} $logrank_{unadj}$ = unadjusted log-rank statistic
^{f} LR_{PH} = partial likelihood ratio under PH
^{g} LR_{exp} = likelihood ratio under constant hazard

Using criterion 3, the split point associated with the highest relative risk is selected.

4. Partial likelihood ratio under proportional hazards:
   Similar to Ciampi (1995), the likelihood ratio (LR) statistic is estimated by maximizing the partial likelihood of the Cox proportional hazards model under the null hypothesis of no effect of the considered split on survival.

5. Likelihood ratio under constant exponential hazard:
   Similar to Davis and Anderson (1989), the likelihood ratio is estimated by maximizing the likelihood of the constant exponential hazard regression model under the null hypothesis of no effect of the considered split on survival.

Using criteria 4 and 5, the split point associated with the highest LR statistic is selected. All computations were carried out using the R software (Ihaka and Gentleman (1996)) in conjunction with the *survival5* package (Therneau (1999)).

3 Results

3.1 Data Splitting Approach

The results of the data splitting approach are given in table 2.

Table 3: Results of internal validation approach

| | resubstitution | | internal validation | | | |
| | point estimates | | point estimates | | 95%-CI | |
split criterion	$-2LL^a$	$R^2_M{}^b$	$-2LL$	R^2_M	$-2LL$	R^2_M
RR^c	3378	.4644	3586	.2072	3579-3592	.2048-.2096
$lr_{adj}{}^d$	3355	.4946	3566	.2381	3560-3573	.2357-.2404
$lr_{unadj}{}^e$	3353	.4971	3574	.2323	3567-3580	.2299-.2348
$LR_{PH}{}^f$	3360	.4885	3573	.2242	3566-3580	.2221-.2265
$LR_{exp}{}^g$	3338	.5146	3567	.2468	3560-3573	.2445-.2491

[a] $-2LL = -2$ log likelihood under Weibull model
[b] R^2_M = Magee R squared
[c] RR = relative risk under PH model
[d] lr_{adj} = adjusted p value based on log-rank statistic
[e] lr_{unadj} = unadjusted p value based on log-rank statistic
[f] LR_{PH} = likelihood ratio under PH model
[g] LR_{exp} = likelihood ratio under constant hazard model

The table shows that, in terms of model likelihood, the best result (-2 LL=1068) in the test set was obtained with the constant hazard likelihood criterion. The split criterion with the second best test set result was the PH likelihood criterion (-2 LL=1072), closely followed by both adjusted and unadjusted log-rank criteria (-2 LL=1073). The worst performance was shown by the relative risk criterion (-2 LL=1076). Similar results were obtained by use of R^2_M.

3.2 Internal Validation Approach

The results of the internal validation approach are given in table 3.

According to internal validation, the best performance in terms of model likelihood was shown by both the adjusted logrank criterion and the constant hazard criterion. The 95% confidence intervals of -2 LL for both criteria were identical (3560-3573). The next best performance was shown by both unadjusted log-rank criterion (95% CI: 3567-3580) and PH likelihood criterion (95% CI: 3566-3580). In terms of R^2_M, the constant hazard criterion showed the best performance (95% CI: 0.2445-0.2491), followed by the adjusted (95% CI: 0.2357-0.2404) and unadjusted (95% CI: 0.2299-0.2348) logrank criterion. Regardless of the measure of explained variation, the performance of the relative risk criterion was found to be worse than that of all other four criteria.

4 Discussion

A substantial number of split criteria for building survival trees has been proposed (Zhang and Singer (1999)). However, no split criterion for survival trees has yet been shown to generally yield trees with superior explained variation. We compared five different splitting algorithms, one out of which uses an adjustment for covariates measured on different scales. As an example, we used data from a prognostic study of outcome after extracorporeal shock wave lithotrypsy in 408 patients. To prevent a possible influence of the choice of pruning algorithm on our results, we decided to compare unpruned trees. In the first step, we used a temporary split-sample approach recommended by Altman and Royston (2000). However, the validity of this approach is limited in smaller samples because the prognostic information in the test sample is not available for model building. In the first approach, the maximum of explained variation in the test sample was found using the constant hazard criterion. However, there was only a small difference between the explained variation of the best and the next three best algorithms. We therefore doubt whether this result is generalisable, as it is based on only one example. In the second step, using the internal validation approach according to Harrell et al. (1996), both the constant hazard criterion and the adjusted logrank criterion showed the best performance in terms of model likelihood, but were again closely followed by the two next best splitting algorithms. Although the confidence intervals reveal significant differences between the splitting algorithms, we doubt whether these results can be generalised, because the results strongly depend on the hazard structure of the analysed sample. Why did the adjusted logrank criterion not perform better than all other split criteria in both approaches? The amount of random variation described by unpruned trees may have outweighed (potentially small) differences between splitting algorithms in terms of explained variation. A potential source of criticism may lie in our predefined maximum terminal node number of 11 (split-sample approach) and 35 (internal validation approach), respectively. These limits may appear arbitrary. Our intention was to construct trees of approximately comparable size, because the explained variation of a tree does not only depend on the information content of a particular split, but also on the total number of splits. We are aware that the resulting trees may still be of different size and that therefore comparability may still be limited. As a possible future consequence, using a measure of explained variation adjusted for the number of tree splits appears to be a potentially fruitful approach. Based on our results, we conclude that (with the exception of the relative risk criterion that generally performed unsatisfactorily) no substantial differences between the studied splitting algorithms in terms of explained variation could be demonstrated using unpruned trees. Therefore, a general recommendation for a best splitting algorithm cannot be given. Simulation studies of pruned trees are necessary to extend our results. Even if a superior splitting algorithm (in terms of explained variation) can be determined, a number of questions will have to be answered before obtained trees can be used in clinical: What are the potential consequences of decisions made on the basis of the tree? Are there competing treatments? What is the (estimated) relationship between risks (and/or costs) and benefits of the considered treatment? Regardless of the statistical method used for prognostic analysis, treatment decisions will always be essentially based on medical and ethical considerations.

References

ALTMAN, D. G., and ROYSTON, P. (2000): What do we mean by validating a prognostic model? *Statistics in Medicine, Vol. 19, 453-473.*

BREIMAN, L., FRIEDMAN, J. H., OLSHEN, R. A. and STONE, C. J. (1984): Classification and Regression Trees. Wadsworth, Monterey.

CIAMPI, A., NEGASSA, A., and LOU, Z. (1995): Tree-structured Prediction for Censored Survival Data and the Cox Model. *Journal of Clinical Epidemiology, Vol. 48, 675-689.*

CLARK, L. A., and PREGIBON, D. (1992): Tree-Based Models, in Chambers, Hastie (Eds.): Statistical Models in S, Chapman and Hall, New York.

COX, D. R. (1972): Regression Models and Life-Tables (with discussion). *Journal of the Royal Statistical Society, Series B, Vol. 34, 187-220.*

DAVIS, R. B. and ANDERSON, J. R. (1989): Exponential Survival Trees. *Statistics in Medicine, Vol. 8, 947-961.*

EDWARDS, A. W. F. (1992): Likelihood. John Hopkins University Press, Baltimore.

EFRON, B., and TIBSHIRANI, R. J. (1993): An Introduction to the Bootstrap. Chapman and Hall, New York.

HARRELL, F. E., LEE, K. L., and MARK, D. B. (1996): Tutorial in Biostatistics: Multivariable Prognostic Models: Issues in Developing Models, Evaluating Assumptions and Adequacy, and Measuring and Reducing Errors. *Statistics in Medicine, Vol. 15, 361-387.*

IHAKA, R., and GENTLEMAN, R. (1996): R: A Language for Data Analysis and Graphics. *Journal of Computational and Graphical Statistics, Vol. 5, 299–314.*

LAUSEN, B. and SCHUMACHER, M. (1996): Evaluating the Effect of Optimized Cutoff Values in the Assessment of Prognostic Factors. *Computational Statistics and Data Analysis, Vol. 21, 307-326.*

MAGEE, L. (1990): R^2 measures based on Wald and likelihood ratio joint significance tests. *American Statistician, Vol. 44, 250-253.*

SEGAL, M. R. (1988): Regression Trees for Censored Data. *Biometrics, Vol. 44, 35-47.*

THERNEAU, T. M. (1999): A package for Survival Analysis in S. *Department of Health Science Research, Mayo Clinic, Rochester, Minnesota, http://www.mayo.edu/hsr/sfunctions.*

ZHANG, H., and SINGER, B. (1999): Recursive Partitioning in the Health Sciences. Springer, New York.

A Type of Bayesian Small Area Estimation for the Analysis of Cancer Mortality Data

U. Schach

Fachbereich Statistik,
Universität Dortmund, D-44221 Dortmund, Germany
uschach@statistik.uni-dortmund.de

Abstract: The distribution of cancer mortality in Germany is collected in two different data sets, one with a high spatial resolution but aggregated data over time, the other with yearly data on a coarse spatial scale. This is due to privacy protection laws, as the data become nearly individual when analyzing rare cancer types or strata of age groups. The aim of this paper is to present a modeling approach to estimate the missing data from the given spatial and temporal marginals. Parameters of spatial and temporal autocorrelation, dispersion, and temporal trend parameters are estimated simultaneously within the Bayesian model, using MCMC techniques based on the Metropolis-Hastings algorithm.

1 Introduction

1.1 Small Area Estimation

The general idea of a small area estimation in its original sense is an intra-polation of information collected on a larger spatial scale to local areas within the study region. Additional variables with a high correlation to the variable of interest are used to improve the estimation, see Rao and Yu (1994). For the purpose of this paper, the small area estimation is necessary as the frequency and distribution of cancer mortality in Germany is published in two different types of resolution. One data set has a high spatial resolution, but aggregated data over time, the other one is based on yearly data, but consists of aggregated data over space. This mode of data presentation is required by privacy protection laws and tabulation procedures. When regarding rare cancer types or further subdivision by age group the time by location cell frequencies become too small. As knowledge about this data is desirable, however, the estimation of the missing data will be performed in this analysis. The small area estimation in this context uses spatial and temporal dependence structures, based on a Bayesian hierarchical modeling approach. The underlying size of the population at risk is available with the highest spatial and temporal resolution.

1.2 Data Sets

The spatio-temporal small area estimation will be illustrated using data on stomach cancer mortality among men in Germany. As described above, the estimation is performed by combining two types of marginal data sets. Data set I has a coarse spatial structure of the 30 regions ("Regierungsbezirk") of former West Germany, but displays yearly data from 1976 to 1990. Data set II consists of stomach cancer mortality figures with the high spatial resolution of 327 districts ("Landkreis") within the study area, but temporaly aggregated over five year periods from 1976-1980, 1981-1985, and 1986-1990. We will use the following notation:

D_{ti} number of cancer cases at time t in district i (unknown)
N_{ti} population size at time t in district i
r_{ti} mortality rate at time t in district i,

where $i = 1, \ldots, I$ denotes the districts within the study region, and $t = 1, \ldots, T$ represents time points for the analysis. Using the notation introduced above, the aim of the small area estimation is to obtain $\hat{D}_{ti}$, using the given marginals $D_{.i}$ and $D_{t.}$, where a dot indicates summation over the dotted index. Additionally, the underlying population size N_{ti} is known and available for the analysis. The region of Braunschweig, located in Lower Saxony in the center of Germany, will be used to illustrate the spatio-temporal small area estimation. Thus, we consider the following table, with the given marginals and the aim is to fill up the missing data, modeled as unknown parameters.

year \ distr.	$i = 1$	$i = 2$	...	$i = I$	Σ
t=1					$D_{1.}$
t=2					$D_{2.}$
$\vdots$					$\vdots$
t=T					$D_{T.}$
Σ	$D_{.1}$	$D_{.2}$	...	$D_{.I}$	$D_{..}$

Table 1: Marginal data for the small area estimation.

For the study area of Braunschweig, we have 11 districts (I=11) and consider five years (T=5) from 1986 to 1990.

2 Spatio-temporal small area estimation

2.1 Spatial Dependence Structures

The spatial structure of the observed area is based on an irregular lattice structure. We imply stochastic dependence of neighboring sites. Two districts are considered to be neighbors, if they share a common border. We model the spatial dependence as a Markov type departure from independence, where the time series definition of a Markov dependence is transferred to spatial data as described by Cressie (1993), pp. 402-410. This means that the observation in district i is dependent on its neighboring sites, denoted by $\{-i\}$. However, it is independent of the remaining sites on the lattice, given the values at the neighboring sites. Markov processes of this type are said to have a conditional autoregressive (CAR) structure. The resulting neighborhood dependencies of the study region of Braunschweig are indicated through the graph in figure 1.

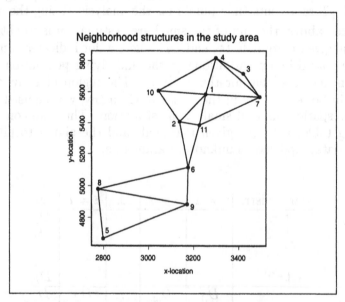

Figure 1: Plot of the neighborhood structures within the study area.

Each of the eleven points displays the corresponding district within the region of Braunschweig, and two points are connected, if they are neighbors, i.e. if they share a common border.

2.2 Spatio-Temporal Model

As described above, a major aim of our method is to estimate the missing parameters of cancer mortality, using the given marginals. Additional to the estimation of the unknown cell frequencies, we use a hierarchical Bayesian model that simultaneously estimates parameters of spatial and temporal autocorrelation, dispersion, and temporal trend parameters, as described by Schach (2000). Based on the idea of conditional independence, given neighboring sites in space and time, we introduce the following three-stage hierarchical spatio-temporal model.

Stage 1: Poisson model

Functional relation:
$$\lambda_{ti} = r_{ti}\,N_{ti} \qquad t = 1,\ldots,T,\ i = 1,\ldots,I$$
$$t = 1: r_{1i} = \mu_1 + b_i \qquad i = 1,\ldots,I$$
$$t > 1: r_{ti} = \mu_t + \alpha\,(r_{t-1,i} - \mu_{t-1}) + \beta\,(\bar{r}_{t-1,-i} - \mu_{t-1}) \qquad i = 1,\ldots,I$$

$$\hat{D}_{ti} \mid \lambda_{ti} \sim \mathrm{Poi}(\lambda_{ti})$$

Stage 2: Prior distributions

$$
\begin{aligned}
b_0 &\sim \mathrm{MVN}(\gamma, V) \\
\gamma &\sim \mathrm{MVN}(0, U),\ U = 11 \times 11 \text{ unity matrix} \\
V &\sim \mathrm{Wishart}(U, I) \\
b &\quad \text{sum to zero transformation of } b_0 \\
\mu_t &\sim \mathrm{N}(\bar{\mu}_t, 1000),\ \bar{\mu}_t = D_{t.}/N_{t.} \\
\alpha &\sim \mathrm{N}(0, 0.0001) \\
\beta &\sim \mathrm{N}(0, 0.0001)
\end{aligned}
$$

Stage 3: Indirect adjustment

$$
\begin{aligned}
D_{t.} &\sim \mathrm{N}(\hat{D}_{t.}, 1000) \\
D_{.i} &\sim \mathrm{N}(\hat{D}_{.i}, 1000)
\end{aligned}
$$

On stage 1 we model the missing values of cancer frequency, dependent on time and space trough an overall level μ_t, dependent on time t and temporal and spatial autocorrelation parameters α and β. We begin with a Poisson model for the estimated number of cancer deaths in district i at time t. The Poisson parameter λ_{ti} is obtained as the product of the mortality rate dependent on space and time and the size of the corresponding underlying population at risk. Here, we must differentiate between the rate at time $t = 1$ and the rate for the remaining points in time. For $t > 1$ the difference of the rate at site i compared to the average is influenced by the corresponding difference at time $t - 1$ and by the difference between the neighbors of i at time $t - 1$, denoted by $\bar{r}_{t-1,-i}$. On stage 2 of the model we specify the prior distributions for time $t = 1$, where we begin to model spatial heterogeneity with the vector of spatial random effects b_0. b_0 is of dimension I, the number of districts within the study region. Instead of assigning a conditional autoregressive dependence structure to b_0 directly, we are giving it a noninformative multivariate normal prior distribution. We expect that the spatial dependence between neighboring sites arises through the model assumption and the data. The vector γ and the precision matrix V are hyperparameters, necessary for the multivariate normal distribution of the vector b_0. The resulting vector b is restricted with a sum-to-zero constraint on b_0 to assure identifiability at time $t = 1$, see also Besag and Kooperberg (1995). The overall level μ_t is assigned a Gaussian prior, depending on $\bar{\mu}_t$ with a high precision. α and β, the parameters of temporal and spatial autocorrelation, are assigned non-informative Gaussian prior distributions with mean 0 and low precision. On stage 3 we use an indirect adjustment of the sum of the estimated mortality figures to the observed marginals. As we have to account for spatial and temporal marginals, the adjustment is two-dimensional. The parameter estimation is invariant under the change of the order of the two adjustments. This computational trick avoids to assign given data (i.e. marginals) to a sum of estimated parameters, which is cumbersome in this type of Bayesian framework.

3 Results

3.1 Parameter Estimation for the Bayesian Model

Before we begin with the presentation of the estimated parameters of the model for the study region of Braunschweig for the years from 1986 to 1990, we will take a look at model diagnostics and convergence. Due to the complexity of the model, a Metropolis-Hastings algorithm has been used for the generation of the Markov chains, as illustrated by Brooks (1998) and Chen et al. (2000),pp. 23-26. We have chosen a burn-in of 4,000 iteration steps and 8,000 additional recorded updates keeping each 20th iteration for the estimation. Figure 2 shows the satisfactory acceptance rates of the sampler. We have used the WinBUGS software for the simulation according to Spiegelhalter et al. (2000).

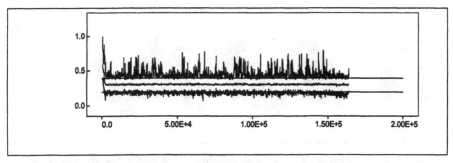

Figure 2: Metropolis-Hastings acceptance rates.

Considering the traces, e.g. for the overall level at time $t = 5$ we can see low autocorrelation of the chain in figure 3.

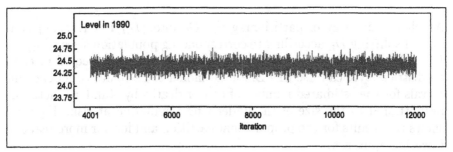

Figure 3: Trace for the level μ_5 in 1990.

Having checked the model diagnostics, we can look at the resulting parameter estimates of the mortality rates per 100,000 persons at risk. Remembering table 1, it has been the aim to fill the missing cell frequencies. The given yearly data had only a coarse spatial structure of single figures for the whole region. After the application of the spatio-temporal small area estimation, we obtain a spatial structure on the basis of the districts for every year, where the aggregated data over five years has been split up into yearly data. To demonstrate the parameter estimates, the resulting rates per 100,000 inhabitants at risk are displayed in figure 4.

It is worth mentioning that the parameters of temporal (α) and spatial (β) autocorrelation, dispersion, and temporal trend are estimated simultaneously within the model. The resulting mean estimate for α is 0.68, and 0.09 for β. Thus, the temporal dependence between observations seems to be much stronger than the spatial correlation.

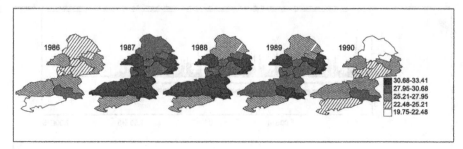

Figure 4: Spatial and temporal resolution of the rates for the study
area for the years from 1986-1990.

3.2 Proportional Partition

An elementary way of partitioning the $D_{t.}$ into $\{\hat{D}_{ti}, i = 1, \ldots, I\}$ consists of splitting $D_{t.}$ according to corresponding population sizes $\{N_{ti}, i = 1, \ldots, I\}$. This can be justified by a Binomial model with equal rates in all districts. We can use this Binomial model to calculate confidence intervals for the estimated number of cancer deaths by standard methods, proportional to the size of the underlying population at risk. Figure 5 shows the results for the proportional partition and for our more realistic Bayesian approach.

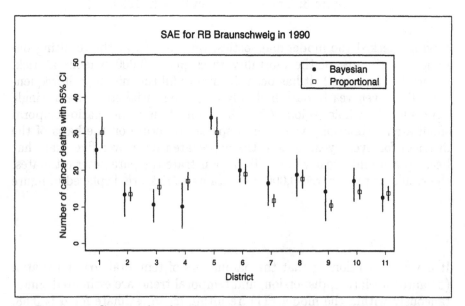

Figure 5: Cancer death estimates and confidence intervals of the
Bayesian approach and the Binomial model in 1990.

4 Discussion

The idea of small area estimation has been applied to combine two different kinds of marginals, in order to obtain data with the highest spatial and temporal resolution, based on underlying population figures. The spatio-temporal small area estimation has been performed with an approach that accounts for spatial and spatio-temporal dependencies within the data. When comparing the resulting parameter estimates $\hat{D}_{ti}$ for the small area with those obtained with a proportional partition, the parameter estimates are nearly identical. However the resulting confidence intervals of the Bayesian model are larger but more realistic, as they account for the dependence structure within the data. When additional covariables are to be included in the model, the proportional partition is no longer valid. Our model can be extended to several temporal blocks, as well as to clusters of regions, as described by Knorr-Held and Raßer (2000) by increasing the number of spatial neighbors. So far, we have explained the procedure without an adjustment for age group. The model can also be refined in that direction. The method is well applicable to the analysis of multi-directional trends within different study regions, for the analysis of temporal trends within small spatial units, and it can easily be extended to age groups and additional covariables.

References

BESAG, J. and KOOPERBERG, C. (1995): On conditional and intrinsic autoregressions. *Biometrika 82, Vol. 4, 733–746.*

BROOKS, S. (1998): Markov chain Monte Carlo method and its application. *The Statistician 47, Vol. 1, 69–100.*

CHEN, M.; SHAO, Q. and IBRAHIM, J. (2000): *Monte Carlo Methods in Bayesian Computation.* Springer, New York

CRESSIE, N. A. (1993): *Statistics for Spatial Data.* Wiley, New York

KNORR-HELD, L. and RAßER, G. (2000): Bayesian Detection of Clusters and Discontinuities in Disease Maps. *Biometrics, Vol. 56, 13–21.*

RAO, J.N. and YU, M. (1994): Small-area Estimation by Combining Time-series and Cross-sectional Data. *The Canadian Journal of Statistics, Vol. 22, 511–528.*

SCHACH, U. (2000): Spatio-temporal models on the basis of innovation processes and application to cancer mortality data. *Technical Report, Vol. 16/2000, SFB 475, Univ. of Dortmund*

374

SPIEGELHALTER, D.; THOMAS, A. and BEST, N. (2000): *WinBUGS Version 1.3 User Manual*. Institute of Public Health, Cambridge and Imperial College School of Medicine, London.

DEA-Benchmarks for Austrian Physicians

M. Staat

Lehrstuhl für VWL, insbes. Mikroökonomik
Universität Mannheim, D-68131 Mannheim

Abstract:

At present, there is virtually no mechanism of cost containment or cost control implemented w. r. t. the outlays for the treatment of patients by general practitioners in Austria. Due to asymmetric information about the true cost of treatment physicians are able to provide an excessive level of services or provide services at excessive cost. We argue that benchmarking physicians' treatment strategy w. r. t. cost efficiency by means of Data Envelopment Analysis makes it possible to implement a reimbursement system that ensures service provision at an efficient cost level by minimizing the ability of physicians to appropriate information rents.

1 Introduction

In Austria, there is no effective mechanism of cost control for the treatment of patients by general practitioners (GPs). The GPs receive a per capita fee for each patient. This covers basic diagnostic measures and therapy. Variable fees apply for additional services provided. These variable fees decrease in services provided per case. This low powered regulation scheme[1] does not prevent physicians claiming higher fees or providing more services than necessary. Therefore, inefficiency w. r. t. the provision of services may well exist.

A number of studies in the health care sector (see, for instance, Hollingsworth et. al (1999) or Staat (2000)) have found marked differences for the efficiency of hospitals. For an analysis of the efficiency of physicians in hospitals, see Chilingerian (1997). It is the goal of this analysis to determine whether this holds true for the service provision by GPs in Austria and if so to which degree.

If there is inefficiency under the present system it is obvious that in order to ensure a cost efficient provision the reimbursement rules for the services which physicians provide will most likely have to be re-engineered, so that only services which are necessary are provided while at the same time underprovision is avoided.

---

[1] See Laffont, Tirole (1999).

We argue that Data Envelopment Analysis (DEA), a nonparametric method to estimate production frontiers, can be used to derive efficiency scores, which can form the basis of a reimbursement scheme with the above mentioned desirable features. The scheme would consist of a reimbursement rule in the spirit of yardstick competition ensuring cost efficient provision of services without the risk of underprovision.

DEA has by now become a standard tool to evaluate the performance of hospitals and other public services.[2] Often, however, some form of benchmarking is used to assess the efficiency of service provision in the health care sector. Hospital rankings, for instance, are frequently based on some form of benchmarking. These rankings, based on simple benchmarking approaches, have led to complaints that they are misleading w. r. t. the relative performance of providers facing different tasks.

As will be shown below, DEA does only compare observations with a similar structure of tasks. We will use cost data on GPs working in the same region of Austria. These data contain the necessary information to ensure that only physicians with nearly identical case load and case mix are compared. Thus, the concerns that benchmarking results in misleading comparisons do not apply to the results of a DEA.

This paper is organized in 6 sections. The following section will illustrate the problems of benchmarking using a simple example. Then DEA is briefly introduced and interpreted as a generalized form of benchmarking. This is followed by a brief section on yardstick competition regulation based on DEA. We then present our results. A final section summarizes the findings.

2 Relative Performance Evaluation Using DEA

2.1 Some Problems with Benchmarking

In this section we focus on the advantages and disadvantages of DEA vis-a-vis benchmarking for the purpose of relative performance evaluation using a simple example. Figure 1 illustrates the difficulties when benchmarking is applied on the basis of more than one parameter. The two benchmarking criteria can be thought of as treatment cost claimed by the physician and cost that was claimed by hospitals to which the physician referred his patients. For this example, we assume that the two types of cost are substitutes and that all GPs face identical tasks.

---

[2]See Hollingsworth et al. (1999).

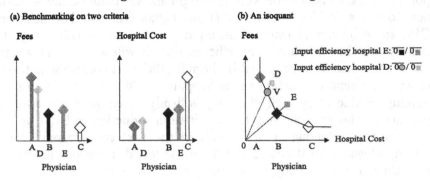

Figure 1: Limits to benchmarking

Inspecting the two figures headed (a) "Benchmarking on two criteria" it would be hard to justify any ranking of the treatment strategies of the GPs A, B and C (i. e. referring a patient at an early or at a later stage). None of the GPs dominates any of the others on both criteria. With standard benchmarking, the efficiency assessment would therefore depend on the relative weight assigned to the cost parameters. The higher the weight put on the GP's cost, the better C will be ranked; the higher the weight put on hospital cost the better A's performance will be ranked.

Figure (b) "An isoquant" illustrates the possibility to benchmark on two criteria simultaneously. Shifting the figure on hospital cost from (a) by 90° and moving it over the physician cost figure will produce the isoquant on the two types of cost for a standard case. The three GPs A, B and C are all located on the isoquant and are therefore viewed to be efficient producers. Physicians D and E are located above the isoquant and are therefore rated as inefficient producers.

The degree of inefficiency can be calculated in a straightforward manner. In case of physician E who is dominated by B w. r. t. both parameters, efficiency is calculated as follows: the closer to the origin any input (cost) combination for a given output is, the more efficient the production of this output (service) is considered.[3] Consequently, one may compare the relative efficiency of any service provider (GP) and a reference point by comparing the relative distance of their cost combinations to the origin. This relative efficiency is calculated as the ratio of the two distances, i. e. $\overline{OB}/\overline{OE}$.

---

[3]This is the logic of an input-oriented approach. Alternatively, one could measure relative efficiently by comparing how much output different DMUs produce, given a comparable input structure.

D's efficiency is measured by the ratio of the distance from the origin to point D and the distance between the origin and the point on the isoquant next to D, i. e. V. The efficiency of D thus equals $\overline{OD}/\overline{OV}$. For inefficient GPs, the efficiency measure of an input-oriented model will always be strictly less than 1, whereas for efficient ones it will equal 1. This ratio of distances has been proposed by Farrell (1957). It corresponds to the inverse of Shepard's distance (see Shepard, 1970). Thus, this way of ranking producers by their efficiency is firmly based on microeconomic production theory. Following this intuitive introduction to DEA we will describe the technique more formally in the next two sections. Since a comprehensive treatment of DEA can be found in various literature, e. g. Cooper et al. (2000), we will keep this very brief.

2.2   Basic Principles of DEA

As outlined above, with standard benchmarking one would be unable to compare physicians taking different approaches to handling certain cases in a sensible way. Instead of applying the same vector of parameter weights to the parameters of all physicians, a DEA assigns a separate vector of weights to the parameters of each observation or decision making unit (DMU). Below, input weights are denoted v_i and output weights are denoted u_r with index i for inputs, index r for outputs, and index j for the observations. The weighted output-input ratio h_0 is then maximized under the restriction that no other observation achieves a score greater than 1 (or any other number chosen for normalization) with the weights that maximize the score of the GP who is evaluated. This results in the following expression (see Cooper et al. (2000), p. 70f):

$$\max_{u,v} h_0 = \sum_r u_r y_{r0} / \sum_i v_i x_{i0} \quad \text{s. t.} \quad \sum_r u_r y_{rj} / \sum_i v_i x_{ij} \leq 1 \quad (1)$$

The index 0 indicates the GP -or more generally DMU- who is evaluated. The formulation of the problem makes clear that the weights underlying the comparison between DMU_0 and the other DMUs are the weights that maximize the efficiency of DMU_0. Comparatively high y's will therefore have high u's, high x's on the other hand will have low v's. This corresponds to comparing the DMUs to the point on the isoquant next to them. Any other reference point chosen would lead to a lower estimate for their efficiency and a suboptimal set of weigths. Proceeding in this way, the three physicians A, B and C in the above example would be rated efficient. Note that it would be impossible to rank two different

GPs with a fixed set of weights as efficient unless they had identical parameter values.

2.3 Formal Representation

The optimization problem (1) can be transformed into a common linear programming (see Cooper et al. (2000), p. 23f). Formula (2) below (in vector notation)[4] is the primal program derived from formula (1). Here, efficiency is measured as the proportional reduction possible for all inputs of an inefficient producer if he produced by means of an efficient technology.

$$
\begin{aligned}
\max_{\theta, \lambda, s^+, s^-} \quad & z_0 = \theta - \epsilon s^+ - \epsilon s^- \\
\text{s.t.} \quad & Y\lambda - s^+ = Y_0 \\
& \theta X_0 - X\lambda - s^- = 0 \\
& \vec{1}\lambda = 1 \\
& \lambda, s^+, s^- \geq 0
\end{aligned}
\tag{2}
$$

The formulation above is the one on which the results to be presented are based. The efficiency score θ is augmented by input slacks s^- and output slacks s^+ multiplied by a non-Archimedian ϵ. Slacks indicate that an input would have to be reduced by more than $(1 - \theta)$ to match the corresponding value of the reference unit or that an output of the reference technology exceeds the respective value of DMU_0. An efficient peer unit enters the reference technology for DMU_0 with a factor $\lambda \geq 0$. In case a DMU is efficient its reference technology is its own input-output-combination and in this case $\lambda = 1$.[5]

For an inefficient DMU there exists a reference technology, i. e. a λ-weighted average of efficient DMUs which is able to produce at least as much output $(Y\lambda - s^+ = Y_0)$ with only a fraction θ of its inputs $(\theta X_0 - X\lambda - s^- = 0)$. Furthermore, the results of a DEA consist of the slacks as well as the corresponding virtual multipliers or shadow prices from the solution of the dual problem.

---

[4]Formula (1) is a constant returns to scale formulation which was chosen to illustrate the similarity to benchmarking whereas equation (2) represents a variable returns to scale formulation. The latter will be used in the analysis.

[5]The results shown in section 5 rest on a specific variant of the programs shown in formulae (2) developed by Andersen, Peterson (1993) and referred to as the "superefficiency"-model in the literature. This allows the measurement of the "excess efficiency" efficient DMUs produce when compared to all *other* observations, i. e. $\lambda_0 = 0$

3   DEA and Yardstick Competition

The problem of cost control for the provision of physicians' services is essentially one of asymmetric information. Bogetoft (2000, see also the literature cited therein) proposes DEA as a means of reducing this information asymmetry by providing additional information to the funding body. This follows closely what Shleifer (1985) developed. He proposed to benchmark hospitals by comparing their input and output to the mean input and output of "similar hospitals", a strategy which Shleifer (1985) related to the concept of yardstick competition (see also Laffont, Tirole (1999)).

Bogetoft (2000) replaces the mean of "similar hospitals" with the DEA-benchmark. This implies that the mean is substituted by a weighted average of hospitals which exactly reflects the input-ouput structure of the hospital -or in our case GP- that is benchmarked. Both methods rest on the principle of using cost information that does not come from the DMU that is benchmarked; it looses its power to manipulate cost information in order to maximize his information rent. DEA, by providing a benchmark much more specific than a mean of "similar observations", leads to minimal information rents, i. e. to an efficient cost level for the provision of the services that can be attained in a setting with asymmetric information.

4   Data

We use data on 635 physicians working in the same region of Austria. The data refer to the first quarter in 1999 and include the number of cases, information about the distribution of the age of patients, the number of patients referred to specialists and to hospitals, the number of patients incapacitated for work -describing case load as well as case mix- and cost data. The cost data include the variable fees paid to the physician, the cost for the consultation of specialists and for the hospitals the GPs referred his patients to, the cost caused by patients who are incapacitated for work, transport cost, cost for drugs and for other means of therapy. The distribution of the cost is displayed in figure 2. We use the fees, the hospital cost and the cost for the medication and other therapeutically relevant means in our analysis to model the input side of the physicians' services. This covers almost 80% of total cost (see figure 2).

Descriptive statistics on the variables used in the analysis, and additionally on the total number of cases are given in table 1. In addition to the cost information, we use the number of patients up to 20 years of

Figure 2: Distribution of Cost

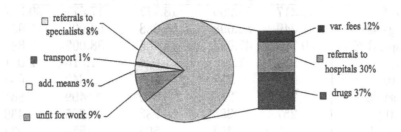

Table 1: Descriptive Statistics

Variable	avg. obs.	min.	max.	avg. ineff.	avg. target (incl. slack)	target /obs.
fees €	18.697	2.294	62.663	18.652	13.486	0,72
drugs €	55.629	7.161	148.174	54.779	40.794	0,74
hospital €	46.764	1.450	129.160	47.401	32.682	0,69
incap.	270,27	27	837	255,50	298,23	1,17
referrals	317,33	13	1173	297,38	337,91	1,14
≤ 20 y	286,61	23	880	280,57	295,28	1,05
21 to 50 y	469,47	82	1192	451,91	492,12	1,09
51+ y	400,23	64	1055	388,53	389,44	1,00
cases	1.156	239	2628	.	.	.

age, the number of patients from 21 to 50 years as well as the number of patients aged 51 and over, the number of cases incapacitated for work and the number of referrals to specialists.

The likely seriousness of cases is described by the number of patients referred to specialists as well as the number incapacitated for work. It is evident that the seriousness of the cases corresponds to the effort (and cost) necessary for treatment. Further information about how cost intensive the treatment is likely to be derived from the number of cases in the three age groups. Patients in the lowest and the highest age cluster require on average more cost intensive treatment (tests not documented here). There is, however, no information on the quality of treatment, for instance a morbidity indicator.

5 Results

The average efficiency of the GPs in the sample (excluding those with an "infinite" efficiency) is 81%, the median is at 77% and the efficiency of

Table 2: Parameter Means by Efficiency

Variable	$z_0 <, 65$	$z_0 <, 85$	$z_0 < 1,05$	$z_0 > 1,05$	$z_0 = \infty$
fees €	20.317	18.296	18.277	17.577	30.317
drugs €	57.249	54.098	55.988	53.906	115.951
hospital €	50.759	47.094	44.955	38.002	87.883
cases	995	1.130	1.234	1.316	2.051
incap.	220	252	303	342	683
referrals	258	287	370	409	861
≤ 20 y	257	288	283	325	439
21 to 50 y	403	453	503	558	934
51+ y	335	390	448	433	786
fees per case	20,91	16,20	14,88	13,02	14,15
no. of GPs	115	316	133	65	6

the inefficient GPs is 75%. Also contained in table 1 are the target values of the reference technologies for each parameter used. The targets for the cost parameters are on average between 26% and 31% lower than the observed values implying potential savings of €2.9 mio for fees, €7.7 mio for drugs and €8.1 mio for hospital cost for the patients of the 635 GPs in the first quarter of 1999. A total of €18,7 mio could have been saved if all physicians had applied fully (cost) efficient treatment strategies. In addition, even if the fraction of patients incapacitated for work rose by 17%, i. e. if case mix became somewhat worse, the full savings potential could be realized.

Table 2 lists the means of the parameters by efficiency ratings of the respective GPs.[6] Note that the GPs with efficiency ratings of less than 0,65 treated on average 995 patients while GPs with an efficiency exceeding 0,65 had on average more than 1192 patients. Even though the most inefficient physicians have only 83,5% percent of the cases compared to the more efficient physicians the average variable fees inefficient GPs earn are higher in absolute terms than those of their more efficient colleagues. This results in a per capita variable fee that is on average 35% higher than that of the other GPs. This is despite the fact that variable fees decrease the more fees are claimed per case.

There is no sign that the patients of these GPs require more costly treat-

---

[6]Colums 3, 4, 5 and 6 refer to all cases not included in column 2 and 3, 4 and 5, respectively. The 6th column describes DMUs for which an efficiency value could not be determined because their structure is very different from that of the *other* observations. The GPs listet in column 6 treat a number of cases which is on average 80% higher than that of the other GPs. Remember that the standard model allows DMUs to be their own reference technology which is not the case with the superefficiency model.

ment than the patients of the more efficient GPs. Neither the number of cases incapacitated for work, nor the number of cases referred to specialists matches that of the more efficient GPs whereas the distribution in the age clusters is the same for all groups of GPs. These findings lend support to the hypothesis that GPs with only a small number of cases may try to compensate the comparatively low income they derive from per capita fees by providing an excessive amount of services to these few patients.

6 Conclusion

The above analysis showed that differences in the efficiency of service provision exist among Austrian GPs. These differences cannot be explained by the fact that some physicians do not provide all the services necessary for their patients and therefore make others appear cost inefficient. This is highly unlikely because under the current system of remuneration there is no incentive for GPs to withhold services or treatment from patients which they consider appropriate. Hence, the differences in efficiency observed must be attributed to the overprovision of services to patients by some physicians. This cannot be remedied with the low powered regulation which currently applies in Austria. It allows the GPs to appropriate information rents from the asymmetric distribution of the information on the true cost of the treatment.

DEA is proposed to resolve the dilemma of having to choose between a low powered regulation, which will ensure provision of services but will not control treatment cost effectively and a high powered regulation, which will control cost but may not ensure the provision of all necessary services. The problem of cost control is essentially one of asymmetric information. By providing a reference technology which is specifically taylored to each observation DEA allows for the maximum possible reduction of the information asymmetry and thus for the reduction of the information rent. We therefore conclude that if proper incentives were provided to the physicians it would be possible to reduce the cost for treatment without depriving patients of necessary services. The present analysis could be extended in various ways. First, it would be desirable to have information on the quality of treatment, even in form of a crude indicator such as morbidity. However, we argued above that there is no reason to suspect that some GPs provide less treatment than appropriate. Therefore, no positive correlation between quality and cost is to be expected. Second, information on more than just one single quarter in 1999 would be useful to check whether the efficiency results will be valid in the longer run or if the results derived are not stable over time. At last, it would be of interest to check the results for their statistical significance. This is possible with a bootstrap approach introduced by Simar and Wilson (2000), for applications of such an approach, see Staat (2001) as well as Staat (2000).

References

ANDERSEN, P. and PETERSEN, F. J. (1993): A procedure for ranking efficient units in Data Envelopment Analysis. *Management Science, Vol. 39, 1261-1264.* BOGETOFT, P. (2000): DEA and Activity Planning under Asymmetric Information. *Journal of Productivity Analysis, Vol. 13, 7-48.*

COOPER, W. W., SEIFORD, L. M. and TONE, K. (2000): Data Envelopment Analysis: A comprehensive Text with Models, Applications, References and DEA-Solver Software. Kluwer.

CHILINGERIAN, J. A. (1997): Exploring why some physicians' hospital practices are more efficient: taking DEA inside the hospital, in Charnes et al. (Eds.) (1997): Data Envelopment Analysis: Theory, Methodology and Applications, 3. Ed. 167 - 193.

HOLLINGSWORTH, B., DAWSON, P. J. and MANIADAKIS, N. (1999): Efficiency measurement of health care: a review of non-parametric methods and applications. *Health Care Management Science, Vol. 2, 161-172.* LAFFONT, J.-J. and TIROLE, J. (1999): A theory of incentives in procurement and regulation. Cambridge.

SHEPARD, R. W. (1970): Theory of Cost and Production Functions. Princeton, NJ.

SHLEIFER, A. (1985): A theory of yardstick competition. *Rand Journal of Economics, Vol. 16, 319-327.*

SIMAR, L. and WILSON, P. W. (2000): Statistical Inference in Nonparametric Frontier Models: The State of the Art. textsl Journal of Productivity Analysis, Vol. 13, 49-78.

STAAT, M. (2001): Bootstrapped estimates for a model for groups and hierarchies in DEA. *European Journal of Operational Research, forthcoming.*

STAAT, M. (2000): Der Krankenhausbetriebsvergleich: Benchmarking vs. Data Envekopment Analysis. *Zeitschrift für Betriebswirtschaft, Ergänzungsheft 4/2000, 123-140.*

Data Mining Tools for Quality Management in Health Care

J. Stausberg, Th. Albrecht

Institut für Medizinische Informatik, Biometrie und Epidemiologie,
Universitätsklinikum Essen, D-45131 Essen, Germany

Abstract: Data mining can be used for the detection of quality deficiencies in health care. Routine data are typically stored in relational databases, which are not easy to understand for end-users. OLAP-tools promise a more intuitive way of analysis and visualization. We evaluated the OLAP-tools from Oracle Corporation with data from our data warehouse, which is also realized in a common relational database management system. Our results reveal that remarkable efforts are necessary to transfer and to prepare data for the multidimensional OLAP-model. Meaningful analyses are not possible without this step. The tools rely on a business view neglecting the granularity and complexity of medical treatment. Once the tools are prepared they offer end-users flexible and graphical analyses in there own terms. We suggest a special adaptation of these tools, which should be tailored for hospital demands.

1 Introduction

Quality management in health care needs complete and accurate data about the provided services and the outcome in terms of the patients. Continuous quality management integrates this task into the routine medical documentation to assure topical analyses and to suffer from the high data quality reached through the multiple use of the same data. The concept of a data warehouse in parallel to the database used in daily practice is proposed as a mean to provide high flexibility for ad-hoc analyses without an increase in the reaction time of the clinical application. A data warehouse can be realized by a relational database management system as well, using reporting functions or assessing the data through an ODBC-connection. This requires some knowledge about the relational model and the implementation at hand, in which the data might be distributed between a lot of tables with non-speaking keys. As an intuitive alternative the multidimensional model for data storage is offered with the catchword online analytical processing (OLAP). The offered tools are dedicated for end-users, whether they are senior physicians, managers or directors. The question is, do standard OLAP products serve the complicated needs of hospital quality management with the ease-of-use expected by today's windows-accustomed users? To answer this question we tried out the features of an evaluation copy of Oracle© Express.

2 How OLAP Works: The Concept of Multidimensional Databases

Instead of organizing data in tables connected by relations, in a multidimensional database the data are organized by dimensions. The dimensions represent the criteria according to which the data are supposed to be analyzed. A dimension is basically a set of values, its "dimension values". Such values could be dates, surgeon IDs or types of operations. Data are assigned to these values. Suppose we have dimensions D_1 through D_n. A tuple $(x_1, ..., x_n) \in D_1$ x ... x D_n would be assigned to one value of the data that is supposed to be analyzed. For instance a date, surgeon ID, and type of operation could be assigned to the length of the operation of that type which the given surgeon performed on that day. But if the surgeon performed more than one operation of one type on one day, only one value can be assigned, which can of course be the average or total of the several values. The assignment of the tuples of dimension values to data is organized by variables. A variable is defined by a name, a data type, and by the dimensions to whose values it assigns data. So in our example, a variable of the type integer, named "Length_of_Operation", dimensioned by date, surgeon ID, and type of operation, would assign an integer (representing the length of the operation in minutes) to each 3-tuple of the three dimensions. So typically, for each column in a relational database, one variable is needed.

When importing from a relational database or a table, the question comes up for each column, whether it should be used as a variable or a dimension. In Express-terminology, a table is dimensioned by the row numbers $\{1, 2, ...\}$ and the column letters $\{A, B, ...\}$. Each pair (e.g. $(C, 5)$) is assigned to one cell. Of course dimensions like that would be of no use. So when importing to Express, certain suitable columns are used as dimensions, i.e. their values are used as dimension values, and the columns that contain the data that is supposed to be analyzed, like duration, revenue etc., are used as the variables, i.e. their values are assigned to the matching dimension values, according to their position in the table. The way in which data are organized in a multidimensional database is very important: unlike in a relational database, where the data can be sorted by any column, the multidimensional data are always sorted by the dimensions, never by the variables. Calculations and aggregations are always performed on the variables' values, never on the dimension values. If the combination of dimensions for one variable permits ambiguity, only one value is imported, resulting in the loss of data and thereby inaccurate analysis. Also, the dimensionality of a variable can't be changed, so it is impossible to add more detail to the data without starting all over by importing it with more detail. On the other hand, too much detail inhibits performance. So it is necessary to know the exact goal of the analysis and to carefully plan the import of the data based on that goal.

In Express, it is possible to define relations between dimensions. In a relation, each value of a dimension, or a tuple of dimension values from several dimensions, is assigned to that of another dimension. A very similar object is the hierarchy, a relation between the different dimension values of one dimension, building a hierarchical structure in the dimension. All these relations are used for aggregation, so that in the assignment of data, one dimension is replaced by the related dimension, and the assigned values are aggregated accordingly.

3 Methods and Material

The data used for the evaluation were collected within the computer-based patient record of the Surgical Center II at the Medical Faculty in Essen, which is in routine use since 1989 (Albrecht et al. (1997)). The medical faculty is a teaching hospital with 1219 beds. The Surgical Center II covers 231 beds, 15 wards, 11 operating rooms and 3 clinics for outpatient care. It comprises functions such as patient admission and discharge, on-line ordering of diagnostic procedures and documentation of clinical data. Covered are the departments of general surgery, neurosurgery, trauma surgery, anesthesiology and intensive care and the blood bank. The computer-based patient record includes basic information on all patients, who stayed in the Medical Faculty since 1989, e.g. name, date of birth, insurance company, insurance ID. Between the laboratory management system and the surgical information system order-entry and result reporting are carried out. The direct input of patient data is routinely used concerning the documentation of surgical procedures for example, which is performed by the surgeons themselves. That documentation is part of the supported care process of surgical patients including operation requests, scheduling, documentation and retrieval. The operative data include performed procedures, diagnoses, complications, times, staff members and so on. In 1997 a data warehouse was brought into routine, which includes a subset of the data from the host (Stausberg et al. (1998)). Every month new data are exported from the computer-based patient record and transferred into the data warehouse. The data warehouse is realized with Microsoft© Access 97. Statistical analyses are performed on a regular basis with different indicators about the structure, processes and outcome of treatment provided by the surgical departments. From this data warehouse, we extracted data concerning the medical operations performed by the department of neurosurgery. The data comprises about 6400 patients and 8000 operations.

The backbone of the Express environment is the Oracle Express Server, the program running the main processes on a server (Oracle Corporation (1999b)). All data reside and are calculated on the server. This server was implemented on a workstation IBM RS/6000 model F50. Several client applications provide the user interface for all actions performed by the server. The client applications were installed on an IBM-compatible personal computer with Microsoft Windows NT 4.0 as operating system. The most important tool is the Oracle *Express Administrator* for the creation and maintenance of databases. A couple of different applications are used to analyze the data: the *Express Analyzer* is the main tool provided by Oracle for analysis of the data, like creating graphs and tables. The *Express Web Publisher* lets the user publish graphs and tables representing Express data on the world wide web, and the *Express Spreadsheet Add-in* provides a link to the Microsoft Office product family by means of an Excel plugin, allowing the exchange of data between Excel and Express (Oracle Corporation (a)). Oracle also provided us with the *Financial Analyzer*, an application that replaces Express Administrator and Express Analyzer in the analysis of financial data (Oracle Corporation (c), Oracle Corporation (d)).

4 Results

Express offers several ways of importing data (Oracle Corporation (b)). The first is setting up an SQL-connection to the source database. However, because our instance of Express Server was running on a Unix workstation, while the Access database was of course running on a Microsoft Windows, we chose to use the second possibility, importing text files. The text files were exported from Access, the columns separated by an assigned character, and then opened by the Express Administrator import utility. Once the columns were correctly recognized by the program, which actually proved to be quite difficult to achieve, dimensions, variables and relations could be mapped with the different columns, and then imported into the mapped objects. Usually, the imported data are not yet suitable for analysis. They have to be aggregated and sometimes various calculations have to be performed. Unfortunately, the possibilities to perform these tasks through the graphical user interface are very limited. Using Express Administrator, only one kind of aggregation is possible without resorting to the command line prompt for the Express Language: the rollup. The rollup is an aggregation based on the hierarchy of a dimension. The sum (never the average or anything else), of the children values are assigned to the parents in the hierarchy tree.

Express Analyzer offers more possibilities. In a table or graph, a right mouse click on a dimension name reveals several different ways of aggregating the data. This feature works only on a limited number of values, so it proved to be useless for aggregating the amount of data we were working with. In order to aggregate our data, we had to use parts of the Express Language (Oracle Corporation (1999a)). An aggregation command in the Express Language consists of the type of aggregation, the variable to be aggregated and, if any, the dimensions by which the result is supposed to be dimensioned. Another classes of functions are the mathematical operations, allowing calculations on a cell by cell basis. The commands can be saved in a formula, an Express object that calculates the data whenever it is needed. Another possibility is to save the results in a variable, which becomes necessary if the calculations take too long. But saving calculations in variables requires an even deeper look into the 500-page Express Language Programming Guide.

Express Administrator offers a tool to schedule and run tasks like data import or calculations in the background. Because a lot of these tasks take a long time and, when executed in the foreground, freeze the application as long as they are processed, this is an important tool to work effectively with Express by continuing to work while the calculations are being performed by the server, or running tasks overnight.

When the data has been imported and treated for analysis, the Oracle Express Analyzer proves to be an easy and comprehensive tool to display the data in tables and graphs. With a right mouse click on a variable or formula, a table or graph is automatically created and then a variety of context-menus and dialog boxes allow the user to select the data which is to be displayed, add different measures, and format the graph or table, all at the click of a mouse button.

Another important tool for the analysis of the data is the Express Spreadsheet Add-In. It adds a menu item to Microsoft Excel which allows the easy import of Express data and the use of a lot of the important features of Express, like drilling down through a hierarchy, from within Excel. As it is also possible to directly transfer the data from Excel to Express via the Spreadsheet Add-In, the whole Microsoft Office product family, which many users are familiar with, can be incorporated into the analysis.

5 Discussion and Conclusion

So do hospitals everywhere need to use specific tools for data mining in quality management? Once the data are organized into measures that represent the data the user wants to analyze, it is easy to analyze and display the data in the Express Analyzer. The analyses can be automatically updated and presented on a regular basis. Unfortunately, the documentation, as well as the software itself, seemed too focused on business management. All examples dealt with the analysis of how much of what was sold where. Most features, like the rollup, and the whole Financial Analyzer, seem tailored only to accustom this kind of analysis. The tools are not yet able to capture the granularity and complexity of medical treatment. Therefor the underlying information has to be aggregated before the analysis with Express.

The problem is organizing it that way. First, the data have to be imported. It is not possible to just import text files from MS Access or Excel, because either the date format doesn't match or the characters separating one column from another are interpreted differently. As described earlier, it is necessary to import variables and dimensions exactly according to ones needs. So it is not possible to just collect data and then worry about the analysis later. One has to think a lot more about what is actually supposed to be analyzed in what way, which columns should be used as dimensions which ones as variables and which ones should be omitted. This contradicts the image of "data mining", mining around in the multidimensional data maybe finding something interesting. When building an Express database, one rather has to think of the mine first, and then build the mountain around it. Express still relies on the Express Language and has not incorporated enough functions into the graphical user interface. Since the data must not be ambiguous, it is usually necessary to import extremely detailed data (e.g. for each and every patient) and later aggregate the data (e.g. the average for all patients). So aggregation has to be used all the time, and at least important functions like this should be accessible through the graphical user interface rather than a command line prompt.

The other product, Oracle Financial Analyzer, was not at all suitable for our purposes. It offers a couple of useful features, like advanced features for data distribution, more spreadsheet-like tables and some forecast functions integrated into the graphical user interface. On the other hand, it has some major limitations: it is impossible to use numbers or days as dimension values, only text and time dimensions with weeks as the most detailed unit are allowed. Variable values can only be numbers. Also, it does not offer a database browser, like the Administrator.

In conclusion Oracle Express and Oracle Financial Analyzer are powerful tools to support an intuitive analyses of business data. They require a carefully configuration through a specialist before providing it to end-users. The domain of health care and the task of quality management are not adequately supported yet. We suggest a special adaptation for hospitals, which should be tailored for the granularity and complexity of medical treatment and could provide a fixed set of views on the data.

6 Acknowledgement

We thank Oracle Corporation for providing us with an evaluation copy of Oracle Express and Oracle Financial Analyzer.

References

ALBRECHT, K., STAUSBERG, J. and EIGLER, F. W. (1997): Five-Year Experience with a Hospital Information System, in Dohrmann, Henne-Bruns, Kremer (eds.): Surgical Efficiacy and Economy (SEE). Proceedings of the 3rd World-Conference, Thieme, Stuttgart, 202.

ORACLE CORPORATION (1999a): Oracle© Express Server. Express Language Programming Guide. 6.2.0.2, March 1999, Part No. 68642-01.

ORACLE CORPORATION (1999b): Oracle© Express Server. Installation and Configuration Guide. Release 6.2.0.2 for Unix, March 1999, Part No. A68169-01.

ORACLE CORPORATION (a): Oracle© Express Spreadsheet Add-In. Release 2.3 User's Guide, Part No. A53721-03.

ORACLE CORPORATION (b): Oracle© Express. Release 6.2 Database Administration Guide, Part No. A59962-01.

ORACLE CORPORATION (c): Oracle© Financial Analyzer. Release 6.2. Installation and Upgrade Guide, Part No. A61298-02.

ORACLE CORPORATION (d): Oracle© Financial Analyzer. Release 6.2. User's Guide, Part No. A61297-01.

STAUSBERG, J., KOLKE, O. and ALBRECHT, K. (1998): Using a Computer-based Patient Record for Quality Management in Surgery, in Cesnik, McCray, Scherrer (eds.): MedInfo 98, IOS, Amsterdam, 80-4.

The Effects of Simultaneous Misclassification on the Attributable Risk

C. Vogel, O. Gefeller

Department of Medical Informatics, Biometry and Epidemiology,
University of Erlangen–Nuremberg, D–91054 Erlangen, Germany

Abstract: Misclassification affects the estimation of measures of association in epidemiologic studies. The paper investigates the situation of non–differential *dependent* misclassification of exposure and disease with respect to its effects on the attributable risk. In addition, the effects on the attributable risk and on the relative risk are compared. Simple situations are presented to illustrate the various distortions of the attributable risk including overestimation, underestimation and cross–over bias, the differences to the effects on the relative risk in these situations are discussed.

1 Introduction

Misclassification affects the estimation of measures of association in epidemiologic studies. Inspired by papers of the last decade considering the effects of non–differential *dependent* misclassification on the relative risk (Brenner et al. (1993), Chavance et al. (1992) and Kristensen (1992)), this paper focuses on the corresponding effects on the attributable risk. After briefly outlining the basic concepts of attributable risk and misclassification using a matrix–based approach, the effects of non–differential misclassification on the attributable risk are discussed. First, the effects of non–differential exposure misclassification *or* non–differential disease misclassification and the effects of non–differential *independent* misclassification of exposure *and* disease – a product of the marginal effects – are presented. Then, the effects of non–differential *dependent* misclassification are derived on the basis of the effects of non–differential *independent* misclassification. Four simple examples illustrate the diversity of effects in the situation of non–differential dependent misclassification, including overestimation, underestimation and cross–over bias. Finally, the results are compared to the corresponding results on the relative risk, followed by an illustration of the differences in two of the examples.

2 Basic Concepts

2.1 Attributable Risk

Levin (1953) introduced the attributable risk which represents the fraction of disease that might be avoided if the entire population contracted disease at the rate of those unexposed to a risk factor. His original definition is equivalent to

$$AR = \frac{P(D=1) - P(D=1|E=0)}{P(D=1)} ,$$

where $P(D=1)$ and $P(D=1|E=0)$ represent the probabilities of disease among the entire population and the unexposed, respectively. Note that the applicability of risk attribution is restricted to situations in which the exposure is associated with an increased risk of disease, i.e. the relative risk exceeds 1. In a public health context the importance of a risk factor depends not only on the relative risk but also on the percentage of the population exposed to this factor: the attributable risk is a function of both the relative risk and the probability of exposure:

$$AR = \frac{P(E=1) \cdot (RR-1)}{P(E=1) \cdot (RR-1) + 1} . \tag{1}$$

2.2 Misclassification

In order to define the concepts of non–differentiality and independence – common but not always suitable assumptions in the context of misclassification – let the variables E' and D' indicate the observed exposure and disease status, respectively.

Misclassification of exposure is called *non–differential* if it is independent of true disease status, i.e. for every $i, j, k \in \{0,1\}$:

$$P(E' = i|E = j \wedge D = k) = P(E' = i|E = j) .$$

Similarly misclassification of disease is called *non–differential* if it is independent of true exposure status, i.e. for every $i, j, k \in \{0,1\}$:

$$P(D' = i|D = j \wedge E = k) = P(D' = i|D = j) .$$

Misclassification of exposure and disease is called *independent* if observed exposure and disease variables are stochastically independent given true exposure and disease variables, that is, if and only if for every $i, j, k, l \in \{0,1\}$ we have:

$$P(D' = i \wedge E' = j|D = k \wedge E = l) =$$
$$P(D' = i|D = k \wedge E = l) \cdot P(E' = j|D = k \wedge E = l) .$$

Otherwise it is called *dependent*.

The observed state probabilities $p'_{ij} := P(D' = i \wedge E' = j)$, or – in other words – the probabilities that the observed disease status is i and the observed exposure status is j, can always be expressed as a homogeneous linear function of the true state probabilities $p_{ij} := P(D = i \wedge E = j)$. It is given by $\vec{p'} = M \cdot \vec{p}$, where $\vec{p} := (p_{11}, p_{10}, p_{01}, p_{00})^T$ and $\vec{p'} := (p'_{11}, p'_{10}, p'_{01}, p'_{00})^T$ and M is the misclassification matrix.

In the following sensitivity Se_E and specificity Sp_E of exposure classification denote the probabilities for correct classification of exposure status among the exposed and the unexposed, respectively. Analogously, sensitivity Se_D and specificity Sp_D of disease classification are defined.

For non–differential exposure / disease misclassification the exposure / disease misclassification matrix M_E / M_D is:

$$M_E := \begin{pmatrix} Se_E & 1 - Sp_E & 0 & 0 \\ 1 - Se_E & Sp_E & 0 & 0 \\ 0 & 0 & Se_E & 1 - Sp_E \\ 0 & 0 & 1 - Se_E & Sp_E \end{pmatrix},$$

$$M_D := \begin{pmatrix} Se_D & 0 & 1 - Sp_D & 0 \\ 0 & Se_D & 0 & 1 - Sp_D \\ 1 - Se_D & 0 & Sp_D & 0 \\ 0 & 1 - Se_D & 0 & Sp_D \end{pmatrix}.$$

The multiplicative structure of non–differential independent misclassification is reflected in the corresponding misclassification matrix, which is the commutative product $M_E \cdot M_D$. The misclassification matrix for non–differential dependent misclassification of exposure and disease is $M_E \cdot M_D + COV$, which has the additional component

$$COV := \begin{pmatrix} Cov_{11} & -Cov_{10} & -Cov_{01} & Cov_{00} \\ -Cov_{11} & Cov_{10} & Cov_{01} & -Cov_{00} \\ -Cov_{11} & Cov_{10} & Cov_{01} & -Cov_{00} \\ Cov_{11} & -Cov_{10} & -Cov_{01} & Cov_{00} \end{pmatrix},$$

incorporating the information about the structure of dependence of classification errors. The parameter

$$\begin{aligned} Cov_{ij} := \ & P(D' = i \wedge E' = j | D = i \wedge E = j) \\ & - P(D' = i | D = i) \cdot P(E' = j | E = j) \end{aligned}$$

describes the group–specific dependence of classification errors in the group with true disease status i and true exposure status j. The maximal possible values have been calculated by Brenner et al. (1993). Note that – in contrast to our definition – in their paper the covariance Cov_{ij} describes the group–specific dependence of classification errors in the group with true *disease* status j and true *exposure* status i.

3   Effects on the Attributable Risk

3.1   Effects of Non–differential / Non–differential Independent Misclassification on the Attributable Risk

The effects of non–differential misclassification – under the assumption that both classifications are better than random – presented by Vogel et al. (2001) can be summarized as follows: (i) under non–differential exposure misclassification the attributable risk is unbiased if the sensitivity is perfect, otherwise it is underestimated, (ii) under non–differential disease misclassification the attributable risk is unbiased if the specificity is perfect, otherwise it is underestimated, and – as a consequence of the multiplicative structure – (iii) under non–differential independent exposure and disease misclassification the attributable risk is unbiased if the sensitivity of exposure and the specificity of disease classification are both perfect, otherwise it is underestimated.

3.2   Effects of Non–differential Dependent Misclassification on the Attributable Risk

The observed attributable risk under non–differential dependent misclassification can be additively decomposed into two components: the observed attributable risk under non–differential independent misclassification AR'_{ind} and a correction term:

$$AR' = AR'_{ind} + \frac{x}{p'_{1.} \cdot p'_{.0}}, \tag{2}$$

where $x := \sum_{i,j=0}^{1}(-1)^{i+j} \cdot Cov_{ij} \cdot p_{ij}$. Note that the components of the denominator of the correction term in equation (2) are not affected by dependence, so that only the numerator of the correction term x varies under dependence. The decomposition (2) can be used to analyse the effects of non–differential *dependent* misclassification on the basis of the effects of non–differential *independent* misclassification.

4   Examples

In this section simple examples are discussed showing the diversity of effects for non–differential dependent misclassification. Our goal is to analyse situations in which errors in exposure and disease classification tend to occur together, which is reflected in positive covariances Cov_{ij}.

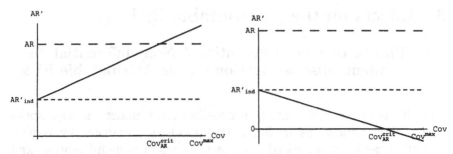

Figure 1: AR' is linearly increasing (situations 1 and 2) or decreasing (situations 3 and 4) in the corresponding covariance.

Following the terminology of Chavance et al. (1992), disease is called *overreported / underreported* if the disease status is correctly classified at least for all diseased / non–diseased subjects, which is equivalent to $Se_D = 1$ / $Sp_D = 1$. Overreporting / underreporting of exposure is defined analogously. Remember that some epidemiologists would give a less extreme definition of over- and underreporting. The following simple situations are to be analysed:

1. Overreporting of exposure and disease

2. Underreporting of exposure and disease

3. Overreporting of exposure, underreporting of disease

4. Underreporting of exposure, overreporting of disease

In all these situation there is only one exposure status ℓ and one disease status k, which might be misclassified. Obviously only the covariance $Cov_{k\ell}$ can take values different from 0. Thus, the observed attributable risk given by equation (2) is a linear function in $Cov_{k\ell}$, linearly increasing in $Cov_{k\ell}$ if $k = \ell$ (situations 1 and 2), otherwise it is decreasing in $Cov_{k\ell}$ (situations 3 and 4). Figure 1 illustrates the observed attributable risk AR' as a linear function of the corresponding covariance. In the case of linear increase there is a critical covariance $Cov_{k\ell}^{crit}$, leading to an unbiased attributable risk. Whenever the covariance $Cov_{k\ell}$ exceeds $Cov_{k\ell}^{crit}$, the attributable risk is overestimated. In the case of linear decrease for a critical covariance $Cov_{k\ell}^{crit}$, the observed attributable risk is 0, larger values of $Cov_{k\ell}$ lead to cross–over bias. The critical covariances for the situations 1–4 defined above can be represented as a function of the relative risk, the probability of exposure and the probability of disease in the unexposed:

$$Cov_{00}^{crit} = \frac{P(E = 1) \cdot (RR - 1)}{(P(E = 1)\,(RR - 1) + 1)\,(1 - P(D = 1|E = 0))} (1 - Sp_D) \cdot Sp_E,$$

$$Cov_{11}^{crit} = \frac{RR-1}{RR} \cdot Se_D \cdot (1 - Se_E) \,,$$

$$Cov_{10}^{crit} = P(E=1) \cdot (RR-1) \cdot Se_D \cdot Sp_E \,,$$

$$Cov_{01}^{crit} = \frac{P(D=1|E=0) \cdot (RR-1) \cdot (1 - P(E=1))}{1 - P(D=1|E=0) \cdot RR} \cdot Sp_D \cdot Se_E \,,$$

which are smaller than the corresponding maximum possible values of the covariances $Cov_{k\ell}^{max}$ if and only if

$$\frac{P(E=1) \cdot (RR-1)}{(P(E=1) \cdot (RR-1) + 1) \cdot (1 - P(D=1|E=0))} < \frac{Sp_D \cdot (1 - Sp_E)}{Sp_E \cdot (1 - Sp_D)} \,,$$

$$\frac{RR-1}{RR} < \frac{Se_E \cdot (1 - Se_D)}{Se_D \cdot (1 - Se_E)} \,,$$

$$P(E=1) \cdot (RR-1) < \text{Min} \left[\frac{1 - Se_D}{Se_D}, \frac{1 - Sp_E}{Sp_E} \right] \,,$$

$$\frac{P(D=1|E=0) \cdot (RR-1) \cdot (1 - P(E=1))}{1 - P(D=1|E=0) \cdot RR} < \text{Min} \left[\frac{1 - Sp_D}{Sp_D}, \frac{1 - Se_E}{Se_E} \right] \,,$$

respectively.

These conditions for the presence of overestimation or cross–over bias can be discussed in more detail. For example, if exposure and disease are both underreported (situation 2) the critical covariance Cov_{11}^{crit} only depends on the relative risk and the sensitivities of exposure and disease classification. Since $\frac{RR-1}{RR}$ is always smaller than 1, the critical covariance is smaller than the maximum possible covariance, whenever Se_E is greater or equal to Se_D. Otherwise, overestimation is more likely for a small relative risk, small values of Se_D and large values of Se_E.

5 Comparison with the Effects on the Relative Risk

Brenner et al. (1993) described the effects of non–differential dependent misclassification on the relative risk: overestimation, underestimation and cross–over bias. Therefore, the following question arises: are there

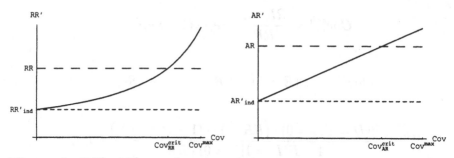

Figure 2: RR', AR' as a function of the corresponding covariance for situations 1 and 2

qualitative differences in the effects on the attributable risk and the relative risk?

Due to the functional relationship between the attributable risk and the relative risk (see (1)) cross–over bias occurs in the same situations; thus, there are no differences in situations 3 and 4.

In contrast, the situations for overestimation do not correspond: a decomposition of the observed relative risk analogously to the decomposition of the observed attributable risk (2) shows that the observed relative risk is also increasing in x. Consequently, both measures are underestimated for small values of x, unbiased for possibly different critical values of x and overestimated for large values of x. Suppose now that RR is unbiased. As a consequence of equation (1), bias in AR can only result from bias in $P(E = 1)$ in this situation. AR is overestimated if and only if $P(E = 1)$ is overestimated, which is equivalent to:

$$P(E = 1) < \frac{1 - Sp_E}{2 - Sp_E - Se_E} . \tag{3}$$

If inequation (3) holds the attributable risk is overestimated for smaller values of x than the relative risk. If $P(E = 1)$ is unbiased, that is if and only if $P(E = 1) = \frac{1-Sp_E}{2-Sp_E-Se_E}$, the critical values of x correspond. If $P(E = 1)$ is underestimated the relative risk is overestimated for smaller values of x.

In the situation of overreporting of exposure and disease (situation 1) the condition (3) is $P(E = 1) < 1$, which is always true. In contrast, in the situation of underreporting of exposure and disease (situation 2) condition (3) is $P(E = 1) < 0$, which is never true. This means that the overestimation of AR starts for smaller / larger values of Cov_{00} / Cov_{11} than the overestimation of RR in situation 1 / 2 , respectively. Figure 2 shows the observed relative and attributable risk for such situations as a function of the corresponding covariance.

6 Discussion

This paper demonstrates that the common assertion that epidemiologic measures are always biased toward the null under non–differential misclassification is not true, when classification errors for exposure and disease are dependent. Overestimation, underestimation and cross–over bias are possible effects on the attributable risk. In two of the examples (situation 1 and 2) starting with small values of the corresponding covariance underestimation is less strong than under non–differential independent misclassification, but larger values can lead to overestimation getting stronger with larger values of the covariance. In the two other examples (situation 3 and 4) underestimation first gets stronger with increasing values of the covariance, sometimes even turns into cross–over bias for larger values of the covariance. While there are no qualitative differences in the kind of bias for the attributable and the relative risk in situations 3 and 4, overestimation of the attributable risk starts for smaller / larger values of the covariance in situation 1 / 2 than overestimation of the relative risk, with the consequence that a separate analysis is necessary for the different risk measures.

The main goal of this paper was to describe the effects on the attributable risk qualitatively, a quantitative investigation of the effects is straightforward: for situations 1 and 2, for example, the bias of AR for the maximal possible covariance $Bias_{max}$ can be calculated, if the maximum possible and the critical value of the covariance, and the bias of AR under non–differential independent misclassification $Bias_{ind}$ are given:

$$\frac{Bias_{ind}}{Bias_{ind} - Bias_{max}} = \frac{Cov^{crit}}{Cov^{max}} .$$

In contrast, a quantitative analysis is more complex for the relative risk because the increase is not linear.

In practice, the epidemiologist has to check, whether the assumption of independent misclassification is realistic. In the case where this assumption is violated a careful analysis of the misclassification structure along the lines of this paper is necessary.

Acknowledgements

This work has been supported by a grant of the Deutsche Forschungsgemeinschaft (grant no. Ge 637/4-1).

References

BRENNER, H., SAVITZ, D.A. and GEFELLER, 0. (1993): The effects of joint misclassification of exposure and disease on epidemiologic measures of association. *Journal of Clinical Epidemiology, 46, 1195–1202.*

CHAVANCE, M., DELLATOLAS, G. and LELLOUCH, J. (1992): Correlated nondifferential misclassification of disease and exposure: Application to a cross-sectional study of the relation between handeness and immune disorders. *International Journal of Epidemiology, 21, 537–546.*

KRISTENSEN, P.(1992): Bias from nondifferential but dependent misclassification of exposure and outcome. *Epidemiology, 3, 210–215.*

LEVIN, M. L. (1953): The occurrence of lung cancer in man. *Acta Unio Internationalis Contra Cancrum, 9, 531-541.*

VOGEL, C., LAND, M. and GEFELLER, 0. (2001): Effects of independent non–differential misclassification on the attributable risk, in Gaul, W., Ritter, G. (Eds.): Classification, Automation and New Media, Springer, Berlin (in press).

Statistical Genetics - Present and Future

A. Ziegler, O. Hartmann, I. R. König, H. Schäfer

Institut für Medizinische Biometrie und Epidemiologie,
Philipps-Universität Marburg, 35033 Marburg, Germany

Abstract: One major task of statistical genetics is the development of efficient methods to identify disease predisposing genes. While genetic mapping and positional cloning of Mendelian disorders can be considered routine nowadays, this systematic approach was successful only for one complex disorder, type II diabetes mellitus. It might hence be questioned whether this route should still be followed for unraveling the genetic background of a disease. One answer to this has been the development of new molecular biological techniques. In this paper, we sketch the recently developed techniques that might be alternatives for the identification of disease susceptibility genes and their function. We illustrate the absolute need for new statistical methods by means of two areas of current research, genome wide association analysis and gene expression arrays. Here, developments are required in study design, quality control and statistical data analysis. We conclude that these manifold challenges can only be accomplished in an interdisciplinary team.

1 Introduction

One major task of statistical genetics is the development of efficient methods for the identification and functional understanding of disease predisposing genes. The methods can be considered routine for disorders following a clear-cut Mendelian mode of inheritance. However, the identification of disease genes is still a laborious business even in these simple situations (Collins (1995)): As a first step, a few families - or even only one - with multiple affected individuals are used to test whether the disease segregates together with specific genetic markers. The latter have known unique chromosomal positions and are spread over the whole genome. If the chromosomal region containing the susceptibility gene is thus mapped, it may be narrowed down in a second step using e.g. association analysis (see section 2). Up to the present, the third step is the sequencing of the fine-mapped chromosomal region allowing the identification of putative disease genes. In the final step, mutations are sought in these candidate genes. This whole procedure is termed positional cloning and has been successfully applied to a series of disorders including cystic fibrosis (OMIM *602421) and Duchenne muscular dystrophy (OMIM *310200).

As an alternative, possible candidate genes might directly arise from genetic databases after the first step, the genetic mapping. These genes might then be screened for functionally relevant mutations. This approach has been successfully applied e.g. to hereditary non-polyposis colon cancer (OMIM *120435). It will be the standard method in a few years because positional cloning will be rendered superfluous by the complete sequence of the human genome.

While these systematic mapping approaches have been successful for many simple Mendelian disorders, the results sound frustrating for complex diseases. Here, positional cloning has so far been successful in only one case: It has been shown that the gene encoding calpain-10 is associated with type II diabetes mellitus (Horikawa et al. (2000)). Other strategies have been responsible for the identification of all other susceptibility genes for complex diseases. As an example, the association between multiple sclerosis and CD45 (OMIM #126200) has been detected by functional cloning, where the fundamental information about the basic biochemical defect is used without reference to a chromosomal position. As a consequence, Schäffer (1998) concluded that diseases might be too complex for us to find a full explanation with current methods.

Currently, researchers pin their hope on new molecular biological techniques for the mapping of complex disease genes. These techniques might be fully exploited together with the findings from the human genome project. In the following, two methodological developments based on chip technology are sketched in addition to their resulting statistical requirements. The first is the analysis of association by using hundreds of thousands of diallelic DNA markers, termed single nucleotide polymorphisms (SNPs). The second are gene expression studies based on microarray technology using mRNA samples.

2 Association Studies Using Single Nucleotide Polymorphisms

The use of SNPs will probably accelerate the identification of disease susceptibility genes by allowing researchers to look for associations between a disease and SNPs in a population. Allelic association is present if a specific allele of a genetic (e.g. SNP) marker occurs more frequently in affected than in unaffected individuals of a population. This might be due to direct causation, where the allele itself is functionally relevant and directly affects the expression of the phenotype. Alternatively, linkage disequilibrium may be present where the allele is found with an allele at another locus that directly affects the expression of the phenotype. This approach differs from the classical linkage method described in the previous section where pedigrees are used to track transmission of a disease

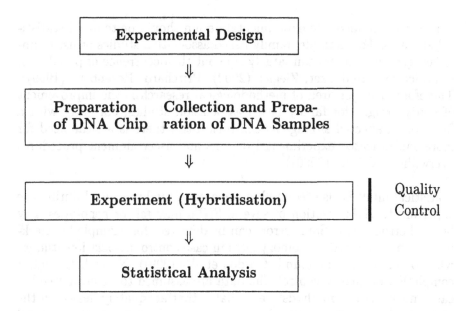

Figure 1: Flow Chart of a Single Nucleotide Polymorphism Association Study

through a family. Obviously, it is much easier to obtain DNA samples from a random set of individuals in a population than from every member of a family over several generations.

With the new chip technologies, these DNA samples are hybridised on a genotyping microarray. Automation is established by robots that distribute the appropriate reagents onto each microarray. Multiple microarray plates can be processed simultaneously providing very high-throughput. Currently, more than 1.42 million SNPs distributed throughout the human genome are known, providing an average density of one SNP every 1.9 kilobases (The International SNP Map Working Group (2001)). Between 10,000 and 100,000 SNPs spread over the genome are usually hybridised in a single experiment nowadays.

The success of these studies depends, however, on the appropriate planning, realization and statistical analysis of the respective experiments. Figure 1 illustrates the different steps of a SNP association study. First of all, statistical input is required to set-up an adequate study design. As a major problem, allelic association may not only occur because of direct causation or linkage disequilibrium but could result from statistical artifact (Horvath, Baur (2000); Strachan, Read (1999)). Thus, the finding could be due to chance, confounded by genetic heterogeneity of the population (population stratification) or selection bias.

One aim of an adequate experimental design should be to avoid statistical artifacts. For example, family based association studies or use of unlinked genetic markers can largely rule out the occurrence of population stratification (Böddeker, Ziegler (2000); Pritchard, Rosenberg (1999)). Therefore, a major area of methodological research is the improvement of study designs for family based association studies, currently with a focus on unaffected siblings as controls. Still, there is a clear need for more sophisticated experimental designs and ascertainment procedures (Terwilliger, Göring (2000)).

Statistical input is also required in the area of quality control during the experiment: Hybridisation of several SNPs may fail or genotypes may be read erroneously. Some errors can be detected, for example by checking for an excess rate of homozygotes in case control association studies, while others cannot be found (Gordon et al. (1999)). Nevertheless, after completion of quality control, the data are assumed to be error free because no statistical methods are available that adequately deal with the undetected genotype errors. It would be even better if quality control were not required in the first place since it is time consuming and expensive. This is especially true in family based studies where checks for Mendelian incompatibilities and double recombinants are usually carried out.

Finally, new statistical methods are required for the analysis of association studies with SNP data because new experimental designs also call for new analysis methods. Here, special consideration should be given to the analysis of masses of data as they arise from SNP experiments. Furthermore, although freely available software packages for statistical genetics exist, for some of them the implemented algorithms and test statistics are not clearly described. In addition, in contrast to conventional statistical programs, most statistical genetic packages do not offer an "all in one"solution but are merely developed for specific applications.

3 Gene Expression Studies Using Comparative Hybridisation

The second methodology based on chip technology are global gene expression studies. The motivation of these studies may be summarized as follows: The behavior of a cell is believed to be controlled by the protein profile, and the protein profile is assumed to correlate with the mRNA profile. While the development of a protein chip is still underway, the global measurement of mRNA is possible and will be routine in a few years. Here, one simultaneously measures the expression profile for thousands of genes for a given mRNA sample from a specific cell type. For

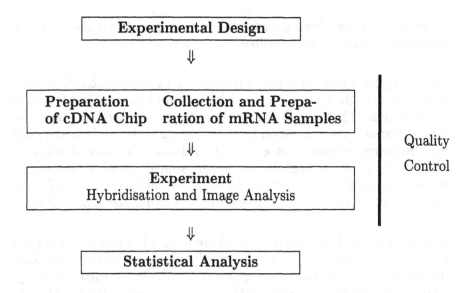

Figure 2: Flow Chart of a Gene Expression Study

the measurement of gene expression profiles, different technologies are used. For example, oligonucleotide arrays measure the absolute amount of mRNA from one sample on a chip. Other techniques like cDNA microarrays directly compare two samples on one chip by using different labeling dyes. Methods also differ in how the expression level is measured.

While the appropriate statistical methods for experiments may vary in detail, the basic methodological problems are generally very similar. What makes microarray experiments so different from conventional methods is the fact that the resulting data structure usually consists of thousands of measured expressed genes on regularly not more then tens or hundreds of different varieties. At the same time measurements underly a number of sources for error paired with confounding factors that add to systematic error: For comparative cDNA microarrays, measured expression heavily depends on fixed factors including dye, chip, and scanner intensity. The quality of the experiment also depends on a series of factors, like mRNA preparation, the quality of the spot, and the spotting process, just to name a few. Finally, not all genes are expected to act independently. Instead, it is very likely that genes with similar function will show similar expression patterns, resulting in multicollinearity between genes.

The methodological problems outlined above clearly point out the absolute need for new statistical methods. Although these heavily depend on the objective of the study, they share some basic properties. Figure

2 depicts the areas where statistical input is required to make successful hybridisation experiments feasible.

Naturally, the success of gene expression studies depends on the experimental design. Here, additional problems need to be considered prior to hybridisation. Thus, the design of the chip may be used to control for random variation in expression levels due to e.g. dust, while at the same time reducing measurement error. In addition, the kind of reference needs to be selected carefully and depends on the applied technology. However, only little has been published about adequate designs for microarray experiments so far (e.g. Kerr et al. (2001), Lee et al. (2000)).

Statistical methods for quality control are required at two stages (Figure 2): a) the production and preparation of the microarray chip, including the setup of the microarray unit, and b) after hybridisation and image analysis to eliminate spurious data. Stage a) is necessary because the protocols for chip production involve numerous steps that might be prone to production errors. While total production failure may be detected easily, a more refined setup of the microarray unit is required for optimal reliability (Hegde et al. (2000)).

Some simple quality control measures at stage b) are already provided by some of the image analysis software. Thus, the size of the spot or the total signal intensity can be used for automated data verification. However, these analyses are rarely performed in practice. Furthermore, data adjustments are necessary to account for systematic error in order to ensure comparability between chips. This normalisation procedure should eliminate differences in labeling and detection efficiency as well as chip quality. Simple methods adjust the total amount of measured intensities between dyes and chips. Others use so-called house keeping genes or control spots. Sometimes linear regression models are applied. Yang et al. (2000) recently proposed to use a locally weighted regression scatter plot smoothing. However, all these normalisation approaches underly some specific model assumptions.

The final part of a gene expression study, depicted in Figure 2, is the statistical analysis. Here, the fundamental problem is the huge amount of measured variables with comparably very few independent observations. This leads to a variety of statistical problems, e.g. multicollinearity or multiple testing (Dudoit et al. (2000), Eilers et al. (2001)). However, for the majority of experiments only descriptive methods have been applied (e.g. Eisen et al. (1998)). These are generally not well-suited to assign confidence to the results. Nevertheless, this will be essential to separate artificial from true findings in complex hybridisation experiments.

4 Discussion

In view of the frustratingly slow progress in unraveling the genetic background of complex diseases, it has been hypothesized that diseases are too complex to be fully explained by current methods. Obviously, most current mapping strategies could be more promising if applied to traits with a clear genetic basis or biologically meaningful genes, as discussed by Weiss and Terwilliger (2000). Nevertheless, advances in molecular biological techniques are also required for a more successful performance of genetic studies for several reasons: In comparison with current methods, cost need to be lowered, results need to be obtained faster, and information from databases need to be used more exhaustively.

In this paper, we have sketched two recently developed molecular biological techniques that may be used with chip technology. We have emphasized that the success of both relies heavily on the statistical input. Both methodologies have been ballyhooed as the future way to map complex disease genes. This attitude might, however, be questioned. Thus, genome-wide association studies using SNP markers typed on DNA chips depend on the degree of association between the investigated alleles and the disease of interest. In addition, it is possible that a SNP study fails to detect an association because the genotyped SNP has not arisen on the disease haplotype.

Different problems have to be faced in microarray experiments. As pointed out by Weiss and Terwilliger (2000), one might work back from a regulated gene with aberrant expression to find its regulator. However, this need not be easier than other mapping approaches. Nevertheless, gene expression studies might be ideally suited to test a variety of hypotheses other than mapping. For example, subsets of genes with similar function can be identified for further investigation (Eisen et al. (1998)).

Summing up, we recommend that future genetic studies focus on genetics with a proper clinical background. Furthermore, if results are based on statistically sound study designs and analyses, one may yield interesting results related to biology, pathobiology and aetiology. As shown, new developments are required that can only be accomplished by joint efforts of geneticists, molecular biologists, clinicians, and statisticians.

References

BÖDDEKER, I. and ZIEGLER, A. (2000): Assoziations- und Kopplungsstudien zur Analyse von Kandidatengenen. *Dtsch Med Wochenschr, Vol. 125, 810–815.*

COLLINS, F. S. (1995): Positional Cloning Moves From Perditional to Traditional. *Nat Genet, Vol. 9, 347–350.*

DUDOIT, S., et al. (2000): Statistical Methods for Identifying Differentially Expressed Genes in Replicated cDNA Microarray Experiments. Tech. Rep. 578, Dept. of Statistics, University of California at Berkeley.

EILERS, P. H. C., et al. (2001): Classification of Microarray Data With Penalized Logistic Regression. *Proceedings of Photonics West 2001: Biomedical Optics Symposium (BIOS), in press.*

EISEN, M. B., et al. (1998): Cluster Analysis and Display of Genome-Wide Expression Patterns. *Proc Natl Acad Sci, Vol. 95, 14863–14868.*

GORDON, D., HEATH, S. C. and OTT, J. (1999): True Pedigree Errors More Frequent Than Apparent Errors for Single Nucleotide Polymorphisms. *Hum Hered, Vol. 49, 65–70.*

HEGDE, P., et al. (2000): A Concise Guide to cDNA Microarray Analysis. *BioTechniques, Vol. 29, 548–562.*

HORIKAWA, Y., et al. (2000): Genetic Variation in the Gene Encoding Calpain-10 is Associated With Type 2 Diabetes mellitus. *Nat Genet, Vol. 26, 163–75.*

HORVATH, S. and BAUR, M. P. (2000): Future Directions of Research in Statistical Genetics. *Stat Med, Vol. 19, 3337–3343.*

KERR, M. K. and CHURCHILL, G. A. (2001): Experimental Design for Gene Expression Microarrays. *Biostatistics, in press.*

LEE, M.-L. T., et al. (2000): Importance of Replication in Microarray Gene Expression Studies: Statistical Methods and Evidence from Repetitive cDNA Hybridizations. *Proc Natl Acad Sci, Vol. 97, 9834–9839.*

OMIM: Online Mendelian Inheritance in Man, OMIM (TM). Johns Hopkins University, Baltimore. URL: http://www.ncbi.nlm.nih.gov/omim/

PRITCHARD, J. K. and ROSENBERG, N. A. (1999): Use of Unlinked Genetic Markers to Detect Population Stratification in Association Studies. *Am J Hum Genet, Vol. 65, 220–228.*

SCHÄFFER, A. A. (1998): Coping With Complexity: Lessons From the Mathematical Sciences. *Hum Genet, Vol. 103, 5–10.*

STRACHAN, T. and READ, A. P. (1999): Human Molecular Genetics. 2nd ed. John Wiley & Sons, New York.

TERWILLIGER, J. D. and GÖRING, H. H. H. (2000): Gene Mapping in the 20th and 21st Centuries: Statistical Methods, Data Analysis, and Experimental Design. *Hum Biol, Vol. 72, 63–132.*

THE INTERNATIONAL SNP MAP WORKING GROUP (2001): A Map of Human Genome Sequence Variation Containing 1.42 Million Single Nucleotide Polymorphisms. *Nature, Vol. 409, 928–933.*

WEISS, K. M. and TERWILLIGER, J. D. (2000): How Many Diseases Does It Take To Map a Gene With SNPs? *Nat Genet, Vol. 26, 151–157.*

YANG, Y. H., et al. (2000): Normalization for cDNA Microarray Data. *Tech. Rep. 589, Dept. of Statistics, University of California at Berkeley.*

Marketing, Finance and Management Science

Market Simulation Using Bayesian Procedures in Conjoint Analysis

D. Baier[1], W. Polasek[2]

[1]Institute of Business Administration and Economics,
Brandenburg University of Technology Cottbus, D-03042 Cottbus

[2]Institute of Statistics and Econometrics, University of Basel, CH-4051 Basel

Abstract: New procedures for predicting market shares of competing products in assumed possible market scenarios are proposed: We use Gibbs sampling in a first phase to obtain the posterior distribution of preference model parameters w.r.t. product choice-relevant attributes and levels. Then, in a second phase, this distribution is used for market simulation. We discuss advantages of the new procedures in an empirical comparison with traditional ones.

1   Introduction

Commercial applications of conjoint analysis for product positioning and pricing usually consist of two major phases: a first phase, where preference model parameters are estimated from observed preferential responses to hypothetical combinations of product attributes and levels, and a second phase, where these estimates are used for market simulation, i.e. for predicting market shares for competing products with assumed attribute-level-combinations in a possible market scenario (see, e.g., Wittink et al. (1994), Baier (1999) for recent surveys).

Various combinations of procedures have been proposed for dealing with the different requirements in these two phases. So, e.g., in the first phase,

$$y_i = X\beta_i + \epsilon_i \qquad (1)$$

is used with ϵ_i as a normally distributed error vector ($\epsilon_{ij} \sim N(0, \sigma_i^2)$) to estimate β_i, the vector of individual preference model parameters. i indexes a sample of N respondents and j a set of n hypothetical attribute-level-combinations (stimuli) which have to be evaluated by each respondent. y_i is the vector of observed preferential responses and X the respective dummy coded design matrix.

In the second phase, the OLS estimates of the parameter vectors are used to predict product choices among a set S of competing products with assumed attribute-level-combinations $x_s^S \; \forall s \in S$. For this purpose, further assumptions have to be made: If we neglect, e.g., the error distribution in equation (1), we get for the probability

$$p_{is|S}^{(1st)} = \begin{cases} 1 & \text{if } (\mathbf{x}_s^S)'\beta_i \geq (\mathbf{x}_{s'}^S)'\beta_i \quad \forall s' \in S \\ 0 & \text{else} \end{cases} \tag{2}$$

that respondent i selects product s if set S is available the so-called first choice rule that respondent i always selects the product that provides him maximum utility on average.

Since equation (2) clearly overstates the choice probability for this product, other – rather ad-hoc – assumptions have been proposed in this second phase. Popular are the BTL (Bradley-Terry-Luce) choice rule or – assuming extreme value distributed errors – the logit choice rule, leading to more equally distributed choice probabilitites:

$$p_{is|S}^{(BTL)} = \frac{((\mathbf{x}_s^S)'\beta_i)^\alpha}{\sum_{s'\in S}((\mathbf{x}_{s'}^S)'\beta_i)^\alpha} \quad \text{and} \quad p_{is|S}^{(logit)} = \frac{\exp(\alpha(\mathbf{x}_s^S)'\beta_i)}{\sum_{s'\in S}\exp(\alpha(\mathbf{x}_{s'}^S)'\beta_i)} \tag{3}$$

(α typically equals 1, see, e.g., Green, Krieger (1988)).

The advantage of combining procedures for estimating individual preference models with ones for predicting product choice probabilities lies in the "richer" information structure of preferential data (compared to choice data) without loosing the possibility to predict market shares, sales volumes or profits for products after aggregation across respondents (see e.g. Gaul et al. (1995), Baier, Gaul (1999)). However, most of these combinations suffer from well-known problems which are often referred (a) to making different assumptions in the two phases (e.g. normally distributed error terms in the first phase and extreme value distributed ones in the second phase), (b) to neglecting prior information about model parameters, and (c) to neglecting products' similarity in market simulation (as discussed in Green, Krieger (1988), Baier, Gaul (1999, 2000), or Huber et al. (2000)).

This paper tries to overcome these problems: As in Allenby et al. (1995) and in Polasek et al. (1998) we derive the posterior distribution for the preference model parameters via Bayesian analysis, and, similar to Baier and Gaul (1999, 2000), we directly use this distribution to predict shares of choices for competing products (without making further assumptions). Consequently: (a) the same modeling assumptions are valid over both phases, (b) prior information is incorporated via rejection sampling, and (c) the products' similarity via the parameter distribution. The new procedures are applied in a real-world application and compared to the above mentioned combinations of procedures.

2 Bayesian Procedures in Conjoint Analysis

Bayesian procedures provide a general framework to combine prior information about model parameters (e.g. non-negativity or ordinal constraints) with the likelihood function of the observed data. The result of this combination, the posterior distribution of the parameters, depends on the modeling assumptions, the observed data, and the assumed prior distributions of the parameters.

In the special case of no prior information the estimation of the regression model (1) will give the usual maximum likelihood estimates

$$\hat{\beta}_i = (\mathbf{X}'\mathbf{X})^{-1}\mathbf{X}'\mathbf{y}_i, \hat{\mathbf{H}}_i = \hat{s}_i^2(\mathbf{X}'\mathbf{X})^{-1}, \hat{s}_i^2 = (\mathbf{y}_i - \mathbf{X}\hat{\beta}_i)'(\mathbf{y}_i - \mathbf{X}\hat{\beta}_i)/n. \quad (4)$$

The more interesting model arises when prior distributions for the parameters are available. A mathematically tractable way is to specify a semi-conjugate multivariate normal and inverse gamma (IG) prior distribution for each respondent i:

$$p(\beta_i) = N(\beta_{i*}, \mathbf{H}_{i*}), \quad p(\sigma_i^2) = IG(s_{i*}^2, n_*). \quad (5)$$

Then the full conditional posterior distribution for β_i and σ_i^2 are given by multivariate normal and inverse gamma distributions

$$p(\beta_i|\mathbf{y}_i, \mathbf{X}, \sigma_i^2) = N(\beta_{i**}, \mathbf{H}_{i**}), \quad p(\sigma_i^2|\mathbf{y}_i, \mathbf{X}, \beta_i) = IG(s_{i**}^2, n_{**}), \quad (6)$$

with parameters

$$\begin{aligned}
\beta_{i**} &= \mathbf{H}_{i**}(\mathbf{X}'\mathbf{y}_i + \sigma_i^2\mathbf{H}_{i*}^{-1}\beta_{i*}), \quad \mathbf{H}_{i**} = (\sigma_i^{-2}\mathbf{X}'\mathbf{X} + \mathbf{H}_{i*}^{-1})^{-1}, \\
s_{i**}^2 &= (\mathbf{y}_i - \mathbf{X}\beta_i)'(\mathbf{y}_i - \mathbf{X}\beta_i) + s_{i*}^2, \quad n_{**} = n + n_*
\end{aligned} \quad (7)$$

(see, e.g., Polasek et al. (1998)). The posterior parameters depend on the prior parameters which have to be carefully specified. A prior-posterior sensitivity analysis will exhibit influential assumptions. Note that equations (6) and (7) can also be used to specify non-informative prior distributions using $\mathbf{H}_{i*}^{-1}=0$, $n_*=0$ and $s_{i*}^2=0$. We will use this special non-informative case in our empirical application.

Rejection sampling within the Gibbs sampler (see, e.g., Allenby et al. (1995)) provides an additional convenient way to incorporate non-negativity or ordinal constraints: Initial estimates $\beta^{(0)}$ and $\sigma_i^{2(0)}$ for the model parameters are arbitrarily set. Then, for $k = 1, 2, \ldots$, indexed versions of equations (4) resp. (5) are used to calculate actual estimates $\beta_{i**}^{(k)}$, $\mathbf{H}_{i**}^{(k)}$, $n_{**}^{(k)}$, and $s_{i**}^{2(k)}$ of the distributional parameters and to draw actual estimates $\beta_i^{(k)}$ and $\sigma_i^{2(k)}$ of the conditional posterior distributions. The drawing is repeated until the pre-specified constraints are fulfilled. We call this approach the restricted Bayesian procedure. Without rejection sampling it is called the unrestricted Bayesian procedure.

The simulated sequence $\{\beta_i^{(k)}, \sigma_i^{2(k)}\}$ ($k=0,1,2,\ldots$) converges under regularity conditions to the posterior distribution of the parameters and can be used in the second phase for market simulation. Similar to Baier, Gaul (2000), a tractable <u>and</u> consistent way could be to approximate the probability that individual i selects product s by

$$p_{is|S} = \frac{1}{K}|\{k \mid (\mathbf{x}_{s|S})'\beta_i^{(k)} \geq (\mathbf{x}_{s'|S})'\beta_i^{(k)} \quad \forall s' \in S\}| \tag{8}$$

(K is a pre-specified number of valid draws), an equation that takes the error distribution in equation (1) into account. It should be mentioned that equation (8) – contrary to equation (3) – provides a way to incorporate the products' similarity into market simulation. In the next section this is shown in an empirical application.

3 Empirical Application and Comparisons

In our example preferential and choice data for a packaged 4 day bus tour to Paris were collected from 100 respondents. Choice-relevant product attributes (traveling time, bus quality, hotel quality, price, additional program, additional visit to Disneyland Paris, each with two or three appropriate levels) were determined using bus-tour catalogues, expert interviews and pre-interviews with potential customers.

The respondents had to perform three different tasks: In the first task, they had to rate directly the importance of the six attributes for their choice decision and the desirability of each of the 2 or 3 levels, all on 11 point rating scales ($\rightarrow$ self explicated data). These data provide later a straight-forward way to get simple estimates of respondents' part-worth functions (so-called $\rightarrow$ self explicated model, SEM) by multiplying the desirability rating of each level with the importance rating of the corresponding attribute. In the second task, 16 stimuli coming from a fractionated factorial design for the above attributes and levels had to be rated ($\rightarrow$ conjoint data). In the third task, they had to decide which of four real-world bus-tours they would buy – if available – and to rank order the remaining ($\rightarrow$ holdout data).

The data for each respondent were analyzed using different traditional procedures: SEM and OLS for parameter estimation, using first choice,

BTL and logit rules (see equations (2) and (3)) for market simulation. Additionally, the individual data were analyzed using the unrestricted Bayesian procedure ('Unrestr. Bayes') and two restricted Bayesian procedures: 'Price-restr. Bayes' with ordinal constraints for the price parameters (i.e. part-worth functions for higher prices are restricted to be lower than part-worth functions for lower prices) and 'Price-hotel-restr. Bayes' with additional ordinal constraints for the hotel quality parameters (i.e. part-worth functions for 1 star hotels are lower than for 2 stars hotels and so on). After a burn-in phase of 500 valid (not rejected) draws, 1.000 valid draws were kept. Rejection sampling did not lead to long sampling times, the estimation time per respondent was lower than 20 seconds.

	SEM	OLS	Unrestr. Bayes	Price-restr. Bayes	Price-ho-tel-restr. Bayes
Traveling time:					
Day	.000	.000	.007 (.019)	.014 (.031)	.015 (.030)
Night	.000	.035	.040 (.038)	.046 (.048)	.043 (.048)
Bus quality:					
Comfort	.290	.327	.304 (.073)	.297 (.097)	.284 (.101)
Luxury	.362	.397	.370 (.064)	.360 (.092)	.347 (.088)
Double-decker	.000	.000	.000 (.005)	.001 (.016)	.001 (.008)
Hotel quality:					
1 star	.000	.000	.003 (.015)	.012 (.034)	.000 (.000)
2 stars	.016	.111	.107 (.053)	.111 (.070)	.100 (.061)
3 stars	.032	.105	.099 (.057)	.105 (.076)	.161 (.067)
Price:					
339 DM	.065	.041	.048 (.049)	.092 (.057)	.096 (.058)
419 DM	.032	.000	.010 (.025)	.034 (.034)	.037 (.037)
499 DM	.000	.129	.123 (.055)	.000 (.000)	.000 (.000)
Additional program:					
Musical (+70 DM)	.282	.199	.183 (.063)	.180 (.088)	.172 (.089)
Sightseeing (+30 DM)	.112	.228	.209 (.055)	.207 (.076)	.203 (.079)
Without	.000	.000	.001 (.009)	.003 (.022)	.004 (.020)
Disneyland Paris:					
With (+95 DM)	.000	.000	.001 (.008)	.003 (.017)	.005 (.020)
Without	.258	.105	.098 (.045)	.097 (.059)	.096 (.060)

Table 1: Estimation results for a selected respondent: Mean standardized part-worth estimates (standard deviation).

Table 1 shows the estimation results for one selected respondent. The estimates are standardized s.t. the part-worth function for the worst level of each attribute becomes zero and the sum of the estimates for the best levels become one. (For SEM in Table 1 the worst levels were 'day', 'double-decker', '1 star', '499 DM', 'without', and 'with (+ 95 DM)'. The best levels were 'night', 'luxury', '3 stars', '339 DM', 'musical (+ 70 DM)' and 'without'.) For the Bayesian procedures the estimates of each draw were standardized resulting in an empirical distribution of the standardized part-worth estimates which is described in Table 1. The most important attribute across all procedures is the bus quality. The constraints for the price parameters are met for SEM and the restricted Bayesian procedures, but not for OLS and 'Unrestr. Bayes'. The same holds for the hotel quality parameters. Spoken in terms of model quality we can say that – for this respondent – the restricted Bayesian procedures have a better face validity since we have no violations of pre-specified ordinal constraints.

	SEM,			OLS,			Price-restr.	Price-hotel-restr.
	first ch.	BTL rule	logit rule	first ch.	BTL, rule	logit rule	Unrestr. Bayes	restr. Bayes	Bayes
Predicted choice probabilities for holdout stimuli:								
Product A	0	.152	.190	0	.186	.209	.006	.015	.018
Product B	0	.357	.301	0	.317	.292	.471	.546	.541
Product C	1	.426	.351	1	.319	.294	.519	.435	.438
Product D	0	.065	.157	0	.179	.205	.004	.004	.003
Spearman's r	.800			.800			.781	.843	.817
Hit rate	0	.357	.301	0	.317	.292	.471	.546	.541
Predicted choice probabilities in market scenario 1:								
Product A	0	.106	.141	0	.141	.162	.006	.015	.018
Product B	0	.251	.224	0	.240	.226	.471	.546	.541
Product C	.500	.299	.261	.500	.242	.227	.260	.218	.219
Product C'	.500	.299	.261	.500	.242	.227	.260	.218	.219
Product D	0	.046	.116	0	.136	.159	.004	.004	.003
Predicted choice probabilities in market scenario 2:								
Product A	0	.112	.140	0	.140	.161	.004	.008	.015
Product B	0	.265	.222	0	.238	.224	.307	.400	.357
Product C	1	.316	.259	0	.240	.225	.236	.260	.260
Product C"	0	.259	.218	1	.248	.232	.450	.329	.366
Product D	0	.048	.116	0	.135	.158	.003	.003	.002

Table 2: Predicted choice probabilities in different market scenarios and predictive validity for the selected respondent.

The data from the holdout task are used in Table 2 to assess the predictive validity of the procedures. Spearman's r measures the correlation between the observed ranking of the holdout stimuli and the ranking of the predicted utility values. The hit rate measures the proportion with which the "bought" holdout stimulus ('product C') shows maximum predicted utility. Here, the restricted procedures also perform best. A little bit surprising is the fact that the procedure with the highest face validity, the 'Price-hotel-restricted Bayes' procedure, has a lower predictive validity than the 'Price-restricted Bayes' procedure. All Bayesian procedures are superior to the traditional procedures.

The superiority becomes even more striking if we have a look at some market simulations in Table 2. We changed the market scenario in the third data collection task by introducing additional products. In market scenario 1 we added a product C' which has the same description as product C. In the market scenario 2 we added a product C" which is improved to product C only for the unimportant attribute additional program. In both market scenarios we would expect that these new products would only or heavily cannibalize choice probabilities from their main competitor, product C. From Table 2 we can see that SEM and OLS in combination with BTL or logit choice rules do not show the desired behavior. In combination with the first choice assumption they clearly overstate the choice probability of product C". The Bayesian procedures show the desired behavior.

	SEM	OLS	Unrestr. Bayes	Price-restr. Bayes	Price-hotel-restr. Bayes
Traveling time:					
Day	.032 (.072)	.030 (.050)	.035 (.055)	.035 (.054)	.035 (.055)
Night	.033 (.063)	.026 (.077)	.031 (.072)	.031 (.073)	.031 (.072)
Bus quality:					
Comfort	.011 (.039)	.023 (.046)	.034 (.056)	.033 (.056)	.034 (.058)
Luxury	.031 (.053)	.058 (.059)	.065 (.066)	.065 (.065)	.066 (.066)
Double-decker	.022 (.044)	.052 (.058)	.059 (.065)	.059 (.064)	.060 (.065)
Hotel quality:					
1 star	.013 (.048)	.012 (.056)	.017 (.058)	.017 (.057)	.000 (.000)
2 stars	.121 (.097)	.014 (.115)	.134 (.111)	.133 (.111)	.122 (.102)
3 stars	.110 (.111)	.169 (.142)	.162 (.137)	.161 (.136)	.187 (.124)
Price:					
339 DM	.285 (.118)	.252 (.134)	.234 (.134)	.247 (.124)	.246 (.123)
419 DM	.152 (.086)	.145 (.084)	.137 (.090)	.139 (.086)	.138 (.085)
499 DM	.000 (.000)	.005 (.026)	.008 (.031)	.000 (.000)	.000 (.000)
Add. program:					
Musical	.170 (.097)	.076 (.128)	.076 (.120)	.075 (.121)	.075 (.120)
Sightseeing	.092 (.141)	.125 (.106)	.121 (.103)	.121 (.104)	.121 (.103)
Without	.148 (.145)	.110 (.131)	.109 (.125)	.109 (.125)	.109 (.125)
Disneyland Paris:					
With	.035 (.081)	.022 (.061)	.022 (.061)	.022 (.061)	.022 (.061)
Without	.174 (.146)	.198 (.164)	.183 (.154)	.183 (.153)	.182 (.153)

Table 3: Standardized estimation results across all respondents: Mean standardized part-worth estimates (standard deviation)

420

Across all respondents we have similar findings. The restricted Bayesian procedure shows improved face validity. Violations of the ordinal constraints cannot be observed. This can easily be seen from the mean standardized part-worth estimates for the worst price level and the worst hotel quality level in Table 3. The superiority of the Bayesian procedure for predictive validity is not so striking as can be seen from Table 4. Mean Spearman rank correlation and the hit rate are slightly worse than with the traditional procedure.

	SEM,			OLS,			Unrestr. Bayes	Price-restr. Bayes	Price-hotel-restr. Bayes
	first ch.	BTL rule	logit rule	first ch.	BTL rule	logit rule			
Hit rate	.54	.347	.298	.80	.358	.310	.747 (.43)	.753 (.43)	.739 (.44)
Spearman's r	.602 (.37)			.752 (.35)			.703 (.38)	.712 (.37)	.684 (.39)

Table 4: Predictive validity across all respondents (standard deviation)

4 Conclusions and Outlook

A new approach for market simulation for conjoint analysis was presented which shows some advantages over the traditional procedures: (a) the same modeling assumptions are used in the parameter estimation phase and in the market simulation phase, (b) an improved face validity through constraints in the estimation phase is obtained, and (c) the modeling of the similarity of products in the market simulation phase is possible.

References

ALLENBY, G. M., ARORA, N., and GINTER, J. L. (1995): Incorporating Prior Knowledge into the Analysis of Conjoint Studies. *Journal of Marketing Research, 32, 152-162.*

BAIER, D. (1999): Methoden der Conjointanalyse in der Marktforschungs- und Marketingpraxis, in: Gaul, Schader (Eds.): Mathematische Methoden der Wirtschaftswissenschaften, Springer, Berlin, 197-206.

BAIER, D., GAUL, W. (1999): Optimal Product Positioning Based on Paired Comparison Data. *Journal of Econometrics, 89, 365-392.*

BAIER, D., GAUL, W. (2000): Market Simulation Using a Probabilistic Ideal Vector Model for Conjoint Data, in: Gustafsson, Herrmann, Huber (eds.): Conjoint Measurement – Methods and Applications, Springer, Berlin, 97–120.

GAUL, W., AUST, E., BAIER, D. (1995): Gewinnorientierte Produktliniengestaltung unter Berücksichtigung des Kundennutzens. *Zeitschrift für Betriebswirtschaft, 65, 835–854.*

GREEN, P. E., KRIEGER, A. M. (1988): Choice Rules and Sensitivity Analysis in Conjoint Simulators. *Journal of the Academy of Marketing Science, 16, 114–127.*

HUBER, J., ORME, B., MILLER, R. (2000): Dealing with Product Similarity in Conjoint Simulations, in: Gustafsson, Herrmann, Huber (Eds.): Conjoint Measurement – Methods and Applications, Springer, Berlin, 393–410.

POLASEK, W., LIU, S., and JIN, S. (1998): Heteroskedastic Linear Regression Models: A Bayesian Analysis, in: Balderjahn, Mathar, Schader (Eds.): Classification, Data Analysis, and Data Highways, Springer, Berlin, 182–191.

WITTINK, D. R., VRIENS, M., BURHENNE, W. (1994): Commercial Use of Conjoint Analysis in Europe: Results and Critical Reflections. *International Journal of Research in Marketing, 11, 41–52.*

Value Based Benchmarking and Market Partitioning

H. H. Bauer[1] M. Staat[2] M. Hammerschmidt[1]

[1]Lehrstuhl für ABWL und Marketing II
Universität Mannheim, D-68131 Mannheim

[2]Lehrstuhl für VWL, insbes. Mikroökonomik
Universität Mannheim, D-68131 Mannheim

Abstract: The paper offers an analytical approach for an integrated treatment of market partitioning and benchmarking within a Data Envelopment Analysis (DEA) framework. Based on an empirical example from the automotive industry we measure product efficiency from the customer's perspective. This is interpreted as customer value, i. e., as a ratio of outputs that customers obtain from a product (e. g., resale value, reliability) and inputs that customers have to invest (e. g., price, running costs). Products offering a maximum customer value relative to all other alternatives represent efficient peers, which constitute benchmarks for different sub-markets. All products benchmarked via the same efficient peer(s) constitute a sub-market including the benchmarks.

1 Introduction

A major tenet of marketing theory is the inclusion of benchmarking into the competitive advantage paradigm. Competitive benchmarks used by firms and managers as reference points evidently affect the choice, direction and implementation of performance-enhancing strategies (Shoham, Fiegenbaum (1999)). The strongest theoretical support for the use of benchmarking is given by strategic reference point theory, which is an extension of prospect theory. In this context the content and risk of marketing decisions are viewed to depend on whether managers perceive their firms as above, below or equal to some given reference point. Arguably, if for example a firm's product position is below the standards set by leading competitors, improvements or strategic action in general may be triggered. Success is viewed as something depending on the position of firms or products relative to competitors.

The reasons that account for the popularity in competition driven companies are twofold: first, with benchmarking performance is evaluated only with regard to other products in the market (relative performance evaluation). Second, benchmarks are real objects ("observed successes") not hypothetical or prescriptive ideals.

Despite these advantages standard benchmarking tools suffer from several deficiencies (Staat, Hammerschmidt (2000)): When benchmarking in a multiple criteria setting a consistent ranking of products necessitates a simultaneous integration of all criteria. Otherwise, it may well happen that a product performs best on one parameter but is inefficient in terms of another. When applied in a multi-criteria setting the results generated by benchmarking exclusively depend on the weights assigned to the parameters. But applying the same vector of parameter weights to all products exogenously would essentially apply one and the same global benchmark to all units. This may lead to extreme performance differences, which are not caused by product inefficiency but by the fact that they may not be comparable to the benchmark.

In addition, with a fixed weighting scheme only one strategy for optimizing products would be rated efficient. In this case, the possibility of alternative approaches (parameter-combinations) to the generation of product value is neglected and at the same time the existence of different consumer segments is implicitly ignored. By using DEA, we develop an approach to derive product reference points to assess and improve product performance in a theoretically and methodologically sound way.

DEA is an exploratory data mining approach, which can be interpreted as a generalized form of benchmarking. The data mining aspect of DEA is given by the fact that the method reveals data structures by grouping observations around efficient units without the necessity of prior knowledge about the factors which determine efficiency. Related studies by Papahristodoulou (1997), Doyle, Green (1991) as well as Kamakura et al. (1988) also use DEA to evaluate product efficiency but not in a marketing context. Bauer et al. (2000) provide a DEA of the car market cast in a marketing context but do not introduce the concept of value based market structuring.

In this paper the efficiency value is measured as an input-output-ratio from the customer's perspective. We show that, in addition to this, DEA achieves market partitioning endogenously. By assigning individual weights to the input-output-parameters different products can be rated as efficient, i. e. serve as benchmarks. All inefficient products located next to the same benchmark(s) have a similar input-output-structure and are clustered in the same market partition. The identification of different benchmarks jointly with similar inefficient products DEA allows us to find "natural" market partitions (product segments). The according segment specific benchmarks are used to assess intra-partition efficiency. Based on an empirical application to the market for compact cars the DEA approach for benchmarking and market partitioning is illustrated.

2 Customer Value as a Basis for Benchmarking and Market Partitioning

Within a value based perspective consumers do not search for products with maximum quality or minimum price but seek to maximize the quality-price-ratio in the sense of value for money (Rust, Oliver (1994)). While forming their judgements about products consumers jointly consider both quality and non-quality related dimensions within an economically oriented decision concept of "higher-order-abstraction" (Sinha, DeSarbo (1998)). This embodies a return to cost trade-off, defined as customer value. This type of sophisticated, two-dimensional purchasing behavior can be expected especially in electronically mediated markets with information driven consumers.

Instead of viewing value solely as a quality-price trade-off, a more systematic, multi-attribute operationalization of customer value is needed (Huber et al. (2000)). Along with these requirements we conceptualize the two basic value dimensions in a more multi-faceted way by measuring customer value as a ratio of weighted outputs and inputs. The principle of modeling the customer value (CV) of a product, which we develop, can be represented as follows:

$$CV = \frac{\text{Outputs}}{\text{Inputs}} = \frac{\sum_r u_r y_r}{\sum_i v_i x_i} \tag{1}$$

Inputs x and respective weights v are indexed by i. They represent an "investment" by the customer necessary to obtain and use a good. In addition to out-of-pocket costs such as price, insurance or running costs inputs could also be non-monetary sacrifices such as time, risk or search costs. Outputs y and respective weights u are indexed by r and represent "outcomes" from a product, i. e. performance attributes from which the customer derives his utility (e. g. reliability, comfort, safety). CV is the customer's economic value derived from the product in the sense of an output to input efficiency value in a customer's perspective. It can be understood as the return on customer's investment.

The analogy of CV and economic efficiency is obvious since products are chosen that offer maximum outputs for given inputs or that demand minimum inputs for a particular output level. This general concept models the customer's trade-off between all received outputs (positive consequences, utility, results) and all inputs (sacrifices, costs) across the entire process of purchasing and using the product. The single input-output-ratios are aggregated into an overall value measure. As a result, we obtain a generalized, broadly applicable measure of customer value because all kinds of customer relevant input and output parameters can be included in our analysis, independent of scale level or dimensionality.

Customer value has often been defined as a higher-order construct to evaluate products, containing much more choice relevant components than one-dimensional approaches do (Sinha, DeSarbo (1998) and Huber et al. (2000)). In spite of this, no empirical attempt has been made to structure product markets on grounds of the customer value. To this end, a market partition is interpreted as a cluster of products that are similar with regard to certain criteria and thus can be considered close substitutes (Bauer, Herrmann (1995)).

Conventionally, only utility or quality related attributes are used as such criteria (see Day et al. (1979)) without connecting them to price-variables within an input-output-function. Typical methods for sub-market identification are MDS, hierarchical cluster analysis or forced switching-methods (DeSarbo et al. (1998) and Day et al. (1979)). Such methods enable researchers to infer, which products belong to one sub-market in terms of similarity w. r. t. particular quality criteria.

Furthermore, these conventional methods do not provide information about which product represents reference units in each of the several sub-markets. We are not aware of any study that jointly treats both market partitioning and benchmarking. Our method enables us to partition the product market and to identify benchmarks for each partition endogenously within a value based view.

3 DEA-based Market Partitioning and Benchmarking

3.1 Basic Principles of DEA

DEA is introduced as a technique which enables the analyst to assess the efficiency value operationalized in the way described above. It is a non-parametric approach to measure the efficiency of observed output-input-structures or decision making units (DMUs), which can be companies, processes or like in this paper products. Thus, we refer to market partitioning in the generic sense as partitioning the product market in several segments (see Bauer, Herrmann (1995)). Product market partitions as well as intra-partition benchmarks and efficiency scores are estimated on grounds of customer value. Thus, results can be transformed in strategy advice for a customer value management (CVM) of products. DEA supports the development of customer value maximizing strategies, i. e. gives directions for the variation of the input and output parameters.

DEA determines the degree of (in)efficiency of a product by measuring its relative distance to an efficient frontier. This frontier (best value line) is made up of all identified "efficient" products. Efficient products offer a particular level and combination of desired outputs demanding the minimum inputs compared to all other products. All products creating a maximum customer value are rated "efficient". This reflects that customers choose the product from which they receive maximum customer value in relation to the relevant alternatives. Thus the efficiency yielded by a DEA is the *relative* customer value. The customer value estimation of a particular DMU_0 is represented in the following expression:

$$\max_{u,v} CV_0 = \frac{\sum_r u_r y_{r0}}{\sum_i v_i x_{i0}} \quad \text{s.t.} \quad \frac{\sum_r u_r y_{rj}}{\sum_i v_i x_{ij}} \leq 1, \quad j = 1, ..., J \quad (2)$$

In the expression input weights are denoted v_i and output weights u_r with index i for inputs and index r for outputs whereas index j runs over the products. DEA assigns an individual vector of weights to each product, optimally adjusted on each specific input-output-structure. High weights are placed on those variables where a product compares favorably, low weights are placed on those variables where it compares unfavorably. These weights give CVM important information about the customer value drivers; these are the parameters that have been assigned high weights by the optimization algorithm.

The sum of the weighted output to input ratios, CV, is maximized under the restriction that no other product achieves a score greater than 1 if the weights that maximize the CV of the product being evaluated are applied to it. Thus, all products with a CV of 1 offer a maximum relative CV in the context of the relevant competition. With a DEA, CV is estimated specifically in relation to the competitive situation in the market, allowing an effective support of competitive advantage management.

3.2 Market Partitioning and Benchmarking: An Overview

We will now illustrate the principle of CV determination in figure 1. It depicts an overall market with seven products (A to G) and two outputs (which can be thought of as comfort, safety) normalized by one input (price). When benchmarking with fixed exogenous weights only one

Figure 1: Illustrating sub-market boundaries

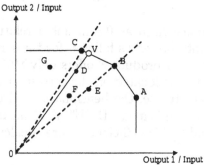

product (A, B or C) could be ranked first. In this case, the ranking would exclusively depend on the relative weights assigned to the outputs. But because all three products create maximum CV in a different way of combining outputs and inputs each product can be considered efficient within a unique value segment. They constitute the efficient frontier of the market as a whole reflecting the best value offered to customers with regard to their specific preferences.

This frontier is extended by two lines branching off horizontally from point A and vertically from point C. This can be justified by the fact that points left to C offer as much of output 2 and less of output 1 than C and therefore can be considered a conservative approximation of the frontier beyond points that are observed. This applies analogously to points below A.

The figure shows that each ray from the origin which intersects with an efficient product forms a boundary of a sub-market. In this example we can partition the overall market into three sub-markets. For each market partition, relative customer value is estimated. Products D, E, and F are all benchmarked against reference units made up of B and C because these are the efficient neighbors. Hence, products B to F belong to one market partition. This is the essence of intra sub-market efficiency evaluation and implies that the efficiency value of D, E, and F is calculated only in comparison to B and/or C. It would result in a quite meaningless comparison if one were to use A as a benchmark for this segment since A with its high level of output 2 and its relatively low output 1 is qualitatively different from products B to F with their more balanced output structure.

The degree of inefficiency for a product is measured by its distance to the origin relative to that of an efficient benchmark. For instance, the benchmark for E is product B as the nearest point on the efficient frontier.

Therefore the inefficiency is calculated as the ratio of the distances of the two output combinations to the origin, i. e. $\overline{OE}/\overline{OB}$.

Assuming the distance ratio as 0.8 means a relative CV for E of 0.8. This value can be interpreted as follows: for the same input (price) that has to be invested for B product E offers only 80% of B's outputs. For D, no existing product is located on the corresponding intersection with the efficient frontier. Hence, the benchmark used to assess the relative value of D is a so called "virtual DMU" V, a linear combination of the efficient peers B and C. The efficiency score is calculated as $\overline{OD}/\overline{OV}$.

3.3   The Formal Model

Continuing the formal discussion of DEA, the optimization problem (2) can be transformed into a linear programming problem. Formula (3) is the primal program of the linearized version of formula (2) above (see Cooper et al. (2000), p. 43). In this input-oriented formulation efficiency is measured as the proportional input reduction an inefficient product would be able to achieve if it applied the same input-output-transformation (strategy) for value creation as the corresponding benchmark on the efficient frontier.

$$\max_{\theta,\lambda,s^+,s^-} z_0 = \theta - \epsilon s^+ - \epsilon s^-$$

$$\text{s.t.} \quad Y\lambda - s^+ - Y_0 = 0 \tag{3}$$
$$\theta X_0 - X\lambda - s^- = 0$$
$$\lambda, s^+, s^- \geq 0$$

The above problem has to be solved separately for each DMU in the sample. It has a number of side conditions that corresponds to the number of input and output parameters $(I + R)$. In contrast, the dual problem (2) in ratio form has a number of side conditions equal to the number of DMUs. The efficiency score θ is augmented by input slacks s^- and output slacks s^+ multiplied by a non-Archimedian ϵ. It is thereby transformed into the so called slack-augmented score z_0. It is determined by comparing actual parameter values of DMU_0, which are denoted X_0 for inputs and Y_0 for outputs with the corresponding values X and Y of the reference unit. In an input-oriented model such as the one in (3) z_0 measures the input reduction possible for DMU_0 when compared to a reference unit.

This unit consists of a linear combination of efficient peers. Inputs x and outputs y of all DMUs are stacked in the vectors X and Y. The factors λ in (3) denote the weights of the efficient peers in the reference unit. It is characterized by outputs $X\lambda$ equal to or greater than outputs Y_0 of DMU_0 and inputs $X\lambda$ less than or equal to X_0.

To recur to figure 1 the input-oriented formulation implies that the value of product E could also be maximized by reducing necessary customer inputs by 20%. That is, the benchmark product B offers the same outputs for only 80% of the inputs (price to be invested) of E. This fraction is denoted by θ. It is equal to CV in formula (2). In the case of product E, the reference unit consists solely of the real product B and therefore $\lambda_B = 1$ and $\lambda_{-B} = 0$. The reference unit V relevant for product D consists of two efficient products, namely B and C. Because V is closer to C the factor λ_C is greater than λ_B.

Non-zero slacks, s^+ and/or s^-, exist for all parameters for which a variation by the proportional factor $1-\theta$ does not suffice to reach the position of the value benchmark. These parameters are weaknesses of the product because on those parameters small variations do not suffice to reach the position of maximum CV. Parameters with zero slacks do contribute to the efficiency of a product and thus indicate strengths of the product.

4   Applying DEA to Data

DEA-based market partitioning and benchmarking is now applied to data from the compact car market. Our analysis includes 30 variants -our observational unit- of the 11 best selling models in the German market in 1994. Compact cars are bought with little emotional involvement. On the output side the value of compact cars arises from technical-functional components (i. e. from basic utility). Thus, we can assume rational, cognitively involved buyers (Papahristodoulou (1997)) at least in a substantial segment of the compact car market.

Our analysis applies to this buyer segment, the data are based on interviews with ADAC-members (German Automobile Drivers Association) that show e. g. an above average road performance. Hence, it is justifiable to model customer value by technical parameters only. We use resale value after 4 years, reliability, safety, comfort, road performance and sufficiency of the catalytic converter as outputs. Price and annual running costs serve as inputs. Instead of reporting on all 30 variants we show only minimum, maximum and average values of these parameters. It is possible, however, to analyze the efficiency of products with higher

Table 1: Descriptive Statistics, ADAC Member Survey, 1996

Parameters	Min.	Max.	Mean
resale value in % of purchase price	0.30	0.56	0.38
reliability on a scale from 0.2 to 1	0.89	0.99	0.95
safety on a scale from 0.2 to 1	0.37	0.45	0.40
comfort on a scale from 0.2 to 1	0.30	0.50	0.40
road performance in km p. a.	15.470	29.200	20.364
advanced catalytic converter (E3 Norm)	0	1	0.57
price in DM	23.100	36.980	26.766
running costs in DM p. a.	2.509	4.727	3.202

emotional involvement by means of DEA. As a prerequisite, data on parameters which adequately describe the product features connected with to emotional involvement, must be available.

40 % of the analyzed model variants are efficient. They create maximum relative value to customers and thus form the efficient frontier. Due to space limitations, we cannot list the results for all 30 variants (available from the authors upon request). We limit our illustration to a few particular models, which suffices to understand the method. The Toyota Corolla, for example, offers below average or average outputs but requires the lowest investments (price, running costs) from customers. Instead, for the VW Golf a customer has to invest above average inputs but receives "market leading" performance on resale value and comfort in return. Both models create maximum value in terms of the output-input-ratio, which is maximal in relation to the market but with different value creating strategies.

In contrast, the Peugeot 306 is dominated by other car models (Corolla, Honda Civic) w. r. t. this ratio provided, thus it does not achieve a CV of 1. The Corolla and the Civic are identified as the nearest efficient neighbors for the Peugeot 306. They serve as benchmarks, from which the reference unit (virtual DMU) is combined. The Corolla and the Civic enter the reference unit with factors $\lambda_{Corolla} = 0,97$, $\lambda_{Civic} = 0,07$. The Corolla is located much closer to the Peugeot therefore it has a much higher importance for the benchmarking of the Peugeot than the Civic. θ is estimated at 0.9, implying that the Peugeot could create maximum CV when reducing price and running cost by 10% $(1 - \theta)$ provided that no slacks exist. But we have non-zero slacks for 5 parameters (price, resale value, reliability, E3 norm, safety). By means of the slacks s^+, s^- and the efficiency score θ DEA provides direct implications for deriving value enhancing strategies. A CVM obtains exact indications of how much any parameter have to be varied to close revealed value gaps.

All other inefficient car models, whose reference point is made up by the Corolla and the Civic are partitioned in the same sub-market.

5 Conclusion

With DEA we propose a method to structure product-markets based on customer value. Since the method measures customer value in a relative way it provides sub-market specific value benchmarks, which has two main advantages. First, DEA estimates intra partition costumer value. An overall market can thus be structured into product segments. Each sub market represents an own, specific approach of creating customer value. Second, target positions are provided for each identified sub-market. On these targets a CVM should be aligned in order to create maximum value for customers. DEA assigns an individual value function to each product, indicating a way to improve (maximize) customer value.

Of course, a better description of the specific advantages of the variants is desirable, including non-technical output parameters like design or brand image. An integration of those parameters into a DEA model is easily handled provided the data are available. Also, several extensions of the DEA specification used in this paper are available which could be applied to our data. For instance, the bootstrap provides a statistical foundation of the DEA approach (for an application, see Staat (2000)).

References

BAUER, H. H. and HERRMANN, A. (1995): Market Demarcation. *European Journal of Marketing, Vol. 29, 18-34.*

BAUER, H. H., STAAT, M. and HAMMERSCHMIDT, M. (2000): Produkt - Controlling - Eine Untersuchung mit Hilfe der Data Envelopment Analysis (DEA). Wissenschaftliches Arbeitspapier W 45 des IMU, Universität Mannheim.

COOPER, W. W., SEIFORD, L. M. and TONE, K. (2000): Data Envelopment Analysis. Kluwer, Boston, Dordrecht, London.

DAY, G. S., SHOCKER, A. D. and SRIVASTAVA, R. K. (1979): Customer - oriented approaches to identifying product-markets. *Journal of Marketing, Vol. 43, 8-19.*

DESARBO, W. S., KIM, Y., WEDEL, M. and FONG, D. K. H. (1998): A Bayesian approach to the spatial representation of market structure from consumer choice data. *European Journal of Operational Research, Vol. 111, 285-305.*

DOYLE, J. R. and GREEN, R. H. (1991): Comparing products using data envelopment analysis. *Omega, Vol. 19, 631-638.*

HUBER, F., HERRMANN, A. and BRAUNSTEIN, C. (2000): Testing the metric equivalence of customer value. Proceedings of the Annual Conference of the AMS, Hong Kong.

KAMAKURA, W. A., RATCHFORD, B. T. and AGRAWAL, J. (1988): Measuring Market Efficiency and Welfare Loss. *Journal of Consumer Research, Vol. 15, 289-302.*

PAPAHRISTODOULOU, C. (1997): A DEA model to evaluate car efficiency. *Applied Economics, Vol. 29, 1493-1508.*

RUST, R. T. and OLIVER, R. L. (1994): Service Quality, in Rust, Oliver (Eds.): Service Quality: New Directions in Theory and Practice, Sage Publications, Thousand Oaks, 1-20.

SHOHAM, A. and FIEGENBAUM, A. (1999): Extending the Competitive Marketing Strategy Paradigm: The Role of Strategic Reference Points Theory. *Journal of the Acacdemy of Marketing Science, Vol. 27, 442 - 454.*

SINHA, I. and DESARBO, W. S. (1998): An integrated approach toward the spatial modeling of perceived customer value. *Journal of Marketing Research, Vol. 35, 236-249.*

STAAT, M. (2000): Der Krankenhausbetriebsvergleich: Benchmarking vs. Data Envelopment Analysis. *ZfB-Ergänzungsheft, No. 4, 123-140.*

STAAT, M. and HAMMERSCHMIDT, M. (2000): Relative Performance Evaluation of Hospitals in Germany, in Albertshauser, Knödler (Eds.): Economics and Political Counseling in Theory and Practice, VWF, Berlin, 221 - 253.

Optimization of Corporate Reorganization Portfolios based on a Genetic Algorithm

R. Bennert[1], M. Missler-Behr[2]

[1]ERNST & YOUNG Deutsche Allgemeine Treuhand AG,
Elisenstr. 3a, D-80335 München, Germany

[2]Institut für Statistik und Mathematische Wirtschaftstheorie,
Universität Augsburg, D-86135 Augsburg, Germany

Abstract: The reorganization of companies threatened by serious existential crises is a basic task of strategic management. Subjects of the strategic considerations are principally all business fields, which have to be audited under aspects of profitability as well as potentials of success. The actual portfolios of business fields have to be optimized as major elements of the reorganization concept with regard to their contributions to profitability and potentials of success in order to ensure the companies economic survival for the future. Due to the fuzzy and uncertain data we have to use, the optimization of reorganization portfolios can be characterized as a non-well-structured problem. Therefore, it requires not the best but pareto-optimal solutions. To find promising portfolio positions an optimization method is presented, which is based on a genetic algorithm.

1   Introduction

A reorganization portfolio does not only cover the business fields which have to be restructured but also those which finance the reorganization by means of liquidation or sale as a whole. Hereby we have to consider quantitative as well as qualitative factors. The quantitative factors especially describe the financial point of view. In this paper the proposed cash-flow and expenditure for the operational functions purchase, production and sale of each business field are used as quantitative factors. Qualitative factors are mainly of relevance to ensure the long-term reorganization success from an economic point of view. As qualitative factors the potentials of success of each business field are here taken into consideration.

Two competitive objectives characterize such an optimization problem: maximizing the cash-flows as well as maximizing the potentials of success. A reduction of the expenditure of the operational functions leads to a higher cash-flow on one hand, but on the other hand it deprives the company of the substantial basis to ensure or to build up future potentials of success.

In the following a crisis-ridden company with four business fields f_1, f_2, f_3 and f_4 is considered. The business fields differ in their expenditure for the operational functions r_{i1} (purchase), r_{i2} (production) and r_{i3} (sale) and in their cash-flows cf_i ($i = 1, \ldots, 4$). We assume that a change of one unit in the expenditure of a operational function (Δr_{ij}) will affect the cash-flow by just 0.8 units. This can be argued for example by cut down problems of personal resources in the connection with social plan requirements.

The potentials of success of each business field differ and will be expressed by a so-called potential index of success e_i ($i = 1, \ldots, 4$) with $e_i \in [0, 1]$. It indicates, how strong the decision-makers of the reorganization estimate potentials of success under certain points of attractiveness as well as risk view (i.e. the innovation potential of products) for each business field.

The potential index e_i will be described by a logistical function depending on the overall operational expenditure r_i. The shape of the function should express that an increase or decrease of the operational expenditure of a business field tends to have a lower affect at a high level of the potential of success factor than on a low one. The marginal utility of the potential index diminishes with incremental expenditure of the operational functions. At a certain level of the overall operational expenditure of a business field the upper value 1 of the potential index is reached. Using these assumptions, we can describe the potential indices by the following general formula (see also Figure 1):

$$e_i(r_i) = \frac{a}{1 + b \cdot e^{-c \cdot r_i}} \tag{1}$$

Calculating the limit of e_i for $r_i \to \infty$ leads to $a = 1$ in equation (1). The constant factors b and c are determined with the help of two estimation points of the index value at minimal expenditure and at a further expenditure value, for example the actual expenditure.

Based on constant sales, the aim of portfolio optimization is to increase the corporate profitability by reallocating the proportions of the expenditure for the operational functions of the business fields by simultaneously placing sufficient potentials of success at the company's disposal in the future.

In the forthcoming example we assume the realistic case that investments are just done in proportions of 100 %, 75%, 50% or 25%. The following input data in Million Euro describe the financial restrictions

Financial restrictions			
purchase	production	sale	$\sum$
company			
r_1	r_2	r_3	
2.5	11.25	6.75	20.5
business fields			
portions	r_{i1}	r_{i2}	r_{i3}
1 = max	1	5	3	9
3/4	0.75	3.75	2.25	
1/2	0.5	2.5	1.5	
1/4 = min	0.25	1.25	0.75	2.25

Table 1: Financial data

for the optimization problem: The company will spend in total not more than 20.5 Mil Euro for all three operational functions. The upper bound for the companywide expenditure for purchase (r_1) will be 2.5 Mil Euro, for production (r_2) 11.25 Mil Euro and for sale (r_3) 6.75 Mil Euro. The expenditure of the business fields for purchase (r_{i1}) are restricted by 1 Mil Euro, for production (r_{i2}) by 5 Mil Euro and for sale (r_{i3}) by 3 Mil Euro. So a business field can mostly invest 9 Mil Euro and at least 2.25 Mil Euro because of the proportion factors. Not all business fields can invest their maximal sum because of the overall restriction of 20.5 Mil Euro. Table 1 summerizes the financial restrictions for the example. The initial cash flow situation of all business fields is negative, given by the following values:

$$cf_1 = -0.75 \qquad cf_2 = -1.25 \qquad cf_3 = -2.25 \qquad cf_4 = -0.75 \qquad (2)$$

1st estimation	2nd estimation	function
$e_1(2.25) = 0.15$	$e_1(2.75) = 0.35$	$e_1(r_1) = \dfrac{1}{1 + 855.86 \cdot e^{-2.23 \cdot r_i}}$
$e_2(2.25) = 0.15$	$e_2(2.75) = 0.45$	$e_2(r_2) = \dfrac{1}{1 + 5665 \cdot e^{-3.07 \cdot r_i}}$
$e_3(2.25) = 0.25$	$e_3(2.75) = 0.40$	$e_3(r_3) = \dfrac{1}{1 + 68.45 \cdot e^{-1.39 \cdot r_i}}$
$e_4(2.25) = 0.10$	$e_4(2.75) = 0.40$	$e_4(r_4) = \dfrac{1}{1 + 28345.33 \cdot e^{-3.58 \cdot r_i}}$

Table 2: Potential indices of success for each business field

The estimation points to state the potential indices of success e_i for the business fields in our example can be seen in Table 2. The shape of the functions is shown in Figure 1.

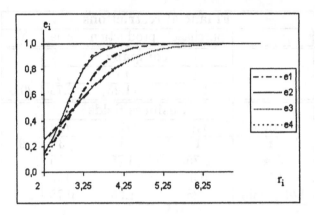

Figure 1: Potential indices of success e_i

The pareto-optimal solution to the decision problem shall be find by applying a genetic algorithm. For the sake of simplicity the liquidation values or fair-market values of the different business fields will not be regarded in decision making. A liquidation or sale of them will be indicated, if the expenditure for the operational functions of a business field converge to minimal investment in a portfolio used for reorganization.

2 Genetic Algorithm

Starting from an initial portfolio position $P_0(f_1, f_2, f_3, f_4)$ of the four business fields f_i ($i = 1 \ldots 4$), which describes the actual situation, the genetic algorithm will model the optimization of the portfolio positioning by means of genetic search operators (crossover and mutation) (see Nissen (1997), Kinnebrock (1994)) without neglecting the above-mentioned restrictions (see above). The resulting portfolio for reorganization does not have to state the global optimum. It is sufficient for the reorganizers to get to know a promising strategic orientation for the company.

Each allocation of the expenditure of the operational functions (r_{ij}) represents a potential reorganization portfolio that will now be coded in a binary way. A two-digit *Gray-Code* (see Table 3) is used to represent the portions of the expenditure as mentioned above (see Table 1).

In Figure 2 the description of a portfolio can be seen. It can be interpreted as follows: Digits 1-6, 7-12, 13-18 and 19-24 describe the resource allocation of the four business fields respectively. The expenditure for purchase of the business fields is given by the digits numbered 1/2, 7/8, 13/14 and 19/20. f_1 and f_3 will spend 4/4 (Gray-Code:10) of 1 Mio. Euro for purchase, f_2 and f_4 will spend 1/2 (Gray-Code:01) of 1 Mio.

portion of expenditure	Gray-Code
1/4	00
1/2	01
3/4	11
4/4	10

Table 3: Used Gray-Code

Euro (see Table 3 and 1). In total 3 Mio. Euro will be spend for the purpose of purchase in this portfolio. But just 2.5 Mio. Euro are disposable for purchase. Therefore the described portfolio is not admissible. In the same way the other digits can be interpreted as expenditure allocation for production and sale.

The starting population for the genetic algorithm contains seven portfolios. This is the initial portfolio position (P^0), which describes the actual situation and shall be improved, as well as six further portfolios, which describe possible future scenarios developed by the reorganizer. The scenario portfolios have to meet the financial restrictions, which means they have to be admissible. To find six realistic starting positions for the different business units will already be a quite demanding task for the reorganizer. Therefore, no larger starting population is considered.

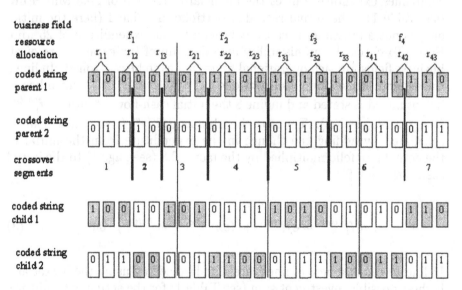

Figure 2: 6-point-crossover between coded portfolio positionings

To create the next generation two parent-strings, representing in our application portfolio positions, will stochastically be selected. Then six crossover points in the parents strings are randomly determined. After interchanging the even numbered string segments at the crossover points, we get two new child-strings (see Figure 2). The children will came in the new population, if they have a higher fitness value as their parents. Otherwise new children are constructed as described above. This procedure will be repeated three times, so that we draw first two parent-portfolios out of seven, then two out of five and finally two out of three. The remaining parent-string is adopted to the new generation. Using six cross-over points means that we get on average 1.5 interchanges in one business field. The number six seems adequate to create sufficient changes in the child-strings. A mutation rate of 1 percent is applied to the gray-bits. As break-off criterion the difference between the highest and the lowest total fitness values of a generation is used. If the difference is less than 0.02 the algorithm is stopped.

The total fitness value of a portfolio is composed by the cash-flow fitness value and the fitness value measuring the potential of success of the company (see line numbered 15, 11 and 14 in Table 4). To calculate the needed indices we start with the actual coded portfolio string (line 1) and the actual invested expenditure of the business fields ex_i^{actual} ($i = 1, ..., 4$) for the three operational functions (line 2) (for the data see Table 1). Adding the expenditure we get the total sum of each business field (line 3) and of the company (line 4). Here we can see, whether the companies expenditure meet the financial restriction of 20.5 Mio. Euro (see Table 1). The actual regarded portfolio in Table 4 (here the initial one) is not admissible, because it provides total expenditure of 25 Mio. Euro. To check the admissibility also the sum of the expenditure of all business fields for each operational function has to be controlled. In line 6 the expenditure of the inital portfolio $ex_i^{initial}$ ($i = 1, ..., 4$), which has to be optimized is stated and in line 8 the initial cash-flow situation $cf_i^{initial}$ ($i = 1, ..., 4$)(see (2)). The actual cash-flow cf_i^{actual} ($i = 1, ..., 4$) can be calculated by adding the change in the expenditure from the initial to the actual portfolio multiplied by the factor 0.8 (see page 1) to the initial cash-flow (line 9):

$$cf_i^{actual} = cf_i^{initial} + (ex_i^{initial} - ex_i^{actual}) \cdot 0.8 \tag{3}$$

In line 10 the numbers of line 9 are standardized by using the lowest and highest possible investment sum (see Table 1) for the actual expenditure. The standardized numbers are the cash-flow fitness values of the business fields, which are summed up to the cash-flow fitness value of the company (line 11). In line 12 and 13 the initial and the actual potential indices of success are given, calculated with the initial and actual expenditure of the business fields. Summing up the actual value of line 13 gives the potential of success fitness value of the company (line 14). The total fitness value of the company (line 15) is calculated by adding the cash-flow fitness weighted by the factor 0.6 and the potential of success fitness value weighted by 0.4. The cash-flow criterion is slightly weighted higher because of the lack of liquidity in reorganization cases.

Figure 3 shows the described initial portfolio in a graphical way. The business fields f_1 and f_4 describe a good portfolio position with few or acceptable expenditure and the best cash-flow position in comparison. For f_2 should be checked, whether a better cash-flow situation can be arranged. The position of f_3 is a worse one. It spends 36% of the overall expenditure of the initial portfolio and its cash-flow situation is ruinous.

	f_1	f_2	f_3	f_4
	Portfolio Positioning P_0			
1	10 | 00 | 01	01 | 11 | 00	10 | 10 | 10	01 | 11 | 10
2	1.00|1.25|1.50	0.50|3.75|0.75	1.00|5.00|3.00	0.50|3.75|3.00
3	3.75	5.00	9.00	7.25
4	25			
5	15.00%	20.00%	36.00%	29.00%
6	3.75	5.00	9.00	7.25
7	0%	0%	0%	0%
8	-0.75	-1.25	-2.25	-0.75
9	-0.75	-1.25	-2.25	-0.75
10	0.78	0.59	0.00	0.26
11	1.63			
12	0.83	1.00	1.00	1.00
13	0.83	1.00	1.00	1.00
14	3.83			
15	2.51			

Legende:

1	coded actual portfolio string
2	actual expenditure of the operational functions
3	actual total expenditure of the business fields
4	actual total expenditure of the company
5	actual relative expenditure of the business fields
6	initial total expenditure of the business fields
7	relative change between the actual and initial expenditure
8	initial cash-flow situation of the business fields
9	actual cash-flow situation of the business fields
10	standardized cash-flow fitness values of the business fields
11	cash-flow fitness value of the company
12	initial potential indices of success of the business fields
13	actual potential indices of success of the business fields
14	potential of success fitness value of the company
15	total fitness value of the portfolio

Table 4: Calculation of the initial portfolio's total fitness

440

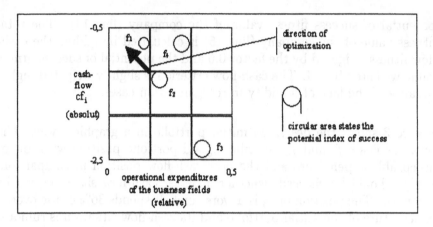

Figure 3: Initial portfolio position

		f_1	f_2	f_3	f_4	TF	TE
			Initial Population				
P_0		100001	011100	101010	011110	2.51	25.00
P_1		010000	010000	000000	111010	2.48	16.00
P_2		011001	010001	010110	100100	2.89	20.50
P_3		101010	000000	000000	000000	2.40	15.75
P_4		010101	000101	101100	110110	2.95	20.50
P_5		011010	100000	000101	000100	2.85	19.25
P_6		010101	000100	000001	101010	2.76	20.00
		Population Fitness				18.83	

Legende:

TF	Total fitness value of the company
TE	Total expenditure of the company

Table 5: Initial population of the genetic algorithm

With the initial population shown in Table 5 the described genetic algorithm is started. Already the inadmissible actual portfolio P_0 can be compared with the other 6 scenario-portfolios. Comparing P_0 and P_1 there seem savings of 9 Mio. Euro possible with just a small change in the total fitness. P_2 and P_4 are the best portfolios with a fitness increase of about 0.4 to the initial portfolio but just 82% of its expenditure. After nine iterations the stop-criterion is satisfied. Table 6 shows the final portfolio population. Graphical results for the two best portfolio constellations can be seen in Figure 4.

In the final population (see Table 6) we find two pareto-optimal portfolios with the highest fitness and the lowest expenditure. These two portfolios should be argued by the reorganization team about realisability and implementation. In both portfolios the percentages of expenditure of all

9ᵗʰ Generation						
	f_1	f_2	f_3	f_4	TF	TE	
P_1	010100	000100	010011	000100	3.2587	14.75	pareto-opt.
P_2	110100	000100	010011	000100	3.2620	15.00	pareto-opt.
P_3	010100	010100	010011	000100	3.2585	15.00	
P_4	010100	010100	110011	010100	3.2559	15.50	
P_5	000101	010100	010101	010100	3.2476	16.25	
P_6	000101	000100	010100	010100	3.2521	15.25	
P_7	010100	010100	000101	000100	3.2570	15.25	
Population Fitness					22.79		

Table 6: Optimization result of the genetic algorithm

business fields are quite close (see Figure 4) but the cash-flow situations differ considerably.

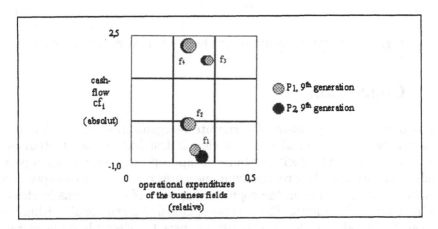

Figure 4: Two pareto-optimal portfolio alternatives

The initial and the pareto-optimal portfolios are graphically compared in Figure 5. The positions of f_3 and f_4 have changed mostly. The assessment of f_3 has extremely improved because its expenditure are cut down more than a half (absolutely), so that the cash-flow become highly positive. But the success index is lower as before. Whereas f_1 has not changed its expenditure absolutely but relatively to the other business fields. Therefore the cash-flow level is the same as in the initial portfolio. The positions of f_4 and f_2 are similar in Figure 3 and 5 but with different evaluations. f_4 holds the best financial position as well as the highest success index. In total also the position of f_2 is highly improved by allocating the financial resources.

The alternative pareto-optimal portfolios serve as a starting point for further analysis in depth with regard to the future situation. Special discussions about the implementation of the modified expenditure proportions of the business fields concerning personnel, equipment, materials, goods and further aspects are needed.

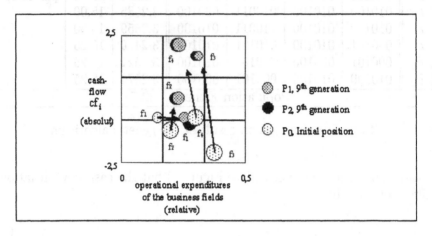

Figure 5: Portfolio positionings before and after optimization

3 Conclusion

The optimization problem of corporate reorganization from a financial as well as an economical point of view is last but not least often not well-structured. The decision about the appropriate reorganization portfolio and corresponding reorganizing measures being under time pressure makes great demands on the cognitive capacity of the persons in charge of reorganization audits. The presented model of a genetic algorithm has been developed in order to support the user to select plausible reorganizing measures by generating pareto-optimal portfolio alternatives.

References

NISSEN, V. (1997): Einführung in evolutionäre Algorithmen: Optimierung nach dem Vorbild der Evolution. Vieweg, Braunschweig

KINNEBROCK, W. (1994): Optimierung mit genetischen und selektiven Algorithmen. Oldenbourg, München

Key Success Factors in City Marketing - Some Empirical Evidence

C. Bornemeyer, R. Decker

Betriebswirtschaftslehre und Marketing,
Universität Bielefeld, D-33615 Bielefeld, Germany

Abstract: In recent years marketing techniques have been applied to solve problems in urban development and urban planning. Despite the growing popularity of city marketing and town centre management little research has been published on the effects of city marketing and possible measures of success. Based on a sample of German city marketing projects a possible procedure to identify key success factors of city marketing is proposed.

1 Motivation

City marketing is "a process whereby urban activities are as closely as possible related to the demands of targeted customers so as to maximise the efficient social and economic functioning of the area concerned in accordance with whatever goals have been established" (Ashworth, Voogd (1988)). The importance of city marketing today can be demonstrated by the large number of cities pursuing marketing activities. According to the surveys by Töpfer (1993) and Grabow, Hollbach-Grömig (1998) 89% and 83% respectively of German cities are engaged in marketing. In the present study 81% of the respondents indicated the use of marketing techniques.

But to ensure the long term support of investors the cities have to demonstrate the success of their marketing activities. However, little research focusses on the analysis of the success of city marketing projects (Bornemeyer et al. (1999)). Therefore, one objective of the research project at hand is to analyse the success of city marketing projects by identifying the factors that determine the success of such projects.

2 Data Description

The empirical investigation is based on a postal questionnaire that was sent to 649 German cities in summer 2000. Figure 1 shows selected questions from the questionnaire. In addition the names of the corresponding variables are indicated.

4. *Which of the following activities have already been part of your city marketing project, which activities are currently carried out, which of them are planned for the future and which ones are not considered? (Please mark **one box per line**!)* (prozess)

	have been part	carried out now	planned	not considered	don't know
Initiation of city marketing activities	☐	☐	☐	☐	☐
Situation analysis	☐	☐	☐	☐	☐
Development of philosophie/role model	☐	☐	☐	☐	☐
Determination of goals	☐	☐	☐	☐	☐
Strategy planning	☐	☐	☐	☐	☐
Planning of measures	☐	☐	☐	☐	☐
Implementation of measures	☐	☐	☐	☐	☐
Evaluation of activities	☐	☐	☐	☐	☐

8. *Which of the following subject areas are treated in your city marketing project? (Please mark all that apply!)* (themen)

☐ General economic aspects   ☐ Leisure and sports
☐ Retailing   ☐ Culture
☐ Catering and hotel trade   ☐ Residence and residential area
☐ Tourism   ☐ Safety
☐ Advertising and public relations   ☐ Traffic
☐ Fares and conferences   ☐ Natural environment
☐ Administration   ☐ Education
☐ Urban design and urban development   ☐ Social aspects
☐ Science and research   ☐ Others: _____

25. *In the following some statements to characterise your city marketing activities are compiled. Please, comment on these statements using the given scale! (Please mark **one box per line**!)*

	completely true				not at all true	don't know
Our city marketing activities are based on a sound strategic basis resulting from a comprehensive situation analysis. (q25_4)	○	○	○	○	○	☐
The municipal authority staff is convinced of the usefulness of city marketing. (q25_8)	○	○	○	○	○	☐
The municipal authority staff is willing to take part in the city marketing activities. (q25_9)	○	○	○	○	○	☐
The mayor is convinced of the usefulness of city marketing. (q25_13)	○	○	○	○	○	☐
The mayor is willing to take part in the city marketing activities. (q25_14)	○	○	○	○	○	☐
The results of target group surveys conducted in course of our city marketing activities transmit important impulses for our activities. (q25_15)	○	○	○	○	○	☐
City promotion is an important element of our city marketing project. (q25_21)	○	○	○	○	○	☐
Public relations have a high priority in our city marketing project. (q25_22)	○	○	○	○	○	☐

Figure 1: Selected parts of the questionnaire

The population comprises German cities with more than 10,000 inhabitants as prior secondary research has shown that only a few smaller cities are pursuing marketing activities. All cities with more than 40,000 inhabitants have been addressed, while for the cities with 10,000 to 40,000 inhabitants a stratified random sample has been drawn. The response rate was 43%, i.e., 280 usable responses. 176 of these cities have at least two years of experience in city marketing. Especially the statements of these cities are relevant in assessing the success of city marketing activities.

3 Key Success Factors and Measures of Success

Key success factors are "those characteristics, conditions, or variables that when properly sustained, maintained, or managed can have a significant impact on the success of a firm competing in a particular industry" (Leidecker, Bruno (1984)). According to Trommsdorff (1993) research in the area of key success factors can be categorised according to the criteria orientation, specificity and causality. The present study aims at identifying key success factors based on a standardised survey and the use of multivariate data analysis techniques. Therefore it has a quantitative orientation, whereas other research on key success factors in city marketing is mainly qualitatively oriented. The study at hand is highly specific as the identified key success factors will only be valid for the marketing activities of German cities. The study aims at approximating the claim for causality by considering successful as well as less successful projects and by using structural equation modelling.

Possible success factors that will be considered in the analysis have been derived by factor analyses of the indicators included in the questionnaire. The results of the two exploratory factor analyses are presented in table 1. The first analysis is based on indicators of action and situational variables. Five factors have been extracted; these can be interpreted as follows. Factor I comprises aspects like the early and steady taking of measures and the offering of additional products and services. Therefore this factor is named "activism". The second factor is labelled "sound (strategic) basis" of the project because indicators like the existence of a general strategic tendency, the existence of a package of measures and the intensity of citizens' participation have high loadings on this factor. Factor III can be named "situation" with respect to the size of the city, as it comprises the indicators number of inhabitants and overnight stays as well as the importance of the city in a region. Factor IV is characterised as "market research" due to the indicators importance of target group surveys and analysis of secondary data. Finally, factor V is named "communication".

The second factor analysis led to the extraction of two factors. Factor VI is named "attitude of municipal groups". Municipal groups comprise the municipal authorities, the town council as well as the public. Factor VII can analogously be characterised as the "attitude of the mayor". It describes the degree of the mayor's conviction of the usefulness of the marketing activities and his willingness to cooperate.

Furthermore, measures of city marketing success have to be derived. First of all, the respondents have been asked to assess the overall success of their marketing activities on a five point scale ("self-assessment"). In addition, two scoring models are used. The respondents were asked to indicate the degree of attainment of different goals (x_{ij}) as well as the importance of these goals (β_{ij}). Using the two scoring models, indices of success can be calculated as follows.

- Additive Scoring Model:
$$s_i^{add} = \sum_{j=1}^{J} \beta_{ij} x_{ij} \quad \forall i$$

- Multiplicative Scoring Model:
$$s_i^{mult} = \prod_{j=1}^{J} \tilde{x}_{ij}^{\tilde{\beta}_{ij}} \quad \forall i$$

with
s_i^{add}, s_i^{mult}	= index of success of city i (additive, multiplicative)
β_{ij}	= weight of indicator j for city i
x_{ij}	= value of indicator j for city i
$\tilde{\beta}_{ij}$	= normed weight of indicator j for city i
$\tilde{x}_{ij}$	= normed value of indicator j for city i

The multiplicative index of success corresponds to the weighted geometric mean of goal attainment. The use of normed values and weights leads to a restriction of the range of this index with respect to interpretation purposes. Both measures are used in discriminant analysis while in structural equation modelling city marketing success is considered as a latent variable and the corresponding degrees of goal attainment serve as indicator variables:

$$\begin{pmatrix} y_1 \\ \vdots \\ y_J \end{pmatrix} = \begin{pmatrix} \lambda_1 \\ \vdots \\ \lambda_J \end{pmatrix} \eta + \begin{pmatrix} \epsilon_1 \\ \vdots \\ \epsilon_J \end{pmatrix}$$

y_j	= degree of attainment of goal j	λ_j	= factor loadings
η	= city marketing success	ϵ_j	= residuals

This approach to measuring success and the results of the two exploratory factor analyses presented above constitute the basis of the measurement model that is used in structural equation modelling.

First exploratory factor analysis					
Indicators of action/	Factors				
situational variables	I	II	III	IV	V
taking measures steadily (q25_19)	0.84*	0.08	0.10	0.20	0.11
taking measures in the beginning (q25_18)	0.71*	0.00	0.03	0.24	0.09
additional services/products (q25_20)	0.60*	0.27	-0.04	0.12	0.24
activism (q25_2)	0.56*	0.53*	0.15	0.07	0.02
sound financial basis (q25_5)	0.50*	0.31	-0.04	0.14	0.10
strategic general tendency (q25_4)	0.11	0.71*	-0.02	0.29	0.03
package of measures (q11)	0.21	0.60*	0.08	0.13	0.21
intensity of citizens' participation (grbbet)	0.24	0.50*	0.00	0.20	0.09
long-term orientation (q25_1)	0.40	0.50*	0.01	0.11	-0.04
analysis of strengths/weaknesses (q25_3)	0.08	0.57*	-0.03	0.54*	0.03
number of subject areas treated (themen)	0.03	0.40	0.17	0.19	0.07
number of participating groups (akteure)	0.12	0.40	0.22	0.08	-0.02
number of process elements (prozess)	0.02	0.35	0.11	0.05	0.12
number of inhabitants (q37_1_1)	0.05	0.12	0.85*	-0.03	0.01
importance of the city in a region (q35)	-0.05	0.06	0.84*	0.05	0.10
number of overnight stays (q37_3_1)	0.00	0.15	0.72*	0.01	-0.04
retail turnover (q37_7_1)	0.02	0.06	0.32	0.10	0.00
target group surveys (q25_15)	0.15	0.14	-0.01	0.63*	0.00
analysis of secondary data (q25_16)	0.11	0.27	0.16	0.74*	0.03
competitive analysis (q25_17)	0.08	0.12	0.07	0.38	0.07
city advertising (q25_21)	0.29	0.16	-0.09	0.09	0.94*
public relations (q25_22)	0.40	0.19	-0.04	0.28	0.47*
number of instr. of citizens' part. (brbbet)	0.18	0.08	-0.07	0.19	0.02
number of capital sources used (brfinakt)	0.31	0.04	0.14	0.00	0.18
establishment of new jobs (q25_6)	0.05	0.15	0.10	0.13	-0.01
integration of town council (q25_12)	0.18	0.18	-0.24	0.18	0.12
experience in city marketing (jahre)	0.13	0.05	0.17	-0.04	0.21
distribution of power (q32)	0.23	0.16	-0.10	0.00	0.07
Variance accounted for	48%	22%	12%	6%	6%

Second exploratory factor analysis			
Indicators of	Factors		Remark: The vari-
intervening variables	VI	VII	ance accounted for
conviction of municipal authorities (q25_8)	0.83*	0.23	by the two factors is
cooperation of municipal authorities (q25_9)	0.80*	0.16	larger than 100 % as
conviction of the town council (q25_10)	0.71*	0.34	common factor analy-
acceptance in public (q25_7)	0.61*	0.25	sis has been used.
cooperation of the town council (q25_11)	0.60*	0.29	
conviction of the mayor (q25_13)	0.26	0.94*	
cooperation of the mayor (q25_14)	0.32	0.82*	
Variance accounted for	83%	19%	

Table 1: Results of the two exploratory factor analyses (varimax rotated)

4 Identification of Key Success Factors

In structural equation modelling a two-step procedure was applied. In a first step the initial measurement model was slightly modified on the basis of the results of a confirmatory factor analysis. Two indicator variables were excluded from the analysis. In a second step the final measurement model was combined with the structural model as depicted in figure 2. The structural model is based on a comprehensive catalogue of hypotheses that have been derived from the relevant literature and several discussions with city marketing experts. Therefore our approach can be charaterised as confirmatory.

The model was estimated using both Maximum Likelihood (ML) and Unweighted Least Squares (ULS). The standardised path coefficients resulting from ML estimation are given in the figure. The coefficients resulting from ULS differ only slightly. The indicated factor loadings are significant at $\alpha \leq 0.05$. The goodness of fit indices for the ML- and the ULS-estimation as well as some "standards" that have been defined in the literature (Homburg, Baumgartner (1998)) are summarised in table 2. The results can be assessed as satisfactory even though the "standards" are not met in all cases. Especially the goodness of fit indices for the ULS-estimation show very good results.

There is a significant influence of the existence of a sound strategic basis, the attitude of municipal groups and the use of communication instruments on city marketing success. The squared multiple correlation (SMC) for city marketing success ist 0.50, i.e. 50% of the variance in city marketing success is explained by the model. The degree of activism and the attitude of the mayor significantly influence the attitude of municipal groups (SMC = 0.43). Other interesting relations refer to the significant influence of the market research activities on the sound strategic basis (SMC = 0.52) and the influence of the sound strategic basis on the attitude of the mayor (SMC = 0.20).

The size of the city ("situation") does not significantly influence city marketing success. This implies that both, large and smaller cities can pursue successful marketing activities.

In order to further investigate the differences between successful and less successful city marketing projects discriminant analyses have been carried out where five different measures of success served as grouping variables (cf. table 3).

The categorisation in the case of the indices of success is based on the mean value of the corresponding index which is used as a kind of benchmark. The factor score results from an exploratory factor analysis of the degrees of attainment of different goals in city marketing. This approach to measuring success can be considered as the counterpart of the measurement model of success that has been used in structural equation modelling. The factor analysis led to the extraction of one factor

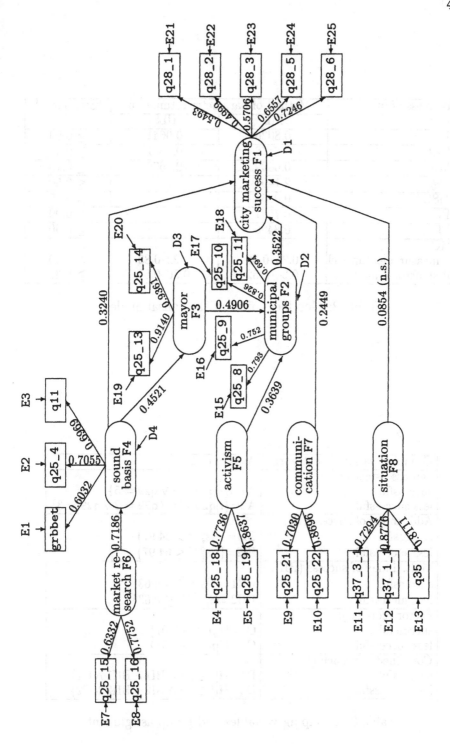

Figure 2: Structural equation model of city marketing success

Goodness of fit index	Method of parameter estimation		"Standards"
	ML	ULS	
GFI	0.8026	0.9641	≥ 0.90
AGFI	0.7478	0.9541	≥ 0.90
RMR	0.0899	0.0670	≤ 0.10
χ^2/df	2.1496	–	≤ 2.50
RMSEA	0.0858	–	≤ 0.05
NFI	0.7561	–	≥ 0.90
CFI	0.8496	–	≥ 0.90
SMC (a) measurement model	0.25–0.88	0.22–0.83	≥ 0.50
(b) structural model	0.20–0.52	0.36–0.89	≥ 0.30

Table 2: Goodness of fit of the causal model

Grouping variables/categories	Assignment
Self-assessment (A): successful	$A1=\{i|q29 = 4 \vee q29 = 5\}$
less successful	$A2=\{i|q29 = 1 \vee q29 = 2 \vee q29 = 3\}$
Add. index of success (BI): successful	$BI1=\{i|s_i^{add} > 64.97\}$
less successful	$BI2=\{i|s_i^{add} \leq 64.97\}$
Mult. index of success (BII): successful	$BII1=\{i|s_i^{mult} > 0.63\}$
less successful	$BII2=\{i|s_i^{mult} \leq 0.63\}$
Factor score (C): successful	$C1=\{i|p_i > -0.01\}$
less successful	$C2=\{i|p_i \leq -0.01\}$
Combined approach (D): successful	$D1=\{i|i \in A1 \wedge BI1 \wedge BII1 \wedge C1\}$
less successful	$D2=\{i|i \in A2 \wedge BI2 \wedge BII2 \wedge C2\}$

Table 3: Grouping variables and group assignment

("success") that accounts for 98% of the variance. Marketing projects have been assigned to group C1 if their factor score (p_i) was larger than the mean of factor scores. This mean does not equal zero as a pairwise deletion of missing values has been carried out for the calculation of the factor loadings while listwise deletion has been applied for the calculation of factor scores.

In terms of the combined approach marketing projects are characterised as successful or less successful if they are considered successful or less successful according to <u>all</u> the criteria of interest (A to C). In this case the remaining projects are excluded from the analysis.

The predictor variables have been derived by one exploratory factor analysis of all indicators presented in table 1. Eight factors that account for 95% of the variance could be extracted. The interpretation of the extracted factors is similar to those discussed earlier:

- Factor I: activism
- Factor II: sound (strategic) basis
- Factor III: situation
- Factor IV: attitude of the town council
- Factor V: market research
- Factor VI: attitude of the mayor
- Factor VII: attitude of municipal authorities
- Factor VIII: communication

Table 4 shows the results of five discriminant analyses using the different grouping variables. The differences between successful and less successful city marketing projects are significant for all grouping variables. The assessment of fit provides evidence of the slightly lower suitability of the additive index of success as a grouping variable.

Factor	Grouping variables				
	A	BI	BII	C	D
I	1.017	0.960	0.937	0.903	1.178
II	0.620	0.187	0.233	0.422	0.519
III	0.535	-0.015	-0.021	0.040	0.274
IV	0.351	0.035	0.159	0.077	0.262
V	-0.011	0.177	0.128	-0.037	-0.024
VI	0.021	0.130	0.220	0.381	0.200
VII	0.431	0.378	0.377	0.393	0.409
VIII	0.197	0.174	0.290	0.448	0.477
Wilks' lambda	0.514	0.805	0.747	0.660	0.454
Roy's greatest root	0.947	0.242	0.338	0.514	1.203
F-values	9.584	2.422	3.380	5.142	7.970
	$(\alpha < 0.01)$	$(\alpha = 0.02)$	$(\alpha < 0.01)$	$(\alpha < 0.01)$	$(\alpha < 0.01)$

Table 4: Discriminant coefficients and assessment of fit

As regards the importance of the factors in discriminating between successful and less successful projects we obtained similar results to those from structural equation modelling. The factors "activism" and "attitude of municipal authorities" have high discriminating power for all grouping variables. The "sound (strategic) basis" is relevant for grouping variables A, C, and D, while the factor "communication" is of high importance in cases BII, C, and D.

Table 5 presents the classification results. The Press's Q statistic shows that the assignment based on discriminant analyses is significantly better than an assignment by chance ($\alpha < 0.01$). The best classification results were obtained by the combined approach (D). This was to be expected because the combined approach integrates the information covered by all other grouping variables.

Finally, the classification results have been cross validated by the leaving-one-out-method (Hand (1997)). It should be noted that the hit rates based on the cross validation sample are only slightly lower than those based on the original sample.

	Grouping variable				
	A	BI	BII	C	D
Hit rate (resubstitution)	79.35%	71.69%	73.69%	72.06%	85.91%
Hit rate (cross validation)	77.53%	64.27%	68.37%	68.01%	83.41%
Proportional-chance criterion	52.47%	54.60%	52.78%	51.42%	54.21%
Press's Q	34.84	17.09	18.89	17.09	34.31
			($\alpha < 0.01$)		

Table 5: Classification results

5 Summary and Outlook

The starting point of our research activities was the need to analyse the success of city marketing projects. On the basis of a comprehensive empirical investigation key success factors of city marketing have been identified using structural equation modelling and discriminant analysis. In addition, several approaches to measuring city marketing success have been presented.

Further research activities are focusing on the development of a classification scheme to evaluate new city marketing projects. Given this scheme, decision support for city management will be provided.

References

ASHWORTH, G. J. and VOOGD, H. (1988): Marketing the city. *Town Planning Review, Vol. 59, 65–80.*

BORNEMEYER, C., TEMME, T., and DECKER, R. (1999): Erfolgs-faktorenforschung im Stadtmarketing unter besonderer Berücksichtigung multivariater Analyseverfahren, in Gaul, Schader (Eds.): Mathematische Methoden der Wirtschaftswissenschaften, Physica, Heidelberg.

GRABOW, B. and HOLLBACH-GRÖMIG, B. (1998): Stadtmarketing, eine kritische Zwischenbilanz. Deutsches Institut für Urbanistik, Berlin.

HAND, D. J. (1997): Construction and assessment of classification rules. Wiley, Chichester.

HOMBURG, C. and BAUMGARTNER, H. (1998): Beurteilung von Kausalmodellen, in Hildebrandt, Homburg (Eds.): Die Kausalanalyse, Schäffer-Poeschel, Stuttgart.

LEIDECKER, J. K. and BRUNO, A. V. (1984): Identifying and using critical success factors. *Long Range Planning, Vol. 17, 23–32.*

TÖPFER, A. (1993): Marketing in der kommunalen Praxis, in Töpfer (Ed.): Stadtmarketing. FBO, Baden-Baden.

TROMMSDORFF, V. (1993): Erfolgsfaktorenforschung über Produkt-innovationen, in Meyer-Krahmer (Ed.): Innovationsökonomie und Tech-nologiepolitik, Physica, Heidelberg.

From Credit Scores to Stable Default Probabilities: A Model Based Approach

S. Höse, S. Huschens

Lehrstuhl für Quantitative Verfahren, insbesondere Statistik,
Fakultät Wirtschaftswissenschaften, TU Dresden,
Mommsenstraße 13, D-01062 Dresden, Germany

Abstract: The default probability of borrowers is one of the crucial inputs for calculating the regulatory capital underlaying credit risk. Currently, default probabilities are estimated in rating grades by using historical default data. In this attempt, information provided by many banks is lost by the classification of credit scores into rating grades. To avoid this loss of information, a direct mapping of credit scores to default probabilities is presented in this paper. This also guarantees stable default probabilities as demanded by the Basel Committee. The model uses an increasing and convex functional relation between credit scores and default probabilities of individual borrowers. Consequently, the approach focuses on the estimation of model parameters instead of the default probability itself. Maximum-Likelihood estimators for the model parameters are derived, which account for heteroskedasticity.

1   Introduction

There are many different kinds of risks against which bank's managements need to guard. For most banks the major risk is credit risk, that is to say the risk of counterparty failure. Therefore, the Basel Accord (1988) requires regulatory capital calculated as credit exposure weighted by risk weights. Recently, the Basel Committee issued a consultative paper (2001a) proposing the Internal Ratings-Based Approach. It suggests the calculation of the regulatory risk weights as a continuous function of stable default probabilities. The Basel Committee demands that these default probabilities are accessed by credit ratings and are calculated as average values over all borrowers within one rating grade. Hence, all borrowers within that grade are treated as having the same probability of default.

Many best-practice banks currently judge borrowers on a finer scale, the credit scoring, using all available information. Borrowers with roughly similar scores are then pooled - under information loss - in one credit rating grade (cp. Carey, Hrycay (2001)). Further, the average default probabilities of the rating grades are estimated from historical default data.

To obtain regulatory risk weights, a model based approach for the estimation of stable default probabilities based on the finer scoring scale is proposed. Therefore, a stable, increasing, and strictly convex functional relation between credit scores and default probabilities of individual borrowers is assumed. This avoids a loss of information and restricts the estimation to few model parameters, here done by Maximum-Likelihood (ML) estimation.

Alternatively, non-parametric regression might be used to specify the relation between credit scores and default probabilities. A fitted non-parametric regression highly depends on the choice of kernel and of bandwidth, especially for small sample size. Further, non-parametric regression guarantees neither the monotonicity nor the convexity functional relation. Thus, it is more promising to use a parametric model based approach, since it intrinsically obeys the desired properties.

2   The Model

A stable, monotone increasing relation between credit scoring

$$s \in S := \{s_{min}, \ldots, s_{max}\} \tag{1}$$

and default probability

$$p(s; \theta) : S \to [0, 1], \quad s \mapsto p(s; \theta) \tag{2}$$

is assumed. This reflects directly the link between credit worthiness, here measured by credit scores, and the probability of default. Hence, small credit scores correspond to high-quality credits with low probability of default and high scores reflect speculative rating grades with high default risk. Although most banks apply scoring systems with an increasing relation between scores and default probabilities (Pfingsten, Schröck (2000)), the model can be easily adapted to the inverted case.

After defining the standardized scores[1],

$$\tilde{s} = \frac{s - s_{min}}{s_{max} - s_{min}} \in [0, 1], \tag{3}$$

the following two alternative functional forms are considered

$$p(s; \theta) = \tilde{s}^{\theta} \quad \text{with} \quad \theta > 0 \tag{4}$$

---

[1]The definition simplifies the comparison between different scoring systems used by different banks.

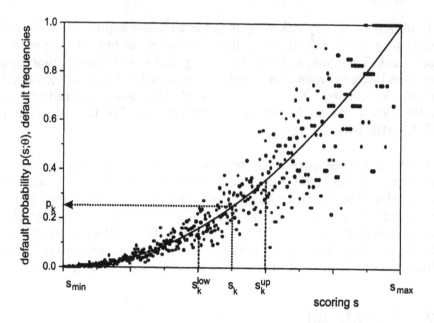

Figure 1: Scores versus default probabilities and default frequencies

and

$$p(s; \theta) = ce^{\theta \bar{s}} \quad \text{with} \quad c \in (0,1), \quad 0 < \theta < \ln(1/c). \tag{5}$$

Note that $c = p(s_{min}; \theta)$ defines a lower bound for all default probabilities.[2] For $\theta > 1$ in (4) both models specify an increasing and strictly convex functional relation between credit scoring and default probability. This functional shape seems to be typical for most scoring systems in practice (cp. Pfingsten, Schröck (2000)).

Illustrating this, a possible relation between scoring and default probability is shown in figure 1. As an example, model (4) is applied with $\theta = 2$. Furthermore, this model and an exponentially decreasing population of scores is used to simulate default frequencies.

In the context of this model based approach, the default probability p_k for the rating grade $k = 1, \ldots, K$ is defined as

$$p_k = p(s_k; \theta) \tag{6}$$

with

$$s_k = \frac{s_k^{low} + s_k^{up}}{2}, \tag{7}$$

---

[2]The Basel Committee on Banking Supervision (2001a) defines the default probability (PD) "... as the greater of the one-year PD associated with the internal borrower grade to which that exposure is assigned, or 3 basis points."

which is also shown in figure (1). Thereby, $[s_k^{low}, s_k^{up})$ denotes the interval of scores belonging to rating grade k with the mean score s_k. Consequently, p_k is independent of the number of borrowers within the rating grade k and is stable over the considered time horizon. This reflects the demands of the Basel Committee (2001a) on the default probability (PD) for corporates: "PD estimates must represent a conservative view of a long-run average PD for the borrower grade in question, ... must be grounded in historical experience and empirical evidence ... over an entire economic cycle ... using a minimum historical observation period of at least 5 years."

For retail portfolios the Basel Committee (2001a) demands: "... as a minimum requirement, a bank must segment by credit scores or equivalent." According the Basel Committee (2001b), the segmentation process must ensure " ... that the risk characteristics of the underlying pool of loans are relatively stable over time, and can be separately tracked."

3   Parameter Estimation

The model parameter θ is determined by Maximum-Likelihood estimation. Therefore, the default events are treated first as binomially distributed using full knowledge about historical defaults in each score (sec. 3.1). On the other hand, defaults can be treated as approximately Poisson distributed (sec. 3.2). Thereby, Poisson approximation on the aggregate level of rating grades allows parameter estimation if information about defaults in each score is missing.

3.1   The Binomial Approach

The default of a single credit i in the scoring class s is described by a binomially distributed random variable

$$X_{i,s} \sim Bin(1, p(s; \theta)) \quad \text{for all} \quad i = 1, \ldots, n_s, \tag{8}$$

where n_s describes the number of credits in scoring class s. Consequently, the total number of defaults H_s in scoring class s is given as

$$H_s = \sum_{i=1}^{n_s} X_{i,s} \sim Bin(n_s, p(s; \theta)), \tag{9}$$

which is binomially distributed under the assumption of mutually independent default events. The probability mass function for the number of defaults H_s in the scoring class s reads as

$$P(H_s = h_s) = \binom{n_s}{h_s} p(s; \theta)^{h_s} (1 - p(s; \theta))^{n_s - h_s} \tag{10}$$

for $h_s = 0, 1, \ldots, n_s$. Note that $P(H_s = 0) = 1$ for $n_s = 0$ or $p(s; \theta) = 0$ and $P(H_s = n_s) = 1$ for $p(s; \theta) = 1$ holds.

The log-likelihood function $l(\theta)$ is then given by

$$l(\theta) = \sum_{s \in S'} \ln P(H_s = h_s) \tag{11}$$

$$= \sum_{s \in S'} \ln \binom{n_s}{h_s} + h_s \ln p(s; \theta) + (n_s - h_s) \ln(1 - p(s; \theta)) \tag{12}$$

with

$$S' := \{s \in S \mid 0 < p(s; \theta) < 1\}. \tag{13}$$

Using $p(s; \theta)$ from (4) the log-likelihood function specializes to

$$l(\theta) = \sum_{s \in S'} \ln \binom{n_s}{h_s} + h_s \ln \tilde{s}^\theta + (n_s - h_s) \ln(1 - \tilde{s}^\theta). \tag{14}$$

From

$$l'(\theta) = \sum_{s \in S'} \frac{n_s \ln \tilde{s}}{1 - \tilde{s}^\theta} \left(\frac{h_s}{n_s} - \tilde{s}^\theta \right) \tag{15}$$

follows that

$$\sum_{s \in S'} \frac{n_s \ln \tilde{s}}{1 - \tilde{s}^{\hat{\theta}}} \left(\frac{h_s}{n_s} - \tilde{s}^{\hat{\theta}} \right) = 0 \tag{16}$$

implicitly defines the ML estimator $\hat{\theta} \in (0, \infty)$ for the unknown parameter θ. Existence and uniqueness can be seen from

$$\lim_{\theta \to \infty} l'(\theta) = \sum_{s \in S'} h_s \ln \tilde{s} < 0, \quad \lim_{\theta \to 0} l'(\theta) = \infty \tag{17}$$

and

$$l''(\theta) = \sum_{s \in S'} \frac{(\ln \tilde{s})^2 \tilde{s}^\theta}{(1 - \tilde{s}^\theta)^2} (h_s - n_s) < 0. \tag{18}$$

With the second model $p(s; \theta)$ from (5) the log-likelihood function (12) specializes to

$$l(\theta) = \sum_{s \in S'} \ln \binom{n_s}{h_s} + h_s(\ln c + \theta \tilde{s}) + (n_s - h_s) \ln(1 - ce^{\theta \tilde{s}}). \tag{19}$$

The ML estimator $\hat{\theta}$ for the unknown parameter θ in this case is implicitly given by

$$\sum_{s \in S'} \frac{n_s \tilde{s}}{1 - ce^{\hat{\theta}\tilde{s}}} \left(\frac{h_s}{n_s} - ce^{\hat{\theta}\tilde{s}} \right) = 0. \tag{20}$$

Existence and uniqueness of the ML estimator $\hat{\theta}$ can be shown analogously to (17) and (18).

In both cases the ML estimator can be numerically determined either by maximizing the log-likelihood function (14) or (19) or by finding the root of the normal equation (16) or (20), respectively. Moreover, the default events show the property of heteroskedasticity, which is already considered by the derived ML estimators.

The default probabilities p_k for the rating grades $k = 1, \ldots, K$ can be estimated from

$$\hat{p}_k = p(s_k; \hat{\theta}), \quad k = 1, \ldots, K, \tag{21}$$

where s_k is the mean score for rating grade k, $\hat{\theta}$ is the ML estimator for θ, and $p(s; \theta) = \tilde{s}^{\theta}$ in the first model and $p(s; \theta) = ce^{\theta\tilde{s}}$ in the second model.

3.2   Poisson Approximation

In this section estimators for the model parameter θ are given, where the binomial distribution for the number of defaults is approximated by a Poisson distribution. The binomial distribution in (9) is replaced by

$$H_s \sim Poi(\lambda_s) \tag{22}$$

with Poisson parameter

$$\lambda_s := n_s p(s; \theta) \tag{23}$$

and probability mass function

$$P(H_s = h_s) = \frac{\lambda_s^{h_s}}{h_s!} e^{-\lambda_s}, \quad h_s = 0, 1, \ldots, \tag{24}$$

where $P(H_s = 0) = 1$ in the degenerate case $\lambda_s = 0$. This approximation is used in some credit risk models (cp. Credit Suisse First Boston (1997)).

Assuming the independence of the variables H_s, the log-likelihood function in this case is

$$l(\theta) = \sum_{s \in S'} h_s \ln n_s + h_s \ln p(s; \theta) - \ln(h_s!) - n_s p(s; \theta) \tag{25}$$

with

$$S' = \{s \in S \mid \lambda_s > 0\} = \{s \in S \mid p(s;\theta) > 0,\ n_s > 0\}. \qquad (26)$$

In the first model, $p(s;\theta) = \tilde{s}^\theta$, the unique ML estimator is implicitly defined by

$$l'(\theta) = 0 = \sum_{s \in S'} n_s \ln \tilde{s} \left(\frac{h_s}{n_s} - \tilde{s}^{\hat{\theta}} \right). \qquad (27)$$

In the second model, $p(s;\theta) = ce^{\theta \tilde{s}}$, the unique ML estimator is given by

$$l'(\theta) = 0 = \sum_{s \in S'} n_s \tilde{s} \left(\frac{h_s}{n_s} - ce^{\hat{\theta}\tilde{s}} \right). \qquad (28)$$

In both models the log-likelihood function is strictly concave in θ, which follows from $l''(\theta) < 0$.

The model parameter θ can still be estimated if detailed information about defaults in scoring classes is missing. The total number of defaults per rating category k is used instead, which is defined as

$$H_{(k)} = \sum_{s \in S_k} H_s, \quad k = 1, \ldots, K, \qquad (29)$$

where $S_k = [s_k^{low}, s_k^{up})$ contains the scores defining rating grade k. The probability mass function of $H_{(k)}$ is complicated, since $H_{(k)}$ is a sum of binomially distributed variables H_s with different parameters n_s and $p(s;\theta)$ for $s \in S_k$. A first approximation for the distribution of $H_{(k)}$ is

$$H_{(k)} \sim Poi(\lambda_{(k)}) \quad \text{with} \quad \lambda_{(k)} = \sum_{s \in S_k} n_s p(s;\theta). \qquad (30)$$

This approximation requires knowledge of the n_s for $s \in S_k$. A second approximation for the distribution of $H_{(k)}$ based on aggregate information is

$$H_{(k)} \sim Poi(\lambda_{(k)}) \quad \text{with} \quad \lambda_{(k)} = n_{(k)} p(s_k;\theta), \qquad (31)$$

where

$$n_{(k)} = \sum_{s \in S_k} n_s \qquad (32)$$

denotes the total number of credits in rating grade k and $p(s_k;\theta)$ is the default probability for rating grade k. This second approximation uses only aggregate information on the level of rating grades. The resulting ML estimators for θ are analogously to (27) and (28) given by

$$\sum_{k=1}^{K} n_{(k)} \ln \tilde{s}_k \left(\frac{h_{(k)}}{n_{(k)}} - \tilde{s}_k^{\hat{\theta}} \right) = 0 \qquad (33)$$

in the first model and by

$$\sum_{k=1}^{K} n_{(k)} \tilde{s}_k \left(\frac{h_{(k)}}{n_{(k)}} - ce^{\hat{\theta}\tilde{s}_k} \right) = 0 \qquad (34)$$

in the second model with

$$\tilde{s}_k = \frac{s_k - s_{min}}{s_{max} - s_{min}}. \qquad (35)$$

These estimators account for heteroskedasticity and can be used to calculate stable default probabilities p_k analogously to (21) for determining regulatory risk weights.

4   Conclusions

Bank's internal credit risk controlling generally uses default probabilities estimated from the very recent history. These do not necessarily obey time stability, which is demanded by the Basel Committee (2001a) for the calculation of regulatory risk weights. In order to guarantee this stability, a model based approach is presented for the estimation of stable default probabilities in rating grades. It uses the known increasing and convex functional relation between credit scoring and default probability of individual borrowers. Consequently, the approach focuses on the estimation of the model parameter instead of the default probabilities. ML estimators for the model parameter are derived for binomially distributed default events as well as for approximately Poisson distributed defaults. In the second case, the model parameter can be estimated even if detailed information about defaults in scoring classes is missing. Aggregate default data on the rating grade level are used instead. In both cases the estimator accounts for the heteroskedasticity. Since default events are treated as being independent, further improvements would also include dependence.

References

BASEL COMMITTEE ON BANKING SUPERVISION (2001a): The Internal Ratings-Based Approach, Consultative Document, January 2001

BASEL COMMITTEE ON BANKING SUPERVISION (2001b): The New Basel Capital Accord, Consultative Document, January 2001

BASEL COMMITTEE ON BANKING SUPERVISION (1988): International Convergence of Capital Measurement and Capital Standards - Basle Capital Accord

CAREY, M.; HRYCAY, M. (2001): Parameterizing credit risk models with rating data. *Journal of Banking & Finance, Vol. 25, 197–270.*

CRDIT SUISSE FIRST BOSTON (1997): CreditRisk+

PFINGSTEN, A.; SCHRÖCK, G. (2000): Bedeutung und Methodik von Krediteinstufungsmodellen im Bankwesen, in Oehler (Ed.): Kreditrisikomanagement - Portfoliomodelle und Derivate, Schäffer-Poeschel, Stuttgart

Dimensions of Credit Risk

R. Kiesel[1], U. Stadtmüller[2]

[1]Department of Statistics
London School of Economics
London WC2A 2AE, England

[2]Abteilung für Zahlen- und Wahrscheinlichkeitstheorie
Universität Ulm
D-89069 Ulm

Abstract: To understand the contribution of various risk factors to the overall riskiness of credit-risky portfolios is one of the most challenging tasks in contemporary finance. Recently, the importance of this issue has been highlighted by the decision of the Basel committee to allow sophisticated banks to use their own internal credit portfolio risk. In this note we use 'default-mode' credit portfolio risk models to study the importance of the factors correlation and granularity for the overall risk of credit risky portfolios.

1 Introduction

The importance of a satisfactory understanding of the contribution of various ingredients of credit risk to overall credit portfolio risk recently has been highlighted by the discussion initiated by the Basel Supervisors Committee (see e.g. BIS (2001) or Hirtle et al. (2001)). The calculation of risk capital based on the *internal rating* approach, currently favored by the Basel Supervisors Committee, relies on an accurate understanding of various relevant portfolio characteristics.

However, few studies have attempted to investigate aspects of portfolio risk thoroughly. In Carey (1998) the default experience and loss distribution for privately placed US bonds is discussed. VaRs for portfolios of public bonds, using a bootstrap-like approach are calculated in Carey (2000). While these two papers utilise a 'default-mode' (abstracting from changes in portfolio value due to changes in credit standing), Kiesel et al. (1999) employ a 'mark-to-market' model and stress the importance of stochastic changes in credit spreads associated with market values.

This empirical uncertainty about the dimensions of credit risk is partly due to the fact that studies on such aspects have to rely on one of the existing credit risk models, see e.g. Ong (1999) or Croughy et al. (2001) for a discussion of such models. As shown in Gordy (2000) results obtained using such models are quite sensitive to assumptions on parameters that are often difficult to observe (or which have to be chosen judgementally by the user). Despite these shortcomings credit risk portfolio models are useful to measure the relative riskiness of credit risky portfolios, not least in the view that other measures such as the spread over default-free interest rate or default probabilities calculated from long runs of historic data, etc suffer from other intrinsic drawbacks.

The aim of this note is to contribute to the understanding of the dimensions of credit portfolio risk and to the role credit portfolio risk models might play in attempts to quantify risk factors. More precisely, we study the effects of diversification within the theory of portfolios of credit risky assets by investigating the impact of correlation of assets and granularity (the extend to which a large fraction of portfolio exposure is to only a few obligors). While correlation is found be an important factor (in line with the standard literature of portfolio theory) the impact of granularity is not so clear cut. Carey (2000) found in a study based on bootstrap-like sampling from a large portfolio of credit risky bonds that differences in granularity have a material effect on risk. However, studies based on credit risk models usually imply that granularity is not a crucial factor, that is, the effect of diversification begins to influence VaR numbers rather early.

To investigate these effects we use a rating-based credit portfolio risk model. We provide an exact formula for the portfolio-loss distribution in the special case of a 'default-mode' model and state an approximative formula for the full model under an (standard) assumption on the underlying risk factor structure.

The structure of this note is as follows: in §2 we deduce an exact formula and state the result of an analytic approximation, in §3 we present results on the effect of correlation and granularity using analytic approximations and discuss the underlying mathematical reason for the results on granularity. §4 concludes with implications for credit risk modeling.

2 Analytic Approximations

We consider a rating-based credit portfolio model, such as CreditMetrics (see J.P.Morgan (1997)). A formal description of such a model consists of N rating categories (one of which corresponds to the default state), a transition matrix of probabilities of rating changes within the time horizon of interest, and some re-evaluation procedure for the exposures within each rating class. To introduce dependencies of the individual exposures a latent factor driving the transitions is assumed. Thus we can describe each individual exposure j by a four-dimensional stochastic vector

$$(S_j, k_j, l_j, \pi(j, k_j, l_j)),$$

where

1. S_j is the (individual) driver for defaults and rating migrations,

2. k_j, l_j represent the initial and end-of-period rating categories,

3. $\pi(.)$ represents the credit loss (end-of-period exposure value).

The loss (or end-of-period value) of the exposure is the given by

$$C_n = \sum_{j=1}^{n} \pi(j, k_j, l_j) \tag{1}$$

Typically, the correlation of the individual transition drivers is modeled assuming common factors (as in the equity-oriented factor models commonly used in portfolio analysis)

$$S_j = \mu_j + \beta_j' f + \sigma_j \epsilon_j,$$

where f (the latent factor) is a p dimensional random vector, ϵ_j (the idiosyncratic risks) are random variables, and β_j are p-dimensional vectors, and $\mu_j, \sigma_j > 0$ are constants. The standard models assume $f \sim N_p(0, \Omega)$ and $\epsilon_j \sim N(0, 1)$ and then the weights can be chosen in such a way that $S_j \sim N(0, 1)$ (compare e.g. Gordy (2000)). Rating transitions of exposure j from rating class k to l are modeled after defining a sequence of cut-off levels

$$-\infty = D_{-1}^k < D_0^k < \ldots < D_N^k = \infty$$

with $I\!P(S_j \in (D_{l-1}^k, D_l^k]) = p_{kl}$ an appropriate transition probability (estimated from historical data or judgementally fixed).

Most applied and theoretical studies use simulation methods to obtain risk measures from credit portfolio models. Recently however there have been attempts to evaluate or approximate the portfolio value distribution given in (1) analytically. In particular, Lucas et al. (1999) show that the loss of the exposure C_n can be approximated (a.s.) by a quantity $g(f)$, which no longer depends on the idiosyncratic risk factors ϵ_j (for related results see also Frey et al. (2001)).

In the following analysis we assume a single common driving factor and a default-mode model, i.e., we have only two rating categories one of which corresponds to default of the obligor. That is, using the above notation, we have

$$S_j = \rho_j f + \sqrt{1 - \rho_j^2} \epsilon_j,$$

where we assume additionally that $f \sim N(0,1)$, $\epsilon_j \sim N(0,1)$ and $f, \epsilon_1, \epsilon_2, \ldots$ are independent. There is a single cut-off level $-\infty < D_j < \infty$ with the event $\{S_j < D_j\}$ corresponding to default of the obligor. So, with p_j the default probability of obligor j, the cut-off level D_j can be found from $p_j = \mathbb{P}(S_j < D_j)$, i.e. D_j is the p_j-quantile of the distribution of default driver S_j. Furthermore, we assume a zero recovery rate, that is, in case of default everything is lost, i.e $\pi(j,1,0) = 1$. Thus the loss of a portfolio (with equal weights) is given by

$$C_n = \sum_{j=1}^{n} \mathbf{1}_{\{S_j < D_j\}}, \tag{2}$$

and we are interested in computing (or approximating) the distribution of C_n.

Now for a single exposure j we have by conditioning on the factor value

$$
\begin{aligned}
q_j &= \mathbb{P}(S_j < D_j) = \mathbb{P}(\rho_j f + \sqrt{1 - \rho_j^2}\,\epsilon_j < D_j) \\
&= \int_{-\infty}^{\infty} \mathbb{P}\left(\epsilon_j < \frac{D_j - \rho_j x}{\sqrt{1 - \rho_j^2}}\right) \phi(x)\,dx \\
&= \int_{-\infty}^{\infty} \Phi\left(\frac{D_j - \rho_j x}{\sqrt{1 - \rho_j^2}}\right) \phi(x)\,dx,
\end{aligned}
$$

where $\Phi(.)$ and $\phi(.)$ are the cumulative distribution function and the density of a standard normal, respectively. Observe, that in case of a constant factor exposure, i.e. $\rho \equiv \rho_j$, and with a homogeneous group of obligors, i.e. $p_j \equiv p$ and so $D_j \equiv D$, we have a constant conditional default probability, i.e. $q_j \equiv q$. In this case, we get for the loss distribution

$$
\begin{aligned}
\pi_k &= \mathbb{P}(C_n = k) = \mathbb{P}\left(\sum_{j=1}^{n} \mathbf{1}_{\{S_j < D\}} = k\right) \\
&= \int_{-\infty}^{\infty} \binom{n}{k} \Phi\left(\frac{D - \rho x}{\sqrt{1 - \rho^2}}\right)^k \Phi\left(-\frac{D - \rho x}{\sqrt{1 - \rho^2}}\right)^{n-k} \phi(x)\,dx.
\end{aligned}
$$

Using the above formula, we can obtain the loss distribution analytically.

On the other hand an application of a strong law of large numbers (Hall et al. (1980), Theorem 2.19) yields an even more convenient approximation:

$$\lim_{n \to \infty} \frac{1}{n} C_n = \Phi\left(\frac{D - \rho f}{\sqrt{1 - \rho^2}}\right) \quad a.s. \tag{3}$$

(Lucas et al. (1999) and Frey et al. (2001) discuss related results and provide extensions).

Approximation (3) makes it particularly easy to calculate (approximative) quantiles of the loss distribution. Indeed, we have

$$\mathbb{P}\left(\Phi\left(\frac{D-\rho f}{\sqrt{1-\rho^2}}\right) \leq y\right)$$

$$= \mathbb{P}\left(\frac{D-\rho f}{\sqrt{1-\rho^2}} \leq \Phi^{-1}(y)\right)$$

$$= \mathbb{P}\left(\frac{1}{\rho}\left(D - \sqrt{1-\rho^2}\Phi^{-1}(y)\right) \leq f\right)$$

$$= 1 - \Phi\left(\frac{1}{\rho}\left(D - \sqrt{1-\rho^2}\Phi^{-1}(y)\right)\right),$$

since $f \sim N(0,1)$. Thus, in order to obtain quantiles for the loss distribution, we have to evaluate the standard normal cumulative distribution function at the corresponding transformed quantile, i.e.

$$y_\alpha = \Phi\left(\frac{\rho q_\alpha + D}{\sqrt{1-\rho^2}}\right), \tag{4}$$

where q_α is the α-quantile of a standard normal.

3 Portfolio Analysis

As mentioned in the introduction correlation is a crucial factor, see e.g. Kiesel et al. (1999), where correlation is found to increase VaR-levels substantially. On the other hand, studies of granularity using credit portfolio models show that granularity is not a crucial factor – somewhat contrary to ones expectation. Below we investigate correlation and granularity using the above results on analytic approximations in our simple one-factor setting. The main tool will be the approximative formulas given in equations (3) and (4).

3.1 Single Grade VaRs

Table 1 reports VaR numbers in percentage points of total exposure obtained by numerical computation and analytic approximation for single grade portfolios of size $N = 500$. Default probabilities for the rating grades are chosen according to the standard transition matrix from the CreditMetrics manual, see J.P.Morgan (1997). (This matrix is also reported in e.g. Ong (1999) or Croughy et al. (2001)). We concentrate on the rating classes BBB down to B, since one-year default probabilities are negligible for the higher classes. The results show an increase of the VaR numbers with increasing correlation. Also, for classes with a significant default probability, exact VaR numbers are reasonably approximated using formula (4). Indeed, for a portfolio of 500 exposures we simply have to multiply the quantile obtained from (4) by 500.

		Exact VaR		Analytic VaR	
Grade	ρ	$\alpha = 1.0\%$	$\alpha = 0.3\%$	$\alpha = 1.0\%$	$\alpha = 0.3\%$
BBB	0.1	0.6	0.8	0.15	0.17
	0.2	0.8	1.0	0.27	0.35
	0.3	0.8	1.2	0.44	0.65
BB	0.1	3.2	3.6	2.28	2.53
	0.2	4.4	5.0	3.66	4.41
	0.3	6.2	7.2	5.53	7.17
B	0.1	9.8	10.6	8.60	9.29
	0.2	13.4	14.8	12.53	14.38
	0.3	18.4	20.0	17.46	21.87

Table 1: Single Grade VaRs

3.2 Correlation

In this section we use an extension of formula (3) to allow for changes of credit quality short of default. We use the standard example portfolios as in Gordy (2000) and Kiesel et al. (1999): An average portfolio containing 15 AAA, 25 AA, 67 A, 156 BBB, 162 BB, 56 B, and 20 CCC rated credits each of the same face value and the same maturity of 5 years. A high quality portfolio which contains 19 AAA, 30 AA, 146 A, 190 BBB, 95 BB, 13 B, and 6 CCC rated credits. Finally, an investment quality portfolio containing 28 AAA, 48 AA, 128 A, 297 BBB rated credits. We base the approximation on a CreditMetrics transition matrix with a one year holding period and assume a single risk factor. VaRs are computed on a mark-to-market basis with deterministic spreads and a fixed recovery rate of 60%. The VaR numbers, reported as percentage of total exposure in Table 2, show that increasing overall correlation increases the VaR levels quite substantially.

Portfolio	ρ	VaR		
		$\alpha = 1.0\%$	$\alpha = 0.3\%$	$\alpha = 0.1\%$
Average Quality	0.1	2.12	2.30	2.46
	0.2	3.16	3.65	4.08
	0.3	4.44	5.39	6.27
High Quality	0.1	1.06	1.16	1.24
	0.2	1.62	1.90	2.16
	0.3	2.34	2.91	3.46
Investment Quality	0.1	0.46	0.51	0.56
	0.2	0.75	0.90	1.05
	0.3	1.13	1.47	1.81

Table 2: VaRs by Analytic Approximation

3.3 Granularity

Finally, Table 3 reports results for single grade portfolios where a significant part of the exposure is given to a few bonds. In particular, we used concentrated portfolios P_1 resp. P_2 of 500 exposures of which 50 resp. 10 make up 40% of the value (later called large loans) and 450 resp. 490 make up the remaining 60% (later called small loans). For reasons of brevity we only report the results for a portfolio consisting of obligors from rating class B with underlying correlation $\rho = 0.3$. Again, the VaR numbers are quoted as percentage of total exposure.

Standard Portfolio P		Concentrated Portfolio P_1		Concentrated Portfolio P_2	
$\alpha = 1.0\%$	$\alpha = 0.3\%$	$\alpha = 1.0\%$	$\alpha = 0.3\%$	$\alpha = 1.0\%$	$\alpha = 0.3\%$
3.68	4.00	3.87	4.19	5.49	5.79

Table 3: Effect of Granularity

Observe that extreme events due to high granularity are caused by defaults of several large loans. However, since we still work under the homogeneous factor structure with normally distributed default drivers (implying that their tail dependence is small), the probability of observing a significant number of defaults of large loans is small. Hence as soon as the number of large loans is increased – hence their relative size is decreased and there are no severe single losses anymore – the diversification effect is observed. This effect is confirmed in Table 3: VaR numbers from portfolio P_1 are already close to the corresponding numbers of the standard portfolio P. The fact that VaR numbers obtained by bootstrapping 'real' portfolios seems to be more sensitive to granularity effects can be seen as an indication that the assumption of normality for the factor structure may be inappropriate.

4 Conclusion

We studied credit risky portfolios with respect to the characteristics correlation and granularity. While correlation clearly plays a major role as a credit risk characteristicum, granularity seems to be less important. However, an analysis of the underlying (probabilistic) model structure shows, that the assumption of normally distributed driving factors contributed significantly to this effect (This would also be true for a simulation based analysis). Thus, there remain concerns on the reliability of risk measures supplied by credit risk models. However, they are clearly a major advance in understanding credit exposures.

References

BIS (2001): Overview of the new Basel capital accord, Working paper, Basel Committee on Banking Supervision.

CAREY, M. (1998): Credit risk in private debt portfolios, *Journal of Finance 53, 1363–1387.*

CAREY, M., (2000): Dimensions of credit risk and their relationship to economic capital requirements, Preprint, Federal Reserve Board.

CROUGHY, M., GALAI, D. and MARK, R. (2001): Risk management, McGraw Hill.

FREY, R. and McNEIL, A. (2001): Modelling dependent defaults, Working paper, ETH Zürich.

GORDY, M. (2000): A comparative anatomy of credit risk models, *Journal of Banking and Finance 24, 119–149.*

HALL, P. and HEYDE, C.C. (1980): Martingale limit theory and applications, Academic Press.

HIRTLE, B.J., LEVONIAN, M.,SAIDENBERG, M., WALTER, S. and WRIGHT, D. (2001): Using credit risk models for regulatory capital: Issues and options, *FRBNY Economic Policy Review 6, 1–18.*

J.P.MORGAN (1997): CreditMetrics.

KIESEL, R., PERRAUDIN, W. and TAYLOR, A.P. (1999): The structure of credit risk: Spread volatility and rating transitions, Preprint, Birkbeck College.

LUCAS, A., KLAASSEN, P., SPREIJ, S. and STRAETMANS, S. (1999): An analytic approach to credit risk of large corporate bond and loan portfolios, Research Memorandum 1999-18, Vrije Universiteit, Amsterdam.

ONG, M.K. (1999): Internal Credit Risk Models. Capital Allocation and Performance Measurement, Risk Books, London.

Cognitive Organization of Person Attributes: Measurement Procedures and Statistical Models

S. Krolak-Schwerdt[1], B. Ganter[2]

[1]Department of Psychology,
University of Saarland, D-66123 Saarbrücken, Germany

[2]Institut für Algebra
TU Dresden, D-01062 Dresden, Germany

Abstract: Theories of social cognition state that person attributes are cognitively organized either by means of a limited number of dimensions or in terms of discrete person types, where traits within a given person type have a unique internal structure. Associated with these two theoretical views are different measurement techniques and statistical models. A methodological study compared trait ratings, trait sortings and subset assignments as typical measurement procedures within the theoretical approaches according to the appropriateness of a dimensional or a categorical statistical model. As expected, the results indicate a dimensional model for rating data and a categorical model for sorting data. In contrast, both statistical models failed to explain the structure of subset assignments. It is shown that trait ratings and sortings differ from subset assignments in that only the first allow for data aggregation and dimension reduction, whereas subset assignments induce a much more complex structure and require a more finely grained data analysis methodology which is provided by Formal Concept Analysis.

1 Introduction

In this paper, we are concerned with methods to analyze how information about persons is cognitively organized. The investigation focuses on the cognitive representation of trait-descriptive terms which are used in ordinary language to characterize people. Current theories of social cognition (e.g. Kunda & Thagard (1996)) assume that such traits and corresponding behaviors which exemplify a given trait are the main constituents of person memory. However, there are very different assumptions regarding the internal structure of this memory system. In empirical research these divergent assumptions are tested by different measurement techniques and statistical models. Thus, the question arises (1) whether a given procedure to measure cognitive relations between traits implies a specific statistical model, (2) whether the structural representation, which is empirically derived, depends on the measurement technique and, if this is true, (3) which statistical models are more adequate approaches in data analysis. These issues will be investigated in the following.

2   How are Traits Organized in Memory?

Though theories of social cognition converge in their assumptions concerning the contents of person memory, there are at least two different views on its internal structure, the dimensional and the typological view (Anderson & Sedikides (1991)).

The dimensional view assumes that the structure of representing social stimuli such as people and their attributes is dimensional. Trait relations and inferences about others can be described using a small number of personality dimensions. Examples are social evaluation (i.e. 'desirable' versus 'undesirable'), activity ('active' versus 'passive') or conscientiousness versus undirectedness (see, e.g., Anderson & Sedikides (1991)).

Typological theories postulate that person attributes are organized in terms of discrete categories or 'types'. Examples of this position are stereotypes like 'tourist' or 'depressed person' (see, e.g., Andersen, Klatzky & Murray (1990)) and the prototype approach of social cognition (Cantor & Mischel (1979)). Regardless of some minor differences betweeen stereotypes and prototypes, the essence of the typological view is that people hold beliefs about how certain sets of traits tend to go together or cluster. Each set of clustering traits constitutes a certain type. In contrast to the dimensional view which states, that the cognitive representation of traits can be described by dimensions regardless of any internal structure of categories, the typological approach assumes that person attributes, which constitute a category, have an internal coherence that differs from trait relations between categories.

These two theoretical approaches involve different operationalizations and statistical models. The dimensional view is typically operationalized by trait-inference procedures (Powell & Juhnke (1983)). Subjects are confronted with a trait term as desciptive of some hypothetical individual and their task is to make inferences about the presence of other traits in that individual. To do so, subjects receive a set of rating scales like 'friendly-unfriendly' and they have to mark on each scale the degree to which one of scale poles characterizes the individual. In the statistical analysis of these rating data, the covariance or correlation matrix among trait terms is analyzed by a factor analytic model, mostly by principal component analysis. The resulting dimensions constitute the required structural representation of the traits.

The typological approach is frequently operationalized by some trait sorting technique in conjunction with a cluster analytic model (Anderson & Sedikides (1991)). In trait sorting, the subject receives a set of trait terms and has to partition the entire set into different categories with each category representing a different person. In the typical sorting task, subjects are required to use each trait exactly once in their sorting such that the categories are mutually exclusive, and the sorting of each subject has to be exhaustive. By counting the frequency with which subjects put each pair of traits into the same category, a similarity matrix of trait terms is obtained. This matrix is analyzed by hierarchical-agglomerative clustering methods. Theoretically postulated categories or 'types' are obtained by the trait clusters.

A variant of the trait sorting method, called 'subset assignment procedure', requires the subject to successively take a subset of trait terms from the attribute sample which characterizes a person of the subject's own choice. After noting the elements of the subset, the terms are replaced into the entire set, and another subset has to be chosen from the set describing a different person. The procedure of assigning a subset of traits to a particular person is repeated several times. In contrast to the typical sorting task, the subject is permitted to use any trait in any number of categories without being forced to take into account all trait terms (Wing & Nelson (1972)).

There is a variety of other techniques which are designed to illuminate the cognitive relations between person attributes such as free association or reproduction tests. As has been outlined above, each measurement technique is connected with particular asssumptions on theoretical grounds and specific statistical models. However, in the context of representing person attributes hardly any investigation can be found which clarifies the relationship between representations from different measurement procedures and the degree to which a particular measurement method determines a specific form of structural representation like the dimensional or categorical type. Powell and Juhnke (1983, p. 919) hypothesize that the interaction between measurement technique and statistical model is of major importance in answering these questions: "When the experimental task is structured so that subjects categorize rather than dimensionalize stimulus objects, a categorical model may be more appropriate... On the other hand, if the experimental task requires direct dimensionalization rather than categorization, a dimensional model would be more appropriate. To our knowledge, this hypothesis has never been explicitly tested."

Rating tasks such as the trait-inference procedure require a dimensionalization in the sense of grading stimuli along prescribed continua. In contrast, the trait sorting method as well as the subset assignment procedure require a discrete stimulus classification and thus a categorization. Consequently, we expect that rating data should be adequately represented by a dimensional model, but not by a categorical model, whereas for sorting data and subset assignments the reverse should be true.

3 Measurement Procedures

In order to test the hypotheses of this investigation, we conducted three studies, each encompassing a different measurement procedure. Stimulus materials in each study consisted of a set of 117 trait descriptive adjectives in everyday language such as 'creative', 'conceited' or 'polite'. These terms were sampled from linguistic taxonomies as well as thesauri and they reflect all relevant personality domains such as abilities, temperament and behavioral style (cf. Angleitner & Ostendorf (1994)).

Study 1: Trait-inference rating. 40 students from all faculties of the University of Saarland participated in the rating study. They received the set of adjectives along with a personality differential which consisted of 15 bipolar seven-point-rating scales. The scales of the personality differential[1] encompass the main personality domains 'extraversion', 'agreeableness' and 'conscientiousness' (cf. Anderson & Sedikides (1991)) and have comparatively high reliabilities. Subjects' task was to rate each trait term on every personality scale according to the degree that a person who has the critical trait may be characterized by one of the scale poles.

Study 2: Trait sorting. Another sample of 40 students received the materials under the instruction of the sorting method. Thus, each subject had to describe several persons of his own choice by simultaneously partitioning the trait sample into different categories, each category representing a different person, under the restrictions, that (1) each trait must be assigned to only one referent person and (2) in the overall sorting all traits must be used. After completion of the task, subjects had to outline their sorting criteria.

Study 3: Subset assignment procedure. A third sample of 40 students had the task to successively take a subset of trait terms from the materials which characterizes a particular person. For each new subset the subject had to choose another person, but otherwise there were no restrictions like in the sorting technique. Finally, subjects had to outline what their criteria for subset assignments were.

---

[1]Because of space limitations, the scales are omitted here. The reader may be referred to Krolak-Schwerdt (1999) for further information.

4 Comparing Statistical Models of Cognitive Organization

The goal of the present research was to investigate if a specific operationalization induces a particular model of representing person attributes, as Powell and Juhnke state, or if the contrary hypothesis is true that the empirically obtained cognitive representation is invariant under different operationalizations. Therefore, the data set from each study was analyzed by dimensional as well as clustering methods.

To analyze the rating data by a dimensional model, the covariance matrix among traits was computed and analyzed by principal component analysis (PCA). The Scree test indicated three dimensions as an adequate representation which accounted for 88.4% of the data variance. By inspection of the factor loadings, the dimensions could be labelled 'evaluation', 'potency' and 'activity' which strongly suggests that the trait terms were rated according to their semantic components. This interpretation was further confirmed by a Procrustes rotation to maximal agreement with the semantic space introduced by Osgood et al. (1957). Osgood's dimensional conceptualization represents connotative associations of verbal concepts and congruence coefficients between 0.88 and 0.95 indicate that this interpretation is valid for the present data.

For cluster analyses, the profile distance between trait pairs on the personality differential was computed and analyzed by Ward's method, Clustroid (Krolak-Schwerdt, Orlik & Kohler (1991)) and Group-Average. The fusion criteria of all three methods suggest a solution of nine clusters. On this fusion level the partitions were compared by the adjusted Rand index (Hubert & Arabie (1985)). Index values in the range between 0.62 and 0.70 indicate that the three clustering methods induce considerable differences in the cluster allocation of traits and thus imply rather unstable solutions. To further determine their goodness of fit to the profile distance data the modified cophenetic correlation (Oldenbürger (1986)) was computed. The maximum value was obtained for Ward's solution which accounted for merely 3.5% of the data variance. Consequently, the clustering solutions did not yield an adequate representation of the data. Table 1 summarizes the results.

For the sorting data, the similarity scores derived from the frequencies with which subjects put trait pairs into the same category were analyzed by the same clustering methods as the rating data. All methods' fusion criteria suggested a five-cluster-solution, where the adjusted Rand index for each comparison of partitions was 0.84. This implies a sufficiently stable clustering representation where the variance accounted for by Ward's method was 78% (see Table 1).

	Rating task	Trait sorting	Subset assignment
1. Explained variance			
Dimensional model	88.4% (PCA)	76% (nMDS)	55% (nMDS)
Clustering model (Ward)	3.5%	78%	47%
Stability	0.62-0.70	0.84	0.45-0.48
2. Sorting and judgement criteria			
	connotative associations	synonyms, denotative components	stereotypes, self concept, celebrities

Table 1: Results of multivariate data analyses

This solution highly corresponds to a taxonomy of trait descriptive terms proposed by Wiggins (1979), and thus, Wiggins' main cluster labels 'warm-agreeable', 'ambitious', 'introverted', 'cold-quarrelsome' and 'extraverted' also apply to our data. Wiggins developped his classification using a deductive reasoning scheme in which synonymous trait terms were amalgamated and antonymous terms were separated into different categories. The correspondence with this denotative classification in conjunction with subjects' statements to have used mainly synonyms as sorting criteria (see Table 1) implies that verbal concepts, especially their descriptive-denotative components are the major organizing principles in the sorting data.

Non-metric multidimensional scalings (nMDS) of data indicated a fourdimensional solution with a percentage of variance accounted for of 76%. However, the dimensional solution was not readily interpretable, since Procrustes rotations to maximal agreement to either Osgood's semantic space or to the 'Big Five' personality dimensions (see, e.g., Anderson & Sedikides (1991)) yielded congruence coefficients below 0.80.

In summary, for the sorting data the typological representation yields a sufficient goodness-of-fit, whereas for the rating data only a dimensional configuration is valid. Thus the main research hypothesis is confirmed by these two measurement techniques showing that the researcher's operationalization is directly reflected by the final representation.

The subset assignment data were treated similar as the sorting data in that frequencies of co-occurence of traits within the same category were analyzed by nMDS and agglomerative clustering techniques. Ward's method, Clustroid as well as Group-Average did not indicate a best fusion level, so any partition might have been chosen. By computing the adjusted Rand index for a five-cluster-solution, values in the range between 0.45 and 0.48 were obtained. Thus, the solutions were rather instable implying a comparatively weak clustering structure.

But dimensional approaches did not induce acceptable configurations either. The Scree test of stress values from nMDS did not indicate a particular number of dimensions as adequate. By choosing rather arbitrarily a fourdimensional solution, only 55% of the variance was explained. Thus, for the subset assignment data both representations, the dimensional as well as the cluster analytic, are invalid.

On the other hand, subjects' statements concerning their judgement criteria (see Table 1) indicate that this measurement technique might be the method of choice in the current context. That is, considering stereotypes, prototypes or the self-concept as sorting criteria is at the focus of social cognition, whereas unrealistic restrictions like in the sorting task or prescribed judgment continua like in the rating task direct subjects' attention to the mainly linguistic features of stimulus materials.

In the subset assignment technique, subjects made their sortings in such different ways, that the aggregation of data across single assignments and subjects did not yield an adequate representation of the data. Since such an agggregation is implied in the computation of similarities among traits, usual nMDS and clustering techniques may be inadequate methods of data analysis. Another methodological framework, which is able to model categorical data structures without any aggregation procedure, is provided by Formal Concept Analysis.

5   A Conceptual Inspection of the Data

Formal Concept Analysis is an elaborated mathematical theory (see Ganter & Wille (1999)) with a methodology to analyze object-attribute data. The basic approach is to consider all attribute combinations that are present in the data, together with the corresponding object sets. These pairs of sets, called 'formal concepts', endowed with a natural partial order relation, form the 'concept lattice', which then is studied. Concept lattices can be visualized using hierarchical order diagrams, similarly to those of trees but with a much richer structure, allowing informative views at the actual data. Since no data reduction is made at this level, such diagrams can quickly become too large to be shown in one picture, and navigation tools become necessary that support browsing. An impression of such a view at the data obtained from the subset assignment procedure is given in Figure 1. It unfolds the 'artist' stereotype and its fragments.

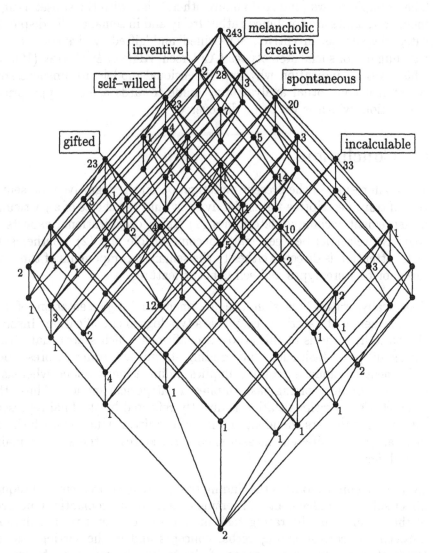

Figure 1: The 'artist' stereotype from the subset assignment data. The nodes correspond to trait combinations that are present in the data. The traits associated with a node are exactly those that can be reached from it via an ascending path. A number attached to a node indicates how often precisely that combination occurs. If no number is given, then these traits occur only together with further ones. We briefly sketch how such a diagram is to be read by explaining the rôle of the leftmost node of the diagram, labelled 2. It represents the trait combination {inventive, self-willed, gifted}, because these are exactly those traits that can be reached via ascending paths from this node. This combination occurs in the data twice by itself and 14 times in combination with further traits; these correspond to the nodes on descendings path from the given one.

The example shows (and so do many others) that with the subset assignment test, traits are combined rather freely and in semantically disparate categories such as 'inventive', 'self–willed' and 'gifted', which constitute a vast number of social stereotypes (Andersen, Klatzky & Murray (1990)). This gives an indication why low dimensional models are unsuccessful. A detailed conceptual analysis may guide towards a more appropriate description, which is yet to be discovered.

6 Conclusions

The main results of the present investigation on the cognitive representation of trait–descriptive terms clearly show that the choice of a particular measurement technique determines (1) which organizational aspects of cognitive relations between traits become visible and (2) whether subjects' attention is directed either to the semantic components of the trait terms or to stereotypes and self concept features.

Specifically, rating tasks induce a dimensional representation of the traits, whereas the sorting task yields a typological classification. Insofar, the basic hypothesis of this study is confirmed which states that structuring subjects' task such that grading stimuli according to prescribed judgement continua is required implies a dimensional model, whereas a discrete stimulus sorting task implies a categorical model. Thus, the researcher's operationalization is directly reflected by the final representation. This renders the identification of cognitive structures, which are invariant under different measurement procedures of the same domain, very difficult.

Even more cumbersome is the finding that these measurement techniques direct subjects' judgement and sorting criteria to the semantic structures of the stimuli - in the rating task in terms of connotative associations between the corresponding verbal concepts and in the sorting task in terms of denotative components of the traits. In contrast, the subset assignment procedure is a more ecologically valid technique in that it directs subjects' attention to stereotypes, the self and other concepts which much more constitute the focus of the present social cognitive domain. However, the subset assignment procedure induces a comparatively complex system of trait categories. To make basic principles of this system visible, Formal Concept Analysis may be the method of choice, since this approach considers the ordering of stimulus–attribute–combinations beyond any data aggregration and is thus appropriate to display the diversity of social categories. As a first demonstration of the advantage of this approach in the current domain, it was shown how semantically incoherent categories corresponding to social stereotypes may be extracted as inherent structures in the data.

References

Andersen, S. M., Klatzky, R. L. and Murray, J. (1990): Traits and social stereotypes: Efficiency differences in social information processing. *Journal of Personality and Social Psychology, Vol. 59, 192–201.*

Anderson, C. A. and Sedikides, C. (1991): Thinking about people: Contributions of a typological alternative to associationistic and dimensional models of person perception. *Journal of Personality and Social Psychology, Vol. 60, 203–217.*

Angleitner, A. and Ostendorf, F. (1994): Von aalglatt bis zynisch: Merkmale persönlichkeitsbeschreibender Begriffe, in Hager, Hasselhorn (Hrsg.): Handbuch deutschsprachiger Wortnormen, Hogrefe, Göttingen, 340–381.

Cantor, N. and Mischel, W. (1979): Prototypes in person perception, in Berkowitz (Ed.): *Advances in experimental social psychology, Vol. 12,* 3–51.

Ganter, B. and Wille, R. (1999): Concept Analysis – Mathematical Foundations. Springer, Berlin et al.

Hubert, L. and Arabie, P. (1985): Comparing partitions. *Journal of Classification, Vol. 2, 193–218.*

Krolak-Schwerdt, S. (1999): Die Wahrnehmung von Persönlichkeitseigenschaften und ihrer Zusammenhänge: ein Methodenvergleich. *Zeitschrift für Sozialpsychologie, Vol. 30, 12–31.*

Krolak-Schwerdt, S., Orlik, P. and Kohler, A. (1991): A regression analytic modification of Ward's method: A contribution to the relation between cluster analysis and factor analysis, in Bock, Ihm (Eds.): Classification, data analysis and knowledge organization, Proceedings of the 14th Annual Conference of the Gesellschaft für Klassifikation e.V., University of Marburg, Springer, Berlin et al., 23–27.

Kunda, Z. and Thagard, P. (1996): Forming impressions from stereotypes, traits, and behavior: A parallel–constraint–satisfaction theory. *Psychological Review, Vol. 103, 284–308.*

Oldenbürger, H. A. (1986): Zur Erhebung und Repräsentation kognitiver Strukturen. Methodenheuristische Überlegungen und Untersuchungen. Braunschweiger Studien zur Erziehungs- und Sozialwissenschaft, Technische Universität Braunschweig, Band 18.

Osgood, C. E., Suci, G. J. and Tannenbaum, P. H. (1957): The measurement of meaning. University of Illinois Press, Urbana, Illinois.

Powell, R. S. and Juhnke, R. G. (1983): Statistical models of implicit personality theory: A comparison. *Journal of Personality and Social Psychology, Vol. 44, 911–922.*

Wiggins, J. S. (1979): A psychological taxonomy of trait–descriptive terms: The interpersonal domain. *Journal of Personality and Social Psychology, Vol. 37, 395–412.*

Wing, H. and Nelson, C. E. (1972): The perception of personality through trait sorting: A comparison of trait sampling techniques. *Multivariate Behavioral Research, Vol. 7, 269–274.*

Manufacturing Branches in Poland - A Classification Attempt

D. Kwiatkowska-Ciotucha[1], U. Zaluska[1], J. Dziechciarz[2]

[1]Department of Forecasting and Economic Analyses
Wroclaw University of Economics
ul. Komandorska 118/120, 53-345 Wroclaw, Poland
(e-mail: ciotucha@credit.ae.wroc.pl, uzaluska@credit.ae.wroc.pl)

[2]Department of Econometrics
Wroclaw University of Economics
ul. Komandorska 118/120, 53-345 Wroclaw, Poland
(e-mail: jdzie@manager.ae.wroc.pl)

Abstract: The goal of the study was the determination of the manufacturing branches attractiveness from the investors' point of view. For the research, a vector of descriptive variables included in the state statistical reports was used. Their normalised values were applied for evaluation of the manufacturing branches. The attractiveness of individual branches on the particular levels of manufacturing sector (*section D* in NACE - Nomenclatures des Activitiees de Communite Europeene) was determined. On the level of *divisions* and *groups* clusters were obtained by k-means method. Additionally the homogeneity of *divisions* was examined. On the level of *class* individual variables have been combined into composite indicator, which has become the evaluation criterion. Finally the ranking of manufacturing branches was made.

1 Introduction

In the newly emerging democracies, among them Poland, some unique managerial problems arise. The most burdening is the lack of the economic consulting and advising infrastructure. This in turn leads to a need for some special research and study assisting management process. As a response for the demand on capital investment decisions support an attempt to establish manufacturing branches attractiveness assessment has been made.

The scope of the research is determined by the data availability. Polish state statistical system collects information according to the NACE Classification. In this system four levels are represented: *17 sections, 60 divisions, 220 groups and 512 classes*. *Section* represents the highest level of classification, *division* is part of *section*, which consist of *groups*. The lowest level of classification is represented by *classes*. For convenience reason the term branch is used. Depending on context it means either *section, division, group* or *class*.

>From the investor's point of view, one of the most important area is the manufacturing sector - the *section D* in the NACE classification. This determines the goals of the study, which consist of above-mentioned attempt to establish manufacturing branches attractiveness assessment and additionally, there is a need to check whether it is enough to assess attractiveness on high level of classification (*division*). It would simplify the analysis. Authors checked whether it would be enough to evaluate performance on *divisions* level instead of analysing much more numerous lower classification levels (*groups* and *classes*).

Ranking of the branches (*divisions, groups* or *classes*) resulting from this research helps at least four groups of decision makers. First of them are investors seeking investment possibilities, among them investors from abroad - they can not apply standard selection procedures used normally in such a situation - capital market is not fully developed, information flow is not effective enough, advisory agencies have no experience. The second group are enterprises managers, who need to evaluate their performance in comparison with other firms in the same branch or with other branches (*divisions, groups* or *classes*). The next type of people, who is interested in such a ranking are bank managers deciding of credit portfolio construction, credit applications evaluation etc. Important users of discussed information are politicians active in local governments. They have to decide which branches to promote in their region, which investments to attract, and which type of skills to teach in local schools and in retraining centres.

In accordance with above, and depending on the end user of the research results (investors, firms managers, credit managers in banks and local politicians), for the evaluation of the Polish manufacturing sector performance - appropriate set of indicators (variables describing firms performance) is to be used.

2 The Data

For the evaluation of the Polish manufacturing sector performance the raw data for the branches from the statistical reports collected by the Polish Central Statistical Office have been applied. The quarterly data, covering period from January 1995 to December 1999 for 21 *divisions*, 97 *groups* and 189 *classes* of *section D* (Manufacturing) were used.

Having in mind that the result of each clustering study depends on the chosen variables, for the purpose of the performance assessment several variables describing all aspects of the firms activity were exploited. It is desirable that the used variables well discriminate objects and simultaneously do not duplicate the contained information. To guarantee this - six variables were chosen, the same for *section, divisions, groups* and *classes*. In the list of variables - in parentheses - their nature is stated:

1. X_1 - the dynamics of incomes from sale in *fixed prices from January 2000 - index on a constant basis - quarterly average from 1995 = 100* (stimulant).

2. X_2 - the cost of obtaining income from total activity (destimulant).

3. X_3 - the profitability rate of net turnover in percent - *the relation of net financial results to income from total activity* (stimulant).

4. X_4 - the liquidity ratio of the second degree - *the relation of current assets decline of stocks to short-term liabilities* (nominant with recommended value range [1.0; 1.5]).

5. X_5 - the liquidity ratio of the third degree - *the relation of current assets to short-term liabilities* (nominant with recommended value range [1.2; 2.0]).

6. X_6 - the ratio of investment - *the relation of investment outlays to income from sale* (stimulant).

In order to avoid negative values, the values of the variable X_3 have been modified by adding the constant value equal to opposite to minimal observed in the dataset.

Depending on the variable nature, an appropriate normalisation formula was used. Normalisation formulas were constructed in such a manner, that they fulfil following requirements: normalised values are in the range [0; 1], normalised variables have a stimulant nature (i.e. the higher value is preferable), normalised variables sustain discrimination ability. Table 1 shows the normalisation formulas.

A few words of argument are needed to discuss the choice of the normalisation formulas used for stimulant and destimulant. Most commonly used transformation formula resulting in [0; 1] range of the normalised values is ratio transformation where for the scaling factor (value) - the maximal value (for the stimulant) or minimal value (for the destimulant) is used. The extreme values are taken from the analysed dataset. Unfortunately - in the situation where extreme values in the dataset vary substantially from the rest of the values in this dataset, such a transformation results in very low variation of the normalised values. In the

Table 1: The normalisation formulas

The nature of the variable	Normalisation formula
Stimulant with values from R_+	$z_{ijt} = \begin{cases} \frac{x_{ijt}}{d_9(x_{it}^k)} & \text{for } x_{ijt} < d_9(x_{it}^k) \\ 1 & \text{for } x_{ijt} \geq d_9(x_{it}^k) \end{cases}$
Nominant with recommended value range $[x_{i,min}; x_{i,max}]$ from R_+	$z_{ijt} = \begin{cases} 1 & \text{for } x_{i,min} \leq x_{ijt} \leq x_{i,max} \\ \frac{x_{ijt}}{x_{i,min}} & \text{for } x_{ijt} < x_{i,min} \\ \frac{x_{i,max}}{x_{ijt}} & \text{for } x_{ijt} > x_{i,max} \end{cases}$
Destimulant with values from R_+	$z_{ijt} = \begin{cases} \frac{d_1(x_{it}^k)}{x_{ijt}} & \text{for } x_{ijt} > d_1(x_{it}^k) \\ 1 & \text{for } x_{ijt} \leq d_1(x_{it}^k) \end{cases}$

x_{ijt} - value of the variable X_i in j-th *division, group* or *class* in period t,
z_{ijt} - value of the normalised i-th variable in the period t,
$d_9(x_{it}^k)$ - ninth decile of the variable X_i in the period t for the k-th level,
$d_1(x_{it}^k)$ - first decile of the variable X_i in the period t for the k-th level,
k = 1, 2, 3 for *divisions, groups* and *classes* levels,
$x_{i,min}$ - the value of lower limit of the recommended value range,
$x_{i,max}$ - the value of upper limit of the recommended value range,
i - variable's number,
j - code of *division, group* or *class* (NACE Classification),
t - period's number

dataset used for the Polish manufacturing branches attractiveness analysis, described phenomenon occurred quite often.

After some experiments with different quantiles, as a remedy for this problem ninth decile for stimulants, and the first decile for the destimulants were chosen as the scaling factor. Apart from the fact, that applied solution guarantee fulfilment of the requirements, which have been formulated earlier, the problem with some data which may be false (errors in data?) is avoided.

3 The Analysis

For each branch, the yearly average values, calculated as an arithmetic mean of quarterly, normalised values for all variables were determined. By doing so, possible seasonality was eliminated.

Branches' clustering was obtained by using k-means method (Hartigan (1975); Grabinski (1992); Gnanadesikan (1997)). For the calculations, the variant with following options was applied:

- The procedure starts with k = 3, 4, 5, 6, 7 clusters. Then objects were moved between those clusters with the goal to minimise variability within clusters and to maximise variability between clusters.

- Initial classification has been selected on the basis of constant interval (STATISTICA (1997)).

Classification with 5 cluster was chosen. This decision was based on the quality of the classification evaluated by the F statistic, clustering stability and interpretability. The five clusters containing manufacturing branches were ordered according to declining performance assessment (attractiveness). The ordering criterion was arithmetic mean of all variables average values in each cluster. Ordered clusters (obtained on the data from 1999) are shown in table 2 and 3.

Table 2: The classification results for the *division* level

Cluster	Number of *divisions* included	*Divisions* codes (NACE Classification)	The value of the ordering criterion
I	7	20 21 22 25 26 31 34	0.907
II	3	28 32 36	0.854
III	1	33	0.818
IV	5	15 16 18 24 29	0.774
V	5	17 19 23 27 35	0.679

Table 3: The classification results for the *groups* level

Cluster	Number of *groups*	*Groups* codes (NACE Classification)	The value of the ordering criterion
I	15	17.3 20.3 20.4 20.5 21.2 22.1 22.2 25.2 26.6 28.2 28.5 31.6 34.1 34.3 36.2	0.890
II	11	15.9 21.1 24.4 25.1 26.1 26.3 26.4 26.5 26.8 31.3 36.5	0.867
III	43	15.1 15.2 15.3 15.5 15.7 15.8 16.0 17.4 17.5 18.1 18.2 19.2 20.1 24.2 24.3 24.5 24.6 26.7 27.4 28.1 28.3 28.4 28.6 28.7 29.1 29.2 29.4 29.5 29.7 31.1 31.2 31.5 32.1 32.2 32.3 33.1 33.2 33.3 34.2 35.1 35.4 36.1 36.6	0.748
IV	10	17.1 20.2 23.2 24.1 26.2 27.2 27.3 29.6 31.4 33.4	0.728
V	18	15.4 15.6 17.2 17.6 17.7 18.3 19.1 19.3 23.1 24.7 27.1 27.5 29.3 35.2 35.3 35.5 36.3 36.4	0.562

The F statistic values analyses indicate significant differences between clusters. Furthermore - on *divisions* level the most discriminating variables were X_1 (the dynamics of incomes from sale) and X_6 (the ratio of investment). On *groups* level all variables appeared to be very discriminative. Additionally high values of the F statistic prove that the descriptive variables and the normalisation formulas choice were appropriate.

Table 4: *Divisions* homogenity analysis

Divisions		Number of *groups* in cluster				
Code	Number of *groups* included	Cluster I	Cluster II	Cluster III	Cluster IV	Cluster V
15	9		1	6		2
16	1			1		
17	7	1		2	1	3
18	3			2		1
19	3			1		2
20	5	3		1	1	
21	2	1	1			
22	2	2				
23	2				1	1
24	7		1	4	1	1
25	2	1	1			
26	8	1	5	1	1	
27	5			1	2	2
28	7	2		5		
29	7			5	1	1
31	6	1	1	3	1	
32	3			3		
33	4			3	1	
34	3	2		1		
35	5			2		3
36	6	1	1	2		2

For *divisions* level cluster I and II were characterised by highest average normalised values (>0.8) of variables X_1, X_2, X_3, X_4, X_5, but in the first cluster there were situated branches which had significantly higher value of the ratio of investment (X_6). For *groups* level - in cluster I there were located the most expansive branches - theyAs the next step of analysis the homogeneity of *divisions* was conducted. The homogeneity was defined as the consistence of individual *group* performance assessment. In other words the *division* homogeneity is manifested by the fact that all *groups* contained in the *division* belong to the same cluster. Provided the *division* homogeneity one may suggest analysis restriction to *divisions* level which is easier, cheaper and faster. Otherwise - the attractiveness analysis on *groups* level is necessary. The homogeneity analysis is presented in table 4. In most cases *groups'* performance assessment differs substantially from that of *divisions*. It means that the analysis on *divisions* level is not sufficient, and exploration on *groups* level is inevitable and necessary. had the most dynamic growth of incomes from sale (X_1). Cluster III have low investment ratio value. Cluster IV is sharply distinguished - has second lowest value of each variable. Branches in cluster V were the least attractive - average normalised values of all variables in this cluster were the smallest.

As the next step of analysis the homogeneity of *divisions* was conducted. The homogeneity was defined as the consistence of individual *group* performance assessment. In other words the *division* homogeneity is manifested by the fact that all *groups* contained in the *division* belong to the same cluster. Provided the *division* homogeneity one may suggest analysis restriction to *divisions* level which is easier, cheaper and faster. Otherwise - the attractiveness analysis on *groups* level is necessary. The homogeneity analysis is presented in table 4. In most cases *groups'* performance assessment differs substantially from that of *divisions*. It means that the analysis on *divisions* level is not sufficient, and exploration on *groups* level is inevitable and necessary.

The analysis shown for the *division* and *group* level has sufficient managerial power. On the *class* level, the obtained clusters are numerous and the heterogeneity inside clusters makes it impossible to use them directly for decision-making. This yields a necessity to establish special framework for that classification level. As a good tool for performance assessment of *classes* a composite indicator may be proposed (Strahl (1998)).

The composite indicator Z, which has the nature of a stimulant, with values in range [0; 1], may be calculated according to the following formula:

$$z_{jt} = \sum_{i=1}^{m} z_{ijt} \cdot w_i \tag{1}$$

where: z_{jt} - value of the composite indicator in period t for *class* j,
z_{ijt} - value of the normalized i-th variable in period t for *class* j,
w_i - weight ascribed to i-th variable, $w_i \in (0,1)$, $\sum w_i = 1$,
j - code of *class* (NACE Classification),
component variables: i = 1, ..., 6, periods: t = 1, ..., 20.

Table 5: Ranking of manufacturing branches

Rank in 1999	Class' code	Description	$\bar{Z}$	Rank in:			
				1995	1996	1997	1998
1	26.12	Shaping and processing of flat glass	0.967	37	16	139	72
2	28.22	Manufacture of central heating radiators and boilers	0.962	11	2	1	5
3	25.22	Manufacture of plastic packing goods	0.961	15	24	2	4
4	22.22	Printing n.e.c.	0.932	8	10	4	1
5	26.30	Manufacture of ceramic tiles and flags	0.923	3	3	3	6
6	26.61	Manufacture of concrete products for construction purposes	0.914	103	144	71	31
7	15.87	Manufacture of condiments and seasonings	0.902	122	47	26	11
8	20.51	Manufacture of other products of wood	0.900	108	38	52	65
9	25.23	Manufacture of builder's ware of plastic	0.899	33	13	21	2
10	34.30	Manufacture of parts and accessories for motor vehicles and their engines	0.895	25	19	12	22
...	...	...	...	...	...	...	...
180	17.14	Preparation and spinning of flax-type fibres	0.500	183	189	180	172
181	15.83	Manufacture of sugar	0.497	153	143	156	180
182	36.40	Manufacture of sports goods	0.488	187	175	179	177
183	29.51	Manufacture of machinery for metallurgy	0.483	171	164	169	178
184	23.10	Manufacture of coke oven products	0.471	123	159	182	186
185	17.22	Woolen-type weaving	0.465	166	180	184	187
186	18.30	Dressing and dyeing of fur; manufacture of articles of fur	0.434	160	173	176	169
187	17.74	Manufacture of knitted and crocheted underwear	0.422	186	179	188	189
188	36.30	Manufacture of musical instruments	0.421	172	125	187	188
189	27.52	Casting of steel	0.383	155	172	186	184

The value of the composite indicator Z for each *class* and period was obtained according to the proposed formula (with equal weights). Next, for each *class* the yearly average values $\bar{Z}$ were calculated as an arithmetic mean of quarterly Z values. Finally, the values of $\bar{Z}$ were taken as an ordering criterion of manufacturing branches. The ranking shown in table 5 has been established due to the 1999 data. Additionally, the past ranking placements (in years 1995, 96, 97 and 98) are shown.

4 The Recommendations

Decisions concerning development and/or investment directions need to be based on the knowledge on the attractiveness of individual economic activity area. Initial suggestion may be acquired from analyses conducted on high level of the economic activity classification (division level in the NACE system). Further steps of the decision making process needs research on the lower level of classification (groups). Hierarchical ordering on the class level in composition with the investors preferences and expertise makes it possible to choose the best possible decision.

References

NACE - Nomenclatures des Activites de Communite Europeene, Central Statistical Office, Warszawa 1991.

GNANADESIKAN, R. (1997): Methods for Statistical Data Analysis of Multivariate Observations, 2nd ed., Wiley, New York.

GRABINSKI, T. (1992): Taxonometric Methods, AE, Krakow.

HARTIGAN, J.A. (1975): Clustering Algorithms, Wiley, New York.

STATISTICA PL for Windows (Volume III): Statistics II (1997), Stat-Soft Inc., Tulsa.

STRAHL, D. (1998): Taxonomy of Structures in Regional Analysis, AE, Wroclaw.

Market-Segments of Automotive Brands: Letting Multivariate
Analyses Reveal Additional Insights

M. Löffler

Marketing Planning, Dr.Ing.h.c. F. Porsche AG,
Porschestraße 15-19, 71634 Ludwigsburg, Germany

Abstract: Market segmentation is of key interest in addressing specific groups of consumers with similar needs. One of the main departures from straightforward cluster-based market segmentation is the "tandem approach", where a factoring step is followed by clustering. The paper presents a modification which takes into account known shortcomings of the approach. The suggested procedure is applied for revealing additional insights into the German automotive market. The approach allows for detecting groups of competing brands, estimates of underlying preference structures and positioning (dis-)advantages of nameplates.

1 The Automotive Market: A Customers & Manufacturers Point of View

To **customers** purchasing an automobile is a task of long-lasting financial and significant social impacts. Purchase decisions of automobiles gain in overall complexity due to the rapidly increasing number of automotive concept segments (e.g., sedan, station wagon, compact, etc.) and models per manufacturer. Customers face this situation often in using decision heuristics. Well-known is e.g. the use of country-of-origin stereotypes (see, e.g., Chao, Gupta (1995)). Another decision heuristic is to classify automobiles respectively nameplates according to their key properties. Following the findings reported by Meffert (1998, p. 1242 ff) purchase decisions are mostly reduced to dimensions like price, brand image, high reliability, styling, and engine power.

Manufacturers face an increasing competition in all automotive concept segments. Custom-tailoring of solutions for mobility and addressing specific needs of customers serve to gain a competitive advantage. Questions of concern include among others

a) characterizing the key dimensions relevant for purchase decisions, after-sales customer satisfaction and future brand loyalty,

b) revealing interchangeable proportions or interrelationships within these dimensions, where changes do not negatively affect the overall brand/product positioning, and

c) identifying the relevant set of competitors and the relative competitive positioning among them.

To answer the above mentioned questions, several techniques in the area of market segmentation have been suggested: they include among others **clustering** (e.g. Bock (1974), Bacher (1992), Gaul and Baier (1994)), **latent class analysis** (Kamakura and Russell (1989)), and **conjoint-based benefit segmentation** (Gaul et al. (1994)). A comprehensive discussion of market segmentation techniques is presented by Wedel and Kamakura (2000).

The paper focuses on aspects b) and c): an approach is presented which allows for the identification of competitive brands. Furthermore, a first glance on the interchangeable proportions of key features within the clusters of competing brands is shown. These proportions can be visualized as gradients of adequately determined cluster-specific functions which indicate areas of same preference or utility: a specific degree of scarcity with respect to one key feature can be compensated by a superiority with respect to another feature and therefore yielding to a brand or product of the same attractiveness. Thus, even if automotive brands are dissimilar with respect to some attributes they may address the same groups of customers. Straightforward clustering assigns automotive brands with significantly dissimilar single attributes into separate clusters and not always provides satisfying results. This motivates in particular the presented modifications of the tandem approach.

Section 2 illustrates the modifications which turn to be appropriate for this situation. In section 3 straightforward clustering and the modified approach are used for analyzing real data in the automotive area. The external validity of the results is tested in section 4. Managerial implications of the findings are discussed in section 5.

2 Modified Tandem Approach

One of the most important departures from straightforward clustering is the so-called **tandem approach** (Arabie, Hubert (1994)): After factoring the inter-variable correlation matrix the objects are clustered in their (e.g.) Varimax-rotated factor space. Despite the reported widespread usage of the tandem-approach, empirical studies by Green, Krieger (1995) as well as Schaffer, Green (1998) report a lack of external validity in comparison to straightforward *k-means*-clustering.

In situations as described in section 1, a modification of the tandem approach performs very well: the modified approach comprises of **step**

494

1, where the data are factored and subsequently Varimax-rotated. In **step 2** a priori-classification is used in order to determine an initial cluster solution. In the final **step 3**, the objects are clustered in the reduced factor-space based on the initial solution of step 2 and using a modified distance measure.

The following notations are used to define the modifications of the tandem approach:

Let $O := \{o_1, \ldots, o_m\} \subset \mathbb{R}^d$ be the (set of the) observations under consideration.

The modifications concentrate mainly on the distance measure $d(i,j)$, which reflects the distance of object o_i from cluster $\mathcal{C}_j \subset O$. Given the vectors $\vec{e}_1, \ldots, \vec{e}_n \in \mathbb{R}^d$, by $[\{\vec{e}_1, \ldots, \vec{e}_n\}] := \{\sum_{s=1}^n \alpha_s \vec{e}_s \,|\alpha_s \in \mathbb{R}\}$ a vector space is defined with dimension $dim[\{\vec{e}_1, \ldots, \vec{e}_n\}]$. Together with the vector $\vec{u} \in \mathbb{R}^d$, $\vec{u} + [\{\vec{e}_1, \ldots, \vec{e}_n\}]$ describes an affine subspace of $\mathbb{R}^d$ according to $\vec{u} + [\{\vec{e}_1, \ldots, \vec{e}_n\}] := \{\vec{u} + \vec{e} \,|\vec{e} \in [\{\vec{e}_1, \ldots, \vec{e}_n\}]\}$.

Step 2 (or preceding iterations of the modified approach in step 3) results in a total of l clusters $\mathcal{C}_1, \ldots, \mathcal{C}_l$. Moreover

$$\mathbf{C}_j := \{\vec{u}_j + [\{\vec{e}_{j,1}, \ldots, \vec{e}_{j,r_j}\}]\}, \tag{1}$$

$j = 1, \ldots, l$, denotes an affine subspace of $\mathbb{R}^d$ as the solution of the minimizing problem

$$\sum_{o \in \mathcal{C}_j} d_{eucl}(o, \mathbf{C}) \rightarrow \min_{\substack{\mathbf{C} := \vec{u} + [\{\vec{e}_1, \ldots, \vec{e}_r\}], \\ \vec{u}, \vec{e}_1, \ldots, \vec{e}_r \in \mathbb{R}^d \\ r < \min\{dim[\{o|o \in \mathcal{C}_j\}], d\}}} . \tag{2}$$

In other words: $\mathbf{C}_j$ is the affine subspace of $\mathbb{R}^d$ which best fits the objects of cluster $\mathcal{C}_j$ in terms of the euclidean distance d_{eucl}.

Now, the distance measure used for the modified tandem approach is defined by

$$d(i,j) := \| pr_{\mathbf{C}_j}(o_i) - o_i \|_{eucl}, \; i = 1, \ldots, m, \; j = 1, \ldots, l. \tag{3}$$

In equation (3) the projection of o_i to the affine subspace $\mathbf{C}_j$ is denoted by $pr_{\mathbf{C}_j}(o_i)$. Distance measures in the manner of equation (3) can be found, e.g., in the area of fuzzy-c-varieties clustering approaches (e.g., Höppner et al. (1997), Bezdek (1981)).

To each of the clusters $C_1, \ldots, C_l$ the affine subspace according to (1) is assigned and used to verify the cluster membership of the objects. Objects (nearly) on a straight line are combined in the same cluster, independent of the Euclidean distance between them.

The following section focuses on testing the modified approach based on two important dimensions ($d = 2$). The solution of problem (2) can be derived by well-known clusterwise regression techniques for $d = 2$.

3   Testing and Empirical Results

Data for testing were kindly made available by Motor-Presse Stuttgart. The underlying survey was carried out between 1999 and 2000. Average values of a total of 126.000 respondents were available. For testing the 20 most important nameplates with respect to market shares were selected. The brands cover 90.9% of the German automobile market in 1999 according to official registration figures. To keep confidentiality, letters "A" to "T" were randomly assigned to the nameplates. The percentage values of brands driven by respondents did not show any meaningful deviation from actual registration figures.

The participants had to write down the makes which best fit to statements like "good looks/styling", "makes sporty cars", "very reliable cars" and "well made". It was possible to mention more than one brand. An extensive listing of all relevant makes was part of the questionnaire.

The survey was conducted in the same manner the first time more than twenty years ago and is repeated on an annual basis. Although the importance of several variables changed over time (e.g., fuel consumption), the questionnaire remained nearly the same. The survey addresses people highly interested in automobiles and does not reflect the German car drivers as a whole. The participants are well-informed concerning automotive topics and act as opinion-leaders.

Originally, one has a three-mode data matrix, but Motor-Presse Stuttgart provides only average values. Therefore, a two-mode matrix is available for further investigations, where the brands are the objects and the average values of the respondents are the treats (variables).

3.1   Straightforward Clustering

Straightforward clustering of data results not only in a "traditional" market segmentation but also provides a benchmark for the modified tandem approach. For comparison purposes four methods were tested: *k-means*, *average linkage*, *Ward* and *single linkage*. 8 variables of actual interest from a total of 13 were selected. Clustering based on 9 up to 13 variables resulted in unchanged cluster membership. The subsequent discussion focuses on a 3 cluster solution, as this reflects the straightforward classification of the car market in the segments "low", "medium", and "luxury" (for different segmentation schemes see Diez (2000)). There was the single solution

$$C_1 = \{Q, R, S, T\}, \ C_2 = \{P\}, \ C_3 = O \setminus \{C_1 \cup C_2\}.$$

From a market orientated point of view the isolation of brands P and S as well as the assignment of brand T to the same cluster as brands Q, R, and S is problematic: the brands are addressing clearly different segments with respect to price sensitivity and expectations concerning emotional product attributes (for details see section 4). Straightforward clustering results in a misleading determination of the existing competitive relationships.

3.2 Clustering by the Modified Tandem Approach

In **step 1** factoring is used in order to analyze the data. The variables were the same as for straightforward clustering (see subsection above). Scree plot as well as a selection based on Kaiser eigenvalue criterion resulted in **two factors**. Eigenvalues are 6.34 (factor 1) and 1.17 (factor 2). The factors explain 93.9 cumulated per cent of the variance. Factor 1 addresses the more quality orientated aspects of automotive brands ("very reliable cars","environmentally compatible cars", "well made", "high safety standards", "advanced technology"), factor 2 focuses on emotions-related dimensions ("makes sporty cars", "I like this make", "good looks/styling"). The factors are subsequently referred by "quality" respectively "emotions". The dimension "quality" consists of a rich set of variables, whereas "emotions" consists of a more focussed one. This is in part due to the fact, that the questionnaire used in the survey remained unchanged for nearly twenty years. It is desirable to reflect the most recent changes in the importance of attributes for brand evaluation (increasing importance of emotional aspects; adjustment of technical product quality). Therefore, the influence of "emotions" should exceed the extent indicated by the corresponding eigenvalue for segmentation purposes.

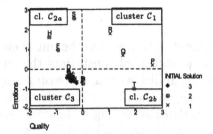

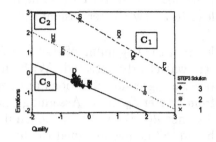

Figure 1: Initial cluster structure.

Figure 2: Final cluster struct. (step 3).

For the present data set Bartlett's test statistic $b = 305.02$ is highly significant ($p < 0.01$), implying that the correlation matrix is not orthogonal (Basilevsky, 1994). The Kaiser-Meyer-Olkin measure of sampling adequacy was MSA=0.705. Although MSA is away from the "marvelous" range of MSA > 0.9, it is in line with values reported to be adequate for factor analysis (see, e.g., Sharma (1996, p. 116)). Altogether, the data were appropriate for the factoring step of the modified tandem approach.

In **step 2** a priori-clustering based on the rotated factor scores was used to derive an initial solution. Following suggestions of Dibb and Simkin (1996) as well as Ragge (2000) for practical orientated classification resp. clustering schemes, a two-dimensional matrix was used. The bounds of the matrix arrays are depicted by dotted lines in figure 1.

Some of the brands under consideration are in a favorable position with respect to the dimensions "quality" as well as "emotions". These brands are combined in cluster C_1. The brands belonging to cluster C_3 are in a reverse situation: factoring resulted in negative (rotated) factor scores for both dimensions. The findings were mixed for the brands combined in clusters C_{2a} as well as C_{2b}. The brands which take such a medium position as in C_{2a} and C_{2b} were combined in step 2 to cluster C_2. Höppner et al. (1997, p. 97ff) report situations of this kind, where separate clusters are combined due to a similar underlying linear structure.

The initial cluster solution is illustrated in figure 1. According to their cluster membership, objects are marked by a cross (cluster 1), a circle (cluster 2) or a rhomb (cluster 3).

In **step 3** brands are clustered using the distance measure described in equation (3). The clustering step starts with clusterwise linear regression using the initial cluster structure plotted in figure 1. Note that, for example, brand "S" is close to regression line $\mathbf{C}_1$ of cluster C_1 although it belongs to cluster C_2 (see figure 2). In step 3, situations of this kind lead to a new cluster assignment.

Even at the beginning of step 3, a linear structure within the initial clusters was clearly detected. Without any additional constraints the clusterwise regression functions are nearly parallel with a decreasing slope from cluster C_3 to C_1. Assigning brands o_i to the cluster C_j with minimal distance $d(i,j)$ results in the **final solution** according to figure 2: e.g., brand "S" is now assigned to cluster $C_1 = \{P, Q, R, S\}$ and nearly coincides with regression line $\mathbf{C}_1$. Due to this reassignment the regression line $\mathbf{C}_2$ of cluster $C_2 = \{F, H, T\}$ changes, too. The **modified tandem approach** results in

$$C_1 = \{P, Q, R, S\}, \ C_2 = \{F, H, T\}, \ C_3 = O \setminus \{C_1, C_2\}$$

which significantly differs from the outcome of **straightforward clustering**.

4   External Validity

This section puts special attention to the relationship of clustering results and exogeneous background variables. The topic is especially emphasized, as the (straightforward) tandem approach is sometimes reported to lack of external validity.

Main differences concerning the (straightforward) clustering results and the results of the modified tandem approach are the inclusion/exclusion of brands P and T to/from cluster $\{Q, R, S, T\}$.

Brands P, Q, R, S, and T altogether account for approx. 45% of the German automobile market, whereas brand P contributes approx. 9.3%, brand T 23.1%. Brand S takes an exceptional price- as well as image positioning. The model range is focused on specific concept segments and addresses a more specific group of customers than the other brands do. The market share of brand S is approx. 0.2%, resulting in a comparatively small number of people with direct product- or driving experience.

For validation purposes the background variables listed in table 1 were evaluated. Additional market research results (which are independent from data used in section 3) like *Fame 2000/2001* and average car prices by makes were considered. They illustrate the high similarity of brands P, Q, R and S as well as their dissimilarity with respect to brand T-supporting the results of the modified tandem approach.

Fame 2000/2001 investigates brand images of durables as well as consumer goods on a statistically representative base. For *Fame 2000/2001* figures of table 1 indicate the percentage values of agreement of the survey participants to the statements "singular, unimitable brand".

No.	validation by	P	Q	R	S	T	remarks
1	*Fame 2000/2001*	71.7	(40.6)	58.6	69.3	37.2	6.508 interviews
2	average price ('000 DM)	62	54	59	>62	35	details for S not available

Table 1: External validation of the clustering results for brands T to S.

Concerning the findings based on the representative survey *Fame 2000/ 2001* (p. 9) brand T has a clearly different level of agreement to the characteristic feature "singular, unimitable brand" in comparison to brands P, R, S. The main reason might be, that brand T addresses a broader group of customers and is therefore seen to be more "common" and popular. Brand Q belongs to company T which results in only slightly higher ratings of brand Q over brand T. Combining brands Q, R, S, and T into one cluster (see the results of straightforward clustering) is not suggested. The results of straightforward clustering are not supported. The same holds true with respect to the inclusion of brand P into the same cluster as brands Q, R, S. A quite similar situation can be found based on the representative survey "Markenprofile 8" (2000).

An additional hint which underlines the validity of the results are the average prices of the brands (see validation no. 2, table 1). Brand T clearly addresses more price-sensitive customers respectively customers with an income situation different from the ones of brands P to S. A structure according to straightforward clustering which combines brands Q, R, S, and T in one cluster and excludes brand P proves to be not appropriate.

Altogether, there is significant external evidence for the cluster structure revealed by the presented approach.

The findings extend recent research results of Schaffer, Green (1998) as well as Green, Krieger (1995). They found the results of straightforward clustering more convincing than the outcomes of the tandem approach. The presented paper illustrates a modification of the tandem approach, which outperformes straightforward clustering in terms of validity. Therefore, based on the underlying data as well as the questions to be answered, both techniques should be taken into consideration.

5 Managerial Implications

The presented approach provides for the situations discussed in section 1 and at least for empirical data currently available two main advantages: first, the cluster structure proved to gain in external validity and second, the approach uses and provides simultaneously clusterwise regression results which allow for additional insights.

The advantageous external validity *preserves from misleading clustering results*, which may indicate inappropriate cluster membership and incorrect competitive structures. As illustrated in section 4, brand T is far away from competing with brands P to S as indicated by straightforward clustering. The modified approach correctly identifies the competitive relationship of brands P to S and the dissimilar positioning of brand T.

The modified approach determines cluster C_j and the clusterwise regression results $\mathbf{C}_j$ simultaneously. Therefore the *mutual importance of the underlying factors* for the brand positioning can be judged on average for the cluster according to the slope of the regression line. In the presented case, a slope $-1 < \sigma_j < 0$ is a first indication for the dominating importance of emotional factors for brand positioning, and vice versa: in order to remain on the regression line $\mathbf{C}_j$, a change in "emotions" by an amount $|\sigma_j \cdot \Delta|$ has to be compensated for by an amount of $|\Delta|$ with respect to the dimension "quality".

For all three clusters, the slope of the clusterwise regression lines fulfill $-1 < \sigma_j < 0, j = 1, \ldots, 3$. When interpreting the regression lines as areas of a specific level of brand attractiveness ("utility"), the *suggestions to improve the competitiveness* of a brand are as follows: it is more advantageous to increase the emotional strength of a brand by a specific amount in comparison to improvements of quality-related aspects. Figure 3 illustrates the situation for brand T: the quality-related improvement of amount Δ_2' leads to a brand attractiveness of level C_2'. An improvement of the same amount Δ_2'' concerning "emotions-related" aspects of brand positioning leads to the superior level C_2''. These findings are in line with common automotive practice for facelifting activities.

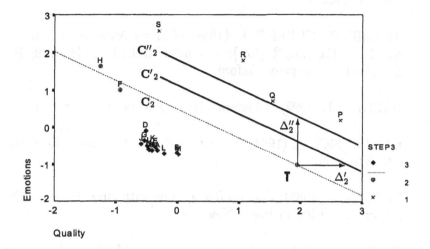

Figure 3: Positioning improvement of brand T.

R&D activities serve to fulfill directly or indirectly specific needs of actual or future customers. As described above in the situation of positioning improvements, the results of the modified approach can be used to *optimize the mix of R&D activities* focusing on more emotion- respectively quality-orientated consumer expectations.

An interesting finding is that the slope of the clusterwise regression lines decline from cluster C_3 to cluster C_1. For the cluster C_1 of luxury cars, emotions-related aspects are more important than for cluster C_3. This underlines again, that especially premium brands have not only to fulfill basic needs of safe and reliable mobility, but also to fit into the self-perception and "lifestyle" of their customers.

6 Conclusion and Outlook

The modified tandem approach provides reliable and valid results for the German car market. The findings gain in external validity compared with outcomes of straightforward clustering and allow for additional insights into the positioning of automotive brands. Based on the convincing results it is indicated to test the approach based on additional data sets. The presented evaluation was made on the brand level. The question concerning the competitive structure on the model level is left for further investigation, too.

References

ARABIE, P., HUBERT, L. (1994): Cluster Analysis in Marketing Research, in: Bagozzi, R. (Ed.): Advanced Methods in Marketing Research. Blackwell & Company, Oxford.

BACHER, J. (1992): Clusteranalyse. Oldenbourg, München.

BASILEVSKY, A. (1994): Statistical Factor Analysis and Related Methods. Wiley, N.Y.

BEZDEK, J. (1981): Pattern Recognition with Fuzzy Objective Function Algorithms. Plenum Press, New York.

BOCK, H.H. (1974): Automatische Klassifikation. Vandenhoeck & Ruprecht, Göttingen.

CHAO, P., GUPTA, P. (1995): Information Search and Efficiency of Consumer Choices of New Cars. *International Marketing Review, 12, 47-59*.

DIBB, S., SIMKIN, L. (1996): The Market Segmentation Workbook. Routledge, London.

DIEZ, W. (2000): Automobil-Marketing. mi, Landsberg.

Fame 2000/2001 (2000), Verlagsgruppe Milchstrasse, Hamburg.

GAUL, W., BAIER, D. (1994): Marktforschung und Marketing Management. 2nd ed. Oldenbourg, Munich.

GAUL, W., LUTZ, U., AUST, E. (1994): Goodwill towards Domestic Products as Segmentation Criterion: An Empirical Study within the Scope of Research on Country-of-Origin Effects, in: Studies in Classification, Knowledge Organization, and Data Analysis, Springer, Heidelberg, *415-424*.

GREEN, P., KRIEGER, A. (1995): Alternative Approaches to Cluster-Based Market Segmentation. *Journal of the Market Research Society, 11, 221-239*.

HÖPPNER, F., KLAWONN, F., KRUSE, R. (1997): Computational Intelligence: Fuzzy-Clusteranalyse. Vieweg, Wiesbaden.

KAMAKURA, W., RUSSELL, G. (1989): A Probabilistic Choice Model for Market Segmentation and Elasticity Structure. *Journal of Marketing Research, 26, 1, 379-390.*

WEDEL, M., KAMAKURA, W. (2000): Market Segmentation, 2nd ed. Kluwer, Boston.

Markenprofile 8 (2000), Gruner + Jahr, Hamburg.

MEFFERT, H. (1998): Marketing. 8nd ed. Gabler, Wiesbaden.

RAGGE, H. (2000): Strategische Geschäftsfeld- und Branchenanalyse, in: Zerres, M. (Ed.), Handbuch Marken-Controlling, Springer, Heidelberg, *41-74.*

SCHAFFER, C., GREEN, P. (1998): Cluster-Based Market Segmentation: Some Further Comparisons of Alternative Approaches. *Journal of the Market Res. Soc., 40, 155-163.*

SHARMA, S. (1996): Applied Multivariate Techniques. Wiley, New York.

Sequence Mining in Marketing

M. Meyer

Seminar für Empirische Forschung und Quantitative Unternehmensplanung,
Ludwig-Maximilians-Universität München, D-80539 München, Germany

Abstract: This paper deals with two different points of view concerning the idea and the profit of sequence mining algorithms: Firstly, from a computational view it is important to develop fast and efficient mining algorithms that discover sequential patterns. In the past, several algorithms were published which have been successfully tested for special purposes. Secondly, besides the computational view the paper deals with the marketing point of view. Generally, marketing researchers try to build marketing models describing and explaining behavioral aspects in the market (e.g. buying behaviour). This is where the marketing point of view differs from the computational approach. While sequence mining algorithms are mainly data driven without considering causes and effects, marketing researchers try to investigate relationships between relevant variables. Those differences are pointed out in order to derive demands on future research.

1 Introduction

The problem of sequence mining is to discover all sequential patterns with a user-defined minimum support in a database of sequences (Agrawal et al. 1995). In the past, several researchers developed so-called sequence mining algorithms which are discussed in the following.

The paper is divided into four parts. The first section deals with some relevant definitions, some database aspects and some formal aspects and finishes with a short comparison of the mentioned computational and marketing point of view. The second section presents important features of some well-known sequence mining algorithms. Some applications of sequence mining algorithms in marketing are discussed in section three. The paper finishes with some conclusions on future work.

1.1 Problem

A sequence is defined as a set of events in chronological order (Agrawal/Srikant 1995; Zaki 2001), which can be represented by an acyclic graph (e.g. figure 1). Typical examples are purchases in retail trade, web access patterns or series of insurance contracts (Zaki 2001; Pei et al. 2000).

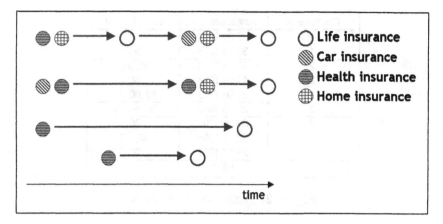

Figure 1: Illustration of sequences

Figure 1 shows series of insurance contracts for four persons during their customer relationship. After concluding a car insurance contract and a health insurance contract the second person decides to contract another health insurance and a home insurance. Later, the person concludes in addition a life insurance. The examples in figure 1 show that many different sequences of contracts are possible. From a marketing point of view it may be interesting to know which sequences are typical, or – with other words – frequent. This is exactly the purpose of sequence mining: Discover all frequent sequence patterns. Such an information may help to offer customers the right product at the right time.

1.2 Database Aspects

Usually, sequence mining algorithms require a special form of the data base. It is important to have customer identifiers, items – for example bought products – and the time when the items were bought. (see figure 2; Agrawal et al. 1995)). Such a representation is very simple so that nearly every database can be transformed into this form. It should be noted that other information is not necessary for the mining task.

1.3 Formal Aspects

This subchapter deals with some relevant terms and formal aspects (according to Zaki 2001). Let $\mathcal{I} = \{i_1, i_2, ..., i_m\}$ be a set of m distinct items. In this context an event is a non-empty unordered collection of items, denoted as $(i_1 i_2 ... i_k)$ where i_j is an item, e.g. the purchase of a

Customer Id	Transaction Time	Items Bought
1	25.06.1993	30
1	30.06.1993	90
2	10.06.1993	10, 20
2	15.06.1993	30
2	20.06.1993	40 ,60, 70
3	25.06.1993	30, 50, 70
4	25.06.1993	30
4	30.06.1993	40, 70
4	25.07.1993	90
5	12.06.1993	90

Figure 2: Example of database

certain product. As defined earlier, a sequence $(\alpha_1 \to \alpha_2 \to \ldots \to \alpha_q)$ is a list of events α_i in chronological order, which is denoted by arrows between the events. We say k-sequence, if the sequence contains exactly k items, e.g. $(A \to BC)$ is a 3-sequence. $\alpha_i < \alpha_j$ means that α_i occurs before α_j. A sequence β contains another subsequence α, if there exists a one-to-one order-preserving function f that maps events in α in events in β, so that $\alpha_i \subseteq f(\alpha_i)$ and if $\alpha_i < \alpha_j$ then $f(\alpha_i) < f(\alpha_j)$.

Given a database $\mathcal{D}$ consisting of a collection of so-called input-sequences, $\alpha \preceq \mathcal{C}$ means that input-sequence $\mathcal{C}$ contains α. In order to discover the frequent sequences we calculate the so-called *support* for a sequence α, i.e. the percentage of input-sequences in the database $\mathcal{D}$ that contain α. If the support of α is at least a minimal user-defined support, then we call α a frequent sequence. Besides the support we calculate the *confidence* of sequence α, which is defined as the probability of α_q given the prefix sequence $\alpha_1 \to \alpha_2 \to \ldots \to \alpha_{q-1}$: $conf(\alpha, \mathcal{D}) = \frac{sup((\alpha_1 \to \alpha_2 \to \ldots \to \alpha_q),\mathcal{D})}{sup((\alpha_1 \to \alpha_2 \to \ldots \to \alpha_{q-1}),\mathcal{D})}$. Or with other words: We calculate the conditional probability.

Figure 3 shows a simple example, which illustrates the relationship between support and confidence. The database consists of five sequences, three of them contain the sequence "Health insurance" followed by "Life insurance". The support of this sequence is 60%. Note that the second sequence contributes to the support, although there is an intervening event. Note as well, that the third sequence does not contribute to the support, because the time interval exceeds a user-defined maximal gap (additionally, a time window is defined as a time interval in which events are united to one event). For this reason the two events do not build a sequence. Generally, support and confidence are simple indicators for the significance of sequences but it is up to the user to decide whether a sequence is relevant or not. Nonetheless, support and confidence lead to a rough preselection of potentially useful sequences.

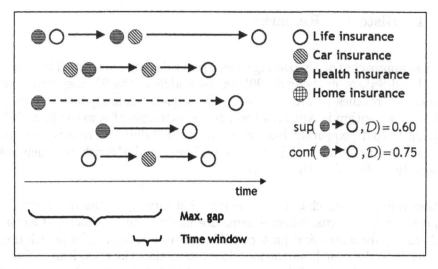

Figure 3: Calculation of support and confidence

1.4 Marketing and Computational Approach

In the past, most researchers focused on the development of efficient and versatile algorithms, whereas the usefulness of these algorithms for realworld applications was less important. Though there are some applications like the analysis of buying behaviour or the investigation of web access patterns. But those applications are mainly data driven, i.e. theoretical foundations of the behaviour were less important. This point of view characterizes the so-called computational approach which differs from the purpose of marketing researchers who try to investigate relationships between relevant variables, i.e. they try to explain the behaviour of web users or customer by building statistical models. This approach can be characterized as being more theory driven.

2 Sequence Mining Algorithms

Generally, the purpose of sequence mining is to discover all frequent sequential patterns. This purpose is of high complexity because we discover sequences of itemsets and not just single item sequences. Further complexity is caused by accepting arbitrary gaps among events with intervening events (see figure 3). For those reasons several algorithms were developed and shall be discussed in the following.

2.1   Historical Remarks

First approaches of discovering association rules were developed by Agrawal and Srikant in the early 90'. In the middle of the 90' they extended their approaches in order to solve the sequence mining task. A well-known algorithm is AprioriAll which was extended afterwards into GSP (Generalized Sequential Patterns). Since 1999 different researchers, for example Han et al. and Zaki, proposed improved algorithms which are shortly explained in the following.

Generally, we can characterize association mining as the discovery for intra-event patterns, whereas sequence mining can be described as the discovery for inter-event patterns. The purpose of mining for association rules is to discover frequent unordered itemsets. The complexity of this task, that is the size of the search space, is $\binom{n}{k}$, where n is the number of items and k is the number of items per set. Mining for sequential patterns is much more complex, because we discover sequences of itemsets (not single items!). This task has a complexity of n^k, where n is again the number of items and k is the length of the sequences.

2.2   AprioriAll and GSP

One of the first sequence mining algorithms and a very popular one is AprioriAll (Agrawal et al. 1995). Without going into detail we can distinguish three phases: In the first phase we scan the database in order to find all frequent itemsets, i.e. to find all itemsets with at least minimum support. In the second phase each transaction is replaced by its frequent itemsets. Finally, the algorithm discovers all frequent sequential patterns by repeated database scans. The repeated database scans are the reason for inefficiencies, because their number depends directly on the length of the sequences. Some limitations, for example AprioriAll can not consider sliding time windows, caused the authors to develop the GSP-algorithm (Srikant et al. 1996) which includes time constraints like the maximum gap between events and sliding time windows, i.e. not all items in an element must come from the same transaction. The number of database scans of GSP increases with the length of the sequences which is comparable with AprioriAll.

Summarizing, AprioriAll and GSP can be described as generate-and-test-algorithms because both firstly generate a set of candidate patterns and then calculate the support of those patterns. This is the reason for the inefficiency of both algorithms.

2.3   WAP-mine-algorithm and SPADE

The WAP-mine-algorithm (Pei et al. 2000) is one of the newer approaches. The main idea is to avoid candidate generation and repeated database scans. Instead of the original database the algorithm uses a special tree data structure by constructing the so-called WAP-tree, which is used for counting the sequences. The main features result from this innovative data structure. It allows a real fast mining process, because there are only two database scans required. Additionally, the algorithm uses conditional search in order to narrow the search space. It should be noted that the researchers developed this algorithm especially for mining web access patters. For this reason it could be demanding to adapt the algorithm to other tasks.

Another new approach is the SPADE-algorithm (Zaki 2001). The main idea of this algorithm can be compared with the WAP-mine-algorithm: Repeated database scans should be avoided. The main step and feature of the algorithm is the decomposition of the problem, i.e. a parallel processing in sub-search spaces takes place. All in all SPADE is a fast and efficient algorithm that only needs three database scans.

Summarizing, AprioriAll and GSP are comparatively simple but inefficient algorithms, whereas WAP-mine and SPADE are much more sophisticated in theory and implementation.

3   Sequences in Marketing Applications

In marketing there are many possible applications of sequence mining algorithms like series of purchases in retail trade and mail order business, series of insurance contracts, web access patterns during (online-)sessions and online buying behaviour during several sessions.

At present the analysis of web accesses and online buying behaviour is of particular interest, especially for content management and adaptive web sites. For this reason the first of the following two examples deals with a web mining application.

3.1   Example: Web mining

By looking at a typical online shop, some important features can be observed:

- It is usual to create a session-id at the beginning of a session for identifying users during their sessions. This ensures the tracking of orientation on the web site for each user.

- If users sign in as customers, it becomes possible to observe their buying behaviour during several sessions, e.g. if different types of users prefer special offers.

Usually data mining starts with the preprocessing of the data which typically requires more than 70 % of the time. Especially, the web mining process includes the transformation of "not-easy-to-read-logfiles" (newer approaches try to avoid this very complex and time-consuming task). The next main step before analyzing the data may be the combination with personal data.

Generally, web sites can provide user-specific information about the orientation through itself, especially about the response to banners or special offers. Combined with an user account the buying behaviour over several sessions can be integrated and analyzed.

On the other hand some problems should be noticed:

- Little customer loyalty and a hard competition in the web, so-called "shop hopping" is very frequent and popular. It is unknown whether customers use other shops as well.

- Many customers give wrong personal information – except their addresses – if they do not see any sense for giving correct answers.

- Finally, by analyzing buying behaviour or web access patterns it could be a problem, that web sites and contexts differ from time to time and from user to user.

For those reasons it is uncertain whether results can be generalized without considering additional information. In figure 4, different layers of information are distinguished:

1. First layer: In general, sequence mining algorithms discover frequent sequential patterns without considering other information. But sequences represent a series of reactions and not a series of independent actions.

2. Second layer: For this reason some additional information should be taken into account. For example the orientation in a web site and the buying behaviour depends on banners and special offers.

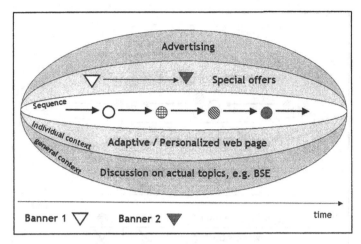

Figure 4: Example: Web mining

3. Third layer: Additionally, the behaviour possibly depends on further context variables, which vary over time.

This leads to an "onion model" with three different layers with information being more or less time dependent (figure 4). While current sequence mining algorithms focus on sequences of the first layer, further relevant information besides the pure web access information should be taken into account for a better understanding of the behaviour.

3.2 Example: Insurances

This example supports the arguments of the preceding example. Figure 5 shows a sequence of conclusions of insurance contracts. As in the preceding example the behaviour must be seen within a context. For example, information like mailings or visits by sales representatives cannot be considered by current sequence mining algorithms. This statement holds as well for general context information.

4 Conclusions

In general sequences provide additional and interesting insights into data. The mentioned computational approach focuses on the development of efficient algorithms which led to several algorithms with significant improvements since 1995. On the other hand the research in marketing is mainly theory driven where data driven approaches can help to find

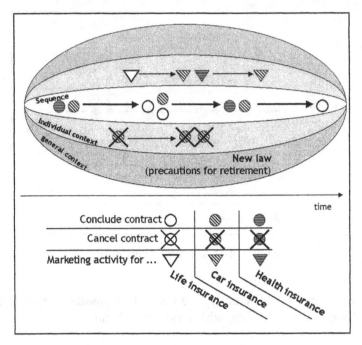

Figure 5: Example: Insurances

answers to open questions. As a conclusion future research should integrate the methodical, data driven computational approach and the theory driven marketing approach because current sequence mining algorithms are not suitable for real world applications in marketing.

It is necessary to take events of different types into account. For this reason future research has to deal with extended data structures and respectively extended or modified algorithms.

References

AGRAWAL, R. and SRIKANT, R. (1995): Mining Sequential Patterns, in Proc. of the Int'l Conference on Data Engineering (ICDE), Taipei, Taiwan, March 1995.

JOSHI, M., KARYPIS, G. and KUMAR, V. (1999): A Universal Formulation of Sequential Patterns. Technical Report No. 99-21, University of Minnesota, Minneapolis.

PEI, J., HAN, J., MORTAZAVI-ASL, B. and ZHU, H. (2000): Mining Access Patterns Efficiently from Web Logs, in Proc. Pacific-Asia Conference on Knowledge Discovery and Data Mining (PAKDD'00).

SRIKANT, R. and AGRAWAL, R. (1996): Mining Sequential Patterns: Generalizations and Performance Improvements, in Proc. of the Fifth Int'l Conference on Extending Database Technology (EDBT), Avignon, France.

ZAKI, M.J. (2001): SPADE: An Efficient Algorithm for Mining Frequent Sequences. *Machine Learning Journal, special issue on Unsupervised Learning, Vol. 42 Nos. 1/2, 31-60.*

Portfolio Management Using Multivariate Time Series Forecasts

M. Pojarliev, W. Polasek

University of Basel
Institute of Statistics and Econometrics,
Bernoullistrasse 28, CH-4051 Basel, Switzerland

Abstract: We use a multivariate VAR-GARCH model to predict the monthly returns and the variance matrix of the MSCI North America, MSCI Europe and MSCI Pacific indices from February 1990 until September 1999. We are interested in the following questions: First, can time series forecasts be successfully transformed into portfolio weights, and second, what kind of forecasts improves the portfolio performance? We compare two minimum-variance portfolios: the first portfolio is based only on the forecasted variance matrix, while the second portfolio uses the predicted returns and the predicted variance matrix. As a benchmark we choose the MSCI World index.

1 Introduction

Many financial analysts have been brought up by the view that stock returns are unpredictable and follow a random walk. What does this random walk hypothesis means for portfolio management? Active portfolio management could not improve the portfolio performance in the long run and, even worse, it increases transactions cost. Gains and losses could be viewed as a consequence of a random walk. On average portfolio managers are not able to beat the benchmark. Thus, the random walk hypothesis favors passive portfolio strategies: Follow the market and minimize the tracking error.

Contrary to the view that stock market are efficient and therefore random walks, empirical studies have shown that the returns of assets are predictable (Ferson and Harvey (1991)). Furthermore, the volatilities of stock returns can exhibit a rich interaction pattern (see e.g. Polasek and Ren (2001)). This paper shows that multivariate time series forecasts can improve the portfolio performance. Using a classical mean-variance framework we investigate the extent to which optimal holdings for the three regions (North America, Europe and Pacific) depart from the benchmark weights. We determine the optimal weights for the MSCI North America, MSCI Europe and MSCI Pacific indices with two different strategies. Portfolio one is the GMV portfolio and uses the predicted variance matrix, portfolio two (the μ-fixed portfolio) is based on the predicted returns and variance matrix.

The paper is organized as follows: In section 2 we estimate a multivariate VAR-GARCH model with exogenous variables for the returns vector of the three MSCI indices. Section three describes the methodology of constructing the portfolios and compares the performance. In the last section we summarize our findings.

2 Multivariate Time Series Forecasts

Let r_t^l be a N-dimensional vector of returns at time t and $l = 1, ..., N$.

$$r_t^l = \mu^l + \epsilon_t^l \tag{1}$$

where μ^l is a constant mean vector of dimension N and the heteroskedastic errors ϵ^l are multivariate normally distributed

$$\epsilon_t^l | I_{t-1} \sim N(0, H_t) \tag{2}$$

where H_t is the conditional variance matrix and I_{t-1} is the information set until time $t-1$. Each element of H_t depends on p lagged values of the squares and cross products of ϵ_t^l as well on q lagged values of H_t. Using the 'half-vectorisation' of symmetric variances, i.e. $h_t = vechH_t$ and $\eta_t = vech(\epsilon_t^l \epsilon_t^{ll})$ the GARCH-equation for the variance matrix is given as

$$h_t = A_0 + A_1 \eta_{t-1} + ... + A_p \eta_{t-p} + B_1 h_{t-1} + ... + B_q h_{t-q}. \tag{3}$$

This parameterization is called *vec* representation of the multivariate GARCH model. Bollerslev et al. (1988) have proposed a *diagonal* representation where each element of the variance matrix $h_{jk,t}$ depends only on past values of H_t and past values of η_t. This means that the variances depend on past own squared residuals and covariances depend on past own cross products of residuals. The *diagonal* model can be obtained from the *vec* representation by assuming a diagonal structure for the coefficient matrices A_i and B_i.

The diagonal representation is preferred in order to reduce the number of parameters. Using four exogenous US economic variables, the mean equation 1 is specified as

$$
\begin{pmatrix} r^A \\ r^E \\ r^P \end{pmatrix} = \begin{pmatrix} \mu^A \\ \mu^E \\ \mu^P \end{pmatrix} + \begin{pmatrix} \alpha_{11} & 0 & 0 \\ 0 & \alpha_{22} & 0 \\ 0 & 0 & \alpha_{33} \end{pmatrix} \begin{pmatrix} r^A_{-1} \\ r^E_{-1} \\ r^P_{-1} \end{pmatrix} +
$$
$$
+ \begin{pmatrix} \beta_{11} & \beta_{12} & \beta_{13} & \beta_{14} \\ \beta_{21} & \beta_{22} & \beta_{23} & \beta_{24} \\ \beta_{31} & \beta_{32} & \beta_{33} & \beta_{34} \end{pmatrix} \begin{pmatrix} x_1 \\ x_2 \\ x_3 \\ x_4 \end{pmatrix} + \begin{pmatrix} \epsilon_A \\ \epsilon_E \\ \epsilon_P \end{pmatrix}. \qquad (4)
$$

Table 1 gives the list of the exogenous variables which were used for model estimation. We have choosen these variables because in Pojarliev and Polasek (2001) we have shown that these variables are useful to explain stock market returns. Using the information criterion AIC (or BIC) we found as best model a VAR(1)-X(0)-GARCH(1,1) model. The 3-dimensional return vector (r^A, r^E, r^P) denotes the MSCI North America, MSCI Europe and MSCI Pacific indices of monthly data for the time period from February 1990 until September 1999.

Symbol	Exogenous Variables
B3m	3-month treasury bill yield
B10y	US benchmark 10 years bond index yield
P/E	price-earnings ratio for the US market
Yen/$	exchange rate Yen to US $

Table 1: Exogenous variables of the VARX model

The GARCH(1,1) part of the model is parameterized as

$$
vechH_t = A_0 + diag(\Psi_1)vech(\epsilon_{t-1}\epsilon'_{t-1}) + diag(\Phi_1)vechH_{t-1}, \qquad (5)
$$

where $diag(\Psi_1)$ and $diag(\Phi_1)$ are the diagonal coefficient matrices of order 1.

For the estimation of the model we use the GARCH module of S-Plus. The program uses the BHHH method of Berndt et al. (1974) for maximum likelihood (ML) estimation. All returns are multiplied by 100.

The VAR(1)-X(0) part specifies the mean equation of the model VARX-GARCH model. In our case we have estimated a VAR model of order 1 together with exogenous variables having lag order 0 (standard deviations are parenthesis):

$$\hat{r}^A = \begin{array}{ccccc} 1.212^* & - & 0.312^* r^A_{-1} & + & 0.041 B3m \\ (0.318) & & (0.104) & & (0.0767) \\ + \ 0.076 B10y & + & 0.448 P/E & - & 0.001 Yen/\$ \\ (0.1107) & & (0.126) & & (0.487) \end{array}$$

$$\hat{r}^E = \begin{array}{ccccc} 0.852^* & - & 0.101 r^E_{-1} & + & 0.345^* B3m \\ (0.362) & & (0.124) & & (0.203) \\ + \ 0.031^* B10y & - & 0.1588 P/E & + & 0.316^* Yen/\$ \\ (0.102) & & (0.163) & & (0.250) \end{array} \qquad (6)$$

$$\hat{r}^P = \begin{array}{ccccc} -0.455 & - & 0.028 r^P_{-1} & + & 0.606^* B3m \\ (0.663) & & (0.087) & & (0.106) \\ + \ 0.316^* B10y & - & 0.134^* P/E & - & 0.731^* Yen/\$ \\ (0.292) & & (0.233) & & (0.235) \end{array}$$

All estimates with a t-value greater than 1 are marked with a star (*). From the first equation in (6) we see that although we have used only fundamental variables from the USA as exogenous variables, the bond yields are not significant for the US stock returns. (Note that USA has a 97% weight in the MSCI North America index). But US bond yields are important for the MSCI Europe and the MSCI Pacific returns. Furthermore we see that a devaluation of the Yen/US $ exchange rate has a negative impact on the MSCI Pacific returns and a positive influence on the MSCI Europe returns.

3 Portfolio Construction

Markowitz (1952) mean-variance efficiency is the classic paradigm of the modern finance theory for asset allocation. The Markowitz efficient frontier represents all efficient portfolios in the sense that all other portfolios have lower expected returns for a given level of risk (measured by the standard deviation). The mean-variance portfolio approach assumes that investors prefer portfolios with high mean returns and low risk (measured by the standard deviation). More risk is only accepted if higher returns can be expected. This means that all investors should hold portfolios on the mean-variance frontier. In our empirical work we will investigate if the benchmark weights, i.e. the proportion of MSCI North America, MSCI Europe and MSCI Pacific indices in the MSCI World index are good choices or if an investor should search for optimal active weights.

If w is the vector of the holding, μ the vector of the expected returns and H the variance matrix of the returns, then the portfolio variance is $\sigma_p^2 = w'Hw$ and the portfolio return is $\mu_p = w'\mu$. Recall that the mean-variance portfolio in absence of a risk-free asset is formulated as (see e.g. Campbell et al. (1997))

$$min \quad w'Hw$$

subject to

$$w'\mu = \mu_{fix} \quad and \quad w'\iota = 1,$$

where ι is a vector of ones. We call the solution of this problem a μfix portfolio, because μ_{fix} is pre-specified as a target. The optimal weights are found to be

$$w_{\mu fix} = g + h\mu_{fix}, \tag{7}$$

where g and h are $(N \times 1)$ vectors,

$$g = \frac{1}{D} \left[B(H^{-1}\iota) - A(H^{-1}\mu) \right]$$

$$h = \frac{1}{D} \left[C(H^{-1}\mu) - A(H^{-1}\iota) \right]$$

and $A = \iota'H^{-1}\mu$, $B = \mu'H^{-1}\mu$, $C = \iota'H^{-1}\iota$, and $D = BC - A^2$. The weight vector of a *global minimum-variance* (GMV) portfolio is computed by

$$w_{gmv} = \frac{1}{C} H^{-1}\iota. \tag{8}$$

We construct two different portfolios base on the predicted conditional variance matrix in equation (5) and the predicted return vector in equation (6).

Portfolio 1 is a GMV portfolio and is based only on the forecasted variance matrix. We compute the portfolio weights for the in-sample period from February 1995 until September 1999 using the formula (8). The weights of portfolio 2 are computed with the formula in equation (7) where we need the predicted variance matrix H, the mean vector of past asset returns μ and the desired portfolio return μ_{fix}. The forecasted variance matrix is the same as for the GMV portfolio 1, but the mean μ is computed as a 12-month moving average of the predicted returns in (6). As desired portfolio return μ_{fix} we choose 12% per year which was approximately the average stock return in the 90ies. To avoid negative weights we use the restriction $minw = 5\%$.

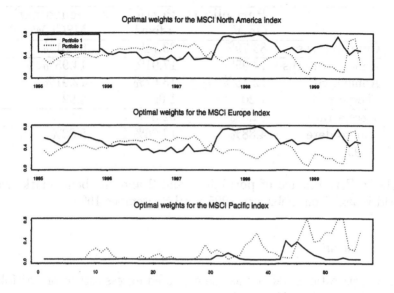

Figure 1: The weights for portfolios 1 and 2 from February 1995 until September 1999

Figure 1 shows the weights of the two portfolios for the in-sample period. In the GMV portfolio the MSCI Pacific index has a lower weight than in the benchmark, while the μ_{fix}-portfolio 2 has decreasing holdings for Europe and increasing Pacific weights for the first nine months of 1999. The average benchmark weights for this period are 54%, 35% and 11% for North America, Europe and the Pacific area, respectively. This is because the MSCI Pacific index was more volatile than the other two series.

3.1 Portfolio Comparison

We compare portfolio 1 and portfolio 2 with the MSCI World index from February 1995 until September 1999 by using the following criteria:

(a) annualized average return (in percent)

(b) annualized standard deviation (SD) (in percent)

(c) cumulative return for 56 months, 3 years and year to date (in percent)

(d) *Sharpe* ratio

	Portfolio 1 GMV	Portfolio 2 μ-fixed	benchmark MSCI World
Cum. returns	83.79%	83.74%	69.51%
Annual. returns	17.95%	17.91%	14.95%
Annual. SD	12.45%	13.67%	12.91%
Sharpe ratio	1.20	1.09	0.92
success rate	64.3%	62.5%	
year to date (1999)	4.88%	13.96%	5.82%

Table 2: Performance of portfolio 1 and 2 and the benchmark (MSCI World index) from February 1995 until September 1999.

(e) *success rate*

The *Sharpe* ratio is defined as the expected excess return of portfolio P divided by the risk of portfolio P:

$$S_P = \frac{r_P - r_{riskfree}}{\sigma_P} = \frac{r_{excess}}{\sigma_P},$$

The risk-free rate $r_{riskfree}$ is assumed to be 3% per year.

The *success rate* is the number of months (in percent) where the portfolio returns are better than the benchmark returns.

Table 2 shows the performance of the two portfolios for the evaluation period.

The GMV portfolio 1 assigns negative active weights for the Pacific index in the first nine months of 1999 and has a larger *Sharpe* ratio than portfolio two: It will be preferred by risk-averse investors. It dominates on the benchmark because it has greater annualized average return and smaller SD. Both strategies lead to success rates of more than 62%, which shows that active portfolio management is rather unlikely a matter of good or bad luck induced by a random walk.

4   Conclusion

This paper has shown how multivariate time series forecasts can be used for active portfolio management. Portfolio 1 (the global minimum variance portfolio) is based only on the predicted variance matrix while portfolio 2 (μfix portfolio) uses variance and mean forecasts. The returns are forecasted using US macro-economic variables as exogenous variables.

We found that both portfolios perform better than the benchmark, i.e. the MSCI World index for in-sample period (from February 1995 until September 1999).

The disadvantage of the two strategies for portfolio construction is the sensitivity of the weights to changes in the forecasts of the variance-covariance matrix: small differences in the predictions can lead to big differences in the active weights. A Bayesian approach where prior information is included in the model can be used to reduce the variability of the optimal portfolio weights (see Polasek and Pojarliev (2001)).

References

BERNDT, E., HALL, B., HALL, R. and HAUSMAN, J. (1974): Estimation and inference in nonlinear structural models. *Annals of Economic and Social Measurement, 653-665.*

BOLLERSLEV, T., ENGLE, R. and WOOLDRIDGE, J. (1988): A capital asset pricing model with time-varying covariances. *Journal of Political Economy, 96, 116-131.*

CAMPBELL, J., LO, A. and MACKINLAY, A. (1997): The econometrics of financial markets. Princeton University Press, Princeton NJ.

FERSON, W. and HARVEY, C. (1991): Sources of predictability in portfolio returns. *Financial Analysts Journal, May-June, 49-56.*

MARKOWITZ, H. (1952): Portfolio selection. *Journal of Finance, 7, 77-91.*

POJARLIEV, M. and POLASEK, W. (2001): What components determine the stock market returns in the 1990's?, in Gaul, Ritter (Eds.): Studies in Classification, Data Analysis, and Knowledge Organization, Springer-Verlag, Heidelberg.

POLASEK, W. and POJARLIEV, M.(2001): Portfolio construction with Bayesian GARCH forecasts, in Fleischmann, Lasch, Derigs, Domschke, Rieder, (Eds.): Operations Research Proceedings 2000, Springer-Verlag, Heidelberg.

POLASEK, W. and REN, L. (2001): Volatility analysis during the Asia crises: A multivariate GARCH-M model for stock returns in the US, Germany and Japan. *Applied Stochastic Models in Business and Industry, 17, 93-109.*

Value-at-Risk for Financial Assets Determined by Moment Estimators of the Tail Index

N. Wagner

Department of Finance and Financial Services
Dresden University of Technology, D-01062 Dresden, Germany

Abstract: In financial risk measurement, and particularly for value-at-risk calculations, the focus of statistical inference lies on the tails of some assumedly stationary, fat-tailed return distribution. Extreme value theory offers estimation approaches based on the asymptotic tail behavior of the underlying distribution. Discussing dependence in financial data and the problem of optimal subsample selection, the paper provides empirical results on the characteristics of the lower tail of two financial return distributions, namely those of daily S&P 500 and DAX returns. The results obtained are compared to those from conventional methods which reveals potential differences and similarities in risk assessment.

1 Introduction

Early empirical research in finance revealed that speculative asset returns are typically drawn from unconditional distributions which show fat-tails when compared to the normal density; see e.g. Blattberg and Gonedes (1974). It is therefore essential to account for fat-tailedness in financial risk management applications (see Huschens and Kim (1999), for example). A popular risk measure is "Value-at-Risk", frequently denoted as VaR, which has the economic interpretation as the amount of capital required in order to insure against bankruptcy under a given probability during some given time horizon. Statistically, VaR involves estimating quantiles of some unknown return distribution. A tool for quantile estimation that allows us to model the extremes of a very general class of distributions and to extrapolate beyond the range of given sample observations is extreme value theory as outlined for example in Embrechts et al. (1997) and Bassi et al. (1998). It turns out that the distribution tails can be characterized by a single parameter, the so-called "tail index". Apart from the application of moment estimators following Hill (1975), estimation of the generalized Pareto distribution is an alternative (see e.g. Emmer et al. (1998), Borkovec and Klüppelberg (2000), and Frey and McNeil (2000)). The latter approach is based on an asymptotic limiting distribution, which is assumed as a parametric model in small samples. Any of the approaches relies on an appropriate threshold selection.

In this paper we discuss the approach to tail index estimation based on the Hill-estimator and make inferences about the lower tails of two financial return distributions, namely those of daily S&P 500 and DAX index returns. The Hill estimator is based on the assumption that the underlying distribution is fat-tailed (i.e. has regularly varying tail at infinity). This explains its use for distributions as typically analyzed in finance. However, there are several notable limitations for practical applications: (i) Although the existence of the second moment of the underlying distribution is not a theoretical requirement, the results e.g. by Wagner and Marsh (2000) indicate that small sample inference has to be treated with caution when the underlying distribution is stable under addition.[1] Another crucial point is that Hill-type estimators are only optimal under independent draws from an exact Pareto distribution. In applications, (ii) the choice of an appropriate subsample of observations from the tail of the return distribution is therefore essential to the estimator's bias/variance trade-off. We use the bootstrap approach to subsample selection proposed in Hall (1990). Furthermore, (iii) as returns typically show non-linear time-series dependence –commonly modeled as "autoregressive conditional heteroskedasticity" (ARCH)– it is important to address the question of robustness with respect to the independence assumption.

Previous empirical studies of the tail index of return distributions include Danielsson and de Vries (1997) and Lux (2001), both considering the problem of optimal subsample selection in estimation. The simulation results in Wagner and Marsh (2000) indicate that ARCH effects increase small sample estimation error. Frey and McNeil (2000) suggest to take dependence into account by fitting an ARCH model to the return data first and then carrying out tail index estimation based on the fitted model residuals. The approach chosen in the present paper is to carry out Hill estimation ignoring dependence in the first place, i.e. by relying on the theoretical robustness results in Hsing (1991) and Resnick and StAricA (1998). The latter results show that consistency of the Hill estimator is given not only under independence but also under quite general forms of dependence including ARCH-type dependence. Ignoring dependence has the advantage that the tail estimator is derived from the original data-set, without the disturbing effects of potential model misspecification, estimation error and questions of robustness of the model estimation results in the presence of outliers. Furthermore, the vast empirical findings on ARCH modelling in finance point out that the usefulness of those models lies in predicting periods of increased, decaying

---

[1]Fortunately however, emprical investigations do generally not support infinite second moments; see Blattberg and Gonedes (1974) and subsequent studies. Also more recently, studying German single stock returns, Runde and Scheffner (1999) find evidence for the existence of the second moment.

return variance *after* exogenous shocks to return variance have occurred. However, risk management in face of extremal events will particularly be concerned with unpredictable, unconditional, shocks to return variance that may cause large portfolio losses. Here we discuss the unconditional approach as an alternative to conditional modelling and focus on the tail of the marginal distribution (see e.g. in Kiefersbeck (1999) and the literature therein), while considering empirical evidence which supports its existence.

The remainder of the paper contains three sections. Section 2 outlines the return model and the quantile estimators derived from extreme value theory. The empirical investigation is carried out in Section 3, providing estimates of the tail index as well as quantiles of the lower tail return distribution for S&P 500 and DAX. The results are compared to those from the normal model and the empirical distribution function. The last section gives a brief discussion.

2 The Model, Extreme Value Theory and Quantile Estimation

We start with a simple discrete time model for the prices S_t of some risky asset

$$S_t = \exp(R_t)\, S_{t-1}, \qquad 1 \le t \le T, \quad S_0 > 0, \tag{1}$$

where the random continuously compounded returns $(R_t)_{1 \le t \le T}$ are drawn from some unknown stationary distribution function F, which typically is fat-tailed.

A commonly used return generating process is the GARCH(1,1) model with constant return expectation μ and time-varying variance σ_t^2

$$
\begin{aligned}
R_t - \mu &= \sigma_t Z_t, & (2)\\
\sigma_t^2 &= \omega_0 + \omega_1 (R_{t-1} - \mu)^2 + \omega_2 \sigma_{t-1}^2, & \omega_0, \omega_1, \omega_2 \ge 0,
\end{aligned}
$$

based on given start random variables $(\sigma_0^2,\ R_0)$. The random variables Z_t are standardized iid draws from a symmetric distribution function. We assume that a stationary marginal distribution F for the return series exists, or more precisely, that the corresponding sufficient condition holds, i.e.:

$$\mathbf{E}\ln\left|\omega_1 Z_t^2 + \omega_2\right| < 0, \quad \omega_0 > 0. \tag{3}$$

In the following we are concerned with losses occurring from holding the asset during a single time period. A loss $L_{t+1} = S_{t+1} - S_t$ is given by model (1) as

$$L_{t+1} = (\exp(R_{t+1}) - 1)S_t \tag{4}$$

where $S_t > 0$ and hence $L_{t+1} < 0$ iff $R_{t+1} < 0$. The concept of Value-at-Risk is generally concerned with

$$\Pr\left(L_{t+\Delta t} \leq -\text{VaR}_{t,\Delta t}^p \,\middle|\, \mathcal{F}_t\right) = p, \tag{5}$$

where $\mathcal{F}_t$ denotes information in time t. Using the unconditional loss distribution which refers to the marginal return distribution F and setting $\Delta t = 1$ in (5) yields

$$\Pr((\exp(R_{t+1}) - 1)S_t \leq -\text{VaR}_{t,1}^p) = \Pr(R_{t+1} \leq r_p) = F(r_p) = p,$$

with $r_p = \ln[(-\text{VaR}_{t,1}^p/S_t) + 1]$. Referring to $F^{\leftarrow}$ as the generalized inverse distribution function of F which is defined by $F^{\leftarrow}(p) \equiv \inf\{r \in \mathbb{R} : F(r) \geq p\}$, we have

$$r_p = F^{\leftarrow}(p). \tag{6}$$

It then follows:

$$\text{VaR}_{t,1}^p = -[\exp(F^{\leftarrow}(p)) - 1]S_t. \tag{7}$$

Extreme value statistics can be used to determine the p-quantile $F^{\leftarrow}(p)$ of the stationary marginal return distribution function F. It can be shown, that F is fat-tailed if and only if the tail $\overline{F}(r) = 1 - F(r)$ of the distribution function F is regularly varying at infinity with parameter $-1/\xi < 0$,

$$\overline{F}(r) = L(r)r^{-1/\xi}, \quad r > 0, \tag{8}$$

where the function $L(r)$ satisfies $\lim_{r \to \infty}[L(tr)/L(r)] = 1$, $t > 0$, i.e. it is slowly varying at infinity. From (8) it follows that F behaves asymptotically like the Pareto distribution with $G(r) = 1 - cr^{-1/\xi}$, $c > 0$, $\xi > 0$. This result forms the starting point of so-called semi-parametric approaches to tail index estimation.

Let $(R_1, \ldots, R_T)$ be a sample of returns from the common marginal distribution function F. Without loss of generality, let $-R_{k,T} \leq \cdots \leq$

$-R_{1,T}$, $k \leq T$, denote k positive upper order statistics, where k is a realization of $K = \#\{i : -R_{i,T} > u\}$ based on a known threshold $u > 0$ and where (8) holds for $r > u$. Then, the so-called Hill estimator is a MLE of ξ conditional on $K = k$. It is given by:

$$\widehat{\xi}_{k,T} = \frac{1}{k} \sum_{i=1}^{k} \ln \left(\frac{R_{i,T}}{R_{k,T}} \right). \tag{9}$$

Consistency follows given that $k(T) \longrightarrow \infty$ and $k(T)/T \longrightarrow 0$ as $T \longrightarrow \infty$ (see e.g. Resnick and StAricA (1998) for details). Considering k lower tail observations $R_{i,T} < 0$ as above, it follows that the corresponding estimate of the lower part of the distribution function is

$$\widehat{F}(r) = \frac{k}{T} \left(\frac{r}{R_{k,T}} \right)^{-1/\widehat{\xi}_{k,T}}, \quad r < -u, \quad R_{k,T} < -u \tag{10}$$

and the estimator for the p-quantile of the distribution is (cp. Embrechts et al. (1997), p. 334-348):

$$\widehat{r}_{k,T,p} = \left(\frac{T}{k} p \right)^{-\widehat{\xi}_{k,T}} R_{k,T}, \quad R_{k,T} < -u. \tag{11}$$

In empirical applications the threshold u and hence the number k of smallest (or largest) sample observations is generally unknown. One practical approach is to determine u by explanatory analysis of the data with graphical methods as outlined in Chapter 6 of Embrechts et al. (1997). Adaptive approaches which aim at selecting an optimal k include the bootstrap approach by Hall (1990) studied e.g. in Danielsson and de Vries (1997) and Wagner and Marsh (2000).

3 An Investigation of Tails and Value-at-Risk

3.1 The Financial Data

The dataset in the empirical investigation of return distributions consists of daily S&P 500 ("SPX") and DAX index levels, $(s_t)_{t=0,\ldots,8262}$, in the period January 1st, 1969 to August 31st, 2000, as obtained from the Datastream database. Since our interest lies in the tail of the distributions only, reinvestment considerations as well as different holiday schedules in each country do not require further attention. The return series $(r_t)_{t=1,\ldots,8262} = (\ln(s_t/s_{t-1}))_{t=1,\ldots,8262}$ are plotted in Figure 1.

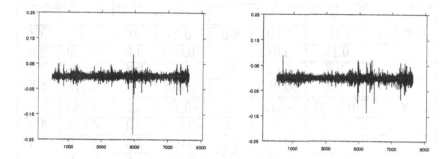

Figure 1: Continuously compounded daily returns of the S&P 500 (left hand side) and the DAX (right hand side)

An exploratory analysis of the data with graphical methods confirms substantial fat-tailedness compared to the normal distribution. Both series in Figure 1 also clearly exhibit periods of varying return variability. Fitting GARCH(1,1) models to both series yields point estimates of the coefficients ω_0, ω_1, and ω_2 in model (2) which are nonnegative at high confidence levels. Based on the parameter estimates, the stationarity condition (3) is satisfied under $Z_t \overset{d}{=} t(9)$ for the S&P 500 and $Z_t \overset{d}{=} t(11)$ for the DAX. One may hence assume the existence of a stationary marginal distribution function F for the return series. While stock market structure definitely changed over the last thirty years, the returns proposedly come from an overall fairly stable generating process.

3.2 Tail and Quantile Estimation Results

Setting the endowment S_T in period T equal to an arbitrary level of 100, $r_p = F^{\leftarrow}(p)$ in equation (7) yields:

$$\widehat{\text{VaR}}_{T,1}^p = -[\exp(\widehat{r}_p) - 1] \cdot 100.$$

In the following we choose three different approaches to estimate the return quantile $\widehat{r}_p$: (i) Fitting a Pareto-like tail by the extreme value approach outlined above. (ii) Using the empirical distribution function. (iii) Assuming a normal distribution.

(i): The tail index estimator (9) is based on a variable number k of largest absolute lower tail sample returns. The Hill estimator typically exhibits a variance-bias trade-off, i.e. high variance for lower values of k and

Estimates:		$\widehat{r}_{k,T,p}$	$\widehat{\text{VaR}}$	$\widehat{r}_{T,p}$	$\widehat{\text{VaR}}$	$\widehat{r}_{N,T,p}$	$\widehat{\text{VaR}}$
	p						
S&P 500	0.01%	−0.094	9.0	−0.23	21	−0.035	3.4
	0.1%	−0.047	4.6	−0.049	4.8	−0.029	2.9
	1 %	−0.024	2.4	−0.024	2.4	−0.022	2.2
	5 %	−0.015	1.5	−0.014	1.4	−0.015	1.5
DAX	0.01 %	−0.12	11	−0.14	13	−0.040	3.9
	0.1 %	−0.058	5.6	−0.064	6.2	−0.033	3.2
	1 %	−0.029	2.8	−0.029	2.8	−0.025	2.5
	5 %	−0.018	1.7	−0.016	1.6	−0.018	1.8

Table 1: Quantile estimation results for the S&P 500 and the DAX return distribution.

increasing bias for larger values of k. The minimal mean squared error bootstrap approach with parameter settings as in Wagner and Marsh (2000) and 500 bootstrap repetitions is used for a data-driven selection of k. The results yield an estimate of the tail index $\widehat{\xi}_{144,T} = 0.298$ for the S&P 500 and $\widehat{\xi}_{177,T} = 0.304$ for the DAX. The p-quantile estimate then follows according to equation (11).

(ii): Based on the empirical distribution function

$$F_T(r) = \frac{1}{T} \sum_{t=1}^{T} I_{\{R_t \le r\}},$$

an empirical quantile estimate $\widehat{r}_{T,p}$ is derived.

(iii): Under the normal model the quantile estimate is

$$\widehat{r}_{N,T,p} = \widehat{\sigma}_T \, \Phi^{\leftarrow}(p) + \widehat{\mu}_T,$$

where $\widehat{\mu}_T$ and $\widehat{\sigma}_T^2$ denote sample mean and variance, respectively.

The resulting quantile and VaR estimates for the S&P 500 and the DAX are summarized in Table 1. The probability p is chosen as 0.01, 0.1, 1 and 5 percent, respectively. Note that a probability of e.g. 0.1 percent relates to a one-day loss which, on average, roughly occurs once every four years.

4 Discussion

Although not obvious from the plots in Figure 1 and the empirical distribution function, the tail index estimation results for the two return series studied indicate quite similar lower tail behavior. While S&P 500 negative returns exhibit one dominating outlying observation (the crash of 1987), DAX negative returns show not as huge, but rather more frequent observations of smaller magnitude. Extreme value analysis reveals tail indices of about 0.3 for both return series. While estimated VaR under the normal model is relatively downward biased, the results for the fitted Pareto tail and the empirical distribution function are quite similar for all but the smallest probability level. Compared to the empirical distribution, extreme value analysis yields less differences in estimated VaR between the two indices, especially when small probability levels are considered. Altogether the results suggest that both indices belong to a class of assets with very similar downside risk characteristics.

References

BASSI, F., EMBRECHTS, P., KAFETZAKI, M. (1998): Risk Management and Quantile Estimation, in: Adler, R. J., Feldman, R. E., Taqqu, M. S. (Eds.): A Practical Guide to Heavy Tails, Birkhäuser, Boston, pp. 111-130

BLATTBERG, R. C., GONEDES, N. J. (1974): A Comparison of Stable and Student Distributions as Statistical Models for Stock Prices, *Journal of Business* 47: 244-280

BORKOVEC, M., KLÜPPELBERG, C. (2000): Extremwerttheorie für Finanzzeitreihen—Ein unverzichtbares Werkzeug im Risikomanagement, in: Johanning, L., Rudolph, B. (Eds.): Handbuch Risikomanagement, Uhlenbruch, Bad Soden, S. 219-241

DANIELSSON, J., DE VRIES, C. G. (1997): Tail Index and Quantile Estimation with Very High Frequency Data, *Journal of Empirical Finance* 4: 241-257

EMBRECHTS, P., KLÜPPELBERG, C., MIKOSCH, T. (1997): Modell-ing Extremal Events for Insurance and Finance, Springer, New York

EMMER, S., KLÜPPELBERG, C., TRÜSTEDT, M. (1998): Var–Ein Mass für das extreme Risiko, *Solutions* 2: 53-63

FREY, R., McNEIL, A. J. (2000): Estimation of Tail-Related Measures for Heteroscedastic Financial Time Series: An Extreme Value Approach, *Journal of Empirical Finance* 7: 271-300

HALL, P. (1990): Using the Bootstrap to Estimate Mean Squared Error and Select Smoothing Parameters in Nonparametric Problems, *Journal of Multivariate Analysis* 32: 177-203

HILL, B. M. (1975): A Simple General Approach to Inference about the Tail of a Distribution, *Annals of Statistics* 3: 1163-1174

HSING, T. (1991): On Tail Index Estimation using Dependent Data, *Annals of Statistics* 19: 1547-1569

HUSCHENS, S., KIM, J.-R. (1999): Measuring Risk in Value-at-Risk based on Student's t-distribution, in: Gaul, W., Locarek-Junge, H. (Eds.): Classification in the Information Age, Springer, Berlin, pp. 453-459

KIEFERSBECK, K. (1999): Stationarität und Tailindex bei zeitdiskreten Volatilitätsmodellen, Diplomarbeit, TU-München

LUX, T. (2001): The Limiting Extremal Behavior of Speculative Returns: An Analysis of Intra-Daily Data from the Frankfurt Stock Exchange, *Applied Financial Economics* 11: 299-315

RESNICK, S., STARICA, C. (1998): Tail Index Estimation for Dependent Data, *Annals of Applied Probability* 8: 1156-1183

RUNDE, R., SCHEFFNER, A. (1999): On the Existence of Moments – With an Application to German Stock Returns, Working Paper, University of Dortmund

WAGNER, N., MARSH, T. A. (2000): On Tail Index Estimation for Financial Return Models, Working Paper No. RPF-295, U.C. Berkeley

Keywords

Authors

A

Albrecht, Th., 385

B

Baier, D., 413
Bak, A., 104
Bauer, H. H., 422
Becker, C., 3
Bennert, R., 433
Bezdek, J. C., 291
Böhm, W., 229
Bornemeyer, C., 443
Brito, P., 12

C

Csernel, M., 22

D

de Carvalho, F.A.T., 12, 22
Decker, R., 443
Domenach, F., 31
Dugarjapov, A., 41
Dziechciarz, J., 483

E

Ernst, C., 321
Ernst, G., 321

F

Fahrmeir, L., 50, 133
Fried, R., 3
Friesen, K., 240
Frühwirth-Schnatter, S., 157

G

Gallegos, M.T., 58
Ganter, B., 472
Gefeller, O., 346, 392
Geyer-Schulz, A., 229
Gössl, C., 50
Godehardt, E., 67
Grauel, A., 82
Groenen, P.J.F., 90
Guimarães, G., 332

H

Hahsler, M., 229
Hammerschmidt, M., 422
Harris, B., 98
Hartmann, G., 257
Hartmann, O., 401
Haybach, G., 133
Hennerfeind, A., 50
Höpfner, L., 357
Hothorn, T., 346
Huber, K.-P., 309
Huschens, S., 454
Höse, S., 454

I

Itoh, Y., 180

J

Jahn, M., 229
Jajuga, K., 104
Jaworski, J., 67

K

Kiesel, R., 463
Klose, A., 257
König, I. R., 401

Druck und Bindung: Strauss Offsetdruck GmbH